Visible Light Communications
Theory and Applications

T0273448

Visible Light Communications
Theory and Applications

Edited by
Zabih Ghassemlooy
Luis Nero Alves
Stanislav Zvánovec
Mohammad-Ali Khalighi

CRC Press
Taylor & Francis Group
Boca Raton London New York

CRC Press is an imprint of the
Taylor & Francis Group, an **informa** business

Editors

Zabih Ghassemlooy received the BSc (Hons) degree in electrical and electronics engineering from Manchester Metropolitan University, UK, in 1981, and the MSc and PhD degrees from the University of Manchester Institute of Science and Technology, UK, in 1984 and 1987, respectively. During 1987–1988, he was a postdoctoral research fellow at City, University of London, UK. In 1988, he joined Sheffield Hallam University, UK, as a lecturer, becoming a professor in optical communications in 1997. In 2004, he joined the Northumbria University, Newcastle upon Tyne (UNN), UK, as an associate dean for research (ADR) in the School of Computing and Engineering. During 2012–2014, he was an ADR in the faculty of engineering, UNN. Currently, he heads the Northumbria Communications Research Laboratory and Optical Communications Research Group. Dr. Ghassemlooy is a visiting professor at Universiti Tun Hussein Onn Malaysia. His research interests are in optical wireless communications, free-space optics, and visible light communications. He has published over 600 articles in 220 journals and 4 books and supervised 50 PhD students. He was the vice-chair of EU Cost Action IC1101 during 2006–2008. He was the IEEE vice-chairman in 2004–2008, the IEEE chairman in 2008–2011, and the chairman of the IET Northumbria Network from October 2011–2015.

Luis Nero Alves graduated in 1996 and received his MSc degree in 2000, both in electronics and telecommunication engineering from the University of Aveiro, Portugal. In 2008, he obtained the PhD degree in electrical engineering from the University of Aveiro. His PhD thesis was on high bandwidth–gain product amplifiers for optical wireless applications. Since 2008, he has been the lead researcher at the Integrated Circuits Group from the Instituto de Telecomunicações, Aveiro. His current research interests are aligned with the IC1101 COST action (OPTICWISE) on optical wireless communications, where he is an active member. Dr. Alves has also worked on several nationally (VIDAS and EECCO, both from FCT) and internationally (LITES–CIP, PADSIC–FP7, and RTMGear–FP7) funded research projects, and industrial contracts.

Stanislav Zvánovec received his MSc and PhD degrees from the Czech Technical University in Prague in 2002 and 2006, respectively. To date, he works as a full professor and a vice head of the Department of Electromagnetic Field and a leader of the Free-Space and Fiber Optics Group. His current research interests include free space and fiber optical systems and electromagnetic wave propagation issues for quasioptical and millimeter wave bands. Until 2014, he was a chair of the Joint MTT/AP/ED/EMC chapter of the IEEE Czechoslovakia Section, and is currently the head of the Commission F of the Czech National URSI Committee. Research within the frame of international ESA projects, EU COST projects IC1101 OPTICWISE (vice-chair of WP1), IC0802, IC0603, ACE 2, Centre for Quasioptical Systems and Terahertz Spectroscopy, and others, holder of several national projects.

Mohammad-Ali Khalighi received his PhD degree in telecommunications from Institut National Polytechnique of Grenoble, France, in 2002. From 2002 to 2005, he was with GIPSA-lab, Télécom Paris-Tech, and IETR-lab as a postdoctoral research fellow. He joined École Centrale Marseille and Institut Fresnel in 2005, where he currently holds an associate professor position. His main research areas of interest include signal processing for wireless communication systems with an emphasis on the physical layer aspects of free-space, underwater, and indoor visible-light optical communications. So far, Dr. Khalighi has coauthored more than 80 journal articles and conference papers on these topics. He has served on the Technical Program Committee of more than 18 international conferences and workshops in the communications area, and the TPC co-chair of the International Workshop on Optical Wireless Communications 2015. Also, he was the vice-chair of Working Group 2 of the FP7 IC1101 COST Action on optical wireless communications.

Contributors

Tunçer Baykaş works as an assistant professor and the head of the computer engineering department at Istanbul Medipol University, Turkey. From 2007 to 2012, he worked as an expert researcher at NICT, Japan. He served as a co-editor and secretary for 802.15 TG3c and contributed to many standardization projects, including 802.22, 802.11af, and 1900.7. He is the vice director of the Centre of Excellence in Optical Wireless Communication Technologies (OKATEM) and the vice chair of 802.19 Wireless Coexistence Working Group. He contributed to the technical requirements document and the channel models of 802.15.7r1 standardization, which will enable visible light communication.

Hasari Celebi received his BS degree from the Department of Electronics and Communications Engineering at Yildiz Technical University, Istanbul, Turkey, and his MS degree from the Department of Electrical Engineering at the San Jose State University, California. He received his PhD degree from the Department of Electrical Engineering at the University of South Florida, Tampa, Florida, in 2008. He is currently an associate professor in the Department of Engineering at Gebze Technical University, Turkey, and is also the Director of the Institute of Information Technologies, Gebze Technical University. Prior to this, he was with Texas A&M University at Qatar (TAMUQ), Doha, as a research scientist. He received The Research Fellow Excellence Award at TAMUQ in 2010. He was also the recipient of The Best Paper Award at the CrownCom 2009 Conference. His research areas include statistical signal processing, estimation theory, localization, frequency diversity and multiplexing, and cognitive radio.

Yeon Ho Chung received his BEng degree in electronic engineering from Kyungpook National University, Daegu, South Korea, in 1984, his MSc degree in communications and signal processing from Imperial College London, UK, in 1992, and his PhD degree in electrical engineering and electronics from the University of Liverpool, UK, in 1996. He was employed as a technical consultant for Freshfield Communications Ltd, UK, in 1994, in the field of the design of mobile radio networks. He has now been working as a professor at the Department of Information and Communications Engineering, Pukyong

National University, Busan, South Korea. He joined the Mobile Communication Research Laboratory of Plymouth University, UK, as a visiting research fellow in 2004. He was a visiting professor at Pennsylvania State University, University Park, USA, in August 2006 and also at Chiba University, Japan, in September 2015. He served as the executive director of the Office of International Relations, Pukyong National University, from August 2008 to July 2012. He is a member of the Editorial Board for *Wireless Personal Communications*, Springer. He has published over 60 articles in the areas of optical wireless communications and mobile radio communications.

Petr Chvojka received his MSc degree in wireless communications from the Czech Technical University in Prague in 2013. He now works toward his PhD in the Department of Electromagnetic Field at the same university where he is a member of the Free-Space and Fiber Optics Team. He was on internships at Ben Gurion University of the Negev, Israel, and Northumbria University, Newcastle upon Tyne, UK, in 2013 and 2014, respectively. He involved in several research projects such as FP7 EU COST IC1101 OPTICWISE (Optical Wireless Communications—An Emerging Technology) and Research of Ambient Influences on Novel Broadband Optical Wireless Systems (LD12058). His current research interests include visible light communications, organic LEDs, and wireless optical communications.

Tamás Cseh received the BS and MS degrees in electrical engineering from the Budapest University of Technology and Economics, Hungary. He commenced his PhD studies in 2011, and he is currently a research assistant at the Department of Broadband Infocommunication and Electromagnetic Theory, Budapest University of Technology and Economics. His research interest includes optical communication with multimode fibers, radio over fiber systems, and dispersion compensation methods and modulation formats in subcarrier optical networks. He has authored and coauthored about 17 articles. He is a member of the Scientific Association for Infocommunications, Hungary (HTE).

José Luis Cura received his PhD degree in electrical engineering from the University of Aveiro, Portugal, in 1999. He is currently an assistant professor in the Department of Electronics, Telecommunications and Informatics at the same university. Dr. Cura has been a member of the Institute of Telecommunications since its foundation, where he has been involved in various projects in the CMOS analog circuit design area.

 **Çağatay Edemen** has received his BSc degree from Marmara University, Istanbul, Turkey, and his PhD degree from Işık University, Istanbul, Turkey, in electronics engineering. He is currently an academic faculty member at Ozyegin University, Istanbul, Turkey. He has been working on several research projects with a focus on information and communication theory and networks in wireless systems. Another field of interest is optical communication. In particular, he is interested in the applications of visible light communication. He also worked in applied research projects which focused on new generation mobile technology in two pioneer Turkish mobile network operators, Turkcell and Türk Telekom. The main objective of these projects was to contribute to the standards IEEE 802.16m and 3GPP LTE. Dr. Edemen was awarded the IEEE WCNC 2008 Best Paper Award for his work titled "Achievable Rates for the Three User Cooperative Multiple Access Channel."

 **Manuel Faria** recently graduated with a master's degree in electrical and computer engineering, the main area of telecommunications, at the Instituto Superior Técnico, Lisbon, Portugal. He deepened his knowledge in optical wireless communications in his master's thesis, which was themed on "transdermal optical communications."

 **Gábor Fekete** received his BSc degree in electrical engineering from the Budapest University of Technology and Economics, Hungary, in 2011, and his MSc degree from the same university in 2013. He is currently pursuing his PhD degree in electrical engineering from the Department of Broadband Infocommunications and Electromagnetic Theory at the Budapest University of Technology and Economics. His research interests cover indoor visible light communication systems, optically generated millimeter-wave, and optical OFDM modulation.

 **Mónica Figueiredo** concluded her PhD study in electrical engineering at the University of Aveiro, Portugal, in 2012. She started her research activities in 2001 at the Telecommunications Institute as a collaborator in the group of Integrated Circuits. Currently, she is a researcher in the same group and an assistant professor at the Polytechnic Institute of Leiria, Portugal. Her current main research interests include the design of communication circuits and systems in programmable logic devices, clock distribution and

alignment techniques, synchronization, timing circuits, and high-speed integrated electronics.

Chadi J. Gabriel received his PhD degree in physics and materials science at Aix-Marseille University, Marseille, France, in 2013. His work focused on underwater sensor networks, wireless optical communication, performance analysis over fading channels, and modulation and coding techniques. Currently, he is working as an expert signal processing researcher at Netatmo Co. in Boulogne-Billancourt, France.

Paul Anthony Haigh received his BEng and PhD degrees from Northumbria University, Newcastle upon Tyne, UK, in 2010 and 2014, respectively. Between 2011 and 2012, Dr. Haigh was awarded the prestigious Marie Curie Fellowship at the European Fellowship for Nuclear Research (CERN) at the youngest age in the history of the organization. His work at CERN focused on the design and testing of radiation-hard high-speed transmitter optical subassemblies for the ATLAS and CMS. During his PhD, Dr. Haigh invented the topic of organic small molecule and polymer visible light communications. He managed to improve data rates in ultralow organic photonic devices from kb/s up to 55 Mb/s. He joined the High Performance Networks Group at the University of Bristol as a research associate in December 2014. His research interests are reconfigurable and agile interfaces between networks. Over the last 4 years, he has published more than 40 refereed journal articles and conference papers.

Matěj Komanec is a research assistant at the Faculty of Electrical Engineering of the Czech Technical University in Prague. He received his MS and PhD degrees in radio-electronics from the Czech Technical University in Prague in 2009 and 2014, respectively. His current research interests include specialty optical fibers, free-space optics, visible light communication, optical interconnects, and fiber sensing. He is a member of OSA and SPIE.

Ivan Kudláček received his MSc degree in electrical engineering and his PhD degree from the Department of Electrotechnology, Faculty of Electrical Engineering, Czech Technical University in Prague (CTU in Prague). He is currently an associate professor and a senior researcher with the Department of Electrotechnology, CTU in Prague. His current research interests are the reliability of electronics devices and ecology electrical equipment.

Thomas D. C. Little received his BS degree in biomedical engineering from Rensselaer Polytechnic Institute, Troy, New York, in 1983, and his MS degree in electrical engineering and PhD degree in computer engineering from Syracuse University, New York, in 1989 and 1991, respectively. Currently, he is a professor at the Department of Electrical and Computer Engineering in Boston University, Massachusetts. He is also an associate dean for educational initiatives for the college and serves as an associate director of the National Science Foundation Center for Lighting Enabled Systems and Applications (LESA), formerly known as the Smart Lighting Engineering Research Center, a collaboration of Rensselaer Polytechnic Institute, the University of New Mexico, and Boston University. His recent efforts address research in pervasive computing using wireless technologies. This includes video streaming, optical communications with the visible spectrum, and applications related to ecological sensing, vehicular networks, and wireless healthcare. He is a senior member of the IEEE, a member of the IEEE Computer and Communications Societies, and a member of the Association for Computing Machinery.

Nuno Lourenço graduated with an MSc degree in electronics and telecommunications from the University of Aveiro, Portugal, in 2010. He then joined the Instituto de Telecomunicações, Aveiro, participating in several R&D activities in the areas of visible light communication, intelligent LED lighting systems, and wireless sensor networks. In 2013, he joined the Zumtobel Group, in the Austrian city of Dornbirn, initially as a project leader in hardware pre-development, and later as a technology scout/expert in the fields of networks and communications. In 2015, he became a consultant, continuing in his line work of analyzing and evaluating the latest developments of the networking world and their potential benefits to the lighting and automation industries. He currently provides support to multiple R&D activities, also including supervision of MSc candidate students, in the topics of indoor location, sensor network architectures for smart lighting, building automation, and visible light communications.

Jose M. Luna-Rivera received his BS and MEng degrees in electronics engineering from the Autonomous University of San Luis Potosi, Mexico, in 1997 and 1998, respectively. He received his PhD degree in electrical engineering from the University of Edinburgh, UK, in 2003. He is currently an associate professor at the College of Sciences at the Autonomous University of San Luis Potosi. His research focuses on signal processing for wireless communication and visible light communications.

Pengfei Luo received his BEng degree in communication engineering from Beihua University, China, in 2007, and his joint MSc–PhD degree in optical communications engineering from Beijing University of Posts and Telecommunications, China, in 2013. He was a research fellow of the Department of Physics and Electrical Engineering, Northumbria University, Newcastle upon Tyne, UK, from December 2013 to October 2014, and a project assistant at Beijing University of Posts and Telecommunications from November 2014 to March 2016. He now works in the Research Department of HiSilicon, Huawei Technologies Co., Ltd, Beijing, China.

Hoa Le Minh received his BEng degree in telecommunications from Ho Chi Minh University of Technology, Vietnam, in 1999, his MSc degree in communications engineering from Munich University of Technology, Germany, in 2003, and obtained his PhD degree in optical communications from Northumbria University, Newcastle upon Tyne, UK, in 2007. Prior to joining Northumbria University as a senior lecturer in 2010 and subsequently the program leader of BEng (Hons) Electrical and Electronic Engineering (2013), he was a research fellow at the Department of Engineering Science and a tutor at St Edmund Hall College, University of Oxford, UK (2007–2010). He worked at R&D Siemens AG, Munich, Germany (2002–2004), as a research assistant in ultrahigh-speed optical communications networks.

Dr. Hoa's expertise is in communications engineering including photonics systems, the emerging inorganic and organic visible light communications technology, smartphone technology, and intelligent mobile ad hoc networks. He has published over 100 journal articles, conference papers, and book chapters.

Rafael Pérez Jiménez received his MS degree in 1991 from Universidad Politécnica de Madrid, Spain, and his PhD degree (Hons) in 1995 from Universidad de Las Palmas de Gran Canaria, Spain. He is a full professor at the ULPGC, where he leads the IDeTIC Research Institute. His current research interests are in the field of optical indoor channel characterization and the design of robust visible light systems for indoor communications, specially applied for sensor interconnection and positioning. He has been awarded with the Gran Canaria Science Prize (2007) and the Vodaphone Foundation Research Award (2010).

Luís M. Pessoa graduated and obtained his PhD degree, both in electrical and computer engineering, from the Faculty of Engineering of the University of Porto (FEUP), Portugal, in 2006 and 2011, respectively. He is currently a senior researcher at INESC TEC, mainly involved in the conception and management of R&D projects, coordination of research students, and fostering the valorization of research results through new contracts with the industry. He has collaborated in several national and international projects in the areas of optical communications and microwave systems. His research interests include digital signal processing using advanced modulation formats, fiber-supported microwave systems, RF/microwave devices, antennas and propagation, and underwater wireless power/communications.

Wasiu O. Popoola received a first class (Hons.) degree in electronic and electrical engineering from Obafemi Awolowo University, Nigeria, and his MSc and PhD degrees from Northumbria University at Newcastle upon Tyne, UK. During his PhD, he was awarded the Xcel Best Engineering and Technology Student of the Year 2009. He is currently a chancellor's fellow at the Institute for Digital Communications, University of Edinburgh, UK. Previously, he was a lecturer in electronic engineering at Glasgow Caledonian University, UK, between August 2012 and December 2014. He has published well over 70 journal articles/conference papers/patents, and a number of those are invited papers; see http://goo.gl/JdCo3R. He was an invited speaker at the 2016 IEEE Photonics Society Summer Topicals. He coauthored the book *Optical Wireless Communications: System and Channel Modelling with MATLAB*®, published by CRC Press in 2012. His research interests include optical (wireless and fiber) and digital communications.

Jose A. Rabadan-Borges received his MS and PhD (Hons) degrees from the Universidad de Las Palmas de Gran Canaria, Spain, in 1995 and 2000, respectively. Currently, he is an assistant professor at the ULPGC. His research interests are in the field of the wireless infrared communications for both wideband local area networks and narrowband sensors networks, high-performance modulation and codifications schemes for VLC communications, and indoor VLC channel characterization.

Michael B. Rahaim is a postdoctoral researcher in the Department of Electrical and Computer Engineering at Boston University, Massachusetts, working with the National Science Foundation Center for Lighting Enabled Systems and Applications (LESA). His research focuses on software-defined radio, visible light communication, heterogeneous networks, and smart lighting. He received his BS degree in electrical and computer systems engineering from Rensselaer Polytechnic Institute, Troy, New York, in 2007, and his MS and PhD degrees in computer engineering from Boston University in 2011 and 2015, respectively.

Sujan Rajbhandari obtained his BEng degree in electronics and communication engineering from the Institute of Engineering, Nepal, in 2004. He obtained his MSc and PhD degrees from Northumbria University, Newcastle upon Tyne, UK, in 2006 and 2010, respectively. He was awarded the P.O. Byrne prize for his MSc project. He worked at Northumbria University from 2009 to 2012 as a senior research assistant and research fellow. He then joined the University of Oxford, UK, as a postdoctorate research assistant in December 2012 and worked in EPSRC's Ultra-Parallel Visible Light Communications (UP-VLC) project. He is currently working as a lecturer at the School of Computing, Engineering and Mathematics, Coventry University, UK. Dr. Rajbhandari has published more than 100 scholarly articles and is a coauthor of the book *Optical Wireless Communications: Systems and Channel Modelling with MATLAB®*. He was an invited speaker in Information and Communication Technology Forum 2015 at Manchester. He has also served as a local organizing and technical program committee member for a number of conferences and proceeding editor for EFEA 2012 and NOC/OC&I 2011. He is a regular viewer for several publications including the IEEE, OSA, and IET journals. His research interests lie in the area of optical communications and signal processing. He is a member of IEEE.

Carlos Ribeiro received his BSc degree (5-year course) in electronic engineering from the University of Coimbra, Portugal, in 1996. In 2003, he received his MSc degree in electronics and computer engineering from the same university. In 2010, he received his PhD degree in electronics engineering from the University of Aveiro, Portugal. In 1997, he joined the Department of Electronics of the Polytechnic Institute of Leiria, Portugal, where he is currently an assistant professor. He is a researcher in signal processing for communications. His main research topics are PHY algorithms for RF and VLC communication systems and its implementation.

He has published tens of research articles and conference papers in international journals. He has been participating in several national and European projects.

 **Luis Rodrigues** graduated with his MSc degree in electronic and telecommunications engineering from the University of Aveiro, Portugal, and he is currently in the MAP-tele PhD program. His master's thesis theme was Error Correcting Codes for Visible Light Communications, aiming performance improvements of OFDM-based VLC systems using an FPGA. He is currently working with analog LED drivers and optical receivers.

 **Julio F. Rufo Torres** received his MS and PhD degrees from the Universidad de Las Palmas de Gran Canaria, Spain, in 2008 and 2016, respectively. His current research interests are in the field of visual light communications systems for indoor communications applied to sensor networks and Internet of things.

 **Elham Sarbazi** received her BSc degree in electrical and computer engineering from the University of Tehran, Iran, in 2011, and her MSc degree (first class honors) on communication systems from the Department of Electrical and Electronics Engineering, Ozyegin University, Istanbul, Turkey, in 2014. She is currently working toward her PhD degree under the supervision of Prof. Harald Haas at the Institute for Digital Communications, University of Edinburgh, Edinburgh, UK. Her research interests mainly include optical wireless communications and visible light communications.

 **Paulo Sérgio de Brito André** received his bachelor's degree in physics engineering, PhD degree in physics, and Agregação title (habilitation) degree from the Universidade de Aveiro, Portugal, in 1996, 2002, and 2011, respectively. In 2013, he joined as an associate professor at the Instituto Superior Técnico, University of Lisbon, Portugal, lecturing courses on telecommunications. Since 2015, he has been a vice director of the Department of Electrical and Computer Engineering. His current research interests include the study and simulation of photonic and optoelectronic components, optical sensors, optical communications systems, and networks.

Parvaneh Shams received her BSc degree in computer engineering from the University of Tabriz, Iran, in 2004, and her MSc degree in electronics and communication engineering from the Iran University of Science and Technology, Tehran, in 2012. Her research interests include optical wireless communications and visible light communications, mainly Mac layer protocol performance. She is currently a PhD student in communication engineering under the supervision of Prof. Niyazi Odabaşıoğlu at Istanbul University, Turkey.

Martin Siegel studied physics and received his diploma from the University Heidelberg, Germany, in 2005 and his PhD degree from the University of Hannover, Germany, in 2009. From 2010 to 2011, he worked at High Q Laser GmbH in Austria, where he was responsible for the development of a pulsed laser system. In 2011, he began working as a technology scout for the Zumtobel Group, identifying and evaluating new technological developments. Since 2016, he has been the Director of Research and Pre-Development at the Zumtobel corporate headquarters in Dornbirn. He has published a number of scientific publications and has presented papers in numerous conferences worldwide. Topics of interest range from lighting and smart-lighting applications all the way to sensor technology and laser development.

Bernardo Silva obtained his MSc degree in electrical and computer engineering from the Faculty of Engineering at the University of Porto (FEUP), Portugal, in 2015. He majored in automation and control, with the specialization in robotics and systems, and minored in enterprise information systems, licensing projects, and electrical design in industrial installations. His final dissertation project was "Underwater Optical Communication: An Approach Based On LED," under the supervision of Prof. Nuno Cruz and Dr. Luís Pessoa. His fields of interest include programming, economics and management, applied electronics, acquisition and signal processing, industrial informatics, and industrial robotics.

Hsin-Mu (Michael) Tsai is an associate professor in the Department of Computer Science and Information Engineering and Graduate Institute of Networking and Multimedia at National Taiwan University, Taipei. He received his BSE in computer science and information engineering from National Taiwan University in 2002 and his MS and PhD degrees in electrical and computer engineering from

Carnegie Mellon University, Pittsburgh, Pennsylvania, in 2006 and 2010, respectively. Dr. Tsai's recognitions include the 2015 K. T. Li Young Researcher Award, 2014 Intel Labs Distinguished Collaborative Research Award, 2013 Intel Early Career Faculty Award (first recipient outside of North America and Europe), and National Taiwan University's Distinguished Teaching Award. Dr. Tsai served as one of the founding workshop co-chairs for the first ACM Visible Light Communication System (VLCS) Workshop in 2014, and TPC co-chair for IEEE VNC 2016 and ACM VANET 2013. His research interests include vehicular networking and communications, wireless channel and link measurements, vehicle safety systems, and visible light communications.

Xuan Tang is a principal investigator and an associate professor at the Fujian Institute of Research on the Structure of Matter, Chinese Academy of Sciences, Fuzhou, since October 2014. She obtained her BEng (first class with honors) degree in electronic and communications engineering in 2008 and her PhD degree from Northumbria University, Newcastle upon Tyne, UK, in 2013. From October 2012 to July 2014, Dr. Tang worked as a postdoctoral researcher at the Department of Electronic Communications Engineering, Tsinghua University, Beijing, China, and then joined the National Basic Research Program of China (973 Program) as the key researcher. From October 2013 to April 2014, she was the visiting academic at the University of Science and Technology of China, Hefei. She has received funding from the China Postdoctoral Science Foundation and National Science Fund for Young Scholars. She has published 40 articles and is an IEEE member. Her research interests are in the areas of optical wireless communications including high-speed infrared/ultraviolet laser communications, visible light communications and optical MIMO systems, and radio frequency communication technologies.

Eszter Udvary received her PhD degree in electrical engineering from Budapest University of Technology and Economics, Hungary, in 2009. She is currently an associate professor at the Department of Broadband Infocommunications and Electromagnetic Theory, Budapest University of Technology and Economics, where she leads the Optical and Microwave Telecommunication Lab. She currently teaches courses on optical communication devices and networks. Dr. Udvary's research interests are in the broad areas of optical communications, including optical and microwave communication systems, radio over fiber systems, optical and microwave interactions, and applications of special electro-optical devices. Her special research focuses on multifunctional semiconductor optical amplifier application techniques. She is deeply involved in visible light communication, indoor optical wireless communication, and microwave photonics techniques. Dr. Udvary has authored more than 80 journal articles and conference papers and 1 book chapter, and she received more than 60 citations. She is a member of IEEE.

H. Fatih Ugurdag is an associate professor at Ozyegin University, Istanbul, Turkey. He received his BS in electrical engineering as well as physics from Bosphorus University, Istanbul, Turkey, in 1986. He received his MS and PhD from Case Western Reserve University, Cleveland, Ohio, in electrical engineering in 1989 and 1995, respectively. He did an MS thesis on machine vision and a PhD dissertation on parallel hardware design automation. He worked in the industry in the United States between 1989 and 2004 at companies such as GE, GM, Lucent, Juniper, and Nvidia as a machine vision engineer, EDA software developer, and chip designer. In late 2004, he joined academia. He is currently a consultant to several companies including Vestel-Vestek, one of the leading consumer electronics companies in Europe. His research interests include real-time hardware/software design in the areas of communications, video processing, and automotive systems.

Murat Uysal is a full professor and the chair of the Department of Electrical and Electronics Engineering at Ozyegin University, Istanbul, Turkey. Prior to joining Ozyegin University, he was a tenured associate professor at the University of Waterloo, Canada, where he still holds an adjunct faculty position. Dr. Uysal's research interests are in the broad areas of communication theory and signal processing, with a particular emphasis on the physical layer aspects of wireless communication systems in radio, acoustic, and optical frequency bands. He has authored some 250 journal and conference papers on these topics and received more than 5000 citations. Dr. Uysal currently serves as the Chair of IEEE Turkey Section. He serves on the editorial boards of *IEEE Transactions on Communications* and *IEEE Transactions on Wireless Communications*. His distinctions include NSERC Discovery Accelerator Supplement Award, University of Waterloo Engineering Research Excellence Award, Turkish Academy of Sciences Distinguished Young Scientist Award, and Ozyegin University Best Researcher Award, among others.

Wantanee Viriyasitavat is a lecturer in the Faculty of Information and Communication Technology, Mahidol University, Bangkok, Thailand, and also a faculty member in the Department of Telematics, Norwegian University of Science and Technology, Norway. During 2012–2013, she was a research scientist at the Department of Electrical and

Computer Engineering, Carnegie Mellon University (CMU), Pittsburgh, Pennsylvania. She received her BS/MS and PhD degrees in electrical and computer engineering from CMU in 2006 and 2012, respectively. During 2007–2012, she was a research assistant at CMU, where she was a member of General Motors Collaborative Research Laboratory and was working on the design of a routing framework for safety and nonsafety applications of vehicular ad hoc wireless networks. Dr. Viriyasitavat has published over 30 conference and journal articles and received numerous awards such as Dissertation Award from National Research Council of Thailand. Her research interests include traffic mobility modeling, network analysis, and protocol design for intelligent transportation systems.

Dehao Wu received his bachelor's degree in optical and information engineering from the Nanjing University of Post and Telecommunication, People's Republic of China, in 2007. He received his master's degree in microelectrical and telecommunication engineering and his PhD degree in cellular optical wireless communication systems from Northumbria University, Newcastle upon Tyne, UK, in 2009 and 2013, respectively. Since 2014, he has been a postdoctoral research fellow in Nanyang Technological University at Singapore. His research interests include the area of indoor optical wireless communications, optical wireless positioning and localization, visible light communications, optical wireless sensing and detecting, and hybrid free-space optics. Dr. Wu has served as a reviewer for several leading publications, including the *Journal of Lightwave Technology* and *IEEE Transactions on Communications*, and several international conferences. He is a member of IEEE.

Zhengyuan Xu received his BS and MS degrees from Tsinghua University, China, and his PhD degree from Stevens Institute of Technology, Hoboken, New Jersey. He was with University of California, Riverside, from 1999–2010, where he became a full professor with tenure and also a founding director of UC-Light Center. He was selected by the Thousand Talents Program of China in 2010. He is professor at the School of Information Science and Technology, University of Science and Technology of China, Anhui. He is a founding director of Wireless-Optical Communications Key Laboratory of the Chinese Academy of Sciences and a chief scientist of the National Key Basic Research Program of China. His research focuses on wireless communication and networking, optical wireless communications, geolocation, and signal processing. He has published over 200 journal articles and conference papers. He was an associate editor and a guest editor for different IEEE journals and a founding co-chair of IEEE GLOBECOM Workshop on Optical Wireless Communications.

 **Petr Žák** graduated from the Department of Electroenergetics, Faculty of Electrical Engineering, Czech Technical University in Prague (FEE CTU), with specialization in lighting engineering in 1992. In 1993, he began working as a lighting engineer in a private company. In 2003, he completed his doctoral studies at FEE CTU, and from 2004, he has been an assistant professor. He is an editorial board member of the magazine *Světlo* and a board member of Czech national committee of the International Commission on Illumination (CIE) and a Czech representative in Division 5 of CIE. Since 2010, he has been a member of the Czech Chamber of Certified Engineers and Technicians active in construction. He is a coauthor of the book *Světlo a osvětlování* (2013). He participated in many indoor and outdoor lighting projects, pilot projects, urban lighting concepts, and luminaries design, for example, lighting of National Gallery in Prague, LED pilot project in Prague, and concept of public lighting in Prague.

1

Introduction

Zabih Ghassemlooy, Luis Nero Alves, Stanislav Zvánovec, and Mohammad-Ali Khalighi

CONTENT

In the past decade, the world has witnessed a dramatic increase in the traffic carried by the telecommunication networks. The increasing demand for high-speed Internet services (high-definition TV, video calls, and cloud-based computing) has underpinned the need for further innovation, research, and development in new emerging technologies capable of delivering ultra-high data rates to the end users. The existing radio frequency (RF) wireless spectrum is outstripping the supply, thus leading to spectrum congestion, which needs urgent attention. This is currently motivating what is known as the "tragedy of the commons" paradigm, a situation in which all users without any clear intention to do so will contribute to deplete a common resource, in this case, available spectrum. Such situations arise in high-density scenarios such as sport venues, concerts, airport, emergency situations, etc., where user demands may lead to the dramatic situation of limited access. Current RF-based communications suffer in particular from multipath propagation effects in dense urban environments, which reduce the link availability and its performance. The limited bandwidth of these systems together with the spectrum congestion means that relatively very few high-definition channels can be accommodated in a given area. This problem is more acute for indoor applications where there is a lack of adequate bandwidth to be shared among the large number of users who want a lion's share of the channel capacity. It is estimated that more than 70% of the wireless traffic takes place in indoor environments (home, office, etc.). Therefore low-cost and highly reliable technologies are required to enable seamless indoor wireless communications. Squeezing more out of RF technologies or using an alternative such as optical technologies are the only two options available.

Regardless of the wireless technologies (i.e., 3G, 4G, 5G, and beyond) that are being adopted, there are only three approaches to increase the capacity of wireless radio systems: (i) release new spectrum and therefore more bandwidth; (ii) increase the number of nodes; and (iii) improve the spectral efficiency. Acquiring a new spectrum is costly, and finding more bandwidth is not a major problem but it is clearly not enough—it is finite. Adding more nodes is also being achieved via cell splitting, but this is rather costly and such systems also become too complex to manage. Also, two nodes do not offer twice the capacity of one, due to interference issues; the law of diminishing returns is at play. In addition, doubling the infrastructure will not lead to doubling the revenue. So in the long run, what are the solutions?

One possible alternative technology that can address and overcome these restrictions is optical wireless communications (OWCs), which utilizes infrared (IR), visible, and ultraviolet (UV) subbands, and remains mostly unexplored so far. Compared to RF, OWC offers superior features such as ultra-high bandwidth (in the order of THz), not being subject to electromagnetic interference, providing a high degree of spatial confinement bringing virtually unlimited frequency reuse, cost effectiveness with no licensing fee, and inherent physical security. With plenty of spectrum available, spectral efficiency is not as critical as in RF systems; nevertheless, most of the techniques developed for improving the spectral efficiency for RF systems can be applied to the optical domain. Whereas most of the proposed solutions for addressing the spectrum scarcity of RF systems consider scratching higher frequencies, such as millimeter and THz waves, with the major drawbacks of increased path loss and expensive transmitter/receiver components, OWC-based transmission systems draw their advantage due to the maturity of the transmission technology (therefore, relatively low-cost, high-performance components) as well as incomparable energy efficiency. The recent, yet well-known OWC in outdoor applications are the free-space optical (FSO) systems that operate at near IR frequencies (i.e., at wavelengths 750–1600 nm). These offer cost-effective and protocol-transparent links with high data rates (up to 10 Gbps per wavelength) and providing a potential solution for the backhaul bottleneck problem over short to long ranges up to a few kilometers.

OWC systems in the visible band (390–700 nm) are commonly referred to as visible light communications (VLCs), which take full advantage of visible light-emitting diodes (LEDs) for the dual purpose of illumination and data communications at very high speeds. VLC is a sustainable and green technology with the potential to revolutionize approaches to how we will use lights in the near future. It can provide solutions for a number of applications including wireless local area, personal area, and body area networks (WLAN, WPAN, and WBANs), heterogeneous networks, indoor localization and navigation (where current GPS is not available), vehicular networks, underground and underwater networks among others, offering a range of data rates from a few Mbps to 10 Gbps.

1.1 State of the Art

Factoring out molecular means of communications, which are part of all natural life on the planet, VLCs are perhaps the oldest means of communications known to humankind. Since the early days of human history, light has served as an essential means of communication. Possible examples are for instance the use of fire signals to communicate between tribes; use of reflected sunlight for ship-to-ship communications both utilized by ancient Greeks. For long-range communication, fire beacons placed on high points were lit from one point to another to deliver messages. Also, the ancient Chinese used smoke signals to communicate information on enemy movements between the army units along the Great Wall. Remarkably, VLC was proposed in more evolved technological society by Alexander Graham Bell with his photophone in 1880. This device was able to modulate sunlight with vibration caused by speech and transmitted the modulated light to an intended receiver. True advancements in VLC came with the discovery of electroluminescence and the LED in 1927, by the Russian scientist Oleg Losev. Coincidentally, Losev foresaw applications of his electroluminescence devices as communication devices. The story was quite different though. Wireless radio based on electromagnetic radiation has been established as the dominant communication technology over the last several decades. OWC based on IR light beams was originally proposed by Gfeller and Bapst in 1979. Their research work marked the beginning of the globally growing research activities, which has culminated in what we now recognize as OWC. Current research trends on OWC are focused on a range of different wavelengths from UV, through to the visible part of the spectrum and ending at the near IR regions of the spectrum. OWC offer unprecedented bandwidths, freedom from spectrum regulation, and inherently secure communications links, when compared to wireless RF systems. Among the research activities on OWC, one that is particularly attractive to be combined with the wireless RF systems to promote novel design concepts and techniques is VLC. VLC systems are currently attracting attention due to the growing use of LEDs for general lighting in a multitude of applications. With characteristics such as longer lifetime, better controllability, and energy efficiency, future lighting will definitely be based on LEDs replacing the conventional lighting devices on a worldwide scale. The unique characteristic of the LEDs is that they can be used at the same time for lighting and communications with unimaginable implications. This unique dual functionality of LEDs can be fully explored as a means to promote a truly green technology. When compared to FSO, most VLC systems are mainly based on diffuse radiation systems, where the presence of line of sight between terminal equipment is not mandatory. However, for high-speed applications, VLC with line-of-sight configurations can also be used. Current research trends have demonstrated the feasibility to achieve high data rate communications (links with data rates above 7 Gbps have been demonstrated) exploring this dual LED functionality. To achieve this performance landmark, several methods have been adopted to mitigate the slow response of power LEDs, for example, the use of optical discrete multitone modulations. Indeed, power LEDs used for lighting are driven with high forward currents when compared to IR LEDs and laser diodes, which makes them slower. Moreover, most power LEDs employ a yellow phosphorous coating to convert blue light into visible light, which further slows down the device response. Another interesting characteristic of VLC systems is their spatial confinement. When used in indoor scenarios, the communications range is limited by the room size, since no radiation crosses the walls. This makes these systems secure and potentially free from eavesdropping. Spatial confinement may also be explored in multiuser scenarios, where different light sources carry different data, but at the cost of increased system complexity.

The present book is composed of 16 chapters that cover a particularly wide scope of the theoretical and application-related aspects of VLC technology and address the different fundamental and practical considerations of these systems in different application scenarios. Along with 16 dedicated chapters, it provides comprehensive illustrations and performance analyses, without forgetting the future perspectives and technology deployment trends. It is an ideal reference for researchers who wish to initiate working on VLC-related research projects, or wish to deepen their knowledge of the field and gain insight into practical considerations. It is also an excellent reference textbook for graduate courses on the topic. Luis Nero Alves et al. in "Lighting and Communications: Devices and Systems" introduces the front-end devices and systems used for communication establishment in VLC systems. On the transmitter side, the focus is on LEDs which are exploited as both communication and lighting devices. The merger of lighting and communications means the same device builds up synergistic opportunities, but is not exempt from challenges. This chapter describes the challenges associated with the design of efficient LED drivers for future VLC systems. On the receiver side, the focus is on the use of photodiodes as the fundamental device for optical signal detection. Then follows a system overview of the most common techniques for optical amplifier design. The chapter ends with an overview of light regulations, something that is traditionally disregarded in VLC system specifications.

Zabih Ghassemlooy et al. in "Channel Modeling" consider this fundamental prerequisite step for the design of VLC links. Focusing on indoor systems, the chapter first introduces different sources of impairment arising from beam propagation or nonideal optoelectronic devices. Indeed, the latter can be considered as a part of the aggregate (i.e., global) communication channel. Then, different optical signal propagation modes are overviewed and the most relevant methods for numerical channel simulation are briefly explained. The authors then describe the limitations arising from the aggregate channel while focusing on the problem of intersymbol interference and how it can affect the link performance, as well as the signal distortion arising from the LED non-linear characteristics. The chapter ends by addressing channel modeling for multiple-input multiple-output (MIMO) VLC systems. Tamás Cseh et al. in "Modulation Schemes" report on intensity modulation and direct detection (IM/DD)-based VLC systems. The chapter discusses baseband modulations including pulse-amplitude modulation, pulse-position modulations, pulse interval modulations, differential amplitude pulse-position modulation, variable pulse position modulation, and compare them in terms of power and bandwidth efficiencies, and peak-to-average power ratio. The chapter also describes multicarrier modulations (i.e., orthogonal frequency division multiplexing [OFDM] and its variants as well as pulse amplitude discrete multitoned, and the special modulation of color-shift keying.

Murat Uysal et al. in "IEEE 802.15.7: Visible Light Communication Standard" overview this standard, approved by IEEE in 2011, with a focus on the features of the physical and MAC layers. The chapter then provides simulation results to demonstrate the key performance metrics as well as the comparison of the different proposed physical layer schemes. Finally, the authors present the ongoing IEEE standardization activities and the most recent proposed amendments to the standard. Hoa Le Minh et al. in "Techniques for Enhancing the Performance of VLC Systems" start with the discussion of a technique for enhancing VLC system capacity. The chapter outlines parallel data transmissions (i.e., MIMO) using multiple LEDs, that are commonly used in home and office lighting, in order to increase channel capacity. In addition, the chapter discusses the OFDM scheme and outlines a viable scheme to overcome the problem of high peak-to-average power ratio. Also included is the dimming techniques adopted in high-speed VLC systems. Rafael Perez-Jimenez et al. in "VLC Applications for Visually Impaired People" provide descriptions of specific outdoor and indoor applications that can be employed in order to allow universal accessibility by disabled people. The chapter deals with both implementation using street lights as a resource for the mobility of blind people, and indoor positioning and guidance. It also presents a VLC-ultrasound hybrid solution and other specific applications for safety and emergency management tools.

Pengfei Luo et al. in "Car-to-Car Visible Light Communications" give an overview of the need for car-to-car communications as part of the intelligent transportation systems, where research and development, products, and standardizations are mostly focused around the RF-based communication technologies for wireless connectivity in vehicular networking. This chapter discusses the VLC technology inherent advantages over the RF-based dedicated short-range communication (DSRC) technology, as well as its key characteristics and features, which could be adopted for intelligent transportation system (ITS) applications. The chapter also outlines vehicular VLC communication models including both single-input single-output (SISO) and MIMO, noise sources, and road surface. Characterization of VLC-based car-to-car communications in terms of the link duration and channel time variation together the system performances and applications is also presented. In the next chapter "Visible Light Communications Based on Street Lighting," Stanislav Zvánovec et al. discuss the main features of LED systems for public lighting systems especially in connection with VLC. Main functions, control systems, and typical parameters of street lighting are given. Furthermore, main aspects associated with lighting performance and aging are summarized. The chapter also outlines recent studies of public lighting-based VLC including ray-tracing simulations, noise parameters, and delay profiles among other topics. Luis Mauel Faria et al. in "Transdermal Optical Communications" discuss an interesting application of optical communications means to establish communications with implantable medical devices (IMDs), placed under the skin. The chapter presents and validates a channel model suitable for the design of communications systems employing optical communications means. The model is further explored to assess the possibility of building energy harvesting means using the same device used for optical signal detection. Conclusions indicate that it is possible to use optical means to communicate with IMDs. Both the skin depth and the radiation wavelength have a direct effect on the signal attenuation, thus revealing optical windows for communication.

Ali Khalighi et al. in "Underwater Visible Light Communications, Channel Modeling, and System Design" consider optical communication in underwater scenarios, which is among the most significant emerging applications. After presenting the fundamental aspects, the chapter provides a comprehensive description of the aquatic channel properties and modeling. Describing light beam propagation in water and the different processes that can affect it in an aquatic medium, the authors explain how these phenomena can be modeled mathematically and discuss channel characterization using analytical and numerical methods. Considerations in the design of the transmitter and the receiver are then addressed and the chapter ends with a description of the realization of a prototype together with some experimental evaluation results. Nuno Lourenço and Martin Siegel in "VLC for Indoor Positioning: An Industrial View on Applications" present an industrial vision of VLC-based positioning systems. Rather than being grounded in strong theoretical and experimental background, this chapter describes potential use cases for VLC systems. This is a corporate vision of one of the major key players in the field of lighting in Europe—Zumtobel. The key ideas discussed are linked to positioning systems based on VLC and their potential interest for lighting applications, ranging from light commissioning systems to position infotainment applications.

Michael B. Rahaim and Thomas D.C. Little in "Optical Small Cells, RF/VLC HetNets, and Software Defined VLC" provide very interesting descriptions and analyses of the application of VLC within next generation wireless RF/VLC networks. The chapter focuses on practical aspects of VLC utilization in the context of small cells, heterogeneous networks integration, and software-defined systems. It discusses a small cell evolution and utilization of VLC directionality for network densification within RF small cells such as femtocells or wireless local area networks. The chapter provides the requirements for coexistence of RF and VLC within mixed-media environments and describes a software defined VLC implementation in RF/VLC heterogeneous networks. Mónica Figueiredo and Carlos Ribeiro address the issue of an OFDM-based VLC system prototyping using reconfigurable hardware tools. Their chapter, entitled "OFDM-Based VLC Systems FPGA Prototyping," presents the design flow for system design merging MATLAB® system generator tools with Xilinx FPGA prototyping. A design example employing DCO-OFDM is used to illustrate the concepts and establish the link with the MATLAB user. Despite the simplicity of the approach, this methodology enables fast system development means, once the user is proficient with the tools used for prototyping.

Yeon Ho Chung in "Smart Color Cluster Indoor VLC Systems" presents the use of red, green, and blue (RGB) LEDs and color clustering to provide relatively high data rates and bidirectional transmission. Solutions supporting user mobility are also presented and multiple access schemes based on color coding are described to address multiuser scenarios. The considered solutions can ensure seamless coverage over various VLC-based connected devices present in a smart home environment, for example. Last, a method for the prospective application of indoor motion detection is described based on the use of multiple detectors. Finally, we cannot forget technologies which will form the majority of devices in near future, so the last chapter "VLC with Organic Photonic Components" by Paul Anthony Haigh et al. is focused on utilization of new organic technologies within VLC. This chapter gives an overview of organic-based VLC focusing on the organic LED (OLED)-based devices, the organic semiconductors, and visible light photodetectors. To enhance the OLED-based VLC links, a number of equalization schemes are discussed and their performances are compared. Finally, an experimental all-organic VLC system employing both OLED and organic photodetectors employing an artificial neural network–based equalizer is introduced and its performance evaluated.

2

Lighting and Communications: Devices and Systems

Luis Nero Alves, Luis Rodrigues, and José Luis Cura

CONTENTS

2.1 Introduction

Humanity has always relied on light to accomplish daily tasks. Sunlight during the daytime was, and still is, the major lighting source. Prior to the invention of electric lights, sunlight and other sources based on candles and gas were used for lighting. This has enabled the advance of economies, which received a major boost with the invention of electric lights. The globalization of electric lighting enabled the 24/7 economy during the second half of the 20th century—24 hours a day, 7 days a week.

Nowadays, the purpose of lighting systems is diverse [1,2]. It is possible to divide lighting systems into three global classes: indoor lighting, outdoor lighting, and signaling [3]. Each of these classes has its own features. Indoor lighting is used in offices, homes, and public spaces. The purpose of lighting in indoor spaces is diverse, with different rules for different functions depending on the purpose of the space. In working spaces, it is important to assure adequate comfort and visibility conditions, thus depending on the characteristics of the space (offices, hospitals, intensive labor places, amongst others), different sets of rules apply. Also, different illumination devices may prove to be more effective depending on the location and environment. In public spaces and homes, both comfort and decorative aspects of light play an important part. Outdoor lighting, on the other hand, serves different purposes such as public safety and security, as well as better visibility. For outdoor lighting there are stringent rules, which are organized into different lighting classes that apply to different road conditions, types of lighting devices, and minimum light levels. Signaling systems, such as traffic control, signaling lights in vehicles, or even lighthouses serve a specific purpose for road drivers, pedestrian safety, and marine navigation.

2.1.1 Lighting Systems

Figure 2.1 depicts a conceptual overview of the main blocks in a typical lighting system. The four main components are the power source (PS),

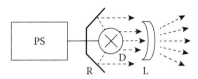

FIGURE 2.1
Lighting system conceptual overview.

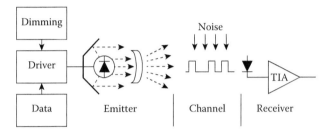

FIGURE 2.2
Combined lighting and communication functions.

the lighting device (D), a reflector (R), and a shaping lens (L). Note that the physical enclosure for the lamp is not shown in Figure 2.1; this will not be covered in this chapter. Each of the components depicted in Figure 2.1 have a specific functionality. The PS is the energy provider, which is responsible for controlling and providing energy to the lighting device. Different lighting devices may have different requirements for the PS, which usually have different names. For instance, a *driver* for a light-emitting diode (LED), *ballast* for fluorescent lamps. The next element is the lighting device where the electrical energy is converted into light. There are several types of lighting devices with different conversion mechanisms, each suitable for different applications. The most common lighting devices available on the market are fluorescent lamps, compact fluorescent lamps (CFL), high-intensity discharge lamps (HIDs), LEDs, and more recently, laser-based visible lights [1,2]. The incandescent lamps with very low energy efficiency are being phased out at a global level. The reflector element is used to confine the light radiation in a specific direction. And finally, a shaping lens is employed to assure uniform lighting conditions.

With the introduction of solid-state lighting devices, for the first time it has become possible to switch these light sources at a high speed (something that is not possible with other lights). Therefore, the possibility exists of dual functionalities of illumination and data communications, thus leading to the emergence of visible light communications (VLC) as shown in Figure 2.2.

When the same device is used for both lighting and data communication purposes, several important system design considerations should be considered. This scenario is particularly relevant when the lighting device is an LED, or an array of LEDs, since other lighting devices offer limited support for communications. On the emitter side, several factors have to be considered starting with the driver itself. LEDs are commonly driven with a constant current [1,3]. Light dimming is preferably attained using the pulse width modulation (PWM) scheme. The current through the LED is switched off by fixed amounts of time on a periodic basis. The emitted light level is proportional to the duty cycle of the PWM signal. PWM offers superior LED light control, since the peak current remains fixed during the on phase. This prevents LED color temperature variation, a feature that will be briefly discussed in this chapter. Combining data transmission with PWM control requires novel circuit design approaches. Most of the market-available LED drivers do not fully support this feature. Apart from the driver, the emitter side also includes a lens and a reflector. The effect of these components on the emitted signal has to be considered. Diffusing lenses open up the radiation pattern of the LEDs, thus producing a more uniform pattern [3]. This is suitable for lighting purposes; however it may induce penalties on the signal quality. The transmitted signal may suffer from signal fading and noise addition while propagating through the wireless channel. The optical power level drops with the square of the transmission distance, thus reducing the signal strength at the input of the receiver. Noise, arising from other light sources (such other lighting devices, or the sun), will interfere with the optical signal thus resulting in degradation of the signal-to-noise ratio (SNR) at the input of the receiver. Noise induced by other light sources has distinct properties from one type of device to another. Gas discharge lamps are known to produce both low- and high-frequency interfering signals periodically on mains power, but with rich spectral contents [3]. Sunlight, on the other hand, produces nearly uniform Gaussian noise on the photodetector (PD).

The second scenario, where lighting and communications rely on different devices, presents quite different design considerations. The impact of the lighting system on communications is limited to the noise contribution of the channel. The lighting device produces noise, which adds to the propagating optical signals between the emitter and the receiver. Mitigating ambient light noise effects is a task that can be performed at the receiver stage. Possible methods may resort to optical filtering, combined with electrical high-pass filtering, or more advanced methods relying on equalization. Particularly relevant is the type of lighting device responsible for inducing noise at the receiver. In this sense, mitigating the interference due to periodic noise sources such as gas discharge lamps of all kinds is a critical task, due to the possibility of high-frequency spectral contents acting on the signal bandwidth. This is however of minor concern for future lighting systems employing LEDs.

2.2 Radiometry, Photometry, and Colorimetry Essentials

Photometry is a branch of the wider field of radiometry. Radiometry can be described as the detector-independent measurement of electromagnetic radiation, while photometry takes into account the detector, more specifically; the detector reflects the response of the human visual system [3]. This difference between photometry and radiometry is essential to understand the scope of both measurement approaches. Radiometry can be regarded as a branch of experimental natural sciences, where the relevant methods are those from experimental physics. The measurement accuracy is generally dependent on the limits of the measuring instruments. On the other hand, photometry has more to do with applied psychology, where the relevant methods are those applied in experimental psychophysics. The accuracy of the measurements is limited by the way the performance of the human visual system is determined. This implies that radiometry is far more precise than photometry. Nevertheless, visual science and lighting engineering are usually expressed in terms of photometric units, since these convey more information about the human vision perception than the analogous radiometric entities [3].

Like photometry, colorimetry describes the human visual perception of color. Color perception is usually described in terms of the response of the three types of cone cells, able to sense the spectral contents of the light with sensitivity peaks in short (S, 420–440 nm), middle (M, 530–540 nm), and long (L, 560–580 nm) wavelengths [4,5]. Usually this is modeled as a tristimulus mapping of color perception for which the CIE chromaticity diagrams are a common tool.

2.2.1 Radiometry

Radiometry is concerned with the energy or the power of optical radiation for a given geometry of propagation. Radiometric measurements cover the entire spectrum from ultraviolet (UV) to infrared (IR) lights, being thus independent of the receiver response. There are four basic radiometric entities to consider: radiant power, radiant intensity, irradiance, and radiance. The following is a description of each these measures [3].

Radiant power or rather radiant flux Φ_e is defined as the total power dQ_e emitted by a light source per unit time dt. The radiant power is expressed in watts (W) and given by:

$$\Phi_e = \frac{dQ_e}{dt}. \tag{2.1}$$

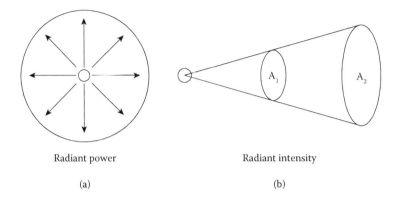

Radiant power   Radiant intensity

(a)   (b)

FIGURE 2.3
Difference between (a) radiant power and (b) radiant intensity.

Figure 2.3a depicts the definition of radiant power. It can be seen that radiant power can be measured as the total power emitted by a radiant source or the power reaching a given surface.

Radiant intensity I_e is defined as the power $d\Phi_e$ emitted per unit solid angle $d\Omega$. It is expressed in watts per steradian (W/sr) and given by:

$$I_e = \frac{d\phi_e}{d\Omega}. \tag{2.2}$$

For configurations described as point sources which hold the inverse square law, a detector with area dA at a distance r from the source defines a solid angle $d\Omega = dA/r^2$. Figure 2.3b depicts the interpretation of radiant intensity.

Irradiance E_e is defined by the ratio of radiant power $d\Phi_e$ and the area of dA of the detector. Irradiance is expressed in watts per square meter (W/m^2) and given by:

$$E_e = \frac{d\Phi_e}{dA}. \tag{2.3}$$

It is possible to establish the relationship between radiant intensity and irradiance for a point source, relating equations (2.2 and 2.3), and the definition of solid angle as given by:

$$E_e = \frac{I_e d\Omega}{dA} = \frac{I_e}{r^2}. \tag{2.4}$$

Radiance is usually employed for extended light sources (sources that cannot be described as point sources). Radiance L_e is defined as the radiant power $d\Phi_e$ emitted from an area dA_e per unit of solid angle $d\Omega$. It is expressed in watts per steradian per square centimeter (W/sr·cm^2) and is given by:

$$L_e = \frac{d^2\Phi_e}{dA_e d\Omega}. \qquad (2.5)$$

2.2.2 Photometry

Photometric measures are analogous to the above-mentioned radiometric quantities. The major difference lies in the fact that in photometry, the measuring equipment takes into account the visual perception of the human eye. The photometric measures are obtained from the corresponding radiometric measures through a weighted average [3]. The weighting function can consider the human visual perception in one of the three defined conditions: under photopic conditions, under scotopic conditions, or under mesopic conditions. The photometric measures analogous to the radiant flux, radiant intensity, irradiance, and radiance are luminous flux Φ_v, luminous intensity I_v, illuminance E_v, and luminance L_v, respectively. These measures are expressed in lumen (lm), lumen per steradian (lm/sr, also known as candela—cd), lumen per square meter (lm/m^2, also known as lux—lx), and candela per square meter (cd/m^2), respectively. Luminous flux is given by:

$$\Phi_v = K_m \int_{380\ nm}^{780\ nm} \Phi_e(\lambda)V(\lambda)d\lambda, \qquad (2.6)$$

where $K_m = 683$ lm/W is a constant establishing the relationship between the (physical) radiometric unit watt and the (physiological) photometric unit lumen. $V(\lambda)$ represents the spectral sensitivity curve of the human visual system. All the other photometric quantities are related to the weighted integral of their corresponding radiometric quantities.

The spectral sensitivity of the human visual system under daytime conditions, also called daytime vision, is described by the $V(\lambda)$ curve. Under daytime visual conditions, only the cones inside the human retina are operational. These cells are responsible for visual perception; their sensitivity to light stimulus varies with the wavelength of the impinging radiation. The sensitivity curve, or $V(\lambda)$ curve, was established as a standard function by the Commission Internationale de l'Eclairage (CIE) in 1924 [3]. It is usually available in either tabulated or graphical forms. Figure 2.4 shows the $V(\lambda)$ curve for photopic vision.

Scotopic vision takes place under low light conditions, when only the rod cells inside the retina are active. Their spectral sensitivity is similar in form to the $V(\lambda)$ curve, for photopic vision. In 1951, the CIE adopted the standard scotopic luminosity function, also available in either tabulated or graphical forms.

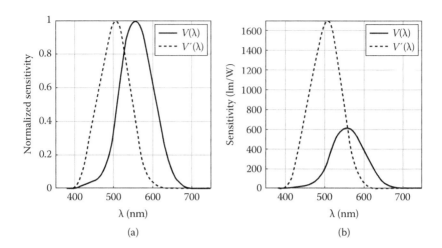

FIGURE 2.4
Photopic and scotopic vision sensitivity curves: (a) normalized version and (b) nonnormalized version.

Scotopic vision sensitivity is expressed by the $V'(\lambda)$ curve; Figure 2.4 depicts both the curves for comparison. It is readily apparent that the major difference is the peak wavelengths. There is also mesopic vision, which relates to intermediate lighting situations. Under these conditions, both the rod and cone cells inside the retina are active. The sensitivity exhibits intermediate values between $V(\lambda)$ and $V'(\lambda)$. Mesopic vision is important for traffic lighting systems, where the road surface luminance stays above the scotopic limit and falls below the photopic limit. Current trends in outdoor lighting are considering mesopic vision for light optimization, due to the fact that photopic vision is a poor predictor of how well humans see at night. The mesopic sensitivity curve is commonly expressed as a linear combination of $V(\lambda)$ and $V'(\lambda)$, which is given as:

$$V_m(\lambda) = (1-x)V'(\lambda) + xV(\lambda), \tag{2.7}$$

where x is a constant between 0 and 1, depending on the photopic luminance.

2.2.3 Colorimetry

Colorimetry is related to the visual perception of color by the human visual system. It also provides qualitative and quantitative descriptions of color. In 1931, the CIE introduced the X, Y, Z tristimulus system, based on the assumption that every color is described as a combination of the three primary colors—red, green, and blue [4]. The tristimulus X, Y, and Z are obtained by the integration of the spectral power distribution $S(\lambda)$ weighted with the three eye-response curves $x(\lambda)$, $y(\lambda)$, and $z(\lambda)$ over the visible wavelength region, which can be expressed as:

$$X = \int_{380\ nm}^{780\ nm} S(\lambda)x(\lambda)d\lambda$$

$$Y = \int_{380\ nm}^{780\ nm} S(\lambda)y(\lambda)d\lambda \qquad (2.8)$$

$$Z = \int_{380\ nm}^{780\ nm} S(\lambda)z(\lambda)d\lambda.$$

Figure 2.5 depicts the tristimulus weighting functions $x(\lambda)$, $y(\lambda)$, and $z(\lambda)$. The X, Y, Z values can be further converted into the color coordinates and represented in a color space. The CIE introduced the xyY color space, depicted in Figure 2.5, currently one of the most used color diagrams. Color coordinates on this map are expressed by the following coordinate conversion:

$$\begin{bmatrix} x \\ y \end{bmatrix} = \frac{1}{X + Y + Z} \begin{bmatrix} X \\ Y \end{bmatrix}. \qquad (2.9)$$

The color map itself is constructed using the weighting functions $x(\lambda)$, $y(\lambda)$, and $z(\lambda)$, and appears calibrated according to the wavelengths, as depicted in Figure 2.5. Coordinates x and y may represent color coordinates for specific devices. Other mappings may adopt a different set of coordinates, such as RGB (red, green, and blue, where the weighting functions are defined in a different manner), amongst others [4]. The Y component has a direct relation with the adaptation of equation (2.6) for illuminance. The fact is that $y(\lambda)$ represents the photopic sensitivity curve $V(\lambda)$ as can be observed in Figures 2.4 and 2.5. Thus, if $S(\lambda)$ is the spectral power distribution (SPD) of the transmitting source, Y represents its associated illuminance measured in lux.

The color gamut or the ability of a given source to represent colors can be adequately represented in these color spaces. For instance, lighting devices are ideally monochromatic sources, however, due to device operational principles, the spectral response may encompass several colors. In this sense, the color gamut is a measure of how "clean" the source can be. The spectral content of lighting devices is of paramount importance, since it has implications for color perception. Two quantitative measures were introduced to characterize these effects, the correlated color temperature (CCT) and the color rendering index (CRI). Black-body radiation changes with temperature starting from red for low temperatures and shifting to blue for high temperatures. The CCT is a measure of the color temperature of a given light source, when

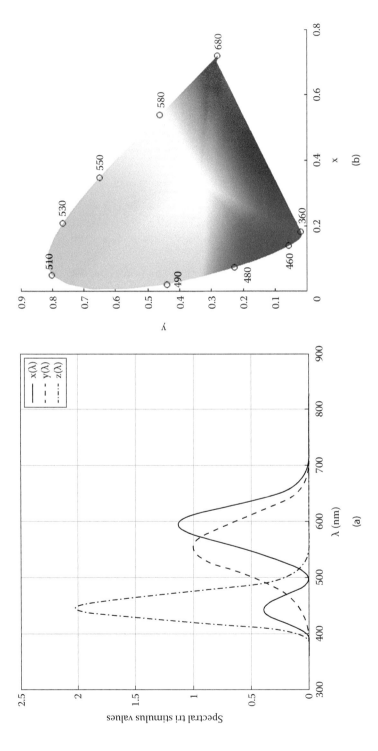

FIGURE 2.5
(a) Tristimulus weighting functions and (b) CIE xyY chromaticity diagram.

compared with a black body. Due to this correlation, the CCT is expressed in Kelvin. This measure is independent of color perception, but rather focused on the light device itself. There are several lighting devices with different CCTs. For instance, for white light sources CCT values may range from 3200 K (typical for incandescent light bulbs) up to 4000 K (normal daylight). The consistency of the CCT with time and operational conditions is an important quality measure of a given lighting device. The CRI, on the other hand, quantifies the color perception enabled by a given light source, when compared with a reference source (usually, the reference is taken as natural daylight). It is expressed as a percentage, with values ranging from 0 to 100. Lighting devices with high CRI enabled more realistic color perception, thus justifying the importance of this parameter on the selection of the device. It is also relevant for device selection to collect information on the variation of the CRI with time and operational conditions.

2.2.4 Black-Body Radiation

As introduced before, the black-body radiation spectrum serves as a basis to define the CCT of lighting devices. The CCT expresses the equivalent temperature of a black body able to produce similar lighting perception [4]. The black-body radiation spectrum is defined by Planck's law given by:

$$S(\lambda) = \frac{2hc^2}{\lambda^5} \left[e^{\frac{hc}{\lambda kT}} - 1 \right]^{-1}, \tag{2.10}$$

where h is Planck's constant, c is the speed of light, k is Boltzmann's constant, T is the temperature in Kelvin, and λ is the wavelength. Figure 2.6 depicts the black-body radiation spectrum for three different temperatures. It also represents the locus of coordinates on the color map, as the black-body temperature increases from 1000 K (near the red–yellow range) to 10,000 K (shifting into the blue range). This locus is also known as Planckian locus. It plays an important role in the characterization of lighting devices, since it covers the range of color temperatures normally used for lighting purposes.

Finding the CCT of a given device involves the determination of the temperature of the black body able to approach the color coordinates of the device. This can be achieved following a suitable optimization problem, able to minimize the distance between a point in the Planckian locus and the point representing the color coordinates of the device. A better approach consists of using the McCamy's formula for the CCT. McCamy's formula applies to color coordinates close to the Planckian locus. Under this condition, it is able to approach the CCT of the device with less than 2% error. McCamy's formula is expressed by the following equation [5]:

$$CCT = 449n^3 + 3525n^2 + 6823.3n + 5520.33, \tag{2.11}$$

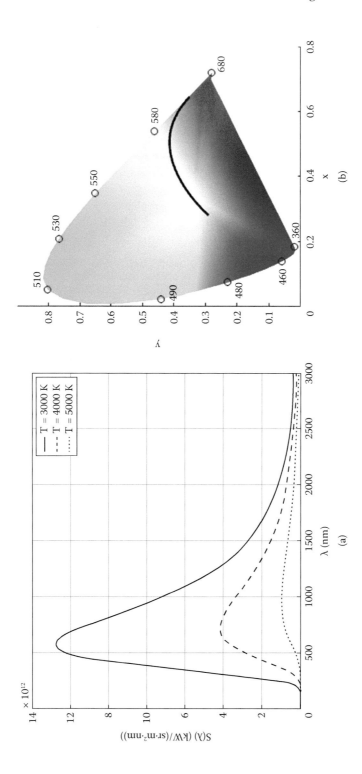

FIGURE 2.6
(a) Black-body radiation spectrum and (b) the Planckian locus on the CIE color map.

with n given by:

$$n = \frac{x - 0.332}{0.1858 - y}.$$ (2.12)

McCamy's formula is derived from a polynomial approximation of the Planckian locus.

2.3 Lighting and Communicating Devices

This section introduces some basic concepts of the physics of LEDs and PDs. It begins with the underlying principles of conduction in semiconductor P–N junctions and photon emission/absorption mechanisms.

2.3.1 Semiconductor Materials and the P–N Junction

To understand the physical principles of LEDs and photodiodes, it is best to start with some basic introduction to semiconductor materials. Semiconductor materials can be broadly categorized as a class of materials between conductor and isolator materials [6]. In this sense, semiconductor materials are not able to conduct current as well as a conductor, such as copper or aluminum. On the other hand, they do not behave as insulators either. Semiconductor materials form crystalline structures, through the sharing of valence electrons between neighboring atoms, in a process known as covalent bonding. Naturally occurring elements, such as silicon (Si) or germanium (Ge), have the special characteristic of forming semiconducting crystals. These are known as the group IV semiconductors. There are other semiconductor alloys formed by composite materials, such as gallium arsenide (GaAs), gallium nitride (GaN), or indium phosphide (InP). These are known as composite semiconductors, or III–V semiconductors. Simple semiconductor materials such as Si and Ge form crystal structures where each atom shares four valence (group IV) electrons with another four neighboring atoms. The crystal structure of composite semiconductors is more complex, since it involves both group III and group V elements, sharing three and five valence electrons, respectively. These are known as bulk or intrinsic semiconductor materials. At thermal equilibrium, the valence electrons are tightly bound to the covalent bonds, and the material exhibits a high resistance. As the electrons acquire energy, which can be either due to thermal agitation or optical radiation, they can jump to the conduction band and the resistivity of the material reduces.

It is possible to add elements of other groups, with a deficit or an excess of valence electrons, to the intrinsic semiconductor. These are the extrinsic

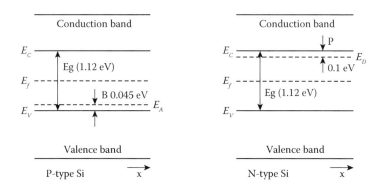

FIGURE 2.7
Energy band diagrams of P-type and N type Si semiconductors.

semiconductors, which can be of two types: type P, having dopant elements with one valence electron less than the intrinsic semiconductor; or type N, having dopant elements with one valence electron in excess. Type P and N dopants are called acceptors and donors. Type P semiconductors have a deficit of electrons on the valence band; as such they can attract free electrons and give rise to hole currents (a hole is the absence of one electron in the crystal lattice). Type N semiconductors have an excess of electrons, which are not bound to covalent bonds and thus free to conduct currents.

Figure 2.7 depicts the energy band diagrams of P and N type Si semiconductors. It is readily apparent how the donor and acceptor dopants improve the conductivity of the sample. For the P-type acceptor elements, such as boron, raise the valence band energy (E_A) above the energy of the intrinsic semiconductor (E_V). Electrons acquiring energy higher than the difference between the conduction band and the valence band energies can jump into the conduction and break free from their native atoms. The same reasoning holds for the N-type semiconductors, with donor elements such as phosphorous, where the conduction band due to the presence of the donor atoms (E_D) is lower than the energy of the intrinsic semiconductor (E_C).

The P–N junction diode is formed when two samples of P and N type semiconductors are brought together into contact. The conduction of the P–N diode depends on the biasing direction of the external field. At thermal equilibrium, excess holes from the P side join together with the excess electrons of the N side, in a process called recombination. The recombination occurs near the junction border leaving the surrounding space empty from free carriers (electrons and holes)—this is the formation of the so-called depletion region. This process gives rise to a built-in potential that prevents holes from the P side from acquiring enough energy to cross and recombine with electrons on the N side. At thermal equilibrium, there is no net current through the junction. Figure 2.8 depicts the energy band diagrams of a P–N junction under the three

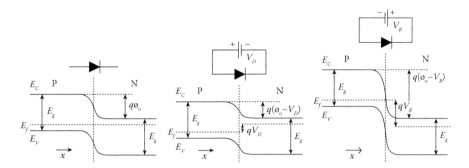

FIGURE 2.8
Energy band diagrams, from left to right: thermal equilibrium, direct biasing, reverse biasing.

possible biasing conditions: thermal equilibrium, direct biasing, and reverse biasing. Under thermal equilibrium, the Fermi level (E_f) of the junction at equilibrium is aligned on both sides of the junction. The energy required to produce this alignment is given by $q\phi_0$, where ϕ_0 is the built-in potential and q the charge of the electron. The separation between valence and conduction bands is the gap energy, E_g. Under direct biasing, the applied external voltage reduces the effect of the built-in potential, making possible current conduction due to minority carriers (electrons from the N side that cross to the P side and holes doing the opposite). On the contrary, reverse biasing adds more energy to the built-in potential, thus contributing to a larger energy separation between the conduction bands of the P and N sides.

As remarked previously, electrons can acquire energy due to other sources, such as thermal agitation or optical radiation [6,7]. Generation of electron–hole pairs due to optical radiation is a process of particular relevance for PDs made from semiconductor materials. On the other hand, recombination of electron–hole pairs near the junction is relevant for light production. Both recombination and generation of electron–hole pairs encompass a process where there are energy-momentum changes. In the process, both energy and momentum are conserved. Recombination occurs when an electron crossing the junction, loses energy quanta given by the gap energy. This energy change involves the emission of a photon with wavelength given by E_g. On the other hand, it may happen that an incidental photon with enough energy produces an electron–hole pair near the junction. This corresponds to an electron in the valence band acquiring enough energy to pass into the conduction band. Generation–recombination of electron–hole pairs depends on the type of semiconductor in use [6]. Formerly, semiconductors were classified as simple and composite, and inside each class, as P and N type according to the presence of a dopant species of known type. There is one further classification, describing how the alignment between valence and conduction bands behaves. Generally speaking, there are two types of

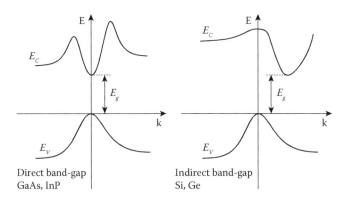

FIGURE 2.9
Direct and indirect band-gap semiconductors E–k diagrams.

semiconductors falling into this classification: direct band-gap and indirect band-gap semiconductors. To better understand the differences between direct and indirect band-gap semiconductors, it is necessary to look into the E–k diagrams (energy versus wavenumber—a representation of momentum). Figure 2.9 depicts the E–k diagrams of direct and indirect band-gap semiconductors [6]. As it can be seen, the conduction and valence energy levels profile changes with the wave vector k. In direct band-gap semiconductors, it happens that the minimal conduction band energy is aligned with the maximal valence band energy. Thus, in direct band-gap semiconductors, the passage of one electron from the valence band into the conduction band involves mostly energy exchange, with a small trade in momentum. On the other hand, for indirect band-gap semiconductors, the same process involves not only energy exchange but also momentum exchange. In the former case, emission or absorption of a photon is enough to produce the energy jump (recall that photons have higher energy when compared to their momentum). In the latter case, a third particle has to be involved—the phonon. Emission or absorption of a photon with the required wavelength is not enough; the phonon has to be involved to compensate the momentum trade and enable the process. Fortunately, phonons are widely available due to thermal agitation of the semiconductor lattice. Nevertheless, the process of photonic emission/absorption in indirect band-gap semiconductors is less efficient [7].

2.3.2 The Light-Emitting Device—LED

LEDs are P–N junction diodes able to take profit from the photonic emission process previously described. LEDs operate in the direct biasing condition, with an external source providing the necessary voltage for current conduction.

Electrons crossing the junction are subjected to a process called radiative recombination. Recombination occurs when an electron loses energy and passes from the conduction to the valence band. In LEDs, this process is accompanied by the emission of a photon with energy given by the difference between the conduction and valence levels. Given the above description, it would appear that LEDs behave as coherent light sources, emitting radiation at a single wavelength given by the band-gap energy of the material they are made of. In fact this is not so, due to the quantum nature of electrons. Electrons are fermionic particles, bonded to obey the Pauli exclusion principle—that is, there cannot be two electrons occupying the same energy level [6]. What happens then, is that the electrons in the conduction band are organized in such a way as to occupy the minimum energy levels available, near the conduction band minimum. Given that the energy levels are very close to each other, as represented in Figure 2.10, when an electron jumps there is a probability of a photon being emitted with energy slightly larger than the band-gap energy E_g. Thus, the emitted photons will exhibit a continuum of wavelengths [7]. Figure 2.10 also depicts the normalized emission intensity $I(E)$ as function of energy, for an AlGaAs LED. The dynamics of $I(E)$ are governed on one side by the density of states in the conduction band (proportional to $(E - E_g)^{1/2}$), and by the Maxwell–Boltzmann distribution on the other (proportional to $e^{-E/k_B T}$, with k_B as the Boltzmann constant and T the temperature):

$$I(E) = I_{peak}\sqrt{E - E_g}e^{-\frac{E}{k_B T}}. \tag{2.13}$$

The spectral line width and the peak wavelength are given by:

$$\Delta\lambda = mk_B T\frac{\lambda_0^2}{hc}, \tag{2.14}$$

and

$$\lambda_0 = \frac{hc}{E_g + k_B T/2}, \tag{2.15}$$

where c is the speed of light, m a constant dependent on the material ($m = 1.8$ for AlGaAs LEDs), and the energy $E = E_g + k_B T/2$ is the energy of the peak in $I(E)$. From (2.15), it is readily apparent that the emission spectrum of the LED depends on temperature. To find the temperature dependence of $I(E)$, it is necessary to know how E_g changes with temperature. Following a result by Varshni, this is given by:

$$E_g = E_{g0} - \frac{AT^2}{B + T}, \tag{2.16}$$

where A and B are Varshni constants dependent on the material ($E_{g0} = 1.932$ eV, $A = 0.658$ meV/K, $B = 248$ K, for AlGaAs LEDs). Figure 2.11 depicts the

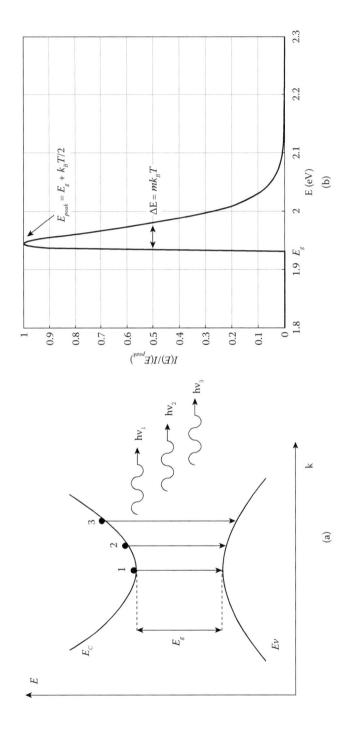

FIGURE 2.10
(a, b) LED emitted spectrum.

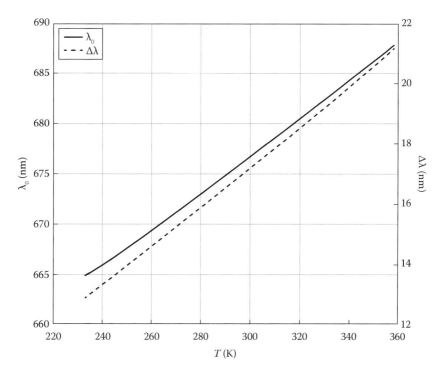

FIGURE 2.11
Temperature dependence of an AlGaAs LED.

peak wavelength and the half-line width against temperature: both are a monotonic functions of the temperature.

Several colors can be synthesized, using an adequate semiconductor with the required band-gap energy. Table 2.1 provides some examples of compound semiconductor materials used to produce LEDs with different colors.

There are several important parameters used to characterize LED performance. Of paramount importance are the internal and external quantum efficiencies of the LED. The internal quantum efficiency, η_{IQE}, measures the effectiveness of the light generation process. It is given by the ratio between the radiative recombination rate (the number of carriers that effectively contribute to photon generation) and the total recombination rate (including the carriers involved in current conduction through the device) as:

$$\eta_{IQE} = \frac{\text{Radiative recombination rate}}{\text{Total recombination rate}} = \frac{\tau_r^{-1}}{\tau_r^{-1} + \tau_{nr}^{-1}}, \qquad (2.17)$$

where τ_r and τ_{nr} are the minority carriers' lifetime constants for radiative and nonradiative recombination, respectively. Alternatively, it could be expressed

TABLE 2.1

LED Semiconductor Materials

Color	Wavelength (nm)	Semiconductor Materials
Infrared	$\lambda > 760$	GaAs, AlGaAs
Red	$610 < \lambda < 760$	AlGaAs, GaAsP, AlGaInP, GaP
Orange	$590 < \lambda < 610$	GaAsP, AlGaInP, GaP
Yellow	$570 < \lambda < 590$	GaAsP, AlGaInP, GaP
Green	$500 < \lambda < 570$	InGaN/GaN, GaP, AlGaInP, AlGaP
Blue	$450 < \lambda < 500$	ZnSe, InGaN
Violet	$400 < \lambda < 450$	InGaN
Ultraviolet	$\lambda < 400$	Diamond, AlGaN, AlGaInN

by the ratio between the number of photons emitted per second and the total number of carriers lost per second as given by:

$$\eta_{IQE} = \frac{\text{Photons emitted per second}}{\text{Carriers lost per second}} = q\frac{\Phi_{ph}}{I_D} = \frac{q}{h\nu}\frac{P_{o(int)}}{I_D}, \tag{2.18}$$

where I_D is the current supplied by the external source, $P_{o(int)}$ is the internally generated radiant flux, and Φ_{ph} represents the number of emitted photons per second. Since not all the emitted photons are effectively emitted as external light due to device construction details, it is usually necessary to introduce another efficiency measure, able to express the ratio between the internally generated photons and the photons that are effectively emitted to the outside. This is introduced by the extraction efficiency, η_{EE}, given by:

$$\eta_{EE} = \frac{\text{Photons effectively emitted}}{\text{Photons generated internally}} = \frac{P_o}{P_{o(int)}}. \tag{2.19}$$

It is possible to express the radiant flux as a function of the device current, using Equations 2.18 and 2.19, resulting in:

$$P_o = \frac{h\nu}{q}\eta_{IQE}\eta_{EE}I_D. \tag{2.20}$$

Equation 2.20 shows that to at large extent, the radiant flux of an LED is proportional to the current supplied to the device. This linear behavior is compromised due to thermal effects and high carrier injection levels. The external quantum efficiency η_{EQE} measures the ratio between the number of effectively emitted photons and the total number of carriers lost due to recombination, which is given by:

$$\eta_{EQE} = \frac{q}{h\nu}\frac{P_o}{I_D} = \eta_{IQE}\eta_{EE}. \tag{2.21}$$

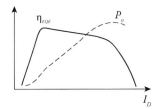

FIGURE 2.12
External quantum efficiency and LED optical power as a function of the supplied current.

Figure 2.12 depicts the general behavior of the external quantum efficiency as a function of the supplied current. For low currents (low carrier injection conditions), the external quantum efficiency increases with the supplied current. For moderate carrier injection, η_{EQE} falls with the I_D; this decrease becomes appreciable at high carrier injection regimes, where the supplied current is high enough to produce heat and degrade LED performance. The optical power, measured in terms of the LED radiant flux, is essentially linear with I_D under moderate carrier injection conditions. Optimizing η_{EQE} is a complex task as it involves not only the device material, but also its architecture, its encapsulation, and its thermal management [7]. Higher power LEDs are in this sense optimized to operate under constant current levels. There is also the power conversion efficiency measuring the ratio between the radiant flux P_o and the externally supplied electrical power, as given by:

$$\eta_{PCE} = \frac{P_o}{I_D V_D} \approx \eta_{EQE}\left(\frac{E_g}{qV_D}\right). \tag{2.22}$$

The power conversion efficiency is approximately proportional to the band-gap energy of the material. All the above efficiency measures are wavelength dependent. It is also usual to use derived quantities able to express the LED efficiency in photometric units, thus the luminous efficacy η_{LE} given by the ratio between the luminous flux, Φ_v, and the supplied electrical power is defined as:

$$\eta_{LE} = \frac{\Phi_v}{I_D V_D} = \frac{K_m}{I_D V_D} \int_{380 \text{ nm}}^{780 \text{ nm}} P_o(\lambda)V(\lambda)d\lambda, \tag{2.23}$$

where K_m is the photometric electric to optical conversion factor with value 683 lm/W, and $V(\lambda)$ is the photopic visual sensitivity curve.

LEDs have different structures [7]. As seen previously, the device structure has a great impact on the external quantum efficient. This is particularly

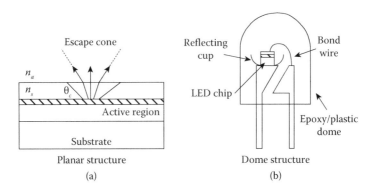

FIGURE 2.13
LED structures: (a) planar LED and (b) dome LED.

evident when considering the extraction efficiency of the device. A good design should improve the light output in terms of the internally generated photons. However, this is not a straightforward task. Two of the most used device structures are the planar and dome, depicted in Figure 2.13. The planar LED is the simplest arrangement. The light extraction is constrained by the reflection coefficients n_a and n_s, due to the change from the LED medium to the exterior. This structure is limited by total internal reflection (TIR). TIR results from the critical angle θ_c due to the different reflection coefficients of the semiconductor–air interface. As a result, some existing directions are forbidden, limiting the extraction efficiency of the device. A more efficient design is the dome structure. Dome LEDs can overcome the TIR effect, placing the LED chip in the center of a plastic or epoxy dome. The reflection index of the plastic/epoxy material is matched to the semiconductor, and the photons reaching the dome surface do not suffer TIR. Dome structures also employ a reflecting cup able to collect photons emitted from nonprivileged directions. There are other LED structures optimized for fiber optic applications which will not be covered here.

2.3.3 LED Circuit Models

The electrical operation of the LED is not much different from a normal P–N junction diode. The current–voltage characteristic of the device has the same form as a normal diode, as given by:

$$I_D = I_o \left(e^{\frac{qV_D}{\eta k_B T}} - 1 \right), \tag{2.24}$$

where I_o represents the device leakage current, when biased in the reverse direction ($V_D < 0$), and η represents a nonideality fitting parameter. As in a normal diode, LEDs also have an intrinsic capacitance. The intrinsic capacitance of a diode as two major contributing effects: (i) the depletion capacitance due to the presence of a depletion region between P and N sides (this component dominates under reverse biasing conditions) and (ii) a second component due to charge diffusion and storage effects under forward bias conditions. A simple model for the capacitance includes these two contributions:

$$C_D = C_j + C_d, \tag{2.25}$$

$$C_j = \frac{C_{j0}}{\left(1 - \frac{V_R}{\phi_0}\right)^m}, \tag{2.26}$$

$$C_d = \frac{q}{k_B T} \tau_T I_D, \tag{2.27}$$

where C_j and C_d stand for the depletion and diffusion capacitance, respectively, V_R is the diode reverse voltage, τ_T is the minority carriers total transit time ($1/\tau_T = 1/\tau_r + 1/\tau_{nr}$), and C_{j0} is the depletion capacitance at zero bias. As can be seen, the behavior of the diode capacitance is strongly dependent on biasing conditions.

When modeling the LED for transmitting applications, two approaches can be considered: digital modulation and analog modulation. Digital modulation of the LED involves two distinct states, which can be on/off or some intermediate configuration. In either case, the device model must be able to handle large signal conditions, possibly covering both forward and reverse operation conditions. This means that Equations 2.25 through 2.27 must be used in their full detail. This is especially relevant for on/off applications, where the rise and fall times are limited by the LED intrinsic capacitance. For analog applications, the device is used in the forward mode, under a defined biasing condition. Assuming the device is biased with (V_Q, I_Q), and that $V_D = V_Q + v_d$, produces small changes in I_D, expanding I_D in a Taylor series around V_Q gives:

$$I_D \approx I_Q \left(1 + \frac{q}{\eta k_B T} v_d + \left(\frac{q}{\eta k_B T} v_d\right)^2 + \dots\right). \tag{2.28}$$

The first term in the expansion gives the linear dependence of I_D on v_d, that is, the small-signal contribution due to v_d, which translates in the device small-signal conductance g_d, given by:

$$g_d = \frac{q I_Q}{\eta k_B T}. \tag{2.29}$$

The linear behavior can be well approximated by the first term in the Taylor expansion as long as v_d remains small, as given by:

$$v_d \ll \frac{\eta k_B T}{q}. \tag{2.30}$$

The small-signal model for the device intrinsic capacitance follows directly from (2.25). Under forward bias conditions, the depletion capacitance is very small when compared to the diffusion capacitance. Thus, the small-signal model includes only the diffusion capacitance evaluated at the biasing condition.

Another important aspect of LED performance is its bandwidth. There are two definitions for the LED bandwidth resulting in different values; these are the optical bandwidth and the electrical bandwidth. The optical bandwidth is the bandwidth to which the LED can be modulated, while the electrical bandwidth is defined as the electrical bandwidth perceived by the photo-detector (PD) [7]. The LED output has the units of power; the power spectrum of the LED is given by:

$$\frac{P_o(\omega)}{P_o(0)} = \frac{1}{\sqrt{1+(\omega\tau_T)^2}}. \tag{2.31}$$

In terms of power, the cut-off frequency, which occurs at 50% of the constant emitted power, is given as:

$$f_{op} = \frac{\sqrt{3}}{2\pi\tau_T}. \tag{2.32}$$

On the other hand, in a PD the detected current is proportional to the incident optical power, thus in terms of power spectrum results we have:

$$\frac{|I_p(\omega)|^2}{|I_p(0)|^2} = \frac{1}{1+(\omega\tau_T)^2}. \tag{2.33}$$

From which the electrical cut-off frequency follows the standard -3 dB definition, given as:

$$f_{el} = \frac{1}{2\pi\tau_T}. \tag{2.34}$$

An interesting observation concerns the power–bandwidth product of an LED. In an LED, the power–bandwidth expresses the trade-off between power and bandwidth, while showing what parameters are involved. In principle, wide bandwidths in LEDs can be achieved through an adequate reduction of the carrier lifetime. However, this impairs the internal quantum efficiency of the device, thus constraining the output power. Using for the definition, the optical bandwidth as expressed in (2.32) together with the output power given by (2.20), and recalling the definition of internal quantum efficiency in (2.17), results in the following power–bandwidth relation:

$$P_o f_{op} = \frac{\sqrt{3}}{2\pi} \frac{h\upsilon}{q\tau_r} \eta_{EE} I_D.$$ (2.35)

The power–bandwidth in an LED is proportional to the injected current [7]. This behavior is maintained for moderate carrier injection conditions. For high carrier injection levels, the output power drops with the external quantum efficiency. As a consequence of this behavior, there is an optimum current level at which the power–bandwidth is maximum. For currents below this value, the power–bandwidth increases linearly with the injected current.

2.3.4 White LEDs

White LEDs are currently seen as lighting devices able to improve efficiency and lifetime in modern lighting scenarios. As such, the complete replacement of conventional lighting devices, like compact fluorescent and HIDs, is currently taking place. This raises the potential to explore new avenues on top of LED lighting scenarios, such as improved lighting conditions tailored for each environment, or even explore the lighting installation to disseminate information through the light [1,2].

The introduction of the white LED was not exempt from problems. For a long time, the problem of generating efficient white light with solid-state devices remained a challenge. White light production can be accomplished using color combination. One possible approach is to combine the emission spectrum of RGB LEDs to form white light. However, achieving the necessary efficacy to make RGB LEDs interesting replacements for conventional lighting devices was another issue. The efficacy of the combined light as well as the color rendering promoted by RGB LEDs was not good enough. Another approach, also exploring color combination, relied on the usage of blue LEDs. Blue LEDs were first introduced by RCA in 1972 [7]. These first blue LEDs were also not efficient enough to be explored as lighting devices. Real advances came in 1994 with the first demonstration of a high brightness blue LED, introduced by Shuji Nakamura. The white LED quickly followed this important landmark. Figure 2.14 compares the SPDs of a typical white LED with the blue, green, and red LEDs [7]. As it can be seen, the spectral emission of white LEDs is able to cover the entire visible range. The photometric characterization of a white LED will be explored later on in this chapter.

The process to convert the blue light into white involves the use of phosphorescence material. Phosphorescence involves a material called phosphor, which absorbs photons with a given energy and re-emits them with a lower energy. The excitation photons have enough energy to free valence electrons within the phosphor. These free carriers undergo a process of non-radiative

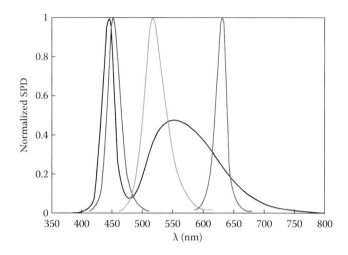

FIGURE 2.14
Normalized spectral power distribution of a white LED (black line) with blue, green, and red LEDs.

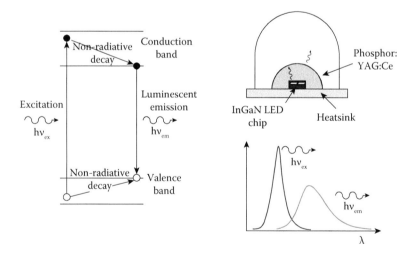

FIGURE 2.15
White light production.

decay by which they lose energy. As depicted in Figure 2.15, the phosphorous emission results in photons of less energy. As energy and wavelength are inversely related, this means that a photon with blue as the dominant wavelength is re-emitted as a photon of less energy and a larger wavelength, in this case being in the yellow region. The combination of blue and yellow photons produces the white light. There are several phosphorescence materials of interest. For white LEDs, one of the most used phosphors is YAG:Ce, yttrium aluminum garnet ($Y_3Al_5O_{12}$), doped with cerium ions (Ce^{3+}). This phosphor is able to absorb blue photons and emit yellow ones, it was also used as scintillator in cathode ray tubes [7]. Figure 2.15 also depicts the phosphor coating on top of the LED chip. It is possible to combine different phosphors and improve the quality of light produced by these devices. The net effect results in tailoring the SPD to achieve some design specifications. Among several possibilities, it is often desirable to control the color temperature of white LEDs (commercial devices are rated for different CCTs); or to improve the LED efficacy (better said, the amount of radiated power that is effectively perceived as white light). Using white lights as general purpose lighting devices has its own problems not seen before. For decades, since their introduction LEDs were seen as indicator devices, with low efficacies and usually used in low-power applications. The use of these low-power devices as means to achieve high luminous output was accomplished using large arrays. These are for instance, the cases of LED advertising panels or even traffic lights. Here, the use of RGB LEDs was a major asset. Lighting applications demanded a different set of characteristics from LEDs, most notable: low cost, high efficacy (at least comparable to state-of-the-art lighting devices), and high luminous output (implying large current handling). This paradigm shift had tremendous implications on device design and specification. Being high-power devices means that these LEDs must handle large operating temperatures. Temperature has a negative effect on the overall performance of LEDs. As previously discussed, it induces spectrum changes, it affects the external quantum efficiency of the device, and furthermore, it reduces the useful lifetime of the device. The lifetime of the LED is measured in terms of its luminous output. Aging traduces in depreciation of the luminous output of the device. Typically, 70% of the maximum luminous output is been used as the cut-off for definition of the device lifetime. Operating LEDs at temperatures higher than the recommended values accelerates aging effects. Device optimization trends take all these into account; for thermal management of the device, high-power LEDs are furnished with specially designed packages able to reduce the thermal resistance. Most often, the LED chips are directly assembled on top of aluminum placeholders; this is a measure to improve thermal extraction from the LED chip to the heatsink.

Another problem related to the usage of high currents is the depreciation of the external quantum efficiency due to high carrier injection levels. This is a source of nonlinear transfer from electrical to optical domains, but more than that it is a negative factor in terms of efficacy. One approach used to circumvent this problem resorts to the use of chip-on-board (COB) LEDs [7]. COB LEDs are arrays of devices assembled in the same package (sometimes, sharing the same phosphor coating). Each device is optimized for maximum efficiency. The output of the combined set of devices delivers higher power and allows an improvement in the efficacy of the overall device, without implying high carrier injection conditions. Currently available COB LEDs operate mostly as single devices, with two terminals and no means of individual device control. Access to individual devices within the COB is an interesting asset. This feature could enable optimized usage for dimming and communication applications, allowing them to achieve higher bandwidths [8].

In lighting applications, LEDs are operated under conditions for which they were optimized, meaning that (i) the LED current should be tuned for highest efficacy, (ii) the operating temperature should not surpass the limits for lifetime specification, and (iii) heatsinks should be employed to improve thermal extraction.

2.3.5 LED's Colorimetric Modeling

A simple approach to model the SPD of white LEDs is to use Gaussian distributions, centered on the device response maxima. Following this approach, the LED's SPD can be approximated by:

$$S(\lambda) = \sum_i w_i S_i e^{-\left(\frac{\lambda - \lambda_i}{\sqrt{2}\sigma_i}\right)^2}, \tag{2.36}$$

where S_i is the spectral power of the device at the peak wavelength λ_i and σ_i represents the power spreading around λ_i. w_i is a weighting factor describing the additive proportions of each peak wavelength. This information can be retrieved from the LED's datasheet. Equation 2.36 can serve as a simple, yet suitable model for both RGB and white LEDs. Taking as an example the case of a white LED, with peak wavelengths on the blue range at 460 nm and yellow range at 555 nm, having a full width half maximum (FWHM) of 25 nm and 150 nm, respectively; the SPD of the device can be described by:

$$S(\lambda) = \xi S_1 e^{-\left(\frac{\lambda - \lambda_1}{\sqrt{2}\sigma_1}\right)^2} + (1 - \xi) S_2 e^{-\left(\frac{\lambda - \lambda_2}{\sqrt{2}\sigma_2}\right)^2}, \tag{2.37}$$

where the weighting factors are taken as ξ and $1 - \xi$, respectively, with $\xi > 0$. Changing ξ between 0.05 and 0.2 produces the results depicted in Figures 2.16 and 2.17. Figure 2.16 shows that indeed, the simulated SPD resembles the SPD of a real white LED. The peak proportions change with ξ and as a consequence the CCT of the simulated device also changes as depicted in Figure 2.16b. This may reflect device fabrication induced variability, with devices from the same lot exhibiting quite different properties. But more interestingly, it can also be used to model possible effects due to driving current changes. As it was previously discussed, current changes may induce temperature variation on the device, and this traduces into shifts

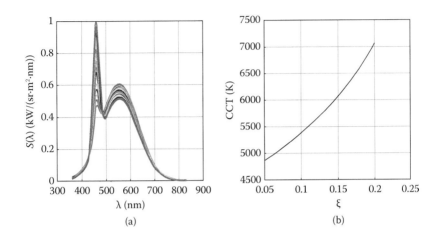

FIGURE 2.16
(a) LED normalized SPD and (b) corresponding CCR.

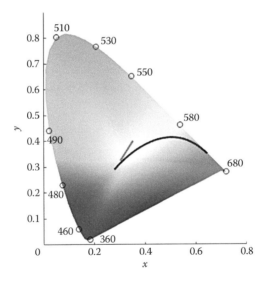

FIGURE 2.17
CIE color map locus.

on the emitted spectrum. Figure 2.17 reflects the SPD changes on the CIE color map, with the curve in gray representing the locus of color coordinates corresponding to the change of parameter ξ. It is also observable that these color coordinates fall close to the Planckian locus (black curve), for the white region of the color map.

2.3.6 Photodetectors

The detection of light can also rely on semiconductors [6]. In the case of LEDs, the annihilation of an electron–hole pair, more precisely the transition of an electron between conduction and valence bands, was accompanied by the release of energy in the form of a photon of given wavelength. In this case, the process of radiative recombination is explored as a means to generate light of a given wavelength. The opposite process is also possible. In a semiconductor sample exposed to an external light source, the impinging photons arriving at the sample may furnish electrons in the valence band the right amount of energy to move into conduction, through an effect known as photon absorption. These photogenerated electrons are free to move in the semiconductor lattice, and as such, they may serve the purpose of charge carriers under the action of an externally applied electric field. Figure 2.18 illustrates this process, recurring once again to band diagrams. The photogenerated current is given by:

$$I_{ph} = P_o \frac{q\lambda}{hc}(1-r)\left(1-e^{\alpha(\lambda)d}\right), \tag{2.38}$$

where P_o is the incident optical power, r is the reflection coefficient at the interface air–semiconductor, $\alpha(\lambda)$ is the absorption coefficient, and d is the length of the sample. The photon absorption properties generally depend on the semiconductor material and wavelength. The quantum efficiency is commonly used to characterize the wavelength dependence of the material in use. For this purpose, the quantum efficiency of the photon absorption process is defined as the ratio between the number of photogenerated carriers and the number of incident photons. The mathematical definition is given by:

$$\eta = \frac{I_{ph}/q}{P_o/h\upsilon} = (1-r)\left(1-e^{\alpha(\lambda)d}\right). \tag{2.39}$$

The quantum efficiency is generally dependent on the absorption coefficient [6]. Figure 2.19 depicts the absorption coefficient for some semiconductor

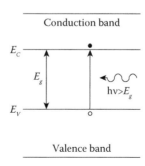

FIGURE 2.18
Carrier generation due to the photon absorption mechanism.

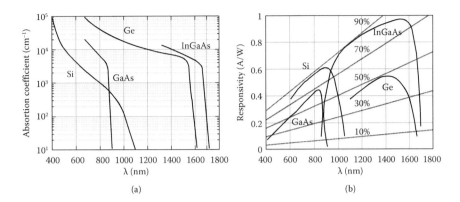

FIGURE 2.19
(a) Absorption coefficient and (b) spectral responsivity.

materials commonly used in PDs. As it can be seen, the absorption coefficient behavior determines the wavelength span to which the PD is sensitive. For all materials, wavelengths above some cut-off wavelength (exhibited by the almost vertical slope region of the absorption coefficient) lead to low values of quantum efficiency, as given in (2.39). This defines the spectral range of the material. Another useful measure is the spectral responsivity of the material [6]. The spectral responsivity is normally used by PD producers as an indication of the device performance. The spectral responsivity is defined as the ratio between the photogenerated current and the incident optical power, which is given by:

$$\mathscr{R}(\lambda) = \frac{I_{ph}}{P_o} = \eta \frac{q\lambda}{hc}. \tag{2.40}$$

Figure 2.19 also depicts the typical spectral responsivity for several semiconductor materials. As it can be seen, indirect band-gap materials such as silicon and germanium have low peak responsivities, as a consequence of the inefficiency associated with the photon carrier generation process, requiring the intervention of a phonon to preserve the energy–momentum balance. Figure 2.19 also represents the theoretical limitations of the spectral responsivity, assuming fixed quantum efficiency values. As it can be seen, both Si and InGaAs exhibit spectral responsivities close to the theoretical limit imposed by 90% quantum efficiency. In visible light radiation detection, Si and GaAs are the most favorable semiconductors, with Si exhibiting a peak responsivity in the IR region.

It is possible to improve the performance of the optical detection process using P–N junctions. As mentioned before, the photon-generated current depends on the existence of an external field. A semiconductor sample of a given type behaves generally as a resistive material. Thus, under the action

of external electric field, the current through the sample is not only due to photogenerated carriers. One possibility to avoid this behavior is to rely on P–N
junction diodes. In a reversed biased P–N diode, there is a strong electric field
applied to the depletion region near the junction. This field can remove any
minority carriers generated within this region. The current passing through
the device is very small. If the depletion region is exposed to external radiation, carriers can be generated by the process previously described. These carriers are then removed from this region by the reverse electric field, giving rise
to a photogenerated current, free from other effects. The current–voltage characteristic of a photodiode combines two contributions: the behavior of a normal junction diode, given by (2.24), and the photogenerated current given by
(2.38). The end result is given by:

$$I_D = I_o \left(e^{\frac{qV_D}{\eta k_B T}} - 1 \right) + I_{ph}. \tag{2.41}$$

Figure 2.20 illustrates the behavior of the current–voltage characteristic of a
typical photodiode. There are three modes of operation that can be employed

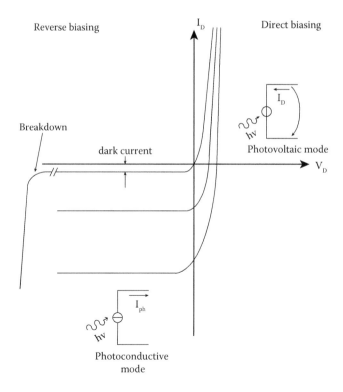

FIGURE 2.20
Photodiode current–voltage characteristic and modes.

for optical signal detection: the photoconductive mode, the photovoltaic mode, and the short-circuit mode. In the photoconductive mode, the photodiode operates under reverse biasing conditions. According to (2.40), the device current is in this case essentially determined by $I_{ph} - I_o$, the last term, I_o, corresponding to the device dark current (the current that flows under no optical radiation exposure). A typical model for this operation mode represents the photodiode as a current source. Photodiodes can also be used in direct biasing conditions. This is known as the photovoltaic mode. In this mode, the device output is represented as a voltage, implying a nonlinear logarithmic relation, dependent on the optical radiation exposure. The photovoltaic mode is mainly explored for energy-harvesting applications. For signal detection, the photoconductive mode is more appropriate due to its linearity and sensitivity. Finally, the short-circuit mode is an intermediate mode where the device feeds a low impedance detector, under absent biasing. This mode is also rarely employed for optical signal detection, due to the fact that the device intrinsic capacitance decreases with the applied reverse biasing. Thus, for large bandwidth applications, the device biasing should be adequately set.

2.3.7 PIN and Avalanche Photodiodes

There are two types of photodiodes normally employed for optical signal detection, PIN (P-type, intrinsic, N type) and avalanche photodiodes (APD). PIN photodiodes employ an intrinsic semiconductor layer (sometimes, a lightly doped P-type layer), between the P and N terminals of the device. This layer acts as the optically active region of the device. The device operates in the reverse biasing mode. The photogenerated carriers drift under the action of the reverse electric field through the intrinsic layer and are collected on the P (holes) and N (electrons) sides. The intrinsic layer acts as an extension of the depletion layer of the device. As such it is exploited as a means to augment optical exposure, but also to reduce the device intrinsic capacitance. Figure 2.21 illustrates the device constitution, its band diagram, and the shape of the electric field.

The APD explores the avalanche effect as a means to improve performance. The avalanche effect occurs under high field regions [6]. Free carriers crossing an intense field region acquire enough energy to generate other free carriers by a process called impact ionization. These generated carriers are on their side accelerated by the electric field and generate more carriers when colliding with the semiconductor lattice. This effect is explored in APD as a current multiplication effect. To achieve this, APDs employ a junction formed by a highly doped N-semiconductor with a P type semiconductor. This junction acts as the current multiplication buffer of the device. The active region, where photon absorption generates the carriers, is formed by an intrinsic layer, terminated with a highly doped P type region. Figure 2.21 illustrates, for comparison purposes, the APD constitution. The current

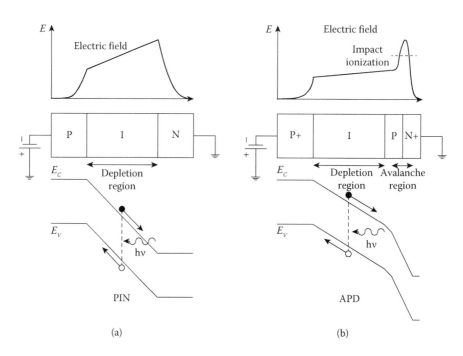

FIGURE 2.21
(a) PIN and (b) APD constitution and working principle.

multiplying effect in an APD is usually modeled by a parameter M, quantifying the ratio between the total generated current I_M and the photogenerated current, which is given by:

$$M = \frac{I_M}{I_{ph}}. \tag{2.42}$$

The spectral responsivity that appears multiplied by M in an APD is defined as:

$$\mathcal{R}(\lambda) = \frac{I_M}{P_o} = \eta \frac{q\lambda}{hc} M. \tag{2.43}$$

2.3.8 Photodiode Electrical Circuit Equivalent Model

Figure 2.22 represents the equivalent circuit model of a photodiode under reverse biasing conditions (photoconductive mode). It contains a current source with value given by (2.38) for a PIN photodiode, or this value multiplied by M for an APD. The device intrinsic capacitance appears in parallel with the current source, with the value given by (2.26). For high-frequency

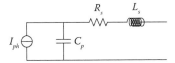

FIGURE 2.22
Photodiode equivalent circuit model.

applications, it is often required to include a series resistance and a series inductance, due to the device bonding wires and terminals. In general, for signal detection applications it is required to include noise sources. Electronic devices employing P–N junctions exhibit shot noise [6]. Shot noise has a quantum nature and is characterized by its uniform spectral distribution, with variance proportional to the signal bandwidth and the average current through the device. More precisely, assuming that the incident optical power is defined as:

$$P(t) = P_o\left(1 + ms(t)\right),$$ (2.44)

where P_o is the average optical power, $s(t)$ is the signal component, and m represents the modulation index. The detected photocurrent in terms of a DC component I_p and a signal component $i_p(t)$ is given by:

$$i_M(t) = M\mathscr{R}(\lambda)P(t) = I_p + i_p(t).$$ (2.45)

The shot noise variance is under this conditions defined by:

$$\sigma_{sh}^2 = 2qI_pBM^2F(M).$$ (2.46)

In (2.46), B is the bandwidth and $F(M)$ is the noise figure. Equation 2.46 is valid for an APD; for a PIN photodiode, the term $M^2F(M)$ is replaced by 1. As it can be seen, the source of noise in a photodiode is linked to the incident optical power. It is also noticeable that noise performance in an APD is worst when compared to PIN photodiodes. This is due to the avalanche multiplicative factor that appears squared in (2.46). Under no light exposure, there is a small current passing through the device, which is known as the dark current. This current is also source of shot noise defined as:

$$\sigma_{sh}^2 = 2qI_oBM^2F(M).$$ (2.47)

The two noise processes are uncorrelated, so the total noise variance due to shot noise is additive.

2.4 LED Drivers for Communications

The design of LED drivers for communications is a challenging task as it entails two distinct areas of electronic design, which often impose conflicting requirements [9,10]. On one side, the power design advocates the usage of power devices and techniques which most often constrain bandwidth. On the other hand, bandwidth optimization relies on linear behavior, demanding that the LED remains close to its operating point. However, for power devices this is generally not a reasonable assumption since the device can be subjected to large signal conditions, imposing large changes from its operating point. The design of the LED driver must also take into consideration the type of modulating signals. Two types of drivers can be considered: On/Off drivers—suited for the transmission of digital modulation formats, and analog drivers—suited for more complex modulation formats demanding continuous or multiple output levels.

2.4.1 ON/OFF Drivers

Digital LED drivers allow the LED modulation in the digital domain (on–off). The common applications require the LED current control from an input signal that usually has a fixed voltage and low current capability. Thus, an active circuit is needed to drive the required current through the LED. Figure 2.23 depicts three configurations used in digital LED drivers. The metal-oxide-semiconductor field-effect transistor (MOSFET) is the preferred active device in the digital domain for its low conduction resistance, R_{DS_ON}. Thus, at low R_{DS_ON} values, the MOSFET can simultaneously handle high currents and achieve low-power dissipation. Nonetheless, bipolar junction transistors (BJTs) can also be used, bearing in mind they require a higher base current to operate (lower input resistance

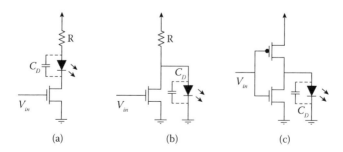

(a)   (b)   (c)

FIGURE 2.23
Digital drivers: (a) single transistor, (b) single transistor inverter, and (c) complementary inverter.

than MOSFETs). Furthermore, they have a lower input maximum voltage specification, which usually requires an additional resistance to limit base voltage [9]. They also have higher saturation voltage, known as V_{CEsat}, leading to higher power dissipation when compared to MOSFETs.

The circuit in Figure 2.23a uses a transistor in series with the LED. As V_{in} increases, the current in the transistor rises, thus the LED current also rises. Considering that the voltage across the transistor is much smaller than the LED forward voltage, the current is limited by the resistor R, according to Ohm's law, which is given by:

$$I_{LED} = \frac{V_+ - V_{LED}}{R}. \tag{2.48}$$

In order to achieve high switching speed, the LED capacitance C_D must be charged and discharged as fast as possible. As the current rises in the transistor, C_D begins to charge, limited by the resistor R. However, when the transistor switches off, C_D sees a high resistance value and has a slow discharge time [10]. To overcome this issue, it is usually connected a second transistor in parallel with the LED to discharge C_D when the primary transistor switches off. However, if a series of several LEDs is used, more than one discharge transistor should be connected to discharge the C_D effectively in all LEDs. An alternative circuit is shown in Figure 2.23b. As opposed to the previous example, the LED is active when V_{in} is in the low state. The on-current is also limited by R as in (2.48). The disadvantage of this configuration is the asymmetry between t_r (rise time) and t_f (fall time) since the LED capacitance is charged through R and discharged through R_{DS_ON}. The third driver circuit uses a complementary metal-oxide semiconductor (CMOS) inverter configuration to drive the LED (Figure 2.23c). Note that only one transistor is on, except during the on/off transitions. The upper transistor charges C_D as well as it feeds current to the LED (current source), and lower transistor drains the charge from C_D (current sink). This circuit allows balancing of the transition times t_r and t_f by adjusting transistor parameters [10,11]. In particular, the upper transistor should be dimensioned to be able to drive LED current plus the C_D charge while the lower transistor must only take care of C_D discharge.

These configurations can drive more than one LED. In order to guarantee the same current for all the LEDs, they are usually arranged in series. However, the transistor must be selected according to the circuit specifications. One of the most important parameters to consider is the maximum allowed voltage across the active device when the control pin is open (V_{DSS} in MOSFET and V_{CEO} in BJT). This voltage is commonly known as the breakdown voltage of the transistor. As said, when the device is on, the voltage drop across it is low. However, in the offstate, the device must handle the maximum voltage drop; in other words, the source power voltage value.

2.4.1.1 Baker Clamps

BJTs are particularly known by their undesirable saturation region. This region is characterized by having the two junctions forward-biased (base-emitter and base-collector). In this region, the collector current is almost directly proportional to the V_{CE} voltage. Removing the transistor from saturation is a slow process which can impair the circuit performance. One technique used to mitigate transistor saturation is the Baker clamp, depicted in Figure 2.24. It is implemented using a Schottky diode connected between the base and the collector. The principle of a Baker clamp is to reduce transistor gain near the saturation region. In other words, in the active region the diode is in the cut-off state. As the collector voltage decreases near the saturation region, the diode starts to be forward-biased, draining current from the base to the collector. This starts to occur when the collector voltage, V_C, is approximately 0.3 V (Schottky diode forward voltage) higher than the base voltage, V_B. Considering a transistor with a V_{BE_ON} of 0.6 V, V_C will be maintained at least 0.9 V higher than the emitter voltage V_E, thus avoiding transistor saturation.

Baker clamps are also used to speed up cut-off time [9]. Assuming the transistor is initially on, when the base current, I_B, falls to zero rapidly, the collector current I_C does not decrease as fast as it is desirable. This occurs due to the charges present in the transistor base that need to be drawn. While the base charges are removed, the transistor is in the saturation mode. Using a Baker clamp, the switch-off time is decreased, due to the fact that base charges are drawn through the collector.

2.4.1.2 Enhancing the Drive Capability

In order to increase the current, either for DC or signal, it is common to connect one or more transistors in parallel. In theory, this connection can be directly implemented. However, due to imperfections in the transistor manufacturing process, it is very unlikely to have two exactly equal transistors. In particular, at the same bias point the current through the transistor is

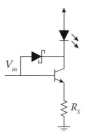

FIGURE 2.24
Baker clamp with Schottky diode.

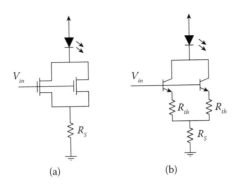

FIGURE 2.25
Enhancing drive capability: (a) parallel combination of MOSFETs and (b) parallel combination of BJTs.

not equal for two devices of the same type. The current imbalance leads to a higher power dissipation in one of the transistors, which naturally increases device temperature [9]. In MOSFETs, higher junction temperatures increase R_{DS_ON} value, decreasing the I_D current. This process is known as negative thermal feedback. Therefore, connecting MOSFETs in parallel as shown in Figure 2.25a usually leads to a current balance between devices. However, opposite to MOSFET, the BJT has positive thermal feedback. This means the current draw increases as the junction temperature increases. Consider the case where the two BJT emitters are connected as in Figure 2.25b without R_{th}. Assuming the transistors' characteristics are slightly different, the same V_{BE} will impose slightly different colector currents. After some time, the transistor that has drawn the higher amount of current will increase in temperature. As a consequence, a higher current value will be drawn with the same V_{BE}. This cycle will repeat until the junction temperature rapidly reaches its breakdown limit and stops operating properly [9]. Thus, BJTs cannot be connected in parallel without some precautions. Figure 2.25b depicts a configuration normally used to combine BJTs in parallel. As it can be seen, R_{th} is connected in series with each emitter. This resistance helps to balance the current in the transistors by creating negative feedback. In this configuration, as the temperature increases, as well as the current through the transistor, the voltage across R_{th} increases. Thus, V_{BE} must also include the voltage drop across R_{th}, as in:

$$V'_{BE} = V_{BE} + I_E R_{th}. \tag{2.49}$$

In other words, assuming the voltage drop across R_s is fixed, then V_{BE} of the transistor is reduced if I_E increases as given by:

$$\Delta V_{BE} = -\Delta I_E R_{th}. \tag{2.50}$$

R_{th} usually has a small value, of the order of few ohms. The result is a decrease in the transistor current which contradicts the thermal runaway effect. In order to minimize the thermal runaway effect, it is common to thermalcouple the parallel transistors, mounting them in the same heat sink. As the temperature increases in one transistor, it will also increase in the other helping to balance the currents. Additionally, using matched transistors, where the characteristics are as identical as possible between transistors, also helps to minimize thermal runway.

2.4.2 Analog Drivers

For more complex modulations such quadrature amplitude modulation (QAM) and orthogonal frequency division multiplexing (OFDM) an analog driver is required [12,13]. Analog drivers should present high linearity. The modulation is assumed for all purposes to be continuous as opposed to the previous case. Two approaches are possible for the design of an analog LED driver: using voltage-mode topologies—where signals are represented by voltages; or using current-mode topologies—where the signals are treated as currents [14]. Regarding circuit theory, current-mode design favors high speed [14]. Using both MOSFETs and BJTs for a given current span, the associated voltage span is a compressed replica. This implies that charging and discharging of the node parasitic capacitances is faster. On another note, concerning LEDs, current-mode drivers seem more appropriate given the linear power-to-current relationship of the LED.

2.4.2.1 *Voltage-Mode Design*

Figure 2.26 shows a voltage-mode analog LED driver [12,13]. A preamplification stage is used to accommodate input signal to LED voltage swing. Later, a push-pull configuration is used in the output stage working as a voltage buffer with high current capability. The output signal is fed into an LED using a bias tee. The circuit shown in Figure 2.26 may work in two possible biasing classes, B or AB (depicted in the figure). Class B is characterized by not having any biasing condition applied to transistors Q_1 and Q_2; the input signal has to drive the transistors on. As it can be seen, for lower V_{in} values the transistor's current is zero, resulting in no output signal. The distortion generated by class B amplifiers is known as crossover distortion, as it occurs when the input signal crosses the zero reference value. This is of limited usage as an LED driving configuration: the voltage span of the LED is usually less than the required voltage to turn on these transistors.

In order to prevent crossover distortion, the output transistors are biased in class AB [10]. Class AB is an intermediate operating class where the transistors are biased close to cut-off. The constant bias current eliminates crossover distortion even at low values of V_{in}, as the transistor's current never reach the zero value. The biasing condition is established by transistor

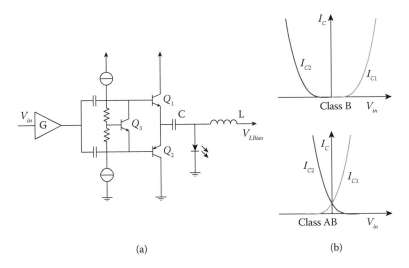

FIGURE 2.26
Typical push-pull stage for LED driving: (a) conceptual circuit and (b) operation classes.

Q_3, operating as a V_{BE} multiplier. As V_{in} increases, the upper transistor starts to draw higher current, and the lower transistor starts to enter into cut-off. The opposite occurs when V_{in} decreases to negative values. Setting the bias voltage on each transistor base slightly higher than V_{BE_ON} often leads to a good compromise between linearity and efficiency. It is also common to use two emitter resistances in series with Q_1 and Q_2. These resistances protect the circuit against overdrive conditions and contribute to linearity by establishing feedback action with the input circuit.

In terms of bandwidth, this circuit operates with high bandwidth given that the voltage gain of an output stage is close to unity. The bandwidth generally depends on the transition frequency of the selected devices. However, as mentioned previously, selecting devices for large current-driving applications may is not exempt from problems: (i) the current handling capabilities generally constrain the transition frequency, merging high power and high bandwidth usually results in increased costs; (ii) high-frequency transistors usually have low breakdown voltages; this is generally true for both BJTs and MOSFETs; and (iii) device combination is normally an option for high current handling; however, it is not suitable for driving a large number of LEDs in series.

2.4.2.2 Current-Mode Design

The LED voltage driving mode is not exempt from linearity problems. The current–voltage characteristic is intrinsically nonlinear, and as such, voltage-to-current conversion implies nonlinear effects. These nonlinear effects tend to have a higher performance impact for high peak-to-average power

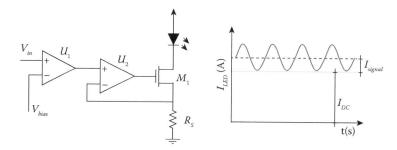

FIGURE 2.27
LED current driving circuit: conceptual circuit and signal plus bias combining.

ratio (PAPR) applications. As established before, the LED output power is linear on the driving current for moderate inversion conditions. Thus, it seems natural that the best choice in terms of linearity is to use current driving alternatives [9].

Figure 2.27 depicts a circuit for analog current driving of an LED. This configuration has the LED in series with the transistor and a current sampling resistor R_S. The transistor input voltage is set by an error amplifier in a feedback loop. By varying its output, the error amplifier will maintain the voltage across R_S equal to the noninverting input port. The DC bias current is obtained by applying a constant voltage to the input signal, V_{bias}. One method to define the LED biasing condition is to set the input signal reference voltage at the preamplification stage. The operational amplifier U_2 has to be carefully selected. First, it must be able to exhibit large bandwidth. On the other hand, it must be stable under small gain operating conditions (for the circuit of Figure 2.27, it will operate with gains close to unity).

Although this configuration is more appropriate for LED luminous flux control, the LED bias current flows through the driving transistor (operating in the linear region), thus increasing power dissipation. This problem is also present in R_S. Furthermore, bearing in mind the driver linearity and intermodulation distortion, the bias current should be selected according to the transistor characteristics.

2.4.3 Pre-emphasis

The circuits above have bandwidth limited by the LED, which creates a pole in the frequency response. Usually, the rest of the driver circuitry can operate at larger bandwidths than the LED optical bandwidth. Thus, from a system design perspective, it is possible to equalize the LED response through a pre-emphasis stage [15,16]. Pre-emphasis can be achieved by providing the necessary gain to compensate the losses introduced by the LED at larger frequencies than its cut-off. This can be achieved using pole-zero compensation networks. In theory, if a zero with a frequency matching the LED's bandwidth is introduced, the compensation would be perfect and the whole system operates without limitation. However, since the compensation must rely on real elements, and most possibly on transistors, it is necessary to consider higher frequency limitation imposed by the compensation network. Assume that the LED frequency response can be modeled as one pole transfer function of the type given by:

$$H_{LED}(s) = \frac{1}{1+s/\omega_{LED}}, \tag{2.51}$$

where ω_{LED} represents the LED optical bandwidth, as seen by the receiver. The compensation network can be adequately represented by:

$$H_c(s) = A_o \frac{1+s/\omega_z}{(1+s/\omega_p)(1+s/\omega_a)}, \tag{2.52}$$

where ω_z, ω_p, and ω_a are the frequencies of the compensation zero, the compensation pole, and the pole due to the amplifier, respectively. Matching the zero to the LED's bandwidth results in the following complete response:

$$H(s) = H_{LED}(s).H_c(s) = \frac{A_o}{(1+s/\omega_p)(1+s/\omega_a)}. \tag{2.53}$$

As it can be observed, with real poles, the best possibility in terms of bandwidth occurs when ω_p and ω_a are equal, corresponding to a first-order critically damped system (an underdamped regime could be achieved using feedback configurations or inductances). In order to study the behavior of the compensation network given by the former equations, it is necessary to define some design strategy. Let the system of (2.53) be represented by the damping coefficient, ξ, and natural frequency, ω_n, defined by:

$$\omega_n = \omega_z\sqrt{\beta\omega_a} \qquad \xi = \frac{\omega_p + \omega_a}{2\omega_n}, \tag{2.54}$$

where β represents the idealized compensation gain defined by the ratio ω_p/ω_z. The cut-off frequency of the complete system is expressed by $\alpha\omega_n$ with α given by:

$$\alpha = \sqrt{(2\xi^2-1)\left(\sqrt{1+1/(2\xi^2-1)^2}-1\right)}. \tag{2.55}$$

Finally, the performance of the compensation network can be analyzed using the bandwidth enhancement ratio (BWER) which is given by:

$$BWER = \frac{\omega_c}{\omega_z} = \alpha\sqrt{\beta\frac{\omega_a}{\omega_z}}. \tag{2.56}$$

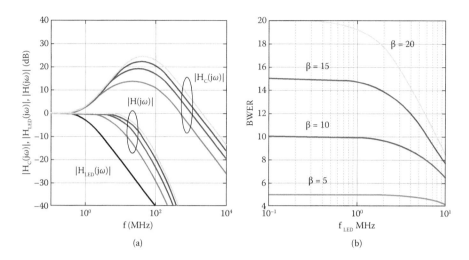

FIGURE 2.28
Pre-emphasis compensation gains: (a) frequency response and (b) BWER.

Figure 2.28 shows the expected performance assuming the LED bandwidth is limited to 1 MHz and that the amplifier's cut-off frequency is 100 MHz. The frequency response plots on the left show the bandwidth improvement for several values of β (in color agreement with the picture on the right). The BWER discloses the limiting effect of the amplifier cut-off frequency for the same values of β, when the LED's bandwidth (the zero of the system) changes. The loss of performance is clear as the LED's bandwidth approaches the amplifier cut-off frequency.

Figure 2.29 depicts two possible circuits for the implementation of a transfer function close to (2.52). In the transistorized circuit, the zero is realized by the time constant imposed by R and C at the input of the compensation stage. The compensation pole and the amplifier depend on the transistors' characteristics and desired gains (which can be set with R_{L1} and R_{L2}). The second circuit relies on an operational amplifier. The frequency limitations are for this case imposed by the operational amplifier. The zero and pole of the compensation network are set by the time constants R_1C_1 and R_2C_2, respectively. The performance of both circuits can be analyzed using the same principles previously described. The definitions of ξ and ω_n are obviously circuit dependent. This circuit's implementations give a more practical view on the parameter β previously described. β is nothing more than a ratio of gains of the idealized pole-zero compensation network. In terms of the second circuit with the operational amplifier, the ratio between the low-frequency gain, given by R_2/R_1, and the high-frequency gain (with an ideal operational amplifier) is set by C_1/C_2.

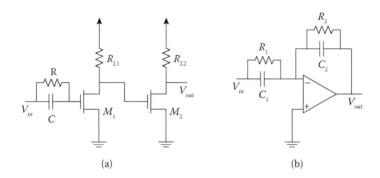

(a)   (b)

FIGURE 2.29
Pre-emphasis circuit implementations: (a) transistorized circuit and (b) operational amplifier-based circuit.

2.4.4 Biasing and Signal Combining

One major benefit for using VLC is the ability to provide simultaneous illumination and data communications using the same lighting device. However, light sources with high (fast) switching features as required for data communications are not always suitable for illumination purposes, since the lighting characteristics would be highly dependent on the communication signal. In order to improve LED lighting and modulation characteristics, it is necessary to combine the communication signal with LED biasing required to support lighting features. This can be achieved using one of two approaches. The simplest one is to rely on a bias tee. The other relies on dedicated circuits able to separate and provide independent control of both functions. The bias tee consists of a passive device with three ports one for signal input, other for biasing, and another providing the combined biasing and signaling functions. The basic operation relies on frequency selective networks. For signaling purposes, the bias tee should provide broadband coverage and provide strong attenuation at low frequencies. For biasing purposes, the bias tee provides strong attenuation for signal frequencies and pass the low frequencies. Achieving good performance relies on the matching quality of the output port. This requires that the load impedance must be matched to the bias tee, otherwise frequency selectivity becomes load dependent. Concerning the LED driving case, the load impedance is the LED, presenting an impedance dependent on the biasing current. There are several design possibilities for bias tee able to maintain performance even with impedance variations. Simple possibilities such as the one in Figure 2.30a, do not support load impedance independence. In this simple case a capacitor is used to feed the signal component and one inductance for the biasing.

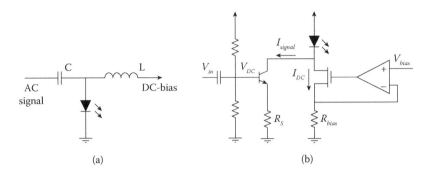

FIGURE 2.30
Biasing and signal combining circuits: (a) bias tee and (b) active combining circuit.

An alternative method is also depicted in Figure 2.30b. It uses a pair of transistors: the first one, on the right, sets the biasing while the second, on the left, drains additional current from the LED, according to the input signal V_{in}. The operational amplifier together with the transistor (which can be a BJT or an MOSFET) creates a transistor with a very high input impedance and no voltage drop. Thus, the biasing current is given by:

$$I_{DC} = \frac{V_{Bias}}{R_{Bias}} + \frac{V_{DC} - V_{BE}}{R_S}, \tag{2.57}$$

and the signal current is given by:

$$I_{signal} = \frac{V_{in}}{R_S}, \tag{2.58}$$

where V_{DC} is the voltage at the signal transistor base. The major design constraint to take into consideration in this circuit relies on the choice of the transistor that makes the signal current source branch. This transistor has to provide enough bandwidth and high breakdown voltages, able to match the application requirements.

2.5 Optical Signal Amplification

Many works on amplifier design for fiber optic communication systems with high speed (i.e., tens or hundreds of Gb/s) data rates have been published in recent years (see references in [17–19]). Although the principle of the design of most of them can be used in VLC systems, the nature of the free-space channel imposes serious limitations on the receiver optical power. In order to maximize the optical signal detection in VLC systems, a PD with a large surface area should be used at the receiver. However, large area PDs have large intrinsic capacitances, which leads to reduced receiver bandwidth provided the amplifier's inputs impedance is high. To overcome the gain–bandwidth product limitations of amplifiers, there are several frequency optimization techniques including inductive peaking [20,21] and capacitive peaking [22,23]. Both of these techniques explore the presence of complex poles within the dynamics of the amplifier as a means to achieve peaking effects on the frequency response. It is possible to extend the amplifier bandwidth if these peaking effects can be made to occur near to the amplifier's cut-off frequency.

2.5.1 Basic Amplifier Topologies

The conversion of the PD current into voltage is achieved by front-end circuits designated as transimpedance amplifiers (TIAs). TIAs can be divided into two major groups: open-loop and feedback [19]. Figure 2.31a shows a generic open-loop TIA. Depending on the architecture, open-loop TIAs can be divided into low input impedance amplifiers and high input impedance amplifiers. Low input impedance amplifiers are suitable for high bandwidth and low noise performance applications. However, they present low sensitivity. On the other hand, high input impedance amplifiers have high sensitivity but low-frequency performance. Feedback TIAs (Figure 2.31b) are usually preferred, mainly because they can overcome the major drawbacks of the others (low sensitivity in low impedance amplifiers, and limited bandwidth in high-impedance amplifiers) while keeping their most attractive features (high bandwidth with small input impedance, and high sensitivity with high gains).

2.5.2 Transimpedance Amplifiers

The TIA design is by itself a challenging design task. There are several design requirements such as gain and bandwidth that conflict with each other. The following discussion highlights some of these problems.

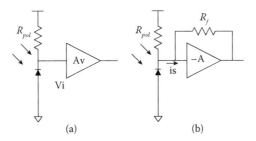

(a)   (b)

FIGURE 2.31
Basic optical receiver amplifying topologies: (a) open-loop configurations and (b) feedback configurations.

2.5.2.1 Gain–Bandwidth Trade-Off

Maximizing both the gain and bandwidth of an amplifier is an old and extensively debated problem in electronic circuit design. Bode was the first to formalize the problem and to deliver a solution, revealing that the product of an amplifier's gain by its bandwidth is generally fixed by the ratio g_m/C, where g_m represents the transconductance of the active devices within the amplifier and C represents a combination of parasitic capacitances at the ports of the amplifier [24]. Figure 2.32a shows a simplified small-signal equivalent circuit of an open-loop front-end receiver. From a circuit design perspective, the photodiode is adequately represented by a current source I_P (representing the optically converted current) in parallel with the intrinsic capacitance C_P. The amplifier input impedance is generally a complex function. However, for illustration purposes it can be assumed as a first-order parallel association of a resistance R_i and a capacitance C_i (in fact, the current-to-voltage conversion can be made by a simple resistor providing a transimpedance gain equal to R, the value of the resistor). Using this simplified model, the pole contribution due to the input circuitry is ruled by $R_i(C_P+C_i)$. Assuming it is possible to design the amplifier in such a way that the other poles in the system have smaller associated time constants, then the input circuit time constant is the dominant one. If a high-impedance amplifier is used, the front-end has a low bandwidth. Thus, in order to meet the high bandwidth requirements, the input circuit time constant should be minimized. There are two possibilities to minimize the input time constant: (i) use PDs with smaller intrinsic capacitances (reducing the active area which, in general, impairs system performance) or (ii) reduce the front-end input impedance. The second method is always preferred since it does not imply drastic changes at the system level nor circuit performance.

A common strategy to reduce input impedance in amplifiers is to use adequate feedback configurations. TIAs can be constructed using feedback configurations, as is the case of shunt-shunt feedback represented on Figure 2.32b. Feedback TIAs represent the best compromise between gain

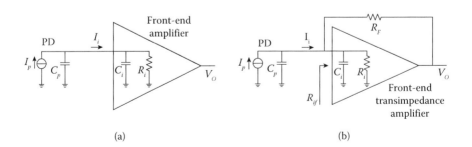

(a)   (b)

FIGURE 2.32
Gain–bandwidth trade-off analysis. Equivalent circuits for (a) open-loop configurations and (b) closed-loop configurations.

and bandwidth. Assuming the simplified model of Figure 2.32 (right), the input time constant is given approximately by $R_{if}(C_P + C_i)$, where R_{if} represents the input resistance with feedback. At a first sight, it seems that the input impedance can be made as small as required, thus reducing the effect of the input time constant. However, a close inspection shows this is not as straightforward as it seems. For a reasonably well-designed TIA, the gain with feedback should be as close as possible to the value of the feedback resistance, R_F. On the other hand, a meaningful approximation of R_{if} is given by R_F divided by the voltage gain (assuming that loading effects posed by input and output circuits are negligible and the voltage gain is large when compared to unity). Both transimpedance gain and input impedance are directly related through R_F, so in general increasing the transimpedance gain implies increasing the input impedance. It is possible to increase voltage gain in order to reduce R_{if}. A simple strategy to increase voltage gain is to include more gain stages on the forward amplifier. This allows larger gains to be achieved but implies an increased complexity of the system dynamics. As the number of stages is increased, the pole-zero contributions due to each added stage become rather evolved, posing serious restrictions to the closed-loop stability.

2.5.2.2 Bandwidth Optimization

Figure 2.33 depicts the detailed small-signal equivalent circuit of a TIA. This circuit employs shunt-shunt feedback through the feedback resistance R_F. The bypass capacitor C_c is used in this circuit as a means to filter low-frequency noise. It can be neglected for high-frequency inspection of the amplifier. Feedback amplifier analysis involves two steps. First, the analysis of the open-loop amplifier, taking loading effects caused by the feedback network into account. The second step addresses the closed-loop transfer function of the amplifier. For open-loop analysis, R_F is replaced by two resistances with a value equal to R_F, one in parallel with the input of the amplifier and another in parallel with the output [11,25]. Straightforward analysis shows that the

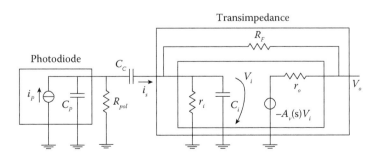

FIGURE 2.33
Transimpedance amplifier small-signal equivalent circuit.

open-loop transfer function, relating the input current from the photodiode to the output voltage of the amplifier, is given by:

$$Z_{ol}(s) = \frac{R_{io}A_v(s)}{1 + sR_{io}(C_P + c_i)},$$ (2.59)

where R_{io} represents the parallel association of R_{pol}, R_F, and r_i. Assuming the forward amplifier has first-order dynamics, given by:

$$A_v(s) = -\frac{A_o}{1 + s/\omega_a},$$ (2.60)

which results in the open-loop transfer function governed by two poles. The closed-loop transfer function is given approximately by:

$$Z_{cl}(s) = \frac{Z_{ol}(s)}{1 + \beta Z_{ol}(s)},$$ (2.61)

where β represents the feedback factor given by $1/R_F$. Straightforward calculations show that:

$$Z_{cl}(s) \approx -\frac{R_F}{1 + 2\xi\, s/\omega_n + (s/\omega_n)^2},$$ (2.62)

$$\omega_n = \sqrt{\frac{A_o\omega_a}{R_F(C_P + c_i)}},$$ (2.63)

$$\xi = \frac{1}{2\omega_n}\left(\omega_a + \frac{1}{R_{io}(C_P + c_i)}\right).$$ (2.64)

The approximation holds for $A_oR_{io} \gg R_F$, that is, the open-loop gain must be larger than the feedback resistance that sets the closed-loop gain. This is usually true for the majority of the design examples employing feedback. It is interesting to note that the natural frequency ω_n is proportional to the square root of the amplifier gain–bandwidth product $A_o\omega_a$. Bandwidth optimization can be achieved if the damping coefficient ξ is chosen such that the denominator of (2.62) becomes a Butterworth polynomial [26]. This occurs for $\xi = \sqrt{2}/2$, when the amplifier will be tuned in the underdamped regime, with a pair of complex conjugate poles. If the value of R_F is set in order to meet this requirement, then the bandwidth of the amplifier is simply ω_n. This simple analysis reveals that frequency optimization of TIAs can be achieved through the improvement of the gain–bandwidth product of the voltage amplifier. This is however limited by the choice of transistor. It is also possible to increase the gain A_o, using cascade amplifiers. This however reduces the stability margins of the closed-loop amplifier.

2.5.2.3 Noise Analysis

Another important aspect of the design of TIAs is noise performance. Noise affects all electronic components. Noise sources have distinct origins, where the most relevant are: (i) thermal noise—originating from thermal fluctuations of charge carriers in conducting materials, (ii) quantum shot noise—usually associated to semiconductor P–N junctions and due to the statistics of the charge transport mechanism, and (iii) flicker noise—present in most electronic devices involving semiconductor materials, in this case associated with lattice defects [6,11,25]. The combined effects of these statistically independent noise sources on a TIA are of paramount importance. The TIA is responsible for signal amplification and conversion. As such, it impairs all the receiver processing chain. If the input noise performance is poor, the reliability of the received data is seriously compromised. The receiver noise performance depends on several factors, one of the most relevant for this discussion is the SNR. The SNR has a direct impact on the achievable bit error rate (BER) performance. System performance can be improved using other processing methods, such as equalization, or employing error detection and correction codes. However, if the received signal has low SNR due to improper design of the amplifying chain, performance will be affected.

In order to better understand how noise affects SNR, it is required to perform noise analysis of the TIA. This can be accomplished by adding all the relevant noise sources to the amplifier small-signal equivalent circuit, as in Figure 2.34. Here, the voltage amplifier noise sources are represented for simplicity by sources i_a and v_a (the associated noise variances are i_a^2 and v_a^2, respectively). The other elements with associated noise sources are resistors, with current noise variance specified by:

$$\sigma_R^2 = 4k_B TB/R. \tag{2.65}$$

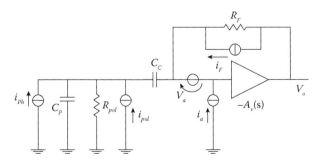

FIGURE 2.34
Transimpedance amplifier with added noise sources.

and the photodiode, with noise variance given by (2.46) and (2.47). The input referred spectral density (neglecting low-frequency behavior due to C_C) is given by [25]:

$$\sigma_n^2 = \int_0^B \left(2q(I_o + I_p) + \frac{4k_BT}{R_{eq}} + i_a^2 + v_a^2 \frac{1 + (2\pi f R_{pol} C_P)^2}{R_{pol}^2} \right) df, \qquad (2.66)$$

where R_{eq} stands for the parallel association of R_{pol} with R_F. Assuming that the signal component detected by the photodiode is $i_p(t)$ as in Section 2.3.8, the input referred SNR is expressed by:

$$SNR_i = \frac{\langle i_p(t) \rangle^2}{\sigma_n^2}. \qquad (2.67)$$

Equation 2.67 is a very general result. The presence of R_F is included in R_{eq}, as such, minimizing noise can be achieved by an adequate choice of the value of R_F. As it has been shown previously, R_F also affects the TIA gain and the bandwidth of the amplifier, implying necessary trade-offs.

2.5.3 Topologies for Improved Performance

2.5.3.1 Electronic Noise Optimization

Silicon technologies—particularly submicron CMOS technologies—have become very attractive due to their low cost and high integration level characteristics. Thermal noise is the dominant noise source in MOS transistors, being dependent on two design parameters: bias condition and transistor dimensions. Increasing transconductance in MOS transistors results in improvements for both frequency performance and thermal noise contributions. However, due to the existence of the photodiode capacitance, the noise minimization problem is slightly more complex. Criteria for defining an optimum value for the design ratio of the input transistor in a TIA have been established. Assuming the input transistor has a minimum length and a maximum bias current (resulting in high transconductance transistors with small parasitic capacitances), the optimum width of this transistor is set in order to match the total amplifier's input capacitance to the PD capacitance C_P and is given by:

$$W_{opt} = \frac{C_P}{C_{ox}L}, \qquad (2.68)$$

where C_{ox} is the oxide capacitance per unit area from gate to channel and L is the channel length. The optimum width for the case where both noise and bias current need to be optimized, as in front-ends for low-power applications, is one-third of the previous value [11,19,25].

With APDs, the current multiplying factor M should be included in the noise analysis. Assuming the noise figure $F(M)$ of APD is essentially proportional to M^{α}, then the optimum value of M that minimizes (2.67) is given by:

$$M_{opt} = \frac{4k_B T/R_{eq} + i_a^2/B + v_a^2/BR_{pol}^2}{\alpha q(I_o + I_p)}. \tag{2.69}$$

2.5.3.2 Differential Topologies

Electromagnetic interference (EMI) is a common source of noise in electronic systems. EMI is caused by surrounding electronic equipment and disturbs the normal operation of highly sensitive circuits (like TIAs). EMI can be reduced using the following strategies: (i) appropriate shielding of the susceptible parts of the receiver and (ii) using differential structures for all the critical circuits as depicted in Figure 2.35 [19]. Figure 2.35a shows a fully differential TIA where the input signal is applied to both its inputs using the same PD [27]. The main advantage of this strategy is the possibility of reducing EMI disturbances at the input stage but has the need for employing differential TIAs with high common-mode rejection ratios (CMRR), which are difficult to design. Pseudodifferential structures (as in Figure 2.35b) have the advantage of avoiding the high CMRR requirement of differential structures. The input stages provide equal gain paths, with phase opposition provided by the PDs. The differential amplifier (with a high CMRR) effectively rejects the signal common-mode components.

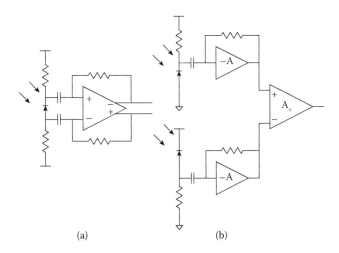

(a) (b)

FIGURE 2.35
Differential TIA configurations: (a) fully differential TIA and (b) pseudo-differential TIA.

2.5.3.3 Dynamic Biasing

When using a PD, the generated photocurrent consists of two components: (i) the signal current (proportional to the incident optical power) and (ii) a noise component. Noise sources in a PD have different origins, namely thermal noise, shot noise, and optical excess noise. Thermal and shot noise contributions are considered white noise sources with small spectral density, while the optical excess noise has its power concentrated in the low-frequency range. This advises the use of high-pass filtering at the input of the front-end, as shown in Figure 2.36a. This high-pass filtering can be realized using bypass capacitors. However, this technique is unsuitable for integration, as it requires large areas to implement the desired capacitor. An alternative design suitable for integration is shown is Figure 2.36b, where a dynamic biasing scheme is applied to the PD [19]. The effect of optical excess noise can be regarded as random fluctuations with a magnitude that can reach 100 times the magnitude of the detected signal. It is possible to eliminate these fluctuations using an error amplifier to detect the output average level and then subtract it (using a controlled current source as shown in Figure 2.36b) from the input, thus removing the noise component from the total generated photocurrent.

2.5.3.4 Automatic Gain Control

There are two quantities which bound the dynamic input range: (i) the front-end sensitivity and (ii) the maximum output signal for which the front-end still exhibits an approximately linear response—strongly affected by the supply voltages. To achieve both high sensitivity and high input dynamic range, the transimpedance gain cannot be fixed and should be adapted to the input signal. Unfortunately, controlling the transimpedance gain while optimizing noise performance, for the typically high sensitivity of these amplifiers, may turn into a difficult task to accomplish. Two strategies have been implemented

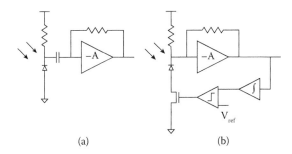

(a)   (b)

FIGURE 2.36
Photodiode biasing schemes: (a) passive biasing with blocking capacitor and (b) active biasing with feedback current control.

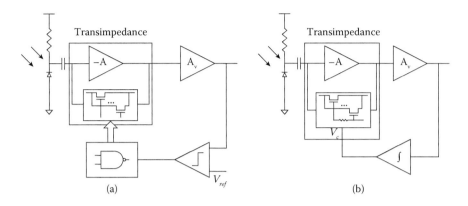

FIGURE 2.37
Automatic gain control schemes: (a) switched gain configuration and (b) continuous gain control configuration.

to circumvent this problem: (i) a switching feedback scheme, see Figure 2.37a [27] and (ii) a controlled feedback scheme, see Figure 2.37b [28].

The switched gain strategy consists of a TIA with a switched feedback network. The transimpedance gain is selected according to a set of previously defined thresholds. If the output level increases (decreases) above a specified limit, the decision circuitry acts on the feedback network in order to decrease (increase) gain. The number and magnitude of the different gains are set to meet the required sensitivity and the required input dynamic range. The overall performance is limited by the design of the front-end with larger gain. This strategy has some shortcomings: (i) the required system bandwidth must be met with all different gains and (ii) the switching scheme must act with a carefully designed time constant in order to prevent both oscillations and signal losses during gain switching.

Some of these disadvantages are overcome using a dynamic gain control scheme. The control circuit acts proportionally to the signal level, varying a set of multiple feedback resistors in order to obtain an easily controlled gain. This control scheme operates in such a way that it is effectively outside the signal path for the largest gains, achieving the lowest internal noise for very low input signals. An advantage of this scheme over the switched approach strategy is the inherent automatic gain control action on the output signal. Furthermore, the absence of the switching unit makes this strategy less prone to oscillatory behaviors.

2.6 Existing Regulation

Regulation related to lighting systems and products is extensive, covering not only the usage of light but also a significant number of other aspects like electrical and chemical safety, energy efficiency, and performance. In fact, a significant part of European Union (EU) regulations for lighting systems focus on the energetic efficiency of market-available products under the eco-design requirements of Energy-Related Products (ErP) Directive 2009/125/EC, the EU Ecolabel, and the joint EU-US Energy Star Agreement. Furthermore, lighting systems and applications are directly or indirectly subject to generic legislation applied to electrical and electronics goods. Among the broadest are the Low Voltage Directive 2006/95/EC, the Restriction of Hazardous Substances (RoHS) Directive 2011/65/EU, and the Registration, Evaluation, Authorization and Restriction of Chemicals (REACH) Directive 2006/121/EC. All the above-mentioned directives and legislation can be found in the Official Journal of the European Union which is available for online consultation on the EUR-Lex portal [29]. In support of legislation is a set of standards (European Norms, EN), for product safety and lighting applications. Conformity with product standards is required and allows the "CE" marking to be used for commercialization of any product within the EU space. Application standards, on the other hand, provide the basic requirements for designing efficient and uniform lighting solutions with the correct brightness, color range, and glare limits, among other factors. Although for VLC systems, product standards may impose limitations particularly regarding energy efficiency; application standards will be responsible for the stringent constraints as they ultimately define the quantity and manner in which light must be used.

As for the future of lighting regulations, the always present economic concerns impose tighter efficiency barriers, thus driving a search for ever more efficient technologies and strategies. However, with material efficiency boundaries being reached it becomes clear that lighting control systems will have an important role to play. Within this scenario, several European lighting industry associations are calling upon the EU regulators to complement the EU ecodesign toward the development of EU-wide lighting system legislation (LSL) [30]. With communication technologies not only being desirable, but an actual requirement, VLC can present advantages under such regulation framework.

In the following sections, we will present a quick overview of existing lighting application standards for indoor, outdoor, and road lighting, as these present the best proliferation prospects for VLC systems.

2.6.1 Indoor Lighting

The lighting requirements for indoor working places are covered by EN 12464 Part 1. Although this is not a legally binding document, it presents the basic guidelines followed by the lighting industry when designing a new solution. The main focus of the document is to provide the lighting requirements for the visual comfort of people under different indoor work environments. The standard specifies quantity and quality figures for the design of lighting solutions. Although it does not give technical details for implementation, it provides best practices, leaving the designers free to use natural, artificial, or a combination of both light sources to achieve the recommended illuminance and visual comfort values [31].

Several criteria are used in the analysis and evaluation of the lighting environment, keeping as a goal the accomplishment of three basic human needs: visual comfort, visual performance, and safety. With these goals in mind, the lighting solution is analyzed based on several factors including luminance distribution, illuminance, directionality and variability of light, color rendering, glare, and flicker. The standard also takes into consideration several other ergonomic parameters that influence the visual performance, such as the size of the task at hand or the ophthalmic capability of the subject. Another important aspect of lighting design is to achieve a uniform luminance distribution, which has a great impact on visual acuity and contrast sensitivity. The standard provides recommended reflectance values of surfaces like the ceiling, walls, and the floor. An interesting value specified in the standard is the average illuminance on surfaces of over 50 lux with a reflectance equal to or over 0.1 on walls and 30 lux when the ceiling has a reflectance over 0.1. The standard also provides the relationship of illuminance on the immediate surroundings of the task area. For task illuminance values over 200 lx, the surrounding area may have an illuminance of less than 25%; however, for values under 200 lx, the surrounding area should maintain the illuminance specified for the task to be executed at that location [31].

There are several other factors used to define the lighting solution in an indoor environment. However, some of the most interesting for VLC applications relate to flicker and stroboscopic effects. Although there is no real description on how to measure these, the standard is very clear in stating that the lighting system should be designed to avoid them, as they are known to cause distractions and other physiological effects such as headaches. On the latest revision of the standard, a new section was introduced, providing guidance for modeling a space in order to show shapes and texture in respect to architectural features, persons, and other objects within a given area [31].

EN12464-1 covers a significant set of interior areas, tasks, and activities. These range from common building locations such as traffic zones, restrooms and first aid, control rooms, and storerooms among others. There are over 20 industrial activities and crafts ranging from agriculture to chemical, electrical, and textile industries. Reference values are also provided for offices and retail premises, places of public gathering, educational premises, healthcare, and transportation areas [31]. These reference values provided should serve as a base for the development of any VLC system in indoor scenarios, providing both a higher and lower limit of the light source power as well as other interesting factors such as directionality or even color rendering. Table 2.2 displays some of the recommended values for different application

TABLE 2.2

Lighting Recommendations for Different Environments

Classification	Recommended Illuminance (lux)	Activity Types
Lighting for spaces with low occupation or without special lighting constraints	20–50	Public areas
	50–100	Spaces with infrequent utilization
	100–200	Spaces of occasional use (storage rooms, locker rooms, etc.)
Lighting for spaces with high occupation or special lighting constraints	300–500	Tasks with low visual demands (classrooms, offices, etc.)
	500–1000	Tasks with common visual demands (amphitheaters, office shops with low precision machinery, etc.)
	1000–2000	Tasks with special visual demands (architecture offices, drawing offices, office shops with high-precision machinery, laboratories, etc.)

scenarios. Another important standard to keep in mind is the EN15193: "Energy performance of buildings—Energy requirements for lighting." Although it only provides the methods to evaluate the energy used in indoor lighting by different buildings, it is a key comparison and evaluation tool in modern building design, thus shaping the amount of lighting power installed.

2.6.2 Outdoor Lighting

For outdoor working places, EN12464 Part 2 provides the same set of basic guidelines for the design of a visually comfortable and safe environment. However, it is clear to see that the ratio of illuminance in the surrounding areas of the task area can be much smaller than in indoor scenarios. Another major difference in this second part of the standard is the definition of obtrusive light in two different time periods, pre- and post-curfew. According to the environmental zones, the light levels may be reduced by a factor of 10 in the post-curfew period. As for the areas of application of the standard, they mention several places from farms, to building sites, fuel stations up to airports [32]. To complement EN12646 Part 2, the light technical performance standard EN13201 for outdoor applications focuses on road lighting. Although it is also not a mandatory standard, it has been adopted and legislated over several countries in the EU. The standard is defined in four parts: Part 1—Selection of lighting classes, Part 2—Performance requirements, Part 3—Calculation of performance, and Part 4—Performance measurements. This standard provides the guidelines for road lighting installations with the main focus on the safe usage or roads and the reduction of obtrusive light especially in residential areas.

2.6.3 LED-Based Traffic Signal Specifications

The two primary factors that determine the luminous intensity requirement of a traffic light head are the luminance of the background L_B and the distance d, from which the signal light is to be seen. The optimum intensity depends on sky luminance and distance from traffic light to driver, according to:

$$I_d = Cd^2L_B, \tag{2.70}$$

where C is a constant equal to 2×10^{-6}. For most traffic signals, the visual range must be at least 100 m under a sky luminance of about 104 cd/m^2, thus allowing safe stopping at a speed of 60 km/h [33]. Under these conditions, and according to Equation 2.70, the optimum intensity of the red light signal is 200 cd.

Fisher noticed that a value of 200 cd was required for a red signal light with a 200 mm diameter to be perceived under standard conditions [34] at a distance of 100 m. Cole and Brown [33] also found that as the angle between the traffic light normal and viewing directions was increased above 3°, the luminous intensity requirements also increased. This effect can be introduced into Equation 2.70, resulting in the well-known Fisher equation for the necessary luminous intensity, given by:

$$I_{d,\alpha} = C\left(\frac{\alpha}{3}\right)^{1.33}d^2L_B, \tag{2.71}$$

where $I_{d,\alpha}$ is the required luminous intensity (cd), α is the viewing angle (in degrees), and L_B is the background luminance (in cd/m^2). The viewing angle is of paramount importance for multilane road scenarios. Usually more than one traffic signal must be employed for these cases. Their orientation toward the road can be a useful design parameter to achieve nearly uniform power distribution for all lanes. This is particular relevant if these traffic lights are designed to support VLC. Furthermore, green and yellow signal lights require higher luminous intensity than the intensity of a red signal light, due to the Helmholtz–Kohlrausch effect [5]. Standard recommendations demand that the intensity ratio for red, yellow, and green uses (R:Y:G) = (1:2.5:1.3). Table 2.3 illustrates the minimum and maximum peak intensity requirements for traffic light signal heads, collected from three current standards [35]. It can be seen that the European standard has minimum requirements when compared to the American.

TABLE 2.3

Traffic Signal Requirements

| LED colors | VTCSH Part 2: LED Vehicle Traffic Signal Modules | | | | European Standard: Traffic Control Equipment—Signal Heads | | | |
| | min | | max | | min | | max | |
	200 mm	300 mm	200 mm	300 mm	200 mm	300 mm	200 mm	300 mm
Red (λ = 620–630 nm)	133	399	800	800	100	339	400	800
Yellow (λ = 580–590 nm)	617	1571	3700	3700	100	339	400	800
Green (λ = 530–560 nm)	267	678	1600	1600	100	339	400	800

Luminous intensity (cd) in the reference axis

References

[1] M. S. Shur and R. Zukauskas, Solid-State Lighting: Toward Superior Illumination, *Proc. IEEE*, vol. 93, no. 10, pp. 1691–1703, 2005.

[2] M. R. Krames, O. B. Shchekin, R. M.-Mach, G. O. Mueller, L. Zhou, G. Harbers and M. G. Craford, Status and Future of High-Power Light-Emitting Diodes for Solid-State Lighting, *IEEE J. Display Technol.*, vol. 3, no. 2, pp. 160–175, 2007.

[3] D. Schreuder, *Outdoor Lighting: Physics, Vision and Perception*, Springer, The Netherlands, 2008.

[4] D. Malacara, *Color Vision and Colorimetry: Theory and Applications*, 2nd edition, SPIE, Bellingham, WA, 2011.

[5] G. Wyszecki and W. S. Stiles, *Color Science: Concepts and Methods, Quantitative Data and Formulae*, Wiley Classics Library Edition, 2nd Edition, Hoboken, NJ, 2000.

[6] S. M. Sze, *Physics of Semiconductor Devices*, 3rd edition, Wiley, Hoboken, NJ, 2008.

[7] E. F. Schubert, *Light-Emitting Diodes*, 2nd edition, Cambridge University Press, Cambridge, UK, 2006.

[8] S. Rajbhandari, H. Chun, G. Faulkner, K. Cameron, A. V. N. Jalajakumari, R. Henderson, D. Tsonev, et al., High-Speed Integrated Visible Light Communication System: Device Constraints and Design Considerations, *IEEE J. Sel. Areas Commun.*, vol. 33, no. 9, pp. 1750–1757, 2015.

[9] N. Mohan, T. M. Undeland and W. P. Robbins, *Power Electronics: Converters, Applications and Design*, 3rd edition, Wiley, Hoboken, NJ, 2003.

[10] J. Millman and C. Halkias, *Integrated Electronics*, 2nd edition, MacGraw Hill, New York, NY, 1972.

[11] P. Gray and R. Meyer, *Analysis and Design of Analog Integrated Circuits*, Wiley, Hoboken, NJ, 1984.

[12] M. Wolf, J. Vucic, D. O'Brien, O. Bouchet, H. Le Minh, G. Faulkner, L. Grobe, et al., *Deliverable 4.2a—Physical Layer Design and Specification: Demonstrator 1*, FP7/ICT 213311, European Commission, Brussels, Belgium, 2010.

[13] O. Bouchet, G. Faulkner, L. Grobe, E. Gueutier, K.-D. Langer, S. Nerreter, D. O'Brien, et al., *Deliverable 4.2b—Physical Layer Design and Specification: Demonstrator 2*, FP7/ICT 213311, European Commission, Brussels, Belgium, 2011.

[14] C. Toumazou, F. J. Lidgey and D. G. Haigh, *Analogue IC Design: The Current-Mode Approach*, IEE Circuits, Devices and Systems Series, Peter Peregrinus Ltd, USA, 1990.

[15] H. Li, Y. Zhang, X. Chen, C. Wu, J. Guo, Z. Gao, W. Pei and H. Chen, 682Mbit/s Phosphorescent White LED Visible Light Communications Utilizing Analogue Equalized 16QAM-OFDM Modulation without Blue Filter, Elsevier, *J. Opt. Commun.*, vol. 354, pp. 107–111, 2015.

[16] X. Huang, Z. Wang, J. Shi, Y. Wang and N. Chi, 1.6 Gbit/s Phosphorescent White LED Based VLC Transmission using a Cascaded Pre-equalization Circuit and a Differential Outputs PIN Receiver, *OSA Opt. Express*, vol. 23, no. 17, pp. 22034–22042, 2015.

[17] R. L. Aguiar, A. Tavares, J. L. Cura, E. de Vasconcelos, L. N. Alves, R. Valadas and D. M. Santos, Considerations on the Design of Transceivers for Wireless Optical LANs, *IEE Electronics & Communications, Colloquium on Optical Wireless Communications*, London, UK, June 1999.

[18] L. N. Alves, High Gain and Bandwidth Current-Mode Amplifiers: Study and Implementation, PhD thesis, Universidade de Aveiro, Aveiro, Portugal, 2008.

[19] L. N. Alves and R. L. Aguiar, Design Techniques for High Performance Optical Wireless Front-Ends, *Proceedings of the Conference on Telecommunications—ConfTele 2003*, Aveiro, Portugal, April 2003.

[20] J. J. Morikuni and S.-M. Kang, An Analysis of Inductive Peaking in Photoreceiver Design, *IEEE J. Lightwave Technol.*, vol. 10, no. 10, pp. 1426–1437, 1992.

[21] S. Shekhar, J. S. Walling and D. J. Allstot, Bandwidth Extension Techniques for CMOS Amplifiers, *IEEE J. Solid-State Circuits*, vol. 41, no. 11, pp. 2424–2439, 2006.

[22] T. Wakimoto and Y. Akazawa, A Low-Power Wide-Band Amplifier Using a New Parasitic Capacitance Compensation Technique, *IEEE J. Solid-State Circuits*, vol. 25, no. 1, pp. 200–206, 1990.

[23] Y.-J. Chan, F.-T. Chien, T.-T. Shin and W.-J. Ho, Bandwidth Enhancement of Transimpedance Amplifier by Capacitive Peaking Design, US Patent No. 6353366, 2002.

[24] H. W. Bode, *Network Analysis and Feedback Amplifier Design*, D. Van Nostrand Company, Princeton, NJ, 1956.

[25] B. Razavi, *Design of Integrated Circuits for Optical Communications*, McGraw-Hill, New York, NY, 2003.

[26] P. Staric and E. Margan, *Wideband Amplifiers*, Springer, The Netherlands, 2006.

[27] E. de Vasconcelos, J. L. Cura, R. L. Aguiar and D. M. Santos, A Novel High Gain, High Bandwidth CMOS Differential Front-End for Wireless Optical Systems, *ISCAS 99—IEEE International Symposium on Circuits and Systems*, Orlando, FL, June 1999.

[28] J. L. Cura and R. L. Aguiar, Dynamic Range Boosting for Wireless Optical Receivers, *ISCAS 2001—International Symposium on Circuits and Systems*, Sydney, Australia, May 2001.

[29] European Union, *EUR-Lex—Access to European Union Law*, Available at: http://eur-lex.europa.eu/ (accessed on February 2017).

[30] Lighting Europe, *Lighting Europe Papers and Publications*, Available at: http://www.lightingeurope.org/library (accessed on February 2017).

[31] European Committee for Standardization, *EN 12464 Light and Lighting—Lighting of Work Places – Part 1: Indoor Work Places*, European Committee for Standardization, Brussels, Belgium, 2011.

[32] European Committee for Standardization, *EN 12464 Light and Lighting—Lighting of Work Places—Part 2: Outdoor Work Places*, European Committee for Standardization, Brussels, Belgium, 2007.

[33] B. L. Cole and B. Brown, Optimum Intensity of Red Road-Traffic Signal Lights for Normal and Protanopic Observers, *J. Opt. Soc. Am.*, vol. 56, no. 4, pp. 516–522, 1966.

[34] A. Fisher, *A Photometric Specification for Vehicular Traffic Signal Lanterns. Part 1: Luminous Intensity Necessary for the Detection of Signals on the Line of Sight*, Institute of Highway and Traffic Research University of New South Wales, Wales, UK, 1969.

[35] C. K. Andersen, New ITE Standards for Traffic Signal Lights, *Presentation at the 2nd Baltimore Regional Traffic Signal Forum*, Baltimore, MD, December 2005.

3

Channel Modeling

Zabih Ghassemlooy, Mohammad-Ali Khalighi, and Dehao Wu

CONTENTS

3.1 Introduction

An important step in the design of a visible light communications (VLC) system is to comprehend the limitations arising from the optical wireless channel. Accurate channel characterization is an important prerequisite to set the system parameters appropriately in order to establish a high-quality link since it permits better exploitation of the available energy and spectral resources in view of optimizing the system design. An accurate channel model is also necessary to precisely predict the performance of VLC systems.

In this chapter, we address channel modeling for VLC systems mainly focusing on indoor systems. We introduce different sources of impairment in VLC systems arising from beam propagation or transmitter (Tx)/receiver (Rx) devices. Indeed, the latter could be attributed to the "global channel" comprising the blocks between the signal Tx and signal detection at the Rx.

After this introduction, we give an overview of different propagation modes in Section 3.2. In addition, we explain methods for numerical channel simulation. Analytical channel modeling for the cases of single- and multiple-source systems is presented in Section 3.3. Then, in Section 3.4, we outline the limitations arising from the aggregate channel while focusing on the problem of intersymbol interference (ISI) and how it affects the link performance particularly in the absence of a line of sight (LOS). This could be due to multipath-induced channel time dispersion, that is, multiple reflections from people and objects within an indoor environment, the lack of sufficient bandwidth of the transmitting device, most commonly a light-emitting diode (LED), the photodetector (PD), and the cabling used for lighting installations. Limitations due to the LED nonlinear characteristics could also be considered as an impairment of the global channel, which is the subject of Section 3.5 where channel distortion modeling is investigated. Finally, channel modeling for multiple-input multiple-output (MIMO) VLC systems will be presented next in Section 3.6.

3.2 Signal Propagation

3.2.1 Propagation Modes

For indoor links, six different configurations have been defined in [1], basically classified depending on the existence/nonexistence of the LOS path between the Tx and Rx. Here we consider some of these configurations that apply to the case of indoor VLCs. For LOS configuration, which is the most basic, the emitter beam angle and the receiver field-of-view (FOV) will specify the transmission channel. For the case of directive links, the Tx and Rx have a small divergence angle and FOV, respectively (see Figure 3.1a), thus requiring very accurate alignment and suffering from blocking due to the movement of people or presence of objects within the room. For the so-called hybrid links, Tx and Rx have different degrees of directionality [1]. In nondirective links, Tx and Rx both have a wide angle—see Figure 3.1b. In the case of the diffuse configuration, see Figure 3.1c (this may be one of the most popular and widely used schemes); the source position plays an important role in the power levels at various points within a room. In this configuration, the Tx pointing up toward the ceiling has a wide beam angle and the Rx has a wide FOV, which collects reflected diffused light from the ceiling, floor, walls, and objects in the room [2]. In general, to establish high data rate links,

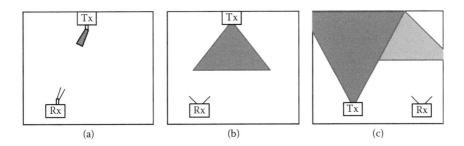

FIGURE 3.1
Link configurations: (a) directed LOS, (b) nondirected LOS, and (c) diffuse.

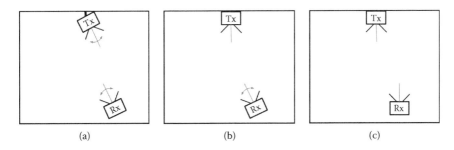

FIGURE 3.2
LOS link classification with respect to mobility: (a) full tracked, (b) half tracked, and (c) nontracked.

the availability of an LOS path is essential since nondirected LOS or diffused configurations will limit the achievable data rate [3]. Indeed, the LOS helps in having a much higher received light intensity (i.e., higher signal-to-noise ratio (SNR) that can be traded to increase the data rate or the link span) and also reduces the risk of ISI, as will be discussed later in Section 3.4, but at the cost of limited mobility or possibility of shadowing/beam blockage.

When considering mobility within an indoor environment, it is essential that a tracking capability is adopted for alignment of the Tx and Rx [4]. Assuming that the Tx is fixed and placed on the ceiling, there are three possible configurations: full-tracking (FT), half-tracking (HT), or nontracking (NT). For the FT link, shown in Figure 3.2a, tracking mechanisms at the Tx and Rx permit the alignment of the Tx and Rx of small apertures, and hence, provide a relatively high SNR at the Rx. For the HT configuration, see Figure 3.2b; only the Tx (or the Rx) tracks the Rx (Tx), thus allowing the use of a lower complexity tracking mechanism, which is more applicable for multiuser systems. For the NT system, see Figure 3.2c; the orientations of the Tx and Rx are fixed and vertical to each other, which permits the implementation of a low-cost system.

In what follows, we will not consider the mobility but focus on the case of nondirected LOS (Figure 3.1b) as the default configuration since it corresponds to a typical indoor VLC system. The non-LOS configuration overcomes the blocking problem by using multiple diffuse reflections from walls and ceiling. In this configuration, the Rx will receive signals from a number of paths, thus ensuring 100% link availability at all times but at the cost of reduced data due to multipath induced ISI. However, it offers much flexibility and ease in the link setup, and hence, it is suitable for point-to-multipoint applications. Note that in general for optical wireless communication (OWC) systems, ISI depends on the data rate and FOV of the Tx and Rx. In VLC systems, the Tx typically has a wide angle of irradiance for the function of lighting in most applications, except the multispot lighting where a Tx with a narrow irradiance angle is normally used.

3.2.2 Channel Simulation

The ideal LOS channel impulse response (CIR) is essentially a time delayed and scaled delta function representing amplitude degradation of the transmitted signal. Therefore, the link attenuation becomes an important parameter that can be derived from the photometric parameters mostly adopted for characterization of LED illumination capability. For the purposes of performance evaluation of VLC systems, one may resort to experimental measurements [5–7] or to numerical simulation of the propagation channel. The second approach is faster and less costly but it can also be helpful prior to the experimental verifications. The propagation channel is fully characterized by the CIR. Whatever the link configuration, the CIR needs to be determined at several points within the indoor environment. Then, the necessity of an accurate and computationally efficient simulation method is obvious. The most critical point in the simulation is the reflections from walls and other objects within the room, which will take time depending of course on the number of reflections taken into consideration.

The reflection characteristics of the surface within an indoor environment depend on a number of factors including the material, operating wavelength of the light source, and the angle of irradiance. The smoothness or roughness of the surface relative to the wavelength will also affect the shape of the reflected patterns. A smooth surface like a mirror or a shiny object reflects the incident beam only in one well-defined direction (i.e., specular reflection), whereas a rough surface reflects the beam in random directions (i.e., diffuse reflection). In practice, most reflections are typically diffuse in nature where the Lambertian model can appropriately be adopted [2,8]. Several measurement campaigns have validated the basic diffuse reflection model, illustrating the importance of the orientation of Tx and Rx as well as the importance of shadowing. There have been several proposed approaches to simulate the diffuse light components. In [2], it was proposed to decompose the room surface into a number of reflecting elements, which scatter the light according to the

Lambertian model, and to sum the reflected light from all these elements at the Rx. However, only single reflections were considered in this method. An extension of the approach of [2] to account for higher-order reflections was proposed in [9] using a recursive algorithm. This method has been widely used in the literature; however, it suffers from high computational complexity. In fact, in order to improve the reliability of the calculated CIR, the number of reflections taken into account should be increased, but this will result in an exponential increase in the computing time [10]. On the other hand, if the number of considered reflections is not sufficient, then the path loss and channel bandwidth will be overestimated [10].

A statistical Monte Carlo approach based on the ray-tracing scheme was proposed in [11] where the ray directions are randomly generated according to the emitter radiation pattern. The trajectories of emitted photons from the Tx are then calculated until they are lost or received in the PD active area. Using a large number of generated photons, the CIR is then estimated. In a more recent work, a lower complexity ray-tracing method was proposed [12]. An iterative method was also presented in [13] that is much faster than the recursive approach of [9]. Using this method, the one-order reflection of tiny surfaces is first calculated, and then used to determine the higher-order reflections in an iterative manner. A so-called integrating sphere method was also proposed in [10] to obtain an analytical channel model that is used to approximate the CIR.

However, most of these works consider the infrared (IR) transmission rather than VLC. The main difference is that for IR communications, a narrowband near-monochromatic IR light source is used and a constant reflectance is considered for the reflectors, independent of the wavelength [14]. It is worth mentioning that most VLC systems use phosphor-based white LEDs with emission in the visible spectrum (from 380 nm to 780 nm) for the reasons of reduced fabrication complexity and cost, compared to red, green, and blue (RGB)-based white LEDs, where all three colors are transmitted simultaneously. However, the bandwidth of these LEDs is very limited (typically several MHz), which is due to the slow time constant of the phosphor. An alternative approach is to use a narrowband blue filter at the receiver in order to filter out the slow yellow component to improve the modulation bandwidth of the LED [15–17]. In this case, most of the results of the work done in the IR band can be exploited as we effectively work in the narrowband. Meanwhile, it should be noticed that the reflectivity of the IR band is higher than that of the visible band [18]. For instance, VLC channel characterization based on the recursive algorithm of [8] is studied in [15,19] where constant reflectance was considered. If the white light is directly used for signal transmission and detection, then we need to modify the channel model to take into account the wavelength-dependent nature of reflectors. This is investigated in [14] where the wideband nature and power spectral distribution of the visible light source are taken into consideration. Indeed, if the number of reflections considered in the simulations is not too high, as a good approximation, the average reflectivity over the entire visible spectrum can be used [14,20].

3.3 Channel Model

3.3.1 Illuminance of LEDs

The illuminance intensity is normally adopted to define the brightness of an LED or an illuminated surface, assuming that the source has a Lambertian radiation pattern, which is given in terms of the spatial angle Ω and the luminous flux Φ as:

$$I = d\Phi/d\Omega. \tag{3.1}$$

Consequently the transmitted (emitted) power P_E is defined as:

$$P_E = \int_{\Lambda min}^{\Lambda max} \int_0^{2\pi} \Phi_e \, d\varphi/d\lambda, \tag{3.2}$$

where Λ_{min} and Λ_{min} are defined by the sensitivity plots of PD, λ is the wavelength, φ is the incident angle, and Φ_e is the energy flux.

The luminous intensity when illuminating a surface defined in terms of the angle of irradiance ϕ is given as [1]:

$$I(\phi) = \frac{m+1}{2\pi} I(0)\cos^m(\phi), \quad \phi \in \left[-\frac{\pi}{2}, \frac{\pi}{2}\right], \tag{3.3}$$

where $I(0)$ represents the center luminous intensity of the LED Tx and m indicates its Lambertian order, given by [21]:

$$m = \frac{-\ln(2)}{\ln(\cos(\Phi_{1/2}))}, \tag{3.4}$$

where $\Phi_{1/2}$ is the semiangle at half illuminance of the Tx. $I(\phi)$ can also be written in terms of the incident power:

$$I(\phi) = \rho \frac{m+1}{2\pi} P_E \cos^m(\phi), \tag{3.5}$$

where ρ is the surface reflection coefficient.

3.3.2 General Transmission Link Model

Like in most OWC systems, intensity modulation with direct detection (IM/DD) is used in most VLC systems [8] for the reasons of reduced cost and implementation complexity. This way, the intensity of the LED, $x(t)$, is modulated by the input signal. Denoting the photocurrent generated

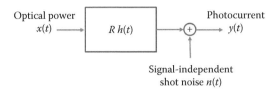

FIGURE 3.3
Baseband-equivalent model of the optical link with IM/DD.

by the PD at the receiver by $y(t)$, the baseband equivalent of the optical link is described by (see Figure 3.3):

$$y(t) = Rx(t) \otimes h(t) + n(t), \qquad (3.6)$$

where R is the PD responsivity, $h(t)$ is the baseband CIR, $\otimes$ denotes convolution, and $n(t)$ is the additive white Gaussian noise. Note that as it represents optical intensity, $x(t)$ is nonnegative.

The receiver noise $n(t)$ is mainly due to the ambient light and in the form of shot noise. The main sources of ambient noise are sunlight and artificial light such as that of incandescent and fluorescent lamps [22,23]. The power spectral density of different ambient light sources and that of the corresponding electrical signals can be found in [1,24]. During daytime, sunlight through windows is typically stronger than the other two sources. In addition, if LEDs are exclusively used for indoor lighting, we are only concerned by the sunlight. Otherwise, to reduce the interference from fluorescent lighting, for example, discrete multitone techniques (DMT) can be used [15] (see Chapter 4). The produced shot noise due to ambient light can degrade the performance of the VLC system. Note that in the case where the blue light is used at the Rx for signal detection by narrow spectral filtering, the influence of ambient light is considerably reduced. If the ambient light is negligible, then the dominant noise source is the Rx preamplifier thermal noise.

3.3.3 Channel Model for Single Source Case

Let us focus on the CIR $h(t)$. If for the sake of simplicity we neglect the diffuse propagation component, that is, consider only the LOS path, the received intensity will depend on the emitter radiation pattern, the receiver optics, and the PD active area. Denoting the emitted optical intensity by P_E, the received optical power P_R is given by:

$$P_R = H(0)P_E, \qquad (3.7)$$

where $H(0)$ is the channel DC gain given by:

$$H(0) = \int_{-\infty}^{\infty} h(t)dt. \qquad (3.8)$$

If we model the emitter by a generalized Lambertian pattern, we have [25]:

$$H(0) = \begin{cases} \dfrac{(m+1)A_{PD}}{2\pi d^2} \cos^m(\phi)T_s(\varphi)g(\varphi)\cos(\varphi), & 0 \leq \varphi \leq \varphi_c \\ 0, & 0 \geq \varphi_c \end{cases}, \qquad (3.9)$$

where A_{PD} is the PD surface area, φ_c is the Rx FOV (semiangle), and d is the distance from LEDs to the Rx point. Also, $T_S(\varphi)$ is the optical filter gain, and the optical concentrator gain $g(\varphi)$ is defined as [8]:

$$g(\varphi) = \begin{cases} \dfrac{n^2}{\sin^2\varphi_c}, & 0 \leq \varphi \leq \varphi_c \\ 0, & 0 \geq \varphi_c \end{cases}. \qquad (3.10)$$

where n is the concentrator refractive index.

Consider Figure 3.4 that shows the geometry of the optical Tx, Rx, and surface reflectors for a typical indoor VLC system. The illuminance at a given point on the receiving plane is given by $I(\varphi)\cos(\varphi)/d^2$ [26].

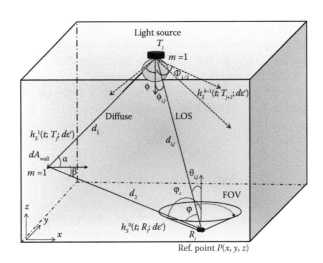

FIGURE 3.4
Geometry of optical Tx, Rx, and reflectors.

Considering power due to the non-LOS paths, the DC channel gain of the reflected path is given by [2]:

$$dH_{ref}(0) = \begin{cases} \dfrac{(m+1)A_{PD}}{2\pi d_1{}^2 d_2{}^2} \rho \, dA_{\text{wall}} \cos^m(\phi)\cos(\alpha)\cos(\beta)T_s(\varphi)g(\varphi)\cos(\varphi), & 0 \leq \varphi \leq \varphi_c, \\ 0, & \varphi \geq \varphi_c \end{cases}$$

(3.11)

where β represents the angle of irradiance from the reflective area of the wall, α is the angle of irradiance to the wall, d_1 and d_2 are the distances between the Tx and the wall, and the wall and a point on the receiving surface, respectively, and dA_{wall} is the size of the reflective area.

Now if we consider both multipath propagation and the LOS component, for the more general case, the total received power P_R is given by [27]:

$$P_R = P_E H_{LOS}(0) + \int_{\text{walls}} P_E dH_{ref}(0).$$

(3.12)

Note that the electrical SNR, which expresses the quality of transmission, can be defined in terms of P_R as:

$$SNR_{ele} = \frac{\left(RH(0)P_R\right)^2}{\sigma_T^2},$$

(3.13)

where σ_T^2 is the total noise variance.

The transmitted intensity of a single light beam undergoes a number of k reflections (bounces) prior to collection at the Rx, which is described by the CIR as [28]:

$$h(t; T_j, R_i) = \sum_{k=0}^{\infty} h_S^{(k)}(t; T_j, R_i),$$

(3.14)

where $h_S^{(k)}(t; T_j, R_i)$ is the impulse response corresponding to the kth reflection. The LOS contribution to CIR is given in terms of the delayed Dirac delta function as:

$$h_S^0(t; T_j, R_i) = VI(\phi_{ij})\left(\frac{A_{Ri}g(\varphi)}{d_{ij}^2}\right) \times \delta(t - d_{ij}/c),$$

(3.15)

where V is the visibility factor $0 < V \leq 1$, with $V = 1$ representing unobstructed LOS path, c is the speed of light, d_{ij} is the distance between the Tx and the Rx, and A_{Ri} is the optical collection area. Also, $g(\varphi)$ is the Rx optical gain function, defined as follows:

$$g(\varphi) = \begin{cases} \cos(\varphi) & \text{if } 0 \leq \varphi \leq \pi/2 \\ 0 & \text{otherwise} \end{cases}.$$

(3.16)

Similarly, the k-bounce response can be calculated using the $(k-1)$-bounce response, which is given by Carruthers et al. in [29]:

$$h_S^{(k)}(t; T_j, R_i) = \int_S \rho d\varepsilon^r \cdot h_S^{(k-1)}(t; T_j, d\varepsilon^r) \otimes h_S^0(t; d\varepsilon^t, R_i), \tag{3.17}$$

where the integral is over the surfaces in S, $d\varepsilon^t$ and $d\varepsilon^r$ represent a differential surface of area dr^2, where the first one acts as Rx with respect to T_j and then as a source with respect to R_i.

Note as $k \to \infty$, $\|h_S^{(k)}(t; T_j, d\varepsilon^r)\| \to 0$ since $\rho < 1$ everywhere; then we can estimate the overall CIR for a number of N-bounce as:

$$h_S(t; T_j, R_i) \approx \sum_{k=0}^N h_S^k(t; T_j; R_i). \tag{3.18}$$

A good approximation can be achieved for $3 < N < 10$, as outlined in [29].

3.3.4 Channel Model for Multiple Sources

Let us consider a general VLC channel with M light sources (or M small elements per each facet) and multiple propagation with N non-LOS paths between a Tx and an Rx. The general link geometry is shown in Figure 3.5. The Rx R_j receives radiation emitted from multiple sources including T_i

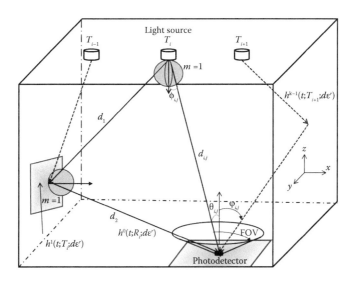

FIGURE 3.5
Geometry of the source, PD, and reflectors.

via the LOS, as well as from k-number of reflections from walls, ceiling, and floor within the room.

For the indoor VLC configuration as shown in Figure 3.5, assuming that T_i emits a unit impulse at $t = 0$ and normalizing P_E to 1, the LOS ($k = 0$) CIR for a particular source T_i and a detector R_j, is given by [30]:

$$h_S^0(t; T_i, R_j) = \frac{I(\phi_{ij})A_{Ri}}{d_{ij}^2} T_s(\varphi_{ij})g(\varphi_{ij})\cos(\varphi_{ij})\text{rect}\left(\frac{\varphi_{ij}}{\varphi_c}\right)\delta\left(t - \frac{d_{ij}}{c}\right), \qquad (3.19)$$

where d_{ij} is the distance between T_i and R_j, and $\delta(\cdot)$ is the Dirac delta function. Also, rect(x) stands for a rectangular function defined as:

$$\text{rect}(x) = \begin{cases} 1 & \text{for } |x| \leq 1, \\ 0 & \text{for } |x| > 1. \end{cases} \qquad (3.20)$$

The CIR for k-bounce ($k \leq 1$) is given by [29]:

$$h_S^k(t; T_i, R_j) = \sum_{n=1}^{M} \rho\, d\varepsilon_n^r \cdot h^{k-1}(t; T_i, d\varepsilon_n^r) \otimes h_S^0(t; d\varepsilon_n^t, R_j). \qquad (3.21)$$

Following the methodology described in [29], all reflective surfaces are represented by a number of small-area elements ε_n, that is, M. In Equation 3.21 and Figure 3.5, both $d\varepsilon_n^r$ and $d\varepsilon_n^t$ represent small reflecting areas that are acting as an Rx with respect to the light source T_i, and then as a source to R_j. The overall CIR taking into account multiple transmitters and multiple reflections can be written as [29]:

$$h(t; T_i, R_j) = \sum_{i=1}^{M}\sum_{k=0}^{\infty} h_i^k(t; T_i, R_j). \qquad (3.22)$$

For the case of high data rate indoor VLC systems that we consider here, because of relatively slow movement of people and fixed objects within a room, the channel can effectively be considered as time invariant. To determine $h_j^k(t; T_i, R_j)$, we carry out the following: (i) calculate the M impulses responses $h_S^0(t, d\varepsilon_n^t, R_j)$; (ii) continue computing $h_S^1(t, d\varepsilon_n^t, R_j)$ until we have $h_S^{k-1}(t, d\varepsilon_n^t, R_j)$; and (iii) use (3.21) to calculate $h_S^k(t, d\varepsilon_n^t, R_j)$ for each receiver. Note that the computation time t_{com} really depends on three key parameters of N and M and the number of Rxs. For a single Rx, $t_{\text{com-1Rx}} = (M^2 \cdot N^2)$ [29]. In a typical room of size 4×4 m^2 with M of 2024 facets, 100 sec $< t_{\text{com}} < 0.03$ sec for a number of Rx elements from 1 to 1000, respectively.

From the CIR, two important items are deducted: channel gain and the root mean square (RMS) delay spread τ. It has been shown that τ and the channel gains are more than sufficient to model diffuse configurations.

The delay spread τ provides a good estimate to how susceptible the channel is to ISI, and can be computed from the CIR using:

$$\tau = \left[\frac{\int (t - \mu)^2 h^2(t) dt}{\int h^2(t) dt} \right]^{\frac{1}{2}}, \tag{3.23}$$

where t is the propagation time and μ is the mean excess delay given by:

$$\mu = \frac{\int t\, h^2(t) dt}{\int h^2(t) dt}. \tag{3.24}$$

Obviously, different room and Tx–Rx configurations can significantly affect τ, and smaller values of τ indicate a higher system transmission bandwidth [13,31]. In OWC systems the most important feature is the channel gain (defined as the ratio between P_R and P_E), which determines the achievable SNR for a given P_E [29].

3.4 Channel Limitations and ISI

3.4.1 Multipath Dispersion

In optical communications, the information carrier signal has a frequency of about 10^{14} Hz. Typically, the PD active area in VLC receivers is about millions of square wavelengths. Since the total generated photocurrent is proportional to the integral of the optical power over the entire PD surface area, this will provide an inherent spatial diversity. Therefore, indoor VLC systems are effectively not subject to multipath fading [8,25]. For the same reason, we are concerned with negligible Doppler spreads and the channel can mostly be considered as time invariant (except when shadowing or beam blockage occurs) [32]. However, multipath propagation of emitted signals in these systems leads to time dispersion and ISI, which will limit the transmission rate [24].

The signal intensity received on the PD surface includes contributions from the LOS (with respect to the transmitters), as well as from reflections of walls or objects within the room [15]. For the LOS contribution, the channel response is modeled by Dirac pulses, whereas for the diffuse part, it is represented by an integrating sphere model [10]. The diffuse component is almost constant and depends on the room properties and the Rx aperture size.

To perform a more detailed analysis, a case study is presented. Consider a room of dimension $5 \times 5 \times 3$ m^3 with a single Tx in the middle of the ceiling. The Rx is placed at a height of 0.5 m that corresponds to a typical desktop. Using equations (3.4, 3.7, 3.9, and 3.10), and for a transmit optical power of 2 W, $\Phi_{1/2}$ of 60°, $I(0)$ of 200 Lux at 700 mA of current, four LEDs, Rx FOV of 60°, PD surface area of 16 mm^2, ρ of 0.8, a unity gain optical filter, and a lens at PD with a refractive index of 1.5, the received optical power distributions corresponding to the LOS path and multipaths (only the first-order reflections) are shown in Figure 3.6a and b, respectively.

We notice that the impact of multipath reflections is most significant at room corners. They have a much lower impact when the Rx is located beneath the Tx, however, which is quite logical as the LOS has the main contribution in the received signal.

In most practical cases, however, the influence of the diffuse component is masked by the strong LOS component. It is shown in [15] that it has no significant influence on the overall channel bandwidth. Indeed, for typical room dimensions, the channel time dispersion corresponding to the LOS component is negligible [15]. For example, considering a room of dimension $5 \times 5 \times 3$ m^3 with Txs configuration as shown in Figure 3.7 the maximum delay between two LOS paths is around 5.5 ns only.

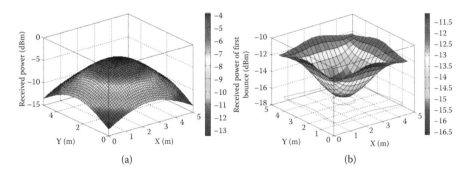

(a)   (b)

FIGURE 3.6
Received optical power distribution corresponding to (a) LOS and (b) first-order reflection multipath. $5 \times 5 \times 3$ m^3 room, Rx height 0.5 m.

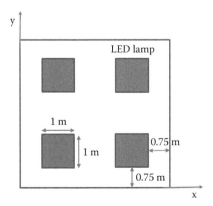

FIGURE 3.7
Example of ceiling lighting design using four LEDs of 1×1 m^2 spaced 7 cm apart with a total number of 900 chips. (Adapted from Grubor, J., et al., *J. Lightwave Technol.*, 26, 3883–3892, 2008.)

In general, the limitation of ISI depends on the transmission scenario and the room properties, and can be quantified by evaluating the channel cut-off frequency. Consider an LED modulation bandwidth of 20 MHz (which is the case when blue filtering is performed, as explained next in Subsection 3.4.2), the Nyquist symbol period is limited to 25 ns, and ISI will occur if transmitted data symbols experience delays larger than 12.5 ns, assuming that LEDs are synchronously driven. Simulation results provided in [15] showed that channel bandwidth limitation corresponds to the minimum bandwidth of 90 MHz at the worst location at the desktop surface, which is significantly above the 20 MHz limitation of the LED itself. As a result, the channel can effectively be considered as frequency nonselective (flat) over the bandwidth of interest.

Note that, alternatively to the cut-off frequency, the transmission bit rate R_b can be used given the RMS delay spread τ as [27]:

$$R_b \leq \frac{1}{10\tau}. \tag{3.25}$$

It should be noticed that in large rooms such as conference halls where we have a noticeable difference between the optical path delays, we may be concerned with ISI for higher transmission data rates. This is specially the case when the Rx is placed in the corner of the room, where the diffuse propagating component becomes predominant [33]. In such cases, advanced modulation schemes should be used.

Note that, in addition to the room dimensions, channel dispersion also depends on the receiver FOV and the distance between the Tx and the Rx. When smaller beam divergence angles are used at the Tx or Rxs of smaller FOV are used, the dispersion is due to multipath scattering and reflections, and hence the ISI is reduced. This is the case for tracked directed links. Such systems can potentially support data transmission speeds of more than 100 Mbps [10,25] but require sophisticated tracking mechanisms to ensure link connectivity. On the other hand, the Tx radiation pattern can be optimized in order to improve link properties. More specifically, we can optimize the Tx Lambertian order through the use of a beam diffuser to maximize the LOS path gain and to minimize the ISI. This is particularly interesting in multicell scenarios. We consider three different room sizes and cell configurations as specified in Table 3.1.

We have shown in Figure 3.8 the normalized CIR for the three cases of Rooms A, B, and C in Table 3.1 where the typical first-order Lambertian order and optimized Lambertian order LEDs are used. We notice that for the Room A in Figure 3.8a, the amplitude of CIR corresponding to LOS increases from 43.5% to 72% by using the Optimum Lambertian order (OLO). Also, the LOS component increases from 35.6% to 81.6% and from 25.6% to 80.3% by using OLO for the two other cases of Rooms B and C, shown in Figure 3.8b and c, respectively. At the same time, the contribution of the reflected paths, and hence the ISI, decreases significantly.

TABLE 3.1

Specification of Studied Indoor VLC Systems

Parameters	Values
LED wavelength (λ)	(500–1000) nm
LED power	200 mW
Half angle FOV of receiver	60 (deg.)
Active area of photodiode	16 mm^2
Gain of optical filter	1.0
Refractive index of a lens at a photodiode	1.5
Reflection coefficient (wall, ceiling, floor)	(0.8, 0.8, 0.3)
Room A (width, length, height)	5 × 5 × 3 m
Number of cells	4
Cell radius (r)	1.77 m
OLO	5.7
Room B (width, length, height)	4 × 6 × 3 m
Number of cells	6
Cell radius (r)	1.41 m
OLO	9
Room C (width, length, height)	5 × 5 × 3 m
Number of cells	9
Cell radius (r)	1.17 m
OLO	13

Note: FOV, field of view; OLO, optimum Lambertian order.

We have also shown in Figure 3.9 the profile of the RMS delay spread for different Rx positions for the nonoptimized and optimized source patterns for the case if a four-cell scenario and 5 × 5 × 3 m room size. The average RMS delay spread by using OLO decreases from ~1.5 ns to ~0.4 ns and the peak RMS delay spread which corresponds to the room corners decreases from ~2.3 ns to ~0.5 ns [34].

In the case of multiple emitting sources, the main factor that impacts the channel frequency selectivity is the asymmetry between the multiple LOS paths rather than multipath reflections [20,35]. As a matter of fact, the RMS delay spread is a useful metric for comparing the degree of frequency selectivity of the different link configurations. However, its absolute value cannot be used to determine the limitation on the transmission rate [20]. One may resort to the channel 3-dB cut-off frequency to determine the degree of channel frequency selectivity. However, this metric is of limited interest in practice except for the case of a purely diffuse channel, that is, blocked LOS. Otherwise, the oscillating behavior of the frequency response due to the contribution of the LOS component makes the 3 dB bandwidth meaningless [20,35]. It is shown in [20,35] that a

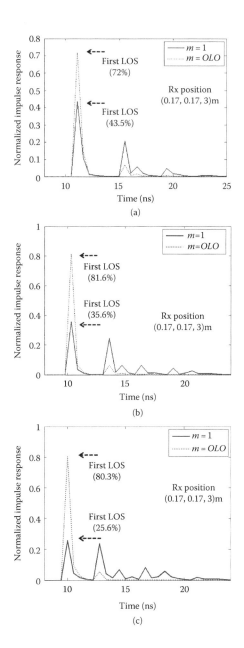

FIGURE 3.8
The normalized CIR when using first-order and optimized Lambertian order light sources in: (a) Room A, (b) Room B, and (c) Room C configurations.

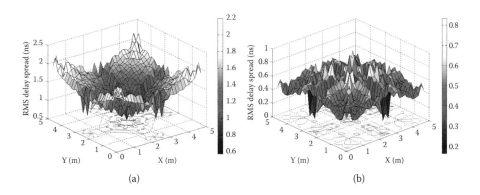

FIGURE 3.9
Spatial distribution of RMS delay spread with and without OLO in a $5 \times 5 \times 3$ m³ four-cell room: (a) with $m = 1$ and (b) with $m = $ OLO.

useful metric in quantifying the amount of ISI is signal-to-ISI ratio (SIR), defined as the ratio of the received powers corresponding to the "desired" signal and ISI, respectively. Defining the ISI on the sampled signal at the Rx, the interest of this metric was demonstrated in [20]. Furthermore, taking into account the effect of the Rx filter, it was shown that a simple Bessel low-pass filter is preferable to the matched filter (assuming perfect time synchronization at the Rx) since it provides a higher SIR at relatively high data rates.

3.4.2 LED Bandwidth Limitation

As a matter of fact, the main limitation on high data rate transmission arises from the limited bandwidth of the LED. This bandwidth limitation is due to the power-bandwidth trade-off and also the parasitic elements in the packaging of the LED [32]. Whereas the trichromatic LEDs have bandwidths of several hundred megahertz and more [17,25], for the reasons of fabrication cost and also color-shift over time, white light LEDs (blue LED plus phosphorous layer) are envisioned for indoor lighting. Typically, the bandwidth of white LEDs is limited to several MHz mainly due to the slow time constant of phosphor. If blue filtering is performed at the Rx to suppress the phosphorescent portion of the optical spectrum, this bandwidth can be extended to about 20 MHz [15].

To increase the transmission rate over this limit, a number of solutions have been proposed so far. One solution is the pre-equalization of the driving circuitry, by which the bandwidth can be increased up to 50 MHz [36,37] but at the price of high SNR penalty (around 20 dB). Another solution is to use multilevel modulation together with DMT modulations, which require complex driving circuitry at the Tx [15,38–40], or to perform frequency-domain equalization (FDE) at the price of increased complexity of the Rx [41–43]. Another alternative is to use MIMO architectures to benefit from spatial multiplexing at the Tx [44–46]. We will consider channel modeling for MIMO systems in Section 3.6.

Note that bandwidth limitation is much more important in the case of using organic devices. The typical bandwidth for organic PDs and organic LEDs (OLEDs) is in the order of 30 KHz and 90 KHz, respectively [47], which is much lower than those for the inorganic counterparts. The use of advanced transmission techniques such as equalization, multilevel modulation, DMT, or MIMO is particularly promising in this case for increasing the data rate beyond the limitations of the components [48–50].

Lastly, although it is not a fundamental issue, cabling can impact the transmission speed in indoor VLC systems. In fact, the difference between the electrical signal paths driving the LEDs on the ceiling can lead to ISI. This obviously depends on the distribution of LED lamps on the ceiling. It is shown in [15] that the critical cabling difference is about 1.6 m considering the Nyquist symbol rate for a bandwidth of 20 MHz. This can be managed easily in installations in medium-size rooms.

3.5 Signal Distortion

3.5.1 Nonlinear LED Characteristics

Another impairment that can affect the performance of a VLC system is the nonlinearity of the LED transfer function, regarding both voltage–current and current-emitted optical power relationships [51–53]. In addition, the signal is also limited due to the limited dynamic range of the LED.

On the other hand, as linear power amplifiers cannot be used for reasons of their high power consumption, we should admit even more nonlinearity [32]. The effect of LED nonlinearity is especially important when high-order constellations are used. It is also problematic when quadrature amplitude modulation (QAM) with DMT is employed in order to increase the data rate or to reduce the impact of the ambient noise from artificial light sources such as fluorescent lighting (see Section 3.3.2). In such a case, the LED nonlinear transfer function causes cross-talk between the subcarriers [53]. It is hence important to consider appropriate modeling of LED nonlinearity in order to evaluate/predict the effective system performance.

Note that, although this impairment does not concern the physical propagation channel, we can consider it as a part of the global channel model, incorporating the imperfect effects of the Tx and Rx devices.

3.5.2 Distortion Modeling

The most common approach to account for LED nonlinearity is to consider memory or memoryless models mostly based on a static model (i.e., neglecting the change in the characteristics of the LED over time) [51]. Considering modulation frequencies well below the LED 3-dB bandwidth, the classical approach is to consider a memoryless model and to use a polynomial fit to the nonlinear transfer function. However, to obtain a realistic model, the polynomial order should be more than five, though a second-order polynomial can provide a fair description of the transfer function.

This way, the output power P_{out} is described as a function of the input current I_{in} as follows:

$$P_{out} = b_0 + b_1(I_{in} - I_{DC}) + b_2(I_{in} - I_{DC})^2, \qquad (3.26)$$

where the coefficients b_0, b_1, and b_2 are the polynomial coefficients and I_{DC} is the DC bias current.

For a more accurate model that can be used for higher modulation frequencies (i.e., large signal bandwidths) the memoryless model is not adequate. Indeed, the frequency-dependency of the current–voltage characteristics of the LED necessitates taking the memory effects of the nonlinearity into consideration [51].

One solution is to use the active region carrier density rate equation [54]. The Volterra series representation of the nonlinearity is the most accurate method but the practical interest of this model is limited due to the high computational complexity for calculating the model parameters that makes it inappropriate for real-time applications [55,56]. As an alternative to this model, a memory polynomial model can be used, as suggested in [57]. Another simplification of this model is to consider two blocks of a linear time-invariant (LTI) system and a memoryless nonlinear system. The order of these two blocks results in Wiener (LTI followed by memoryless nonlinear) [58] or Hammerstein [51] models (otherwise).

3.6 MIMO VLC Systems

3.6.1 Interest of MIMO Structures

In most VLC systems, we have a relatively high SNR available. In order to achieve high data rates despite the limited bandwidth of LEDs, one solution is to perform spatial multiplexing by using the MIMO technique. MIMO systems offer a higher data throughput as well as increased link range without the requirement for additional power or bandwidth. MIMO systems have been widely proposed for optical interconnects between source and detector arrays in order to simplify the source-detector alignment [59]. In VLC systems, however, MIMO systems have attracted attention for their ability of increasing channel capacity [16,17,44–46,60].

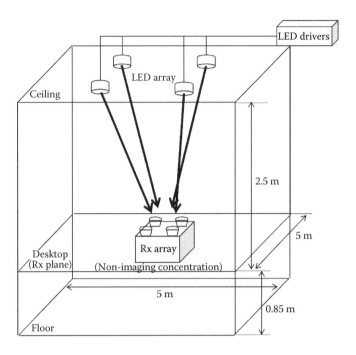

FIGURE 3.10
Example of a nonimaging optical MIMO system.

In contrast to radio frequency [61] or free-space optical communication [62], where MIMO systems are used to achieve increased link reliability by providing spatial diversity gain, in VLC systems we are concerned with a deterministic channel, and hence the only interest of MIMO architectures is for increasing the data throughput.

3.6.2 Channel Modeling for MIMO VLC Systems

For MIMO VLC systems, there are two approaches of nonimaging and imaging receivers. Let us start with the simpler one, that is, the nonimaging system where at the Rx side, nonimaging lenses are used for collecting the transmitted intensity. Figure 3.10 shows an example of such a system where the LEDs and receivers are arranged in a 2 × 2 array. Each PD, through nonimaging concentrators, collects the light from the LEDs with different intensities.

Let us consider the general case of an MIMO architecture with N_T transmitters and N_R receivers. As explained in Section 3.4.1, the LOS propagation component dominates the diffuse one in most practical cases. Furthermore, we can practically neglect the difference between the propagation delays of the different LOS paths [44]. So, the MIMO channel can be described by a matrix H of dimension $N_R \times N_T$, whose entries are the DC channel gains between a pair of Tx–Rx.

$$H = \begin{bmatrix} h_{11} & \cdots & h_{1N_T} \\ \vdots & \ddots & \vdots \\ h_{N_R 1} & \cdots & h_{N_R N_T} \end{bmatrix}. \tag{3.27}$$

For instance, h_{ij} represents the channel gain between the jth PD and ith LED. The channel matrix includes the LOS component as well as the diffuse component arising from multipath reflections. A general form of H is proposed in [63] as follows:

$$H = G_r(D + F_s \cdot \Pi), \tag{3.28}$$

where D represents the contribution of the LOS components, F_s and G_r denote the Tx and Rx profiles, respectively, and Π is the environment matrix representing the contribution of surface reflectors.

If we denote the vectors of transmitted and received signals at a given time reference by $X = [x_1,\ldots, x_{N_T}]^T$ and $Y = [y_1,\ldots,y_{N_R}]^T$, respectively, we can write:

$$Y = RP_{\text{LED}} H X + n, \tag{3.29}$$

where R is the detector responsivity, n is the noise, and P_{LED} is the average transmitted power. Then, the transmitted data are obtained, for instance, based on channel inversion:

$$\hat{X} = H^{-1}Y, \tag{3.30}$$

where it is assumed that H is known at the Rx, which can be realized through the transmission of some pilot signals. Inverse filtering is justified by the relatively high SNR available at the Rx in VLC systems. However, in order to estimate the transmitted signals from Equation 3.30, the channel matrix must obviously be of full rank. For the configuration shown in Figure 3.10, this is not the case when the Rx is situated in the center of the room or along its axes. As a result, by nonimaging MIMO, the channel bandwidth is position dependent; depending on the LED configuration geometry and Rx position, the channel matrix can be ill-conditioned, and in the worst case, rank-deficient. Note that inverting an ill-conditioned H results in a significant noise amplification and consequently a considerable bit error rate (BER) increase. (The reader is referred to [44], Figure 3, which shows the dependence of the BER to the Rx position for a configuration similar to Figure 3.10.)

To circumvent the problem of rank-deficient channel matrix, an imaging lens system [64] can be used at the Rx. Figure 3.11 illustrates the imaging MIMO structure, where the LED arrays are "imaged" to the Rx plane via the imaging lens [44]. This requires a large enough Rx area so that the images of the LED arrays fall on the detectors for all possible Rx positions inside the room [65]. By paraxial approximation, we neglect image distortion due to the dependence of magnification on the angle of incidence of the rays.

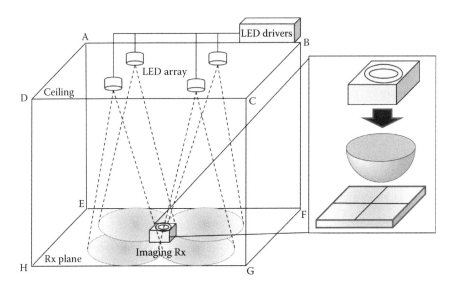

FIGURE 3.11
Schematic of an imaging optical MIMO system.

   This technique is quite efficient but the Rx imaging lens is bulky and introduces additional expense and complexity. Another approach is to use a standard
camera technology [66], but the problem is the limited FOV as such cameras are
designed to produce focused images that match the human eye. Lastly, the use
of a hemispherical imaging lens has been recently proposed in [45], which has
the advantage of providing a very wide FOV with low correlation between the
underlying subchannels (in the case of using several PDs at the Rx).

References

[1] J. M. Kahn and J. R. Barry, Wireless infrared communications, *Proc. IEEE*, vol.
    85, no. 2, pp. 265–298, 1997.
[2] F. R. Gfeller and U. H. Bapst, Wireless in-house data communication via diffuse
    infrared radiation, *Proc. IEEE*, vol. 67, no. 11, pp. 1474–1486, 1979.
[3] L. Grobe, A. Paraskevopoulos, J. Hilt, D. Schulz, F. Lassak, F. Hartlieb, C. Kottke,
    V. Jungnickel and K.-D. Langer, High-speed visible light communication sys-
    tems, *IEEE Commun. Mag.*, vol. 51, no.12, pp. 60–66, 2013.
[4] V. Jungnickel, A. Forck, T. Haustein, U. Kruger, V. Pohl and C. von Helmolt,
    Electronic tracking for wireless infrared communications, *IEEE Trans. Wireless
    Commun.*, vol. 2, no. 5, pp. 989–999, 2003.
[5] J. M. Kahn, W. J. Krause and J. B. Carruthers, Experimental characterization of
    non-directed indoor infrared channels, *IEEE Trans. Commun.*, vol. 43, no. 234,
    pp. 1613–1623, 1995.

[6] A. Yokoi, Samsung Yokoham Research Institute, https://mentor.ieee.org/802.15/dcn/08/15-08-0436-00-0vlc-vlc-channelmeasure ment-for-indoor-application.pdf

[7] K. Cui, G. Chen, Q. He, and Z. Xu, Indoor optical wireless communication by ultraviolet and visible light, *SPIE*, vol. 7464, 2009. DOI: 10.1117/12.826312.

[8] J. R. Barry, *Wireless Infrared Communications*, Kluwer Academic, Boston, MA, 1994.

[9] J. R. Barry, J. M. Kahn, W. J. Krause, E. A. Lee and D. G. Messerschmitt, Simulation of multipath impulse response for indoor wireless optical channels, *IEEE J. Sel. Areas Commun.*, vol. 11, no. 3, pp. 367–379, 1993.

[10] V. Jungnickel, V. Pohl, S. Noenning and C. Von Helmolt, A physical model for the wireless infrared communication channel, *IEEE J. Sel. Areas Commun.*, vol. 20, no. 3, pp. 631–640, 2002.

[11] F. J. Lòpez-Hernàndez, R. Pèrez-Jimènez and A. Santamarìa, Ray-tracing algorithms for fast calculation of the channel impulse response on diffuse IR wireless indoor channels, *Opt. Eng.*, vol. 39, no. 10, pp. 2775–2780, 2000.

[12] M. Zhang, Y. Zhang, X. Yuan and J. Zhang, Mathematic models for a ray tracing method and its applications in wireless optical communications, *Opt. Express*, vol. 18, no. 17, pp. 18431–18437, 2010.

[13] J. B. Carruthers and P. Kannan, Iterative site-based modeling for wireless infrared channels, *IEEE Trans. Antennas Propag.*, vol. 50, no. 5, pp. 759–765, 2002.

[14] K. Lee, H. Park and J. R. Barry, Indoor channel characteristics for visible light communications, *IEEE Commun. Lett.*, vol. 15, no. 2, pp. 217–219, 2011.

[15] J. Grubor, S. Randel, K.-D. Langer and J. W. Walewski, Broadband information broadcasting using LED-based interior lighting, *J. Lightwave Technol.*, vol. 26, pp. 3883–3892, 2008.

[16] D. O'Brien, L. Zeng, H. Le-Minh, G. Faulkner, J.W. Walewski and S. Randel, Visible light communications: Challenges and possibilities, *International Symposium on Personal, Indoor and Mobile Radio Communications, (PIMRC)*, Cannes, France, pp. 1–5, September 2008.

[17] Z. Ghassemlooy, H. Le Minh, P. Haigh and A. Burton, Development of visible light communications: Emerging technology and integration aspects, *International Conference on Optics and Photonics Taiwan (OPTIC2012)*, Taipei, Taiwan, 6–8 December, 2012.

[18] T. Haran, Short-Wave Infrared Diffuse Reflectance of Textile Materials, MS dissertation, Georgia State University, 2008.

[19] T. Komine, J. H. Lee, S. Haruyama and M. Nakagawa, Adaptive equalization system for visible light wireless communication utilizing multiple white LED lighting equipment, *IEEE Trans. Wireless Commun.*, vol. 8, no. 6, pp. 2892–2900, 2009.

[20] S. Long, M. A. Khalighi, M. Wolf, S. Bourennane and Z. Ghassemlooy, Investigating channel frequency selectivity in indoor visible light communication systems, *IET Optoelectronics*, vol. 10, no. 3, pp. 80–88, 2016.

[21] R. Wang, J.-Y. Duan, A.-C. Shi, Y.-J. Wang and Y.-l. Liu, Indoor optical wireless communication system utilizing white LED lights, *15th Asia-Pacific Conference on Communications (APCC2009)*, pp. 617–621, 2009.

[22] A. J. C. Moreira, R. T. Valadas and A. M. de Oliveira Duarte, Performance of infrared transmission systems under ambient light interference, *IEE Proc. – Optoelectronics.*, vol. 143, no. 6, pp. 339–346, 1996.

[23] A. C. Boucouvalas, Indoor ambient light noise and its effect on wireless optical links, *IEE Proc. Optoelectronics*, vol. 143, no. 6, pp. 334–338, 1996.

[24] Z. Ghassemlooy and A. Hayes, *Indoor Optical Wireless Communications Systems—Part 1: Review*, School of Engineering, Northumbria University, 2003.

[25] T. Komine and M. Nakagawa, Fundamental analysis for visible-light communication system using LED lights, *IEEE Trans. Consum. Electron.*, vol. 50, no. 1, pp. 100–107, 2004.

[26] T. Do, H. Junho, J. Souhwan, S. Yoan and Y. Myungsik, Modeling and analysis of the wireless channel formed by LED angle in visible light communication, *International Conference on Information Networking (ICOIN2012)*, pp. 354–357, 2012.

[27] Z. Ghassemlooy, W. O. Popoola and S. Rajbhandari, *Optical Wireless Communications: System and Channel Modelling with MATLAB®*, CRC, Boca Raton, FL, 2012, ISBN: 978-4398-5188-3.

[28] P. L. Eardley, D. R. Wisely, D. Wood and P. McKee, Holograms for optical wireless LANs, *IEE Proc. Optoelectronics*, vol. 143, pp. 365–369, 1996.

[29] J. B. Carruthers, S. M. Caroll and P. Kannan, Propagation modelling for indoor optical wireless communications using fast multi-receiver channel estimation, *IEE Proc. Optoelectronics*, vol. 150, pp. 473–481, 2003.

[30] J. B. Carruthers and S. M. Carroll, Statistical impulse response models for indoor optical wireless channels, *Int. J. Commun. Syst.*, vol. 18, pp. 267–284, 2005.

[31] J. B. Carruther and J. M. Kahn, Angle diversity for nondirected wireless infrared communication, *IEEE Trans. Commun.*, vol. 48, pp. 960–969, 2000.

[32] A. Jovicic, J. Li and T. Richardson, Visible light communication: Opportunities, challenges and the path to market, *IEEE Commun. Mag.*, vol. 51, no. 12, pp. 26–32, 2013.

[33] P. A. Haigh, Z. Ghassemlooy, S. Rajbhandari, I. Papakonstantinou, and W. Popoola, Visible light communications: 170 Mb/s using an artificial neural network equalizer in a low bandwidth white light configuration, *IEEE J. Lightwave Technol.*, vol. 32, no. 9, pp. 1807–1813, 2014.

[34] D. Wu, Z. Ghassemlooy, H. L. Minh, S. Rajbhandari, M.A. Khalighi and X. Tang, Optimization of Lambertian order for indoor non-directed optical wireless communication, *Optical Wireless Communications Workshop, International Conference on Communications in China (ICCC)*, Beijing, China, pp. 43–48, 2012.

[35] S. Long, M. A. Khalighi, M. Wolf, S. Bourenanne and Z. Ghassemlooy, Channel characterization for indoor visible light communications, *International Workshop on Optical Wireless Communications (IWOW)*, Madeira, Portugal, pp. 75–79, September 2014.

[36] H. Le-Minh, D. C. O'Brien, G. Faulkner, L. Zeng, K. Lee, D. Jung and Y. Oh, High-speed visible light communications using multiple resonant equalization, *IEEE Photon. Technol. Lett.*, vol. 20, no. 14, pp. 1243–1245, 2008.

[37] H. Le Minh, D. O'Brien, G. Faulkner, L. Zeng, K. Lee, D. Jung, Y. J. Oh and E. T. Won, 100-Mb/s NRZ visible light communications using a postequalized white LED, *IEEE Photon. Technol. Lett.*, vol. 21, no. 15, pp. 1063–1065, 2008.

[38] J. Vucic, C. Kottke, S. Nerreter, A. Büttner, K.-D. Langer and J. W. Walewski, White light wireless transmission at 200+ Mb/s net data rate by use of discrete-multitone modulation, *IEEE Photon. Technol. Lett.*, vol. 21, pp. 1511–1513, 2009.

[39] J. Vucic, C. Kottke, S. Nerreter, K.-D. Langer and J. W. Walewski, 513 Mbit/s visible light communications link based on DMT-modulation of a White LED, *J. Lightwave Technol.*, vol. 28, no. 24, pp. 3512–3518, 2010.

[40] A. M. Khalid, G. Cossu, R. Choudhury and P. Ciaramella, 1-Gb/s transmission over a phosphorescent white LED by using rate-adaptive discrete multitone modulation, *IEEE Photon. J.*, vol. 4, no. 5, pp. 1465–1473, 2012.

[41] M. Wolf and M. Haardt, Comparison of OFDM and frequency domain equalization for dispersive optical channels with direct detection, *ICTON Conference*, Coventry, UK, pp. 1–7, 2012.

[42] S. Long, M.A. Khalighi, M. Wolf, Z. Ghassemlooy and S. Bourennane, Performance of carrier-less amplitude and phase modulation with frequency domain equalization for indoor visible light communications, *International Workshop on Optical Wireless communications (IWOW)*, Istanbul, Turkey, September 2015.

[43] M. Wolf, S. A. Cheema, M. A. Khalighi and S. Long, Transmissionschemes for visible light communications in multipath environments, *ICTON Conference*, Budapest, Hungary, pp. 1–7, July 2015.

[44] L. Zeng, D. C. O'Brien, H. Le Minh, G. E. Faulkner, K. Lee, D. Jung, Y.J. Oh and E.T. Won, High data rate multiple input multiple output (MIMO) optical wireless communications using white LED lighting, *IEEE J. Sel. Areas Commun.*, vol. 27, no. 9, pp. 1654–1662, December 2009.

[45] T. Q. Wang, Y. A. Sekercioglu and J. Armstrong, Analysis of an optical wireless receiver using a hemispherical lens with application in MIMO visible light communications, *J. Lightwave Technol.*, vol. 31, no. 11, pp. 1744–1754, 2013.

[46] A. H. Azhar, T.-A. Tran and D. O'Brien, A gigabit/s indoor wireless transmission using MIMO-OFDM visible-light communications, *IEEE Photon. Technol. Lett.*, vol. 25, no. 2, pp. 171–174, 2013.

[47] P. A. Haigh, Z. Ghassemlooy, H. Le Minh, S. Rajbhandari, F. Arca, S. F. Tedde, O. Hayden and I. Papakonstantinou, Exploiting equalization techniques for improving data rates in organic optoelectronic devices for visible light communications, *J. Lightwave Technol.*, vol. 30, no. 19, pp. 3081–3088, 2012.

[48] P. A. Haigh, Z. Ghassemlooy and I. Papakonstantinou, 1.4 Mb/s white organic LED transmission system using discrete multi-tone modulation, *IEEE Photon. Technol. Lett.*, vol. 25, no. 6, pp. 615–618, 2013.

[49] P. A. Haigh, Z. Ghassemlooy, S. Rajbhandari and I. Papakonstantinou, Visible light communications using organic light emitting diodes, *IEEE Commun. Mag.*, vol. 51, no. 8, pp. 148–154, 2013.

[50] P. A. Haigh, Z. Ghassemlooy, I. Papakonstantinou, F. Arca, S. F. Tedde, O. Hayden and S. Rajbhandari, A MIMO-ANN system for increasing data rates in organic visible light communications systems, *International Communication Conference*, Budapest, Hungary, pp. 5322–5327, June 2013.

[51] K. Ying, Z. Yu, R. J. Baxley, H. Qian, G.-K. Chang and G. T. Zhou, Nonlinear distortion mitigation in visible-light communications, *IEEE Wireless Commun. Mag.*, vol. 22, no. 2, pp. 36–45, April 2015.

[52] D. Tsonev, S. Sinanovic and H. Haas, Complete modeling of nonlinear distortion in OFDM-based optical wireless communication, *J. Lightwave Technol.*, vol. 31, no. 18, pp. 3064–3076, 2013.

[53] I. Neokosmidis, T. Kamalakis, J. W. Walewski, B. Inan and T. Sphicopoulos, Impact of nonlinear LED transfer function on discrete multitone modulation: Analytical approach, *J. Lightwave Technol.*, vol. 27, no. 22, pp. 4970–4978, 2009.

[54] R. Windisch, A. Knobloch, M. Kuijk, C. Rooman, B. Dutta, P. Kiesel, G. Borghs, G. H. Dohler and P. Heremans, Large signal modulation of high efficiency light

emitting diodes for optical communication, *IEEE J. Quant. Electron.*, vol. 36, no. 12, pp. 1445–1453, 2000.

[55] M. Schetzen, Nonlinear system modeling based on the Wiener theory, *Proc. IEEE*, vol. 69, no. 12, pp. 1557–1573, 1981.

[56] T. Kamalakis, J. W. Walewski, G. Ntogari and G. Mileounis, Empirical volterra-series modeling of commercial light-emitting diodes, *J. Lightwave Technol.*, vol. 29, no. 14, pp. 2146–2155, 2011.

[57] L. Ding, G. T. Zhou, D. R. Morgan, Z. Ma, J. S. Kenney, J. Kim and C. R. Giardina, A robust digital baseband predistorter constructed using memory polynomials, *IEEE Trans. Commun.*, vol. 52, no. 1, pp. 159–165, 2004.

[58] H. Qian, S. J. Yao, S. Z. Cai and T. Zhou, Adaptive postdistortion for nonlinear LEDs in visible light communications, *IEEE Photon. J.*, vol. 6, no. 4, 2014. DOI: 10.1109/JPHOT.2014.2331242.

[59] A. G. Kirk, Free-space optical interconnects, In *Optical Interconnects: The Silicon Approach*, L. Pavesi and G. G. Guillot, Eds. Berlin: Springer, pp. 343–377, 2006.

[60] O. González, Multiple-input multiple-output (MIMO) optical wireless communications, In *Optical Communication*, N. Das, Ed., InTech, pp. 393–414, 2012.

[61] A. J. Paulraj, D. A. Gore, R. U. Nabar and H. Bolcskei, An overview of MIMO communications: A key to gigabit wireless, *Proc. IEEE*, vol. 92, no. 2, pp. 198–218, 2004.

[62] S. Arnon, J. R. Barry, G. K. Karagiannidis, R. Schober and M. Uysal, Eds., Advances *Optical Wireless Communication Systems*, Cambridge University Press, 2012.

[63] Y. Alqudah and M. Kavehrad, MIMO characterization of indoor wireless optical link using a diffuse-transmission configuration, *IEEE Trans. Commun.*, vol. 51, no. 9, pp. 1554–1560, 2003.

[64] J. M. Kahn, R. You, P. Djahani, A. G. Weisbin, B. K. Teik and A. Tang, Imaging diversity receivers for high-speed infrared wireless communication, *IEEE Commun. Mag.*, vol. 36, pp. 88–94, 1998.

[65] K. D. Dambul, D. C. O'Brien and G. Faulkner, Indoor optical wireless MIMO system with an imaging receiver, *IEEE Photon. Technol. Lett.*, vol. 23, no. 2, pp. 97–99, 2011.

[66] S. Hranilovic and F. R. Kschischang, A pixelated MIMO wireless optical communication system, *IEEE J. Sel. Topics Quant. Electron.*, vol. 12, no. 4, pp. 859–874, 2006.

4

Modulation Schemes

Tamás Cseh, Sujan Rajbhandari, Gábor Fekete, and Eszter Udvary

CONTENTS

The visible light communication (VLC) systems use intensity modulation and direct detection (IM/DD). For IM/DD systems, the optical intensity must be a real value and nonnegative. As a result of the constraints of IM/DD, modulation schemes that are advantageous in radio frequency (RF) communications that may not offer the same advantage in VLC.

For example, the orthogonal frequency division multiplexing (OFDM) scheme does not offer the same spectral and energy efficiencies as in the RF domain [1,2]. Depending upon available signal-to-noise ratio (SNR) and system nonlinearity, it is experimentally shown in [2,3] that pulse amplitude modulation (PAM) with decision feedback equalizer (DFE) can outperform complex modulation schemes such as OFDM. As a result, the baseband modulation schemes like PAM and pulse position modulation (PPM) are often the prime candidates for standards and industrial applications. On the other hand, an OFDM scheme is inevitability often the first choice to maximize the system capacity because of the feasibility to match the system spectral profile with the constellation mapping using bit and power loading. This chapter provides an overview of the modulation schemes applied in VLC systems. The description includes baseband modulations, multicarrier modulations, and special color-shift keying (CSK) for VLC.

4.1 Baseband Modulations

The baseband modulation schemes can be classified into: (i) pulse amplitude, (ii) pulse position, and (iii) pulse interval modulation depending upon the method used to encode information into the optical carrier.

4.1.1 Pulse Amplitude Modulation

The PAM scheme is one of the most popular signaling techniques for VLC system because of its simplicity. The information in the PAM scheme is encoded in the amplitude of an optical pulse. In an L-PAM scheme (where $L = 2^M$ and M is a positive integer), a pulse is selected from the following L alphabets to represent the M-bit input symbol.

$$b_k = \{0, 1, 2, \ldots, (L-1)\} \tag{4.1}$$

The time waveform of 4-PAM scheme for different binary input is given in Figure 4.1. The time waveform of PAM can be represented as:

$$s(t) = P_t \sum_{k=-\infty}^{\infty} b_k p(t - kT_{sym}), \tag{4.2}$$

where P_t is the average optical power, $p(t)$ is any unit energy pulse, and T_{sym} is the symbol duration.

The binary PAM (2-PAM), also popularly known as on–off keying (OOK) is the simplest form of baseband IM/DD. The popularity of OOK is due to its simplicity in implementation and power efficiency. In the OOK scheme, a binary "one" is represented by an optical pulse that occupies the entire or part of the bit duration while a binary "zero" is represented by the absence of an

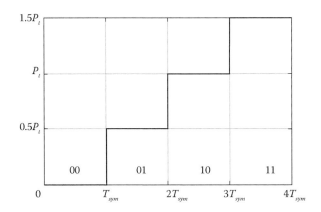

FIGURE 4.1
4-PAM waveforms for combinations of binary input bits.

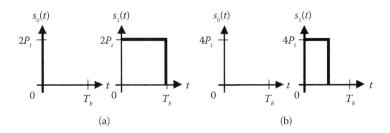

FIGURE 4.2
Time waveforms of (a) OOK-NRZ and (b) OOK-RZ with $\gamma = 0.5$.

optical pulse. The optical pulse can have different duty cycles (γ), which is defined as the ratio between the pulse duration and bit (symbol) duration. The non-return-to-zero (NRZ) OOK has a duty cycle of one and the return-to-zero (RZ) scheme has a duty cycle $\gamma <1$. Figure 4.2 shows the waveforms of OOK-NRZ, and OOK-RZ with a duty cycle γ of 0.5, where P_t represents the average transmit power and T_b is the bit duration. The signal can be described by:

$$p(t) = \begin{cases} (2P_t)/\gamma & \text{for } t \in [0, \gamma T_b) \\ 0 & \text{elsewhere} \end{cases}. \tag{4.3}$$

In additive white Gaussian noise (AWGN) channel without any distortion, the optimal maximum likelihood receiver for OOK is a matched filter matched to the transmitted pulse shape $p(t)$ followed by a sampler and threshold detector. For an independent and identically distributed (IID) system, the probability of occurrence of the zeros and ones are equal. Hence, the optimal threshold level is at the midway between expected one and zero levels [4]. In an AWGN channel, the bit error probability for OOK modulation scheme is given by [4]:

$$P_e = Q\left(\frac{RP_r}{\sqrt{\gamma N_0 B}}\right),$$ (4.4)

where $N_0/2$ is double-sided power spectral density, P_r is the average received optical power, R is the photodiode responsivity, and B is the bandwidth. The bandwidth requirement for baseband modulation is generally defined as the span from DC to the first null in the power spectral density of the transmitted signal. For the rectangular pulse shaping, bandwidth B of OOK-NRZ is equal to bit rate R_b. The error probability is a function of electrical SNR, which is defined as [5]:

$$\text{SNR} = \frac{(RP_r)^2}{N_0 R_b}.$$ (4.5)

Equation 4.4 clearly indicates that reducing the duty cycle improves the power efficiency, that is the average power required to achieve the desired bit error rate (BER). For example, OOK-RZ ($\gamma = 0.5$) requires 3 dB less power in comparison to OOK-NRZ. However, the spectral efficiencies reduced by a factor of duty cycle, that is bandwidth requirement for OOK-RZ ($\gamma = 0.5$) is twice that of OOK-NRZ. Hence, OOK-NRZ is often the preferred option for VLC systems as most of the VLC systems have higher SNR but lower system bandwidth (<10 MHz). Furthermore, a bit stuffing is necessary to recover clock at the receiver for OOK-RZ as it allows a long low signal without any 0 to 1 transition [6]. The bit stuffing further reduces bandwidth efficiency.

The higher level PAM further improves the spectral efficiency but at the cost of power efficiency. The L-PAM scheme encodes M-bit input in a symbol. Hence, the required system bandwidth, noise variance, and sampling rate are reduced by M. On the other hand, in order to maintain same average power as OOK, the minimum decision distance d between the two symbols is reduced by [7,8]:

$$d = \sqrt{\frac{3}{L^2 - 1}}.$$ (4.6)

This reduction in the minimum decision distance means higher optical power are required to achieve the same error probability as OOK. In AWGN systems, the bit error probabilities of L-PAM can be approximated as [8]:

$$P_{b-PAM} = \frac{2(L-1)}{L\log_2 L} Q\left(\sqrt{\frac{M}{(L-1)^2}} \frac{RP_r}{\sqrt{N_0 R_b}}\right).$$ (4.7)

To achieve a data rate of R_b in an AWGN channel, the bandwidth B, and optical power penalty P_{pb} for L-PAM relative to OOK is given by [7,8]:

$$B = 1/M; \qquad P_{pb} = (L-1)/\sqrt{M}. \tag{4.8}$$

This clearly demonstrates that there is a trade-off between the spectral and power efficiencies. For example, 4-PAM improves the spectral efficiency by two but requires in optical power penalty of ~6.5 dB. Higher level PAM further reduces the power efficiencies. Moreover, system nonlinearities, limited dynamic range issues are more pronounced in higher level PAM causing further penalty.

In a bandlimited system, further power penalty occurs due to intersymbol interference (ISI). The ISI is compensated by electronic equalization. One of the advantages of PAM schemes over carrier base modulation schemes is that analog and digital pre-equalization or post-equalization is feasible [9–11]. The power penalty using an equalizer is not negligible and depends on the system bandwidth and the data rate ratio. It is analytically shown in [12] that 4-PAM with DFE will offer an advantage over 2-PAM when the data rate is higher than five times the bandwidth. Higher level PAM with DFE offers improved performance when the data rate is significantly higher than bandwidth. Le Minh et al. demonstrated an 80 Mbps VLC link using a commercial white LEDs and analog pre-equalization [13]. They have achieved a bandwidth of 45 MHz out of few MHz of raw white LED bandwidth using a combination of three LC circuits and hence achieved an error-free transmission rate up to 80 Mbps using OOK. Using passive post-equalization, the same group showed a data rate of 100 Mbps [9]. Li further improved the bandwidth by combining analog active and passive equalizers and demonstrated a 200 Mbps link using a pre-equalizer [14], 340 Mbps link using a post-equalizer [15], and 550 Mbps using pre- and post-equalizers [16] for the OOK-NRZ modulation scheme. An organic VLC system using OOK-NRZ and a post-equalizer demonstrated a data rate of 20 Mbps [17]. A data rate of 512 Mb/s was achieved using complementary metal-oxide semiconductor (CMOS)–controlled high-bandwidth blue micro-LEDs [18]. Using micro-LEDs, error-free (BER < 10^{-12}) transmission rates of 1.6 Gbps and 2 Gbps were achieved using 2-PAM and 4-PAM schemes and a digital feedforward pre-equalizer [11]. By using a PAM scheme in a multiple input multiple output (MIMO) VLC system, a data rate of >1 Gbps was demonstrated [19,20]. These demonstrations clearly show there is a plethora of research activities for practical implementation of PAM for VLC systems.

4.1.2 Pulse Position Modulation

As the name suggests, the information in PPM scheme is encoded in the position of a pulse within a symbol. An L-PPM symbol consists of L time slots of equal duration. Within the symbol, all slots except the information bearing

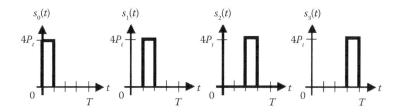

FIGURE 4.3
The time waveforms of all possible 4-PPM symbols for two input bits.

slot are empty. The position of this pulse carries the information about input bit sequence. The position of the pulse corresponds to the decimal value of the M-bit input data. In order to maintain the same throughput as OOK-NRZ, PPM slot duration T_{s_PPM} is shorter than the OOK bit duration T_b by a factor L/M:

$$T_{s\_PPM} = \frac{T_b M}{L}. \qquad (4.9)$$

Figure 4.3 shows the time waveforms of all possible 4-PPM symbols for two input bits. The PPM signal can be expressed as:

$$x(t)_{PPM} = \begin{cases} 1, \text{ for } t \in [(m-1)T_{s\_PPM}, mT_{s\_PPM}) \\ 0, \qquad\qquad \text{elsewhere} \end{cases}, \qquad (4.10)$$

where $m \in \{1,2, \dots L\}$.

The PPM symbols are orthogonal with a ratio of the pulse over the empty slots being $1/L$. Hence, as L is increased, the average power efficiency improves while the bandwidth efficiency is reduced. Due to its poor bandwidth efficiency, PPM is more susceptible to multipath induced ISI compared to OOK-NRZ. Hence, PPM is advantageous over OOK only in a power-limited system. As the information in PPM is in the position of a pulse within a symbol, the PPM receiver requires both slot and symbol synchronization. Hence, the PPM receiver is more complex than the OOK receiver. Like in the case of OOK scheme, a threshold detector is commonly used at the receiver. Since the probability of a zero is greater than the probability of a one for $L > 2$, the optimum threshold level does not lie midway between one and zero levels. It is a complicated function of the signal and noise powers, and L. However, in an AWGN channel with a low error probability, the midway threshold is near optimum and hence slot error probability $P_{sle-PPM}$ can be approximated as:

$$P_{sle-PPM} = Q\left(\sqrt{\frac{LM}{2}}\frac{RP_r}{\sqrt{N_0 R_b}}\right). \qquad (4.11)$$

For IID system, the symbol error rate (SER) and BER can be calculated from slot error rate using following [21]:

$$P_{sye-PPM} = 1 - (1 - P_{sle-PPM})^L \quad P_{be-PPM} = \frac{L/2}{L-1} P_{sye-PPM}, \tag{4.12}$$

where $P_{sye-PPM}$ and P_{be-PPM} are the symbol and bit error probabilities, respectively.

Since PPM symbols are orthogonal, a maximum likelihood detector is not the fixed threshold detection but is a symbol-by-symbol decoding using a "soft" decision scheme. In a soft decision scheme, the amplitude of all slots within a symbol is compared and the slot corresponding to the largest amplitude is assigned a one, and zeros are assigned to the remaining slots. The soft decoding offers improved performance in comparison to the threshold detection in the presence of ISI [22]. In the AWGN channel, the symbol error probability for soft decoding is given by [21]:

$$P_{sye-PPM-S} = \frac{1}{\sqrt{2\pi}} \frac{1}{\sqrt{2\pi}} \int_{-\infty}^{\infty} \{1 - [1 - Q(y)]^{L-1}\} e^{-\left(y - \sqrt{2E_s/N_0}\right)^2/2} dy, \tag{4.13}$$

where $E_S = M(RP_r)^2 T_b$ is symbol energy.

The greatest strength of PPM is its unparalleled power efficiencies in comparison to all other baseband modulation schemes. Hence, PPM is the most popular modulation schemes for handheld devices where lower power consumption is essential. PPM was adopted for IEEE 802.11 standard and IrDA serial data communication. The weakness of PPM, however, is its low spectral efficiency. Hence, a number of variations of PPM have been proposed to improve spectral efficiencies including multilevel PPM [23], differential PPM (DPPM) [24], and differential amplitude pulse position modulation (DAPPM) [25,26]. Some of the popular variations of PPM will be described in the following sections.

4.1.3 Pulse Interval Modulation

Pulse interval modulation (PIM) is an anisochronous modulation technique where empty slots between two pulses carry the information. There are a number of variations of PIM which either improve throughput or reduce power requirement. PIM modulation normally starts with a pulse followed by empty slots, the number of which depends on the information being encoded and hence PIM has built-in symbol synchronization. The simplest form of PIM is the digital PIM (DPIM), where data are encoded as a number of discrete time slots between adjacent pulses. The symbol length is variable and is determined by the information content of the symbol. In *L*-DPIM, each symbol starts with a pulse followed by empty. The number of slots depends on the decimal value of the *M*-bit input data [27]. Figure 4.4 shows

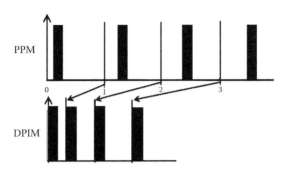

FIGURE 4.4
The symbol structure of DPIM.

the time waveform of 8-DPIM. The minimum and maximum symbol durations are T_s and LT_s, respectively, where T_s is the slot duration with an average number of slots being $(L + 1)/2$. DPIM displays a higher transmission capacity by eliminating all the unused time slots within PPM symbol. Parallels can be drawn between DPIM and DPPM. They have same symbol length for a given M-bit input, same power, and bandwidth requirements. The only difference is DPIM symbols start with a pulse whereas DPPM symbols end in a pulse preceded by empty slots. Hence, DPPM is not exclusively discussed in this chapter as all analysis carried out for DPIM is also valid for DPPM.

Because of variable symbol duration in DPIM, the overall data rate is also variable. In order to achieve the average data rate of R_b, the slot duration T_{S_DPIM} is given as [28]:

$$T_{s\_DPIM} = \frac{M}{R_b \bar{L}_{DPIM}},$$ (4.14)

where $\bar{L}_{DPIM}$ is the average symbol length of DPIM. The minimum and maximum symbol length of DPIM is 1 and 2^M. Hence, the average symbol length $\bar{L}_{DPIM}$ is given as:

$$\bar{L}_{DPIM} = \frac{L+1}{2}.$$ (4.15)

The transmitted optical power is also varied, but the average transmitted optical power is calculated according to the average symbol length.

A DPIM receiver consists of matched filter followed by a threshold detection. Because of the unequal probability of ones and zeros, the threshold level set at the mid-level between expected ones and zeros level is non-optimum. The optimum threshold level is a function system SNR and bit resolution and it is out of the scope of this chapter. The interested reader can refer to [4].

However, at high SNR, the mid-level threshold offers near optimum performance. In AWGN channel with the mid-level threshold, the slot error probability in DPIM $P_{sle-PPM}$ can be approximated as:

$$P_{sle-DPIM} = Q\left(\sqrt{\frac{M\bar{L}_{DPIM}}{2}} \ \frac{RP_r}{\sqrt{N_0 R_b}} \right). \tag{4.16}$$

Above equation indicates that DPIM has better power efficiency than OOK, however, the bandwidth requirement is higher. On the other hand, the bandwidth requirements of the DPIM is approximately half of PPM for the case of $M > 2$ and power requirements is higher than PPM.

Further variations of PIM have been suggested including Dual-Header PIM (DH-PIM) and Multilevel DPIM (MDPIM). Like in DPIM, DH-PIM symbols start with a pulse, followed by a number of empty slots [29]. However, the pulse duration depends on the most significant bit (MSB) of the input word. The pulse duration for MSB = 1 is double that of pulse duration when MSB = 0. This way, the header itself differentiates the MSB, that is, a less number of empty slots (half of DPIM) are required to represent remaining bits, henceforth increasing the throughput. In MDPIM, the pulse amplitude rather than pulse duration depends on the MSB. A symbol starts with an amplitude of A if the MSB is 0 and $2A$ if the MSB is 1, followed by a number of empty slots [4]. MDPIM can be considered as a variation of DAPPM with two amplitude levels.

4.1.4 Differential Amplitude Pulse Position Modulation

By combination of two or more modulation schemes, it is feasible to improve the data throughput, bandwidth efficiency, or peak-to-average power ratio (PAPR). A number of such variations had been suggested which combines PAM with PPM or DPIM. DAPPM, which is a combination of PAM and DPPM, offers advantages over other modulation schemes including PPM, DPPM, and DH-PIM in terms of the bandwidth requirements, capacity, and PAPR [25,26]. In DAPPM, a block of $M = \log_2(A \times L)$ input bits is mapped to one of 2^M distinct waveforms. The symbol length varies from $\{1, 2, ..., L\}$ and the pulse amplitude is selected from $\{1, 2, ..., A\}$, where A and L are integers. A set of DAPPM waveforms are shown in Figure 4.5. The average number of empty slots preceding the pulse can be lowered by increasing the number of amplitude levels A thereby increasing the achievable throughput in the process. When compared with similar modulation techniques, a well-designed DAPPM will require the least bandwidth. DAPPM suffers from a high average power and a large DC component, thus restricting its use to applications where power is not a premium. It is also susceptible to the baseline wander. DPPM (DPIM) and MDPIM can be considered as special cases of DAPPM with $A = 1$ and 2, respectively.

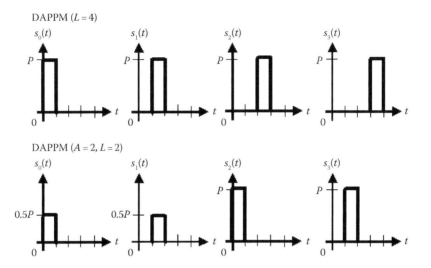

FIGURE 4.5
Symbol structure of DAPPM, M = 2 bits/symbol.

The transmitted DAPPM signal is defined as [25]:

$$s(t)_{DAPPM} = \sum_{k=-\infty}^{\infty} \left(\frac{P_p}{A}\right) b_k p(t - kT_s),$$ (4.17)

where $b_k \in \{0, 1,...A\}$, $p(t)$ is a unit amplitude rectangular pulse shape, T_s is the one-chip duration, and P_p is the peak power.

Assuming IID random data, each symbol is equally likely and the symbol length of DAPPM is given by:

$$\overline{L}_{DAPPM} = \frac{L+1}{2M},$$ (4.18)

where $\overline{L}_{DAPPM}$ is the average length of a DAPPM symbol.

Compared to the other modulation schemes, DAPPM provides better bandwidth efficiency and higher transmission capacity. The capacity of DAPPM approaches 2A times and A times that of PPM and DPPM, respectively, as the number of bits/symbol increases.

In nondispersive channel the slot error probability is given by:

$$P_{se} = p_0 Q\left(\frac{\theta_1 RP_r}{A\sqrt{N_0 R_b}}\right) + p_A \sum_{i=1}^{A-1}\left(Q\left(\frac{(i-\theta_i)RP_r}{A\sqrt{N_0 R_b}}\right) + Q\left(\frac{(\theta_{i+1}-i_i)RP_r}{A\sqrt{N_0 R_b}}\right)\right)$$

$$+ p_A Q\left(\frac{(A-\theta_A)RP_r}{A\sqrt{N_0 R_b}}\right),$$ (4.19)

where p_0 is the probability of receiving the empty slots, p_A is the probability of receiving a pulse, and θ_i are the optimum threshold levels.

As in the case of DPIM, a single chip error affects not only the current symbol but the whole packet. Therefore, packet error rates are often used to compare the variable symbol length modulation schemes. The packet error rate of DAPPM is given by:

$$PER = 1 - (1 - P_{se})^{\overline{L}_{DAPPM}D/M} \approx \frac{\overline{L}_{DAPPM} \cdot D \cdot P_{se}}{M}, \qquad (4.20)$$

where D bit packet is transmitted and the average chip sequence length is $(\overline{L}_{DAPPM}D)/M$.

4.1.5 Variable Pulse Position Modulation

Variable pulse position modulation (VPPM) is a modulation scheme recommended for VLC scheme as outlined in IEEE 802.15.7 Section 8.2. VPPM supports simultaneously illumination with dimming control and communication. It is a combination of PAM and pulse width modulation schemes. This scheme uses binary PPM for communication and the pulse width for dimming control. Bits in VPPM are distinguished by the pulse position during the symbol period. Zero occurs when the pulse is aligned to the left of the symbol period and one occurs when the pulse is aligned to the right of the symbol period. The width of the pulse can be adjusted to reduce the average intensity of the source while maintaining that the mean of one and zero are equal. Since each VPPM symbol, for a given dimming level, has an equal mean, the data are unlikely to produce flicker. In essence, VPPM is Manchester OOK with a variable duty cycle and has equally poor spectral efficiency. Figure 4.6 shows the coding scheme with different dimming control.

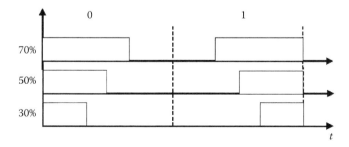

FIGURE 4.6
VPPM coding scheme.

The VPPM signal can be expressed as:

$$s(t) = \begin{cases} \sqrt{E_s \cdot \dfrac{d}{50}} \cdot \varphi_0(t), & b = 0 \\[4mm] \sqrt{E_s \cdot \dfrac{d}{50}} \cdot \varphi_1(t), & b = 1 \end{cases}, \tag{4.21}$$

where E_s is the symbol energy, d is the dimming level ($0 \le d \le 100$), and $\varphi_i(t)(i = 0, 1)$ is the basis function that changed according to dimming level. The basic functions are defined as:

$$\varphi_0(t) = \begin{cases} \sqrt{\dfrac{100}{d \cdot T}}, & 0 \le t \le \dfrac{d \cdot T}{100} \\[3mm] 0, & otherwise \end{cases} \qquad \varphi_1(t) = \begin{cases} \sqrt{\dfrac{100}{d \cdot T}}, & \left(1 - \dfrac{d}{100}\right) \cdot T \le t \le T \\[3mm] 0, & otherwise \end{cases}, \tag{4.22}$$

where $\varphi_i(t)$ and the corresponding template pulse $q_i(t)$ are normalized to have a unit energy:

$$\int_0^T \varphi_i^2(t) = 1 \quad and \quad \int_0^T q_i^2(t) = 1. \tag{4.23}$$

The error probability can be expressed as:

$$BER = P\{b = 0\} \cdot P_{e|b=0} + P\{b = 1\} \cdot P_{e|b=1} = \frac{1}{2} \cdot erfc\left(\gamma \cdot \sqrt{\frac{E_s}{2 \cdot N_0}} \cdot \sqrt{\frac{d \cdot (1 - \alpha)}{50}}\right), \tag{4.24}$$

where priori probabilities of $b = 0$ and $b = 1$ are equal to ½.

4.1.6 Comparisons of Baseband Modulation Schemes

There are a number of criteria against which effectiveness of a modulation scheme is measured. The power and bandwidth efficiencies are benchmarks to compare modulation schemes. Besides, the PAPR, transmission reliability, transmission capacity, and system complexity are other important parameters. The comparison of different baseband modulation schemes in terms of power and bandwidth efficiencies and PAPR is made in the following section. OOK-NRZ is taken as the benchmark modulation scheme against which performance of other modulation schemes is measured.

TABLE 4.1

The Normalized Average Power Requirements

PAM	PPM with Hard Decision	PPM with Soft Decision	DPIM
$\frac{P_{avg\_PAM}}{P_{avg\_OOK}} = \frac{(L-1)}{M}$	$\frac{P_{avg\_PPM}}{P_{avg\_OOK}} = \sqrt{\frac{4}{L\log_2 L}}$	$\frac{P_{avgPPM}}{P_{avgOOK}} = \sqrt{\frac{2}{L\log_2 L}}$	$\frac{P_{avg\_DPIM}}{P_{avg\_OOK}} = \sqrt{\frac{8}{(L+1)\log_2 L}}$

Sources: Kahn, J.M. and Barry, J.R., 1997, *Proc. IEEE*, 85, 265–298; Ghassemlooy, Z., et al., *Presented at the Proceedings of the IASTED International Conference on Wireless and Optical Communications*, Canada, 2001.

TABLE 4.2

The Normalized Bandwidth (Normalized to OOK-NRZ) Requirements

PAM	PPM	DPIM	DAPPM
$B_{req\_PAM} = \frac{1}{M}$	$B_{req\_PPM} = \frac{L}{M}$	$B_{req\_DPIM} = \frac{L+1}{2M}$	$B_{req\_DAPPM} = \frac{(L+1)}{2M}$

Sources: Ghassemlooy, Z., et al., *Optical Wireless Communications: System and Channel Modelling with MATLAB®*, 1st ed., CRC Press, Boca Raton, FL, 2012; Kahn, J.M. and Barry, J.R., *Proc. IEEE*, 85, 265–298, 1997; Tanaka, Y., et al., *12th IEEE International Symposium on Personal, Indoor and Mobile Radio Communications*, vol. 2, pp. 81–85, San Diego, CA, 30 September–03 October 2001.

4.1.6.1 Power Efficiency

The power efficiency is defined as the average optical power required to achieve a desired BER performance in an AWGN channel. The average power requirement for OOK-NRZ is given as:

$$P_{avg\_OOK} = \sqrt{\frac{N_0 R_b}{2R^2}}\, Q^{-1}\left(P_{e\_bit\_OOK}\right). \qquad (4.25)$$

Table 4.1 shows the normalized average power requirements for PAM, PPM, and DPIM [5,30].

4.1.6.2 Bandwidth Efficiency

Table 4.2 represents the normalized bandwidth requirements for PAM, PPM, DPIM, and DAPPM [4,5,31]. The bandwidth requirement B_{req} for a modulation scheme depends upon the minimum slot duration. The transmission capacity of a modulation scheme depends on bandwidth requirement. Considering rectangular pulse shaping, bandwidth requirement can be approximated as:

$$B_{req} = \frac{1}{\tau_{min}}, \qquad (4.26)$$

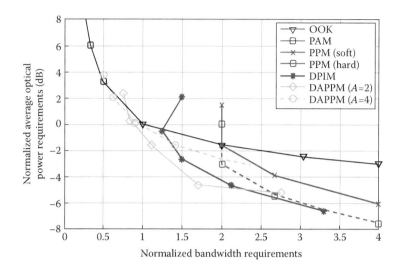

FIGURE 4.7
Optical power requirement normalized to the OOK-NRZ versus bandwidth requirement for
L-PPM, *L*-DPIM, and *L*-DAPPM.

where τ_{min} is the minimum slot duration. The τ_{min} depends on the average number of slots. The bandwidth requirements for OOK-NRZ can be approximated to:

$$B_{req\_OOK} = R_b. \tag{4.27}$$

The normalized optical power requirement versus the bandwidth requirement is shown in Figure 4.7 clearly demonstrates that there is a trade-off between bandwidth and average power efficiencies. Among binary modulation, OOK-NRZ is the most bandwidth efficient whereas PPM is the most power efficient. The power efficiencies of *L*-PAM and *L*-DPIM (*L*-DPPM) improve for higher *L* at the cost of bandwidth efficiencies. Though there is some performance difference at *L* < 16, these modulation schemes tend to offer similar power and bandwidth efficiencies higher bit resolutions.

4.1.6.3 Peak-to-Average Power Ratio

Because of a limited dynamic range of the practical system, and to avoid nonlinearity in optoelectronic devices, a low PAPR is desirable. The PAPR of PAM is 2 irrespective of the bit resolution. However, PAPR of other modulation schemes depends on bit resolutions. Table 4.3 shows the PAPR for PPM, DPIM, and DAPPM modulations.

Figure 4.8 demonstrates the PAPR of PPM, DPIM, and DAPPM for a number of bit resolutions. PPM showed the highest PAPR, and the PAPR

TABLE 4.3

Peak-to-Average Power Ratio

PPM	DPIM	DAPPM
$\text{PAPR}_{\text{PPM}} = 2^M$	$\text{PAPR}_{\text{DPIM}} = \frac{(2^M + 1)}{2}$	$\text{PAPR}_{\text{DAPPM}} = \frac{(2^M + A)}{(A + 1)}$

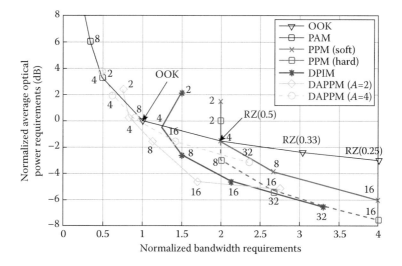

FIGURE 4.8
PAPR of PPM, DPIM, and DAPPM versus M.

value increases exponentially with bit resolution. On the other hand, of DAPPM is relatively low and does not increase significantly with bit resolution.

4.2 Optical OFDM

Multicarrier modulation schemes can be more efficient than the baseband modulation schemes. The VLC has two main challenges: the limited bandwidth of the LEDs and the multipath propagation. The typical modulation bandwidth of LEDs is around couple tens of MHz. In order to achieve a higher data rate, complex modulation schemes such as phase shift keying (PSK), quadrature amplitude modulation (QAM), or OFDM modulations can be used. The most popular and applicable choice in VLC systems is OFDM, since it offers improved spectral efficiency than PSK, QAM and it has a strong robustness against the ISI airing from multipath propagation or limited system bandwidth [31–33]. The use of OFDM was noted first in [32] and its popularity increased significantly as the OFDM is robust against multipath propagation [31]. The multipath propagation causes linear distortion in the channel and it leads to ISI. In order to reduce the linear distortion of the dispersive channel, a cyclic prefix can be inserted to the OFDM symbols. Therefore, the symbol period of an OFDM symbol should be increased. This growth of the symbol period which is called guard interval needs to be higher than the impulse response of the channel [34,35].

The multipath propagation can also cause frequency selective fading in the RF system, and it leads to ISI as well [34]. However, by dividing the channel to N parallel parts, the bandwidth of each channel part is smaller than the coherence bandwidth of these parts. Thus, the linear distortion of the channel may be avoided; therefore, the OFDM system can reduce dispersion effect compared to the single carrier system [34,63]. The multicarrier systems like OFDM have an added advantage: unmodulated pilot tones can be used for characterizing the dispersive channel. The effect of the characterized dispersive channel can be equalized at the receiver [33]. The capacity of a multicarrier system can be increased by partly overlapped subcarriers. In this case, the subcarriers are orthogonal, hence, the nearby channels do not disturb each other. As OFDM does not need to apply guard bands, thus the capacity of the link is as large as the capacity of the single carrier system [34]. The comparison of the spectral efficiency of different modulation types is shown in Figure 4.9. So, the OFDM is capable of combining the high channel capacity with the protection against multipath propagation, frequency

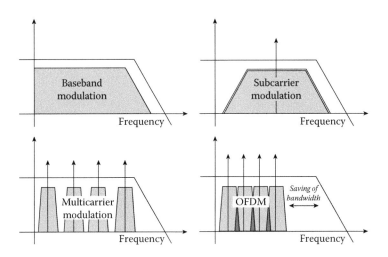

FIGURE 4.9
Spectral efficiency of different modulation types.

selective fading. All of these reasons make the OFDM scheme a suitable modulation type of VLC system [36].

The traditional OFDM signal widely applied to RF system is complex and bipolar. Due to IM/DD, the signaling for VLC system must be a real and unipolar [37,38]. Therefore, the traditional OFDM signal is modified to make them real-valued and unipolar. There are a number of variations of the unipolar OFDM that are proposed for VLC system such as DC-biased optical OFDM (DCO-OFDM), asymmetrically clipped optical OFDM (ACO-OFDM), unipolar OFDM (U-OFDM), pulse-amplitude-modulated discrete multitone modulation (PAM-DMT), and Flip-OFDM.

4.2.1 Properties of Bipolar OFDM

In order to show the properties of OFDM modulation and the general block diagram of the OFDM transmitter and receiver, the mathematical expression for OFDM is given as [35]:

$$s_{OFDM}(t) = \mathrm{Re}\left\{ \sum_{i=-\infty}^{\infty} \sum_{n=-N_{SC}/2}^{N_{SC}/2-1} c_{nki} g(t-iT_s) e^{j2\pi f_0(t-iT_s)} e^{j2\pi \frac{n}{T_s}(t-iT_s)} \right\}, \quad (4.28)$$

where c_{nki} is an element of a symbol sequence, which has the kth value from an M-ary alphabet at the ith time slot and at the nth subcarrier [39]. T_s is the symbol period and $g(t)$ is the pulse shaping function. The element of a symbol sequence c_{ni} is a complex, and it depends on subcarrier modulation of the OFDM symbol. It can be written as:

$$c_{nki} = \begin{cases} e^{j2\pi \frac{k-1}{M}} & \text{for} \quad M\text{-PSK} \quad , \\ a_{ki} + j \cdot b_{ki} & \text{for} \quad M\text{-QAM} \end{cases} \quad (4.29)$$

where a_{ki} and b_{ki} are elements of the symbol sequence of PAM. The higher order modulation schemes such as QAM can be written as a quadrature modulation of two PAM signals. Two separate k-bit symbols from the information sequence on two quadratic carriers give the QAM modulation [39].

To understand better the formula of OFDM, that should be given with the baseband equivalent as well:

$$s_{OFDM}(t) = \mathrm{Re}\{s_{OFDM,Base}(t) \cdot e^{j2\pi f_0 t}\}, \quad (4.30)$$

where

$$s_{OFDM,Base}(t) = \sum_{i=-\infty}^{\infty} g(t-iT_s) \cdot \sum_{n=-N_{SC}/2}^{N_{SC}/2-1} c_{ni} e^{j2\pi \frac{n}{T_s}(t-iT_s)}. \quad (4.31)$$

Equation 4.31 could be rewritten to:

$$s_{OFDM,Base}(t) = \sum_{i=-\infty}^{\infty} s_H(t - iT_s) = \sum_{i=-\infty}^{\infty} s_H[i] \cdot \delta(t - iT_s), \qquad (4.32)$$

where:

$$s_H(t) = g(t) \cdot \sum_{n=-N_{SC}/2}^{N_{SC}/2-1} c_{ni} e^{j2\pi \frac{n}{T_s} t}, \qquad s_H[i] = \sum_{n=-N_{SC}/2}^{N_{SC}/2-1} c_n[i] e^{j2\pi \frac{n \cdot i}{N_{SC}}}. \qquad (4.33)$$

Although the overlapping multicarrier modulation can solve the spectral efficiency, but the arrays of sinusoidal generators and coherent modulators and demodulators still make the OFDM structure unreasonably expensive [40]. Weinstein and Ebert showed that the OFDM modulation and demodulation can be realized by using inverse discrete Fourier transform (IDFT) and discrete Fourier transform (DFT) [41]. For IDFT and DFT operations numerical algorithms exist, which make these operations faster. These are inverse fast Fourier transform (IFFT) and fast Fourier transform (FFT) [40]. These algorithms decrease the number of complex multiplications from N^2 to $(N/2) \cdot \log_2(N)$ [41]. Therefore, by applying IFFT and FFT for OFDM modulation and demodulation, it becomes cheaper and less complicated. In [41], the OFDM modulation and demodulation are studied, and it is proven that IDFT (or IFFT) and DFT (or FFT) can replace complicate up- and down-conversions. As a consequence, the equations which describe OFDM signal (Equation 4.28 and Equations 4.30 through 4.33) are modified to:

$$s_H(t) = g(t) \cdot IFFT\{c_{ni}\}, \qquad (4.34)$$

$$s_{OFDM,Base}(t) = \sum_{i=-\infty}^{\infty} s_H(t - iT_s) = \sum_{i=-\infty}^{\infty} g(t - iT_s) \cdot IFFT\{c_{ni}\}, \qquad (4.35)$$

$$s_{OFDM}(t) = \mathrm{Re}\{s_{OFDM,Base}(t) \cdot e^{j2\pi f_0 t}\} = \mathrm{Re}\left\{ \left(\sum_{i=-\infty}^{\infty} g(t - iT_s) \cdot IFFT\{c_{ni}\} \right) e^{j2\pi f_0 t} \right\}. \qquad (4.36)$$

As it was mentioned before, OFDM can provide a protection against ISI by using guard interval. This guard interval is a cyclic prefix in practice because the end of the symbol is copied to the start of the symbol as a guard interval. The transmitter put this cyclic prefix to start of the symbol and the receiver throws away the cyclic prefix before it processes the received signal. Thus, the receiver makes the signal process just in the steady state. By using a cyclic prefix, which is longer than the impulse response, the ISI can be avoided [41]. The mathematical description of the cyclic prefix is also an

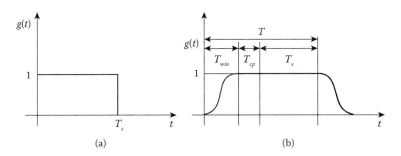

FIGURE 4.10
The simplest representation of (a) pulse shape function and (b) pulse shape function with cyclic prefix and windowing. (From Asadzadeh, K., 2011, Open Access Dissertations and Theses, Paper 6079. With permission.)

important point to describe OFDM signals. The pulse shaping function [$g(t)$] could be more complex than a simple rectangular pulse. To connect cyclic prefix to an OFDM symbol, we should modify the length of $g(t)$ [35]. However, windowing can modify $g(t)$, too. The rectangular pulse shape is not spectral efficient, because the Fourier transform of the rectangular pulse is a sinc function, which has sidelobes [35]. To reduce these sidelobes the pulse shape function could be windowed. A raised-cosine pulse shape may be applied to reduce the sidelobes, and reduce the occupied bandwidth of the OFDM signal [35]. The shape of this windowed pulse shape function, which is extended with a cyclic prefix is plotted in Figure 4.10.

In Figure 4.10 T_{win} is window time, T_{cp} is the duration of cyclic prefix, and T_s is the effective symbol time. T is the total length of an OFDM symbol. The mathematical expression of the pulse shaping function is the following [35]:

$$
g(t) = \begin{cases}
\frac{1}{2}[1 - \cos\pi(t + T_{win} + T_{cp})/T_{win}] & -T_{win} - T_{cp} \leq t < -T_{cp} \\
1 & -T_{cp} \leq t < T_s \\
\frac{1}{2}[1 - \cos\pi(t - T_s)/T_{win}] & T_s < t \leq T_s + T_{win}
\end{cases} \quad (4.37)
$$

After the consideration of OFDM modulation with a cyclic prefix and windowing, the block scheme of OFDM transmitter and receiver is shown in Figures 4.11 and 4.12, respectively [41]. The OFDM transmitter is divided into baseband part and RF part. The signal is produced by the baseband part, which is upconverted to the carrier frequency by RF upconversion block. The data bits stream to a serial to parallel converter, which is the input of the subcarrier symbol mapper. This symbol mapper makes the modulation of each subcarrier, which can be N-QAM or N-PSK, and this module produces the c_n symbols of the subcarriers. Then these symbols are converted to OFDM symbols by an IFFT block, and after that the cyclic prefix is added to the OFDM

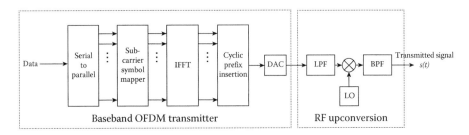

FIGURE 4.11
Block scheme of OFDM transmitter. (From Armstrong, J. and Schmidt, B. J. C., 2008, *IEEE Commun. Lett.*, 12, 343–345. With permission.)

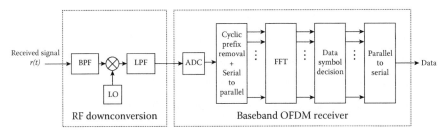

FIGURE 4.12
Block scheme of OFDM receiver. (From Armstrong, J. and Schmidt, B. J. C., 2008, *IEEE Commun. Lett.*, 12, 343–345. With permission.)

symbols. In order to get an analog OFDM signal, the discrete OFDM symbols have to be converted by a digital-to-analog converter. The baseband OFDM needs to be set to the given RF band, which conversion is made by the RF upconverter. This block is a simple upconverter with a mixer and the necessary filters [41]. The block scheme of OFDM receiver is similar to the transmitter. It has an RF downconversion block which converts the OFDM signal from the RF band to baseband. First, the baseband OFDM receiver produces a digital signal from the analog signal with an analog-to-digital converter, and then it removes the cyclic prefix. The sampled and useful r_m symbols have to Fourier transformed with an FFT block to get c_k' received information symbols. The received data are produced from the information symbols by data symbol decision block and a parallel to serial converter [41].

4.2.2 Unipolar OFDM Formats for VLC

Real and unipolar signals are required in IM/DD VLC systems. In order to obtain the real signal, the input vector to the IFFT is constrained to a

Hermitian symmetry [42,43]. To prove that it will be real signal, consider the following [43]:

$$s_H(t) = g(t) \cdot \sum_{n=-N_{SC}/2}^{N_{SC}/2-1} c_{ni} e^{j2\pi\frac{n}{T_s}t} = g(t) \cdot \left[\sum_{n=-N_{SC}/2}^{-1} c_{ni} e^{j2\pi\frac{n}{T_s}t} + \sum_{n=1}^{N_{SC}/2-1} c_{ni}^* e^{j2\pi\frac{n}{T_s}t} \right],$$

(4.38)

$$s_H = g(t) \left[\sum_{n=1}^{N_{SC}/2-1} c_{ni} e^{-j2\pi\frac{n}{T_s}t} + c_{ni}^* e^{j2\pi\frac{n}{T_s}t} \right] = g(t) \left[2 \cdot \text{Re} \left\{ \sum_{n=1}^{N_{SC}/2-1} c_{ni} e^{-j2\pi\frac{n}{T_s}t} \right\} \right],$$

(4.39)

$$s_H(t) = g(t) \left[2 \cdot \text{Re} \left\{ \sum_{n=1}^{N_{SC}/2-1} c_{ni} e^{-j2\pi\frac{n}{T_s}t} \right\} \right] = 2g(t) \sum_{n=1}^{N_{SC}/2-1} a_{ni} \cos(2\pi \frac{n}{T_s} t).$$

(4.40)

The equation gives Fourier polynomial of a periodic even function. With a shift of half symbol period, we get an odd function. That means the information in the first $N/2$ samples is repeated in the second half of OFDM symbol [44]. The OFDM signal with the Hermitian symmetry input is still bipolar, which is not subtle for an IM/DD system [36]. Three methods exist in order to get unipolar signals: (a) DC bias addition (as in DCO-OFDM), (b) clipping negative signal at zero level (as in ACO-OFDM and PAM-DMT) [42,43], and (c) generating unipolar OFDM based on the extraction of the positive and negative part of the bipolar OFDM signals (as in Flip-OFDM and unipolar OFDM [U-OFDM]) [41,45].

4.2.3 DC-Biased Optical OFDM

DCO-OFDM is the easiest way to ensure the non-negativity of OFDM signals. DCO-OFDM adds a DC bias to the bipolar OFDM signal. The required DC bias to satisfy nonnegativity is equal to the maximum negative amplitude of the OFDM signal [36]. The time samples of the bipolar OFDM and DCO-OFDM are compared in Figure 4.13.

The DCO-OFDM can be expressed mathematically as well [42]:

$$s_{DCO-OFDM}(t) = s_{OFDM}(t) + s_{DC} + n(s_{DC}) \approx s_{OFDM}(t) + s_{DC},$$

(4.41)

where s_{DC} is the DC bias and $n(s_{DC})$ is the clipping noise. If the DC bias is high enough, the clipping noise can be neglected. The DC bias is relative to the power of $S_{OFDM}(t)$ [42]:

$$s_{DC} = \mu \sqrt{E\{s_{OFDM}^2(t)\}},$$

(4.42)

where $E\{s_{OFDM}^2(t)\}$ is the power of the signal. With the value of k, the DC bias can be expressed in the terms of decibels, by defining it as $10\log_{10}(k^2+1)$ [42].

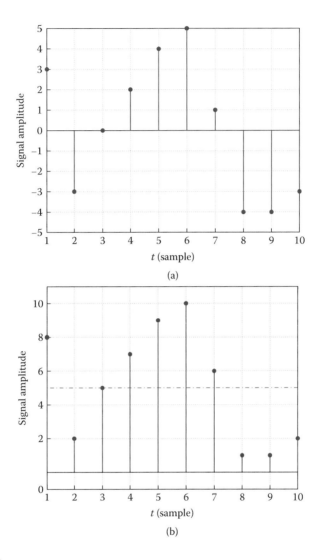

FIGURE 4.13
The samples of (a) bipolar OFDM and (b) DCO-OFDM.

When the number of subcarriers is high enough, the magnitude of the OFDM signal has a Gaussian distribution with zero mean value. According to [36], for the zero mean Gaussian distributed random variable x with a standard deviation of σ, the random variable is in the range of $-2\sigma < x < 2\sigma$ with a 97.8% probability. Therefore, the next equation is satisfied [36]:

$$\Pr\{x + 2\sigma > 0\} \cong 97.8\%, \tag{4.43}$$

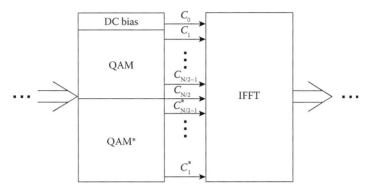

FIGURE 4.14
Block diagram of DCO-OFDM.

where, $P_r\{\}$ is the probability function. This value is very close to 1, so the DC bias may be set to at least twice the standard deviation, thus $\mu = 2$. However, the optimal DC bias depends on the number of subcarriers: for larger constellation such as 256-QAM, the DC bias should be higher to get unclipped DCO-OFDM signal [42]. Figure 4.14 shows the block diagram of DCO-OFDM transmitter [43], where $c_0-c_{N/2}$ are the symbols at the kth subcarrier. The c_0 symbol sets the DC bias and $c_1-c_{N/2}$ symbols modulate each subcarrier. For real signals, Hermitian symmetry has to be constrained:

$$C_n = C_{N-n}^* \quad \text{for} \quad 0 < n < \frac{N}{2}. \tag{4.44}$$

The DCO-OFDM is a simple solution to get unipolar OFDM signal, but the main disadvantage is the lower power efficiency.

4.2.4 Asymmetrically Clipped Optical OFDM (ACO-OFDM)

In order to improve the power efficiency of the unipolar OFDM modulation format, negative signal clipping at zero level is applied. The ACO-OFDM can be expressed mathematically as [38]:

$$s_{ACO-OFDM}(t) = \begin{cases} s_{OFDM}(t) & \text{if} \quad s_{OFDM}(t) \geq 0 \\ 0 & \text{if} \quad s_{OFDM}(t) < 0 \end{cases}. \tag{4.45}$$

The clipped signal is shown in Figure 4.15. When the DC bias is set to zero, the hard clipping might be avoided by applying ACO-OFDM [45]. It is shown in [43,46] that the clipping noise can be avoided by encoding information symbols on only the odd subcarriers as shown in Figure 4.16 [37].

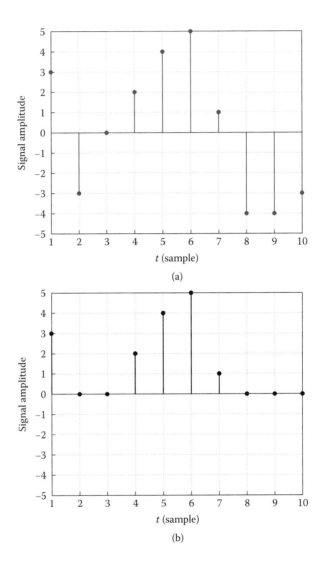

FIGURE 4.15
The samples of (a) bipolar OFDM and (b) clipped OFDM.

Since only odd subcarrier is modulated, the ACO-OFDM has only the half the spectral efficiency of DCO-OFDM. However, there is no information loss when the signal is clipped, because of the antisymmetry of the modulated signal [36,47]:

$$s_H\left[i + \frac{N_{SC}}{2}\right] = \sum_{n=-N_{SC}/2}^{N_{SC}/2-1} c_n[i] e^{j2\pi \frac{n \cdot i}{N_{SC}}} e^{j\pi n}. \tag{4.46}$$

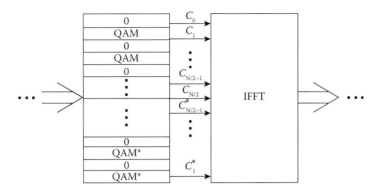

FIGURE 4.16
Modulation scheme of ACO-OFDM.

Therefore, if only the odd carriers are modulated:

$$s_H\left[i + \frac{N_{SC}}{2}\right] = -s_H[i],\qquad(4.47)$$

where $s_H[i]$ is a discrete time sample of the symbol and N_{SC} is the length of the symbol. When the time sample $s_H[i]$ is smaller than zero, and it is clipped during the optical modulation, $s_H[i + (N_{SC}/2)]$ must be positive, and $s_H[i]$ can be recovered by using $s_H[i + (N_{SC}/2)]$. The spectral investigation also showed this robustness against clipping. The clipping noise is noticed only at the even subcarriers, and the information is received at the odd subcarriers [6]. The equation of the clipped ACO-ODM is [36,45]:

$$X_c(k) = \begin{cases} \dfrac{X(k)}{2} & \text{if } k \text{ is odd} \\ X_c^n(k) & \text{if } k \text{ is even} \end{cases},\qquad(4.48)$$

where $X_c(k)$ is the clipped signal at k^{th} subcarrier, $X(k)$ is the original signal at the kth subcarrier, and $X_c^n(k)$ is the clipping noise. Due to the antisymmetry of the ACO-OFDM, the clipping noise is orthogonal to the data [36,45].

4.2.5 Pulse-Amplitude-Modulated Discrete Multitone (PAM-DMT)

PAM-DMT is similar to ACO-OFDM, but the subcarriers are modulated by PAM. Furthermore, the mathematical expression of PAM-DMT is equal to the mathematical expression of ACO-OFDM (Equation 4.45), as it applies asymmetrical clipping. The figure of the clipped signal is also the same for PAM-DMT (Figure 4.15). According to [41], if the data are modulated using PAM only on the imaginary components of the subcarriers, clipping noise

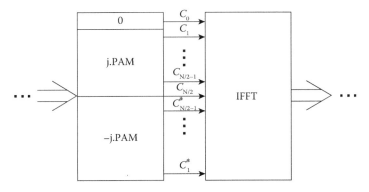

FIGURE 4.17
Modulation scheme of PAM-DMT.

does not affect the system performance since noise is a real value signal, so it is orthogonal to the modulation [43]. The modulation scheme of PAM-DMT is shown in Figure 4.17 [37].

In a PAM-DMT system, there is no DC bias. All of the subcarriers are modulated, but the modulation uses only the imaginary part of the subcarrier, thus the spectral efficiency is the same as ACO-OFDM [43]. Although it has a limited spectral efficiency, it is more power efficient than DCO-OFDM, because it has also an antisymmetry (Hermitian symmetry). It is described in [36,47] as:

$$s_H[i] = \sum_{n=-N_{SC}/2}^{N_{SC}/2-1} c_n[i]e^{j2\pi\frac{n\cdot i}{N_{SC}}} = \sum_{n=-N_{SC}/2}^{N_{SC}/2-1} j \cdot b_n[i]e^{j2\pi\frac{n\cdot i}{N_{SC}}}, \tag{4.49}$$

$$s_H[i] = 2 \cdot \sum_{n=-N_{SC}/2}^{N_{SC}/2-1} b_n[i]\sin\left(\frac{2\pi n i}{N_{SC}}\right), \tag{4.50}$$

$$s_H[N_{SC}-i] = 2 \cdot \sum_{n=-N_{SC}/2}^{N_{SC}/2-1} b_n[i]\sin\left(\frac{2\pi n[N_{SC}-i]}{N_{SC}}\right) = 2\cdot$$

$$\sum_{n=-N_{SC}/2}^{N_{SC}/2-1} -1 \cdot b_n[i]\sin\left(\frac{2\pi n i}{N_{SC}}\right) = -s_H[i], \tag{4.51}$$

$$s_H[N_{SC}-i] = -s_H[i]. \tag{4.52}$$

Similar to ACO-OFDM, the clipping of the negative signals does not lead to the information loss, because the clipped signal components can be reconstructed by using this antisymmetry [32]. The clipping noise is also orthogonal to the modulation, as the data have only imaginary part, but the clipping noise is a real value [43]. The next equation describes the clipped PAM-DMT according to [36] is given by:

$$\mathrm{Im}\{X_c(k)\} = \frac{D_k}{2}, \qquad \mathrm{Re}\{X_c(k)\} = \mathrm{Re}\{X_c^n(k)\}, \qquad (4.53)$$

where D_k is chosen from PAM symbols. The original PAM-DMT is $X(k) = j^*D_k$ [32]. The imaginary part of the clipped signal is half of the original signal, the real part of the signal is only the clipping noise.

4.2.6 Unipolar OFDM (U-OFDM)

The U-OFDM was introduced in [41], and an almost the same concept named Flip-OFDM was suggested in [45]. In U-OFDM (or Flip-OFDM), the negative and the positive part of the real bipolar OFDM signal are extracted. Hence, the Hermitian symmetry is preserved. The polarity of the negative parts of the symbol is inverted before the transmission of both positive and negative parts in a consecutive OFDM symbol [41]. The bipolar OFDM symbol and the U-OFDM symbol are compared in Figure 4.18.

The mathematical formula of the U-OFDM symbol can be expressed as [45]:

$$s_H[i] = s_H^+[i] + s_H^-[i - \frac{N_{SC}}{2}], \qquad (4.54)$$

where

$$s_H^+[i] = \begin{cases} \left(\varepsilon[i] - \varepsilon\left[i - \frac{N_{SC}}{2}\right]\right) \cdot s_H[i] & \text{when} \quad s_H[i] \geq 0 \\ 0 & \text{when} \quad s_H[i] < 0 \end{cases}, \qquad (4.55)$$

$$s_H^-[i] = \begin{cases} \left(\varepsilon[i] - \varepsilon\left[i - \frac{N_{SC}}{2}\right]\right) \cdot s_H[i] \cdot (-1) & \text{when} \quad s_H[i] < 0 \\ 0 & \text{when} \quad s_H[i] \geq 0 \end{cases}, \qquad (4.56)$$

where $\varepsilon[i]$ is step function:

$$\varepsilon[i] = 1 \quad \text{when} \quad i \geq 0. \qquad (4.57)$$

At the demodulator side, the original OFDM can be recombined by subtracting the negative frame [47].

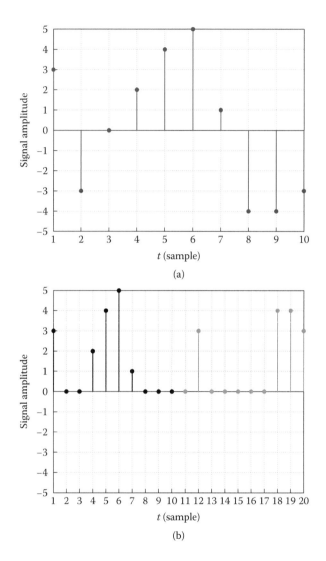

FIGURE 4.18
The samples of (a) bipolar OFDM and (b) U-OFDM.

4.2.7 Performance of the OFDM Formats for VLC

The performance of the OFDM formats for VLC is analyzed in the terms of BER, spectral efficiency, and power efficiency.

4.2.7.1 BER Performance of Bipolar OFDM in AWGN Channel

The perfectly synchronized OFDM can be viewed as a set of parallel Gaussian channels [35]. So the received symbols are the summation of transmitted symbols and the independent noise samples. The received symbols can be analyzed on each subcarrier. Therefore, the BER performance of OFDM should be also analyzed on the subcarriers, thus, the BER performance of subcarrier modulations can describe the BER performance of OFDM as well. The symbol error probabilities P_e are written by several authors before, and in this section, a summary of them is presented for typical subcarrier modulations such as M-PAM, M-PSK, and M-QAM. The symbol error probability is given by [39] for M-PAM:

$$P_e = 2\left(1 - \frac{1}{M}\right)Q\left(\sqrt{\frac{6\log_2 M}{M^2 - 1}\frac{E_b}{N_0}}\right), \qquad (4.58)$$

where E_b is the energy of the bits and N_0 is the noise spectral density. P_e has only an approximate formula for M-PSK.

$$P_e \approx 2Q\left(\sqrt{(2\log_2 M)\sin^2\left(\frac{\pi}{M}\right)\frac{E_b}{N_0}}\right). \qquad (4.59)$$

The approximate P_e formula for M-QAM signal is given by [48,49]:

$$P_e \approx \frac{4(\sqrt{M} - 1)}{\sqrt{M}}Q\left(\sqrt{\frac{3\log_2 M}{M - 1}\frac{E_b}{N_0}}\right). \qquad (4.60)$$

According to [39,48–50], the bit error probability P_b could be also calculated, which is related to BER, as given by:

$$P_b = \frac{2(M - 1)}{M\log_2 M}Q\left(\sqrt{\frac{6\log_2 M}{M^2 - 1}\frac{E_b}{N_0}}\right) \quad \text{for} \quad M\text{-PAM}, \qquad (4.61)$$

$$P_b \approx \frac{2}{\log_2 M}Q\left(\sqrt{(2\log_2 M)\sin^2\left(\frac{\pi}{M}\right)\frac{E_b}{N_0}}\right) \quad \text{for} \quad M\text{-PSK}, \qquad (4.62)$$

$$P_b \approx \frac{4(\sqrt{M} - 1)}{\sqrt{M}\log_2 M}Q\left(\sqrt{\frac{3\log_2 M}{M - 1}\frac{E_b}{N_0}}\right) \quad \text{for} \quad M\text{-QAM}. \qquad (4.63)$$

Based on these equations, it can be noted that the performance of M-PAM is equal to the performance of M^2-QAM [51]. The M-QAM and the M-PAM modulations are frequently used in unipolar OFDM formats, therefore, M-PAM and M-QAM are compared to the most typical baseband modulations.

As shown in Figure 4.19, the bipolar OFDM with 4-QAM has better BER performance than both OOK-NRZ and OOK-RZ, and the bipolar OFDM

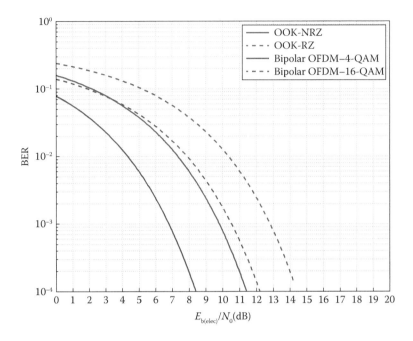

FIGURE 4.19
BER performance of OOK-NRZ, OOK-RZ, and bipolar OFDM with 4-QAM and 16-QAM.

with 16-QAM can approach the same BER with OOK-NRZ, if the bipolar OFDM has 1 dB higher SNR than OOK-NRZ.

4.2.7.2 Comparison of Unipolar OFDM Formats for VLC

The BER performance of the unipolar OFDM formats is compared to the bipolar OFDM formats. The BER performance of the DCO-OFDM depends on the bias level. For a high bias level, the clipping noise can be neglected, and the SNR for a given BER is approximately equal to the SNR of bipolar OFDM plus the bias level in dB [42]. In ACO-OFDM, only the half of the electrical power is used at the odd subcarriers. Hence, ACO-OFDM requires 3 dB higher SNR to approach a given BER than the bipolar OFDM [42]. The comparison of the bipolar OFDM, DCO-OFDM, and ACO-OFDM is shown in Figure 4.20. The bias of the DCO-OFDM is 7 dB as in [42]. It is also important to note that the BER performance of ACO-OFDM with 4-QAM is equal to the BER performance of OOK-NRZ [38]. Furthermore, it can be noticed, that the BER performance of the ACO-OFDM with 4-QAM is similar to the BER performance of bipolar OFDM with 16-QAM. The BER performance of the ACO-OFDM with 16-QAM is slightly better than the BER performance of the DCO-OFDM with 4-QAM.

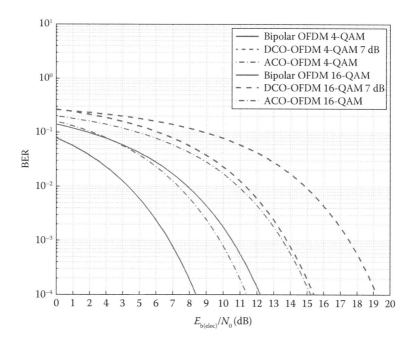

FIGURE 4.20
BER performance of bipolar OFDM, DCO-OFDM at 7 dB biasing, and ACO-OFDM. All of the modulation formats investigated with 4-QAM and 16-QAM.

According to [43], the performance of ACO-OFDM and PAM-DMT is exactly the same, because for ACO-OFDM the half of the subcarriers are filled, and for PAM-DMT the half of the quadrature. Thus, for PAM-DMT the power is half of the bipolar OFDM, similarly to ACO-OFDM [51]. As it was mentioned before, the performance of M-PAM is equal to the performance of M^2-QAM [51], therefore, ACO-OFDM with M^2-QAM is comparable with M-PAM-DMT, and their BER performance is exactly the same. The BER performance of U-OFDM (Flip-OFDM) and the BER performance of the ACO-OFDM is also the same, as their SNR are equivalent [45]. Consequently, ACO-OFDM and U-OFDM with M^2-QAM and M-PAM-DMT have same BER performance. It can be demonstrated in Figure 4.21 by using 4-QAM as a subcarrier modulation of ACO-OFDM and U-OFDM, and 2-PAM-DMT. These results are also compared to bipolar OFDM and DCO-OFDM with 7 dB biasing with 4-QAM modulation. By comparing the unipolar OFDM modulations in the term of spectral efficiency [61,62], it is important to note, that DCO-OFDM is the most efficient modulation type, because DCO-OFDM modulates all the subcarriers. The data rate of the DCO-OFDM is twice that of the ACO-OFDM [13]. The spectral efficiency can be defined by the first null of the spectra, and normalize relative to OOK with the same data rate [42].

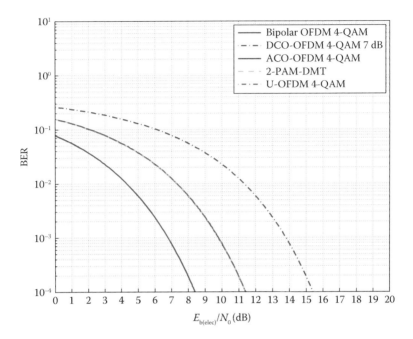

FIGURE 4.21
BER performance of bipolar OFDM, DCO-OFDM at 7 dB biasing, ACO-OFDM, PAM-DMT, and U-OFDM. All of the OFDM formats investigated with 4-QAM and PAM-DMT are investigated with 2-PAM.

Hence, the normalized bandwidth of DCO-OFDM is $(1 + 2/N_{SC}) \log_2 M$ and for ACO-OFDM $2(1 + 2/N_{SC}) \log_2 M$ [42]. The spectral efficiency of the ACO-OFDM and PAM-DMT is the same [43]. According to [45], the spectral efficiency of ACO-OFDM and U-OFDM is approximately the same for a sufficient number of subcarriers.

In terms of power efficiency, ACO-OFDM is more power efficient than DCO-OFDM, because ACO-OFDM does not apply a DC bias. The power efficiency of ACO-OFDM, PAM-DMT, and U-OFDM are the same [43,45]. The DCO-OFDM is inefficient in the terms of optical power especially for the constellation up to 256-QAM, however, the DCO-OFDM is more efficient than ACO-OFDM for larger constellations (such as 1024-QAM), because in this case the spectral efficiency is more important than power efficiency [52].

To conclude, DCO-OFDM is the most spectral efficient and less power efficient than unipolar OFDM modulation. The other three unipolar OFDM modulations (ACO-OFDM, PAM-DMT, and U-OFDM) have the same performance in term of spectral efficiency, power efficiency, and BER. However, 50% receiver complexity can be saved by U-OFDM over ACO-OFDM [16]. PAM-DMT can be an appropriate choice, when the adaptation to the frequency response of the channel is crucial, because PAM-DMT modulates all of the sub-carriers, but ACO-OFDM modulates only the odd subcarriers. PAM-DMT is advantageous especially when the number of the subcarriers is small [51].

4.2.7.3   PAPR Performance of OFDM

OFDM signal is sensitive to nonlinear distortions, which cause intermodulation between the subcarriers. This intermodulation leads to intercarrier interference in OFDM systems, which can cause degradation in the quality of the communication [53]. OFDM signals have a relatively high dynamic range, which is described often in the terms of PAPR. The high dynamic range is a consequence of the multicarrier modulation. As a result, subcarriers can add constructively or destructively, which may cause a large variation in signal power [41]. The high PAPR requires large saturation power for power amplifiers which leads to low power efficiency [41]. The nonlinearity of the optical modulation may cause limitations in the communication. The modulation of direct modulated light sources is described by the bias current-optical power characteristic. It determines the nonlinearity of the modulation. The distortion is more considerable at high powers [36]. As the illumination and communication work together in VLC, high-power optical sources are often used, and it makes the nonlinear distortion significant. Therefore, the issue of the high PAPR of OFDM is still a serious problem. The PAPR can be defined in [41] as:

$$PAPR = \frac{\max\{|s(t)^2|\}}{E\{|s(t)^2|\}}, \quad t \in [0, T_s], \tag{4.64}$$

where $E\{\}$ is the mean value operator.

The theoretical maximum of PAPR of OFDM is $10\log_{10}(N_{SC})$, where N_{SC} is the number of subcarriers [35,41]. According to [41], OFDM system with 256 subcarriers, the theoretical maximum of PAPR is 24 dB, but it is rarely observed. A better way to characterize PAPR is the complementary cumulative distribution function (CCDF), which is described as [41]:

$$P_c = \Pr\{PAPR > \xi_p\}, \tag{4.65}$$

where P_c is a probability that PAPR exceeds a particular value of ξ_p [3].

According to [41], for more likely probability regimes, where CCDF is around 10^{-3}, the PAPR is lower than the maximum value (24 dB) however, this value is around 11 dB, which is still relatively high. The PAPR of an OFDM signal is relatively high for both RF and optical systems; therefore PAPR reduction has been an active researching area [41]. Several techniques to reduce PAPR exist. They were overviewed in [54]. The PAPR reduction techniques can be classified into signal scrambling techniques and signal distortion techniques (signal clipping, peak windowing, envelope scaling, random phase updating, peak reduction carrier, and companding). Signal scrambling techniques contain the techniques with explicit side information (coding based, probabilistic schemes) and without explicit side information (Haddamard transform method and dummy sequence method). Some other techniques try to modify the modulation scheme of the unipolar OFDM modulation. A novel approach applies discrete Hartley transform (DHT) in ACO-OFDM systems in order to reduce PAPR of the signal. There are two groups of unipolar OFDM modulations. DCO-OFDM which is spectrally efficient, but not power efficient, and there is a group of power efficient, but spectrally inefficient OFDM formats (ACO-OFDM, PAM-DMT, U-OFDM). When the power efficiency is crucial (for smaller constellations up to 256-QAM) ACO-OFDM is advantageous, but for larger constellations (such as 1024-QAM), the spectral efficiency is the most important parameter, therefore, DCO-OFDM is the best solution. Alternative spectrally efficient modulation formats are carrierless amplitude and phase (CAP) modulations. Using CAP modulation, two orthogonal signals do not need overhead and carrier. It is effective for band limited VLC environment, where higher spectral efficiencies can be achieved. But, it requires more complex, adaptive bit loading techniques [55].

4.3 Color-Shift Keying (CSK)

CSK is a special modulation method in VLC systems. Most VLC systems use a blue LED and a yellow phosphor layer on LED(s) which converts the blue light into white. However, the phosphor layer has a long relaxation time and it limits the maximum modulation frequency. Using RGB (red, green, and blue) LEDs the former frequency limit can be eliminated from the system. However, the RGB LEDs are more expensive and require a complex control circuit to create white light. Because of these reasons RGB LEDs are rarely used in commercial devices at the moment. CSK modulation schemes are designed to operate with RGB LEDs in order to provide higher order, spectrally efficient modulation. Data are sent on the instantaneous color of the RGB triplet, while maintaining an average perceived chromaticity. The color-based modulation of CSK has several advantages over intensity modulated schemes. The constant emitted light guarantees an absence of flicker at all frequencies. The constant luminous flux of the source leads to near constant current drive, which in turn implies a reduced inrush current when modulating data, strong signal isolation from the power line and a reduction in inductance caused by large switching currents. The bit rate is decided by the symbol rate and the number of color points on the constellation. That means CSK bit rate is not limited by the frequency response of the LEDs. However, the use of RGB LEDs is not prevailing in lighting systems. Phosphor-based visible LEDs are more often used and they are not suitable for CSK. CSK is standardized in Section 12 of IEEE 802.15.7. In the standard, three LEDs are used in the CSK system. However, many papers discussed the four LEDs CSK systems as an improvement of these systems. Four colors provide a larger color band in the CIE 1931 color space, which increases the distances between symbol points, and the constellation can be similar to *M*-QAM.

CSK modulation is supported by the PHY III layer in the IEEE 802.15.7. Table 4.4 shows the possible modulation schemes, the error corrections, and the applied clock frequencies [56]. All CSK devices have to support the 12 MHz clock frequency. The devices send their parameters (e.g., which color bands are supported) to a coordinator, then it selects one color channel, which will be used for CSK communication. PHY II layer is used for this handshaking communication.

The block scheme of the reference CSK transmitter is shown in Figure 4.22 [56]. The incoming data are scrambled to reduce the error effect on the transmission. The scrambled data are always encoded with RS(64,32), except for the two highest data rate, where forward error correction (FEC) is not used.

TABLE 4.4

The Operation Mode of PHY III Layer

Modulation	Clock (MHz)	FEC	Data Rate (Mb/s)
4-CSK	12	RS(64,32)	12
8-CSK		RS(64,32)	18
4-CSK	24	RS(64,32)	24
8-CSK		RS(64,32)	36
16-CSK		RS(64,32)	48
8-CSK		None	72
16-CSK		None	96

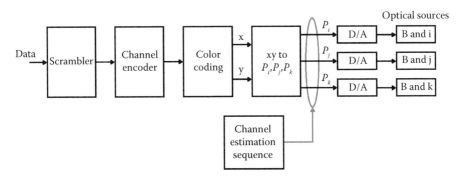

FIGURE 4.22
Block scheme of a PHY III transmitter.

TABLE 4.5

Scrambler Seed Selection

TDP	x_{init}	PRBS output
P1	0011 1111 1111 1111	0000 0000 0000 1000
P2	0111 1111 1111 1111	0000 0000 0000 0100
P3	1011 1111 1111 1111	0000 0000 0000 1110
P4	1111 1111 1111 1111	0000 0000 0000 0010

The color coding block codes the information into the *xy* color coordinate system according to the applied channel. Finally, the colors are transformed into intensities (P_i, P_j, P_k) to generate constant luminous flux.

The equation of the scrambling pseudorandom binary sequence (PRBS) generator is:

$$g(D) = 1 + D^{14} + D^{15}, \tag{4.66}$$

where D is a single bit delay element. The applied PRBS, $x[n]$, can be calculated as:

$$x[n] = x[n-14] + x[n-15], \quad n = 0, 1, 2 \ldots \tag{4.67}$$

In the above equation, '+' is a modulo-2 addition. All PRBS needs an initialization value x_{init}. It is specified in Table 4.5 [56].

Figure 4.23 shows the scrambling process [56]. The initialization value of the PRBS has to correspond to the seed identifier (Table 4.5), corresponding to the topology dependent pattern (TDP). The seed values are increased every transmission (included the retransmission, too). For example, if the seed value was P_3 in the first frame, then it will be P_4 in the second frame and P_1 in the third frame. For the CSK transmission, three color light sources are used, which are out of the defined seven color bands (Figure 4.24). The color bands are selected by the center wavelength and it determines the three vertices of the CSK constellation triangle [56]. The center frequency of some optical sources can differ from the center of the band plan or its spectrum can be distributed among over multiple frequency bands (Figure 4.25).

Band (nm)	Code	Center (nm)	(x,y)
380–478	000	429	(0.169, 0.007)
478–540	001	509	(0.001, 0.597)
540–588	010	564	(0.402, 0.597)
588–633	011	611	(0.669, 0.331)
633–679	100	656	(0.729, 0.271)
679–726	101	703	(0.734, 0.265)
726–780	110	753	(0.734, 0.265)

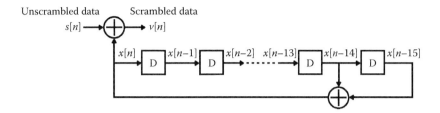

FIGURE 4.23
Block diagram of the scrambling process.

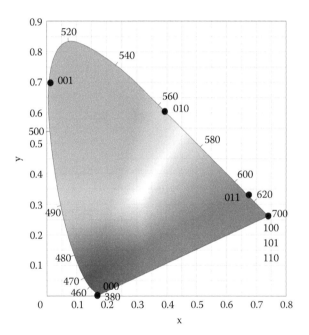

FIGURE 4.24
XY color coordinates of the seven bands and their position in the XY color diagram.

All CSK constellation design rules can be found in Section 12.5 of IEEE 802.15.7. Figure 4.26 shows the constellation design and data mapping of 4-CSK. Three bands I, J, K are selected and they determine the center wavelength of the three colors in the XY coordinate system (table in Figure 4.24). We can get the four symbols easily. One symbol is the center of the triangle and the other three symbols are the center of I, J, K bands, which are the vertices of the triangle. In the case of 4-CSK, two bits are assigned to all symbols.

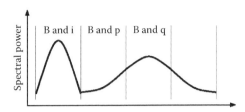

FIGURE 4.25
Optical sources spectrum over the channels.

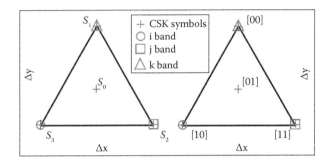

FIGURE 4.26
Design rule and the data mapping of 4-CSK.

The design rule and data mapping for 16-CSK [56] are shown in Figure 4.27. The three bands determine the vertices of the color triangle and its center wavelength. It provides only four symbol points, but 16 points are needed. The original triangle is separated into nine small equilateral triangles. The symbols will be the vertices and the centers of the nine triangles. In the case of 16-CSK, 4 bits are assigned for all symbols.

For CSK modulation, only those color bands can be used together from the seven bands, which allocate a triangle in the xy color coordinate system. At the receiver side, we have to find out which colors were used. For this calculation, we know only the light intensity and the colors of the transmitters. Figure 4.28 shows an example of this process. (x_i,y_i), (x_j,y_j), and (x_k,y_k) are the points of the color sources, while (x_p,y_p) is one of the 4-CSK symbols [56]. This point is created by the intensities (P_i,P_j,P_k) of the light sources. To determine the (x_p,y_p) point we have to know the relation between the intensities and the xy coordinates.

$$x_p = P_i x_i + P_j x_j + P_k x_k \quad y_p = P_i y_i + P_j y_j + P_k y_k \quad P_i + P_j + P_k = 1 \quad (4.68)$$

In the receiver, the above-mentioned equations are used to determine the received *xy* coordinate and it will be transformed into the proper symbol.

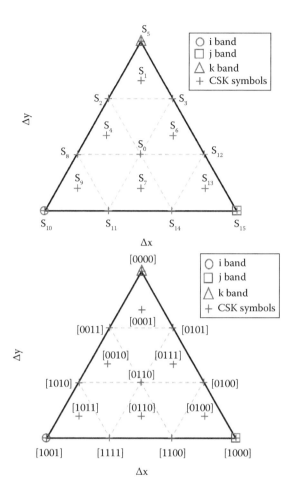

FIGURE 4.27
Design rule and the data mapping of 16-CSK.

The block scheme of a realized CSK system is in Figure 4.29. The incoming data are mapped into symbols and each symbol is coded into a color. The x,y coordinate of the color is then coded to the intensities of the RGB LED. At the receiver side, three photodiodes are necessary, one for each color of the RGB LED. Color filtering is achieved by an optical filter before the photodiode. AWGN is added to all colors after the detection. The decision is made in the color space, therefore the intensity of the detected three colors have to be decoded back to the x,y coordinate. The color-signal symbol decoding is made in the final block. The signal decoding can improve the system performance, if the symbol decision is made in signal space (based on intensities) instead of the color space [57,58]. Signal space detection is better

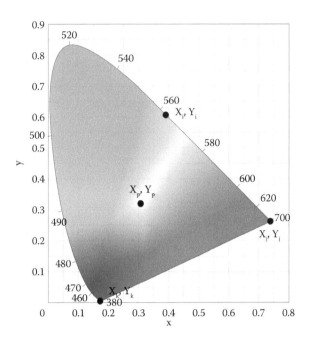

FIGURE 4.28
Example of 4-CSK demodulation.

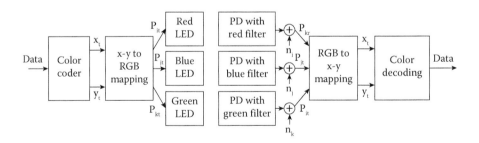

FIGURE 4.29
Block scheme of a CSK system.

because the distance between all symbols is equal. Nine valid color band combinations (CBCs) are defined in the standard from the seven color bands. The valid combinations are shown in Table 4.6.

The nine CBCs cover different color regions therefore their constellations will be different. It can result in their transmission performance not being the same [57]. Figure 4.30 shows the minimum Euclidean distance for different CBCs in the function of the constellation size [59]. CBC-2 and CBC-8 has the

TABLE 4.6

Valid Color Band Combinations

	Band i	Band j	Band k
1	110	010	000
2	110	001	000
3	101	010	000
4	101	001	000
5	100	010	000
6	100	001	000
7	011	010	000
8	011	001	000
9	010	001	000

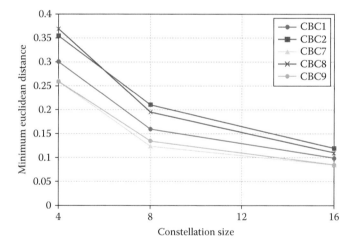

FIGURE 4.30
Minimum Euclidean distance for various constellation sizes and CBCs.

largest distance between the symbols, while the symbols are the closest to each other if CBC-7 or CBC-9 is applied. CBC-1 can be found between them. In the BER examination the same results are expected. BER was calculated in AWGN and line-of-sight channel and the data were uncoded [57]. The BER curve in the function of the optical SNR is in Figure 4.31. The BER curves are similar to the minimum Euclidean distance curve. The CBC-2 has the best performance, while CBC-7 requires more SNR for the same BER performance, like CBC-2. The performance of the CBC has to be taken into consideration at CSK system design. Depending on the selected CBC, there can be 2 dB difference in the requested SNR to achieve the same BER.

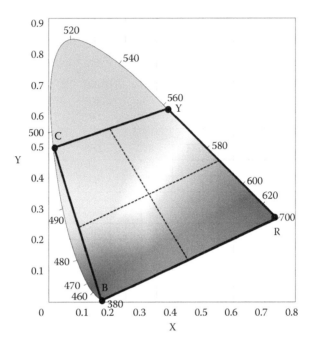

FIGURE 4.31
Color space of a QLED system.

To increase the applied color and intensity spaces efficiency, many papers investigated four LED (Quad LED)-based CSK systems. The idea of the QLED system was to use multiple TLED (Three LED) systems to extend the color space. At any time only one TLED is active and the receiver has to be able to decide which TLED was active. This decision can be avoided if a four-dimensional constellation is made by QLED system. In that case the TLED sets cover only a part of the gamut without any overlapping. It combines four sets of three-dimensional constellations. Because of it, the receiver can detect the received intensity as a point in the four-dimensional constellation and it is not necessary to know which TLED set was active. The color space of a QLED system is in Figure 4.31. The four colors usually applied are the following: blue, cyan, yellow, and red. It makes it possible to create simple symbol mapping and constellation design as in *M*-QAM. The transformation equations between the intensities and chromaticities contain the fourth color.

$$x_p = P_i x_i + P_j x_j + P_k x_k + P_l x_l \quad y_p = P_i y_i + P_j y_j + P_k y_k + P_l y_l \quad P_i + P_j + P_k + P_l = 1$$

$$(4.69)$$

These linear equations do not have an accurate solution and it can give negative values for intensities. Negative intensities are impossible in VLC systems; that is why QLED systems use only three LEDs at the same time. In that case,

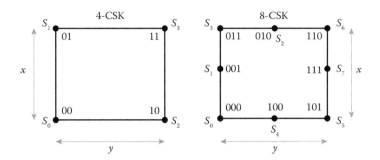

FIGURE 4.32
4- and 8-CSK symbol mapping in a QLED system.

the similar equations have to be solved as in the three LED systems. The color space (Figure 4.31) is divided into four parts [57]. All color of one part can be covered by three LEDs which are in the Blue-Red-Yellow (BRY) points. An example for the simple symbol mapping is in Figure 4.32. In the case of 4-CSK the symbols are in the color points of the four LEDs. At 8-CSK the S_2 and S_4 points are at the halfway point of two color LEDs. The QLED system has 4.4 dB SNR improvement compared to the TLED system [57].

4.4 Conclusion

This chapter provides an overview of the modulation schemes applied in VLC systems. A major limitation of existing VLC systems is the limited modulation bandwidth of LEDs. Using more complex modulation to improve the spectral efficiency and ISI for radio communications has been well studied. The same approaches can be applied in VLC systems, but the difference between radio and VLC has to be taken into account. IM/DD schemes require the modulation signal to be real-valued and positive. The chapter describes real-valued baseband modulations, multicarrier OFDM approaches, and special CSK for VLC. The modulation methods popular in VLC are described and compared based on bandwidth requirement, power efficiency, PAPR, and BER.

References

[1] J. Armstrong, OFDM for optical communications, *J. Lightwave Technol.*, vol. 27, pp. 189–204, 2009.
[2] G. Stepniak, M. Schuppert and C. A. Bunge, Advanced modulation formats in phosphorous LED VLC links and the impact of blue filtering, *J. Lightwave Technol.*, vol. 33, pp. 4413–4423, 2015.

[3] S. Loquai, R. Kruglov, B. Schmauss, C. A. Bunge, F. Winkler, O. Ziemann, E. Hartl and T. Kupfer, Comparison of modulation schemes for 10.7 Gb/s transmission over large-core 1 mm PMMA polymer optical fiber, *J. Lightwave Technol.*, vol. 31, pp. 2170–2176, 2013.

[4] Z. Ghassemlooy, W. O. Popoola and S. Rajbhandari, *Optical Wireless Communications: System and Channel Modelling with MATLAB®*, 1st ed. Boca Raton, FL: CRC Press, 2012.

[5] J. M. Kahn and J. R. Barry, Wireless infrared communications, *Proc. IEEE*, vol. 85, pp. 265–298, 1997.

[6] T. Lueftner, C. Kroepl, M. Huemer, J. Hausner, R. Hagelauer and R. Weigel, Edge-position modulation for high-speed wireless infrared communications, *IEEE Proc. Optoelectronics*, vol. 150, pp. 427–437, 2003.

[7] K. Szczerba, P. Westbergh, E. Agrell, M. Karlsson, P. A. Andrekson and A. Larsson, Comparison of intersymbol interference power penalties for OOK and 4-PAM in short-range optical links, *J. Lightwave Technol.*, vol. 31, pp. 3525–3534, 2013.

[8] J. R. Barry, *Wireless Infrared Communications*. Boston, MA: Kluwer Academic, 1994.

[9] H. Le Minh, D. O'Brien, G. Faulkner, Z. Lubin, L. Kyungwoo, J. Daekwang, O. Yunje and W. Eun Tae, 100-Mb/s NRZ visible light communications using a postequalized white LED, *IEEE Photon. Technol. Lett.*, vol. 21, pp. 1063–1065, 2009.

[10] M. Hoa Le, D. O'Brien, G. Faulkner, Z. Lubin, L. Kyungwoo, J. Daekwang and O. YunJe, High-speed visible light communications using multiple-resonant equalization, *IEEE Photon. Technol. Lett.*, vol. 20, pp. 1243–1245, 2008.

[11] X. Li, N. Bamiedakis, X. Guo, J. McKendry, E. Xie, R. Ferreira, E. Gu, M. Dawson, R. V. Penty and I. H. White, Wireless visible light communications employing feed-forward pre-equalization and PAM-4 modulation, *J. Lightwave Technol.*, vol. 34, pp. 2049–2055, 2016.

[12] S. Randel, F. Breyer, S. C. J. Lee and J. W. Walewski, Advanced modulation schemes for short-range optical communications, *IEEE J. Sel. Top. Quant. Electron.*, vol. 16, pp. 1280–1289, 2010.

[13] H. Le-Minh, D. O'Brien, G. Faulkner, Z. Lubin, L. Kyungwoo, J. Daekwang and O. Yunje, 80 Mbit/s Visible light communications using pre-equalized white LED, *34th European Conference on Optical Communication*, pp. 1–2, 2008.

[14] H. Li, X. Chen, J. Guo, D. Tang, B. Huang and H. Chen, 200 Mb/s visible optical wireless transmission based on NRZ-OOK modulation of phosphorescent white LED and a pre-emphasis circuit, *Chin. Opt. Lett.*, vol. 12, p. 100604, 2014.

[15] L. Honglei, C. Xiongbin, H. Beiju, T. Danying and C. Hongda, High bandwidth visible light communications based on a post-equalization circuit, *IEEE Photon. Technol. Lett.*, vol. 26, pp. 119–122, 2014.

[16] H. Li, X. Chen, J. Guo and H. Chen, A 550 Mbit/s real-time visible light communication system based on phosphorescent white light LED for practical high-speed low-complexity application, *Opt. Express*, vol. 22, pp. 27203–27213, 2014.

[17] P. A. Haigh, F. Bausi, T. Kanesan, L. Son Thai, S. Rajbhandari, Z. Ghassemlooy, I. Papakonstantinou, et al., A 20-Mb/s VLC link with a polymer LED and a multilayer perceptron equalizer, *IEEE Photon. Technol. Lett.*, vol. 26, pp. 1975–1978, 2014.

[18] J. J. D. McKendry, D. Massoubre, S. Zhang, B. R. Rae, R. P. Green, E. Gu, R. K. Henderson, A. E. Kelly and M. D. Dawson, Visible-light communications using a CMOS-controlled micro-light- emitting-diode array, *J. Lightwave Technol.*, vol. 30, pp. 61–67, 2012.

[19] S. Rajbhandari, H. Chun, G. Faulkner, K. Cameron, A. V. N. Jalajakumari, R. Henderson, D. Tsonev, et al., High-speed integrated visible light communication system: Device constraints and design considerations, *IEEE J. Sel. Areas Commun.*, vol. 33, pp. 1750–1757, 2015.

[20] Z. Shuailong, S. Watson, J. J. D. McKendry, D. Massoubre, A. Cogman, G. Erdan, R. K. Henderson, A. E. Kelly and M. D. Dawson, 1.5 Gbit/s multi-channel visible light communications using CMOS-controlled GaN-based LEDs, *J. Lightwave Technol.*, vol. 31, pp. 1211–1216, 2013.

[21] J. G. Proakis, *Digital Communications*, 4th ed., New York: McGraw-Hill, 2001.

[22] S. Rajbhandari, Z. Ghassemlooy and M. Angelova, Bit error performance of diffuse indoor optical wireless channel pulse position modulation system employing artificial neural networks for channel equalisation, *IET Optoelectronics*, vol. 3, pp. 169–179, 2009.

[23] P. Hyuncheol and J. R. Barry, Modulation analysis for wireless infrared communications, *IEEE International Conference on Communications, ICC '95 Seattle, "Gateway to Globalization,"* vol. 2, pp. 1182–1186, 1995.

[24] D. Shiu and J. M. Kahn, Differential pulse position modulation for power-efficient optical communication, *IEEE Trans. Commun.*, vol. 47, pp. 1201–1210, 1999.

[25] U. Sethakaset and T. A. Gulliver, Differential amplitude pulse-position modulation for indoor wireless optical communications, *EURASIP J. Appl. Signal Process.*, vol. 2005, pp. 3–11, 2005.

[26] U. Sethakaset and T. A. Gulliver, Performance of differential pulse-position modulation (DPPM) with concatenated coding over optical wireless communications, *IET Commun.*, vol. 2, pp. 45–52, 2008.

[27] Z. Ghassemlooy, A. R. Hayes, N. L. Seed and E. D. Kaluarachchi, Digital pulse interval modulation for optical communications, *IEEE Commun. Mag.*, vol. 36, no. 12, pp. 95–99, 1998.

[28] Z. Ghassemlooy and A. R. Hayes, Pulse interval modulation for IR communications, *Int. J. Commun. Syst.*, Special issue, vol. 13, pp. 519–536, 2000.

[29] N. M. Aldibbiat, Z. Ghassemlooy and R. McLaughlin, Indoor optical wireless systems employing dual header pulse interval modulation (DH-PIM), *Int. J. Commun. Syst.*, vol. 18, pp. 285–305, 2005.

[30] Z. Ghassemlooy, A. R. Hayes and N. L. Seed, The effect of multipath propagation on the performance of DPIM on diffuse Optical wireless communications, *Presented at the Proceedings of the IASTED International Conference on Wireless and Optical Communications*, Canada, 2001.

[31] Y. Tanaka, T. Komine, S. Haruyama and M. Nakagawa, Indoor visible communication utilizing plural white LEDs as lighting, *12th IEEE International Symposium on Personal, Indoor and Mobile Radio Communications*, vol. 2, pp. 81–85, San Diego, CA, 30 September–03 October 2001.

[32] H. Elgala, R. Mesleh and H. Haas, Indoor broadcasting via white LEDs and OFDM, *IEEE Trans. Consum. Electron.*, vol. 55, no. 3. pp. 1127–1134, 2009.

[33] N. LaSorte, W. Justin Barnes and H. H. Refai, The history of orthogonal frequency division multiplexing, *Proceedings of the IEEE Global Telecommunications Conference (GLOBECOM)*, New Orleans, LA, 30 November–4 December 2008.

[34] R. Prasad, *OFDM for Wireless Communication Systems*, Artech House, Boston, MA, 2004.

[35] K. Asadzadeh, Efficient OFDM signaling schemes for visible light communication systems, Open Access Dissertations and Theses, Paper 6079, 2011.

[36] X. Li, R. Mardling and J. Armstrong, Channel capacity of IM/DD optical communication systems and of ACO-OFDM, *IEEE International Conference on Communications 2007, ICC'07*, Glasgow, pp. 2128–2133, 24–28 June 2007.

[37] J. Armstrong, B. J. C. Schmidt, D. Kalra, H. A. Suraweera, A. J. Lowery, Performance of asymmetrically clipped optical OFDM in AWGN for an intensity modulated direct detection system, *IEEE Global Telecommunications Conference*, (GLOBECOM '06), San Francisco, CA, pp. 1–5, 27 November–01 December 2006.

[38] N. Kumar, *Visible Light Communication Systems for Road Safety Applications*, PhD Thesis, Universidade de Aviero, chapter 3, pp. 31–71, 2011.

[39] S. B. Weinstein and P. M. Ebert, Data transmission by frequency-division multiplexing using the discrete Fourier transform, *IEEE Trans. Commun.*, vol. COM-19, pp. 628–634, 1971.

[40] W. Shieh and I. Djordjevic, *OFDM for Optical Communications*, Elsevier, Burlington, NJ, 2010.

[41] J. Armstrong and B. J. C. Schmidt, Comparison of asymmetrically clipped optical OFDM and DC-biased optical OFDM in AWGN, *IEEE Commun. Lett.*, vol. 12, no. 5, pp. 343–345, 2008.

[42] D. J. F. Barros, S. K. Wilson, Senior Member, IEEE, and J. M. Kahn, Comparison of orthogonal frequency-division multiplexing and pulse-amplitude modulation in indoor optical wireless links, *IEEE Trans. Commun.*, vol. 60, no. 1, pp. 153–163, 2012.

[43] R. Mesleh, H. Elgala and H. Haas, An overview of indoor OFDM/DMT optical wireless communication systems, *7th International Symposium on Communication Systems Networks and Digital Dignal Processing (CSNDSP)*, July 2010, Newcastle upon Tyne.

[44] N. Fernando, Y. Hong and E. Viterbo, Flip-OFDM for unipolar communication systems, *IEEE Trans. Commun.*, vol. 60, no. 12, pp. 3726–3733, 2012.

[45] D. Tsonev, S. Sinanovic and H. Haas, Novel unipolar orthogonal frequency division multiplexing (U-OFDM) for optical wireless, *IEEE 75th Vehicular Technology Conference (VTC Spring)*, 6–9 May, Yokohama, 2012.

[46] V. Vijayarangan and R. Sukanesh, An overview of techniques for reducing peak to average power ratio and its selection criteria for orthogonal frequency division multiplexing, *J. Theor. Appl. Inform. Technol.*, vol. 5, no. 9, pp. 25–36, 2009.

[47] A. R. S. Bahai and B. R. Saltzberg, *Multicarrier Digital Communications, Theory and Applications of OFDM*, Kulwer Academic, New York, NY, 2002.

[48] P. Singhal, Multicarrier OFDM system performance in AWGN Channel, International, *J. Recent Trends Math. Comput*, vol. 1, no. 1, pp. 18–24, 2012.

[49] H. Schultze and C. Lüders, *Theory and Applications of OFDM and CDMA*, Wiley, Chichester, UK, 2005.

[50] S. C. Jeffrey Lee, S. Randel, F. Breyer and A. M. J. Koonen, PAM-DMT for intensity-modulated and direct-detection optical communication systems, *IEEE Photon. Technol. Lett.*, vol. 21, no. 23, pp. 1749–1751, 2009.

[51] M. S. Islim, D. Tsonev and H. Haas, Spectrally enhanced PAM-DMT for IM/DD optical wireless communications, *IEEE 26th Annual International Symposium on Personal, Indoor, and Mobile Radio Communications (PIMRC)*, Hong Kong, pp. 877–882, 30 August–02 September 2015, 2015.

[52] D. Tsonev, S. Sinanovic and H. Haas, Pulse shaping in unipolar OFDM-based modulation schemes, *2012 IEEE Globecom Workshops (GC Wkshps)*, Anaheim, CA, pp. 1208–1212, 3–7 December 2012.

[53] U. S. Jha and R. Prasad *OFDM towards Fixed and Mobile Broadband Wireless Access*, Artech House, Boston, MA, 2007.

[54] S. D. Dissanayake and J. Armstrong, Comparison of ACO-OFDM, DCO-OFDM and ADO-OFDM in IM/DD systems, *J. Lightwave Technol.*, vol. 31, no. 7, pp. 1063–1072, 2013.

[55] P. A. Haigh, Multi-band carrier-less amplitude and phase modulation for band-limited visible light communications systems, *IEEE Wireless Commun.*, vol. 22, no. 2, pp. 46–53, 2015.

[56] IEEE P802.15.7 Working Group. *IEEE Standard for Local and Metropolitan Area Networks—Part 15.7: Short-Range Wireless Optical Communication Using Visible Light*, IEEE Computer Society, 2011.

[57] R. Singh, T. O'Farell and J. P. R. David, An enhanced color shift keying modulation scheme for high-speed wireless visible light communications, *J. Lightwave Technol.*, vol. 32, no. 14, pp. 2582–2592, 2014.

[58] K.-I. Ahn and J. Kwon, Color intensity modulation for multicolored visible light communications, *IEEE Photon. Technol. Lett.*, vol. 24, no. 24, pp. 2254–2257, 2012.

[59] R. Singh, T. O'Farell and J. P. R. David, Performance evaluation of IEEE 802.15.7 CSK Physical Layer, *Globecom 2013 Workshop—Optical Wireless Communication*, 2013.

[60] D. Tsonev and H. Haas, Avoiding spectral efficiency loss in unipolar OFDM for optical wireless communication, *Proceeding of the International Conference on Communications (ICC)*, Sydney, Australia, IEEE, 10–14 June 2014.

[61] M. Islim, D. Tsonev and H. Haas, A generalized solution to the spectral efficiency loss in unipolar optical OFDM-based systems, *Proceeding of the International Conference on Communications (ICC)*, London, UK, IEEE, 8–12 June 2015.

[62] T. Cseh and T. Berceli, Optimum modulation for radio-over-fiber links transmitting OFDM NQAM RF signals, *2012 IEEE International Topical Meeting on Microwave Photonics*, pp. 1–4, Paper P29, Noordwijk, 2012.

5

IEEE 802.15.7: Visible Light Communication Standard

Murat Uysal, Çağatay Edemen, Tunçer Baykaş, Elham Sarbazi, Parvaneh Shams, H. Fatih Ugurdag, and Hasari Celebi

CONTENTS

5.1 Introduction

Visible light communications (VLC) use the visible spectrum (wavelengths of 390–750 nm or frequency band of 400–790 THz) and provide wireless communication using omnipresent light-emitting diodes (LEDs). Since the human eye perceives only the average intensity when light changes fast enough, it is possible to transmit data using LEDs without a noticeable effect on the lighting output and the human eye. Simultaneous use of LEDs for both lighting and communications purposes is a sustainable and energy-efficient approach that has the potential to revolutionize how we use light. VLC can be used in a wide range of short- and medium-range communication applications including wireless local, personal, and body area networks (WLAN, WPAN, and WBANs), vehicular networks, and machine-to-machine communication among many others. Besides energy efficiency, VLC offer several other inherent advantages over radio frequency (RF)-based counterparts, such as immunity to electromagnetic interference, operation on unlicensed bands, additional physical security, and a high degree of spatial confinement allowing a high reuse factor.

There is growing academic interest in VLC, which has resulted in a rich literature spanning from channel modeling to physical layer design and upper layer issues (see, e.g., Lee et al. 2011; Miramirkhani and Uysal 2015; Mesleh, Elgala, and Haas 2011; Acolatse, Bar-Ness, and Wilson 2011; Fernando, Hong, and Viterbo 2012; Gancarz, Elgala, and Little 2013; Jovicic, Li, and Richardson 2013; Hong et al. 2013; Bykhovsky and Arnon 2014; Hsu, Chow, and Yeh 2015; Elgala and Little 2015; Nuwanpriya et al. 2015; Kizilirmak, Narmanlioglu, and Uysal 2015; Hussein and Elmirghani 2015; Li, Zhang, and Hanzo 2015; Kashef et al. 2015). Along with academic interest, industrial attention to VLC has triggered related standardization activities to avoid fragmentation of proprietary vendor solutions in this emerging market. In Japan, the Visible Light Communications Consortium (VLCC) (www.vlcc.net) championed the standardization activities and proposed two standards known as the visible light communication system standard and the visible light ID system standard. These two standards were accepted by the Japan Electronics and Information Technology Industries Association (JEITA) in 2007 and became known as JEITA CP-1221 and JEITA CP-1222, respectively. More recently, in June 2013, the JEITA CP-1223 visible light beacon system standard was approved as an improved version of the JEITA CP-1222.

The Institute of Electrical and Electronics Engineers (IEEE) also recognized the potential of this emerging technology and produced IEEE Standard 802.15.7, which was approved in June 2011 (IEEE, 2011). This standard defines a physical layer (PHY) and a medium access control (MAC) layer for VLC and promises data rates sufficient to support audio and video multimedia services. In this chapter, we first provide an overview of this IEEE standard describing the main features of PHY and MAC layers. Then, we present simulation results to demonstrate the key performance metrics such as bit error rate (BER), throughput, and latency. The last section before the conclusions is reserved for the most recent standardization activity, which will be amended to IEEE Standard 802.15.7.

5.2 Overview of IEEE Standard 802.15.7

A personal area network (PAN) is the connection of information technology devices within a short distance. IEEE Standard 802.15.7 defines visible light communication personal area network (VPAN) as its network type. In a VPAN, a coordinator is responsible for starting and maintaining a network. The coordinator also assigns new devices to an existing VPAN. Three different network topologies, namely peer-to-peer, star, and broadcast, are introduced for VPANs, as shown in Figure 5.1.

- **Peer-to-peer topology:** Figure 5.1a illustrates the simple network infrastructure of peer-to-peer topology. The peer-to-peer networking topology is designed to support connectivity between two nodes that normally can be used for both sending and receiving, and act as both a device and a coordinator.

- **Star topology:** Figure 5.1b illustrates the network infrastructure for star topology. In this topology, a coordinator controls the network communications and can communicate with all the devices within the network.

- **Broadcast:** Figure 5.1c illustrates the network infrastructure for broadcast topology. The coordinator transmits data that will be received by each device in the network. The communication in this mode is unidirectional, and the destination address is not required.

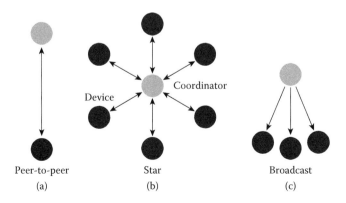

FIGURE 5.1
Network topologies: (a) peer-to-peer, (b) star, and (c) broadcast.

TABLE 5.1

VLC Device Classification

	Infrastructure	Mobile	Vehicle
Fixed coordinator	Yes	No	No
Power supply	Ample	Limited	Moderate
Form factor	Unconstrained	Constrained	Unconstrained
Light source	Intense	Weak	Intense
Physical mobility	No	Yes	Yes
Range	Short/long	Short	Long
Data rates	High/low	High	Low

In the IEEE 802.15.7 standard, three classes of VLC devices are considered, namely infrastructure, mobile (portable), and vehicle. The main features of each class are summarized in Table 5.1.

5.2.1 Network Architecture

The IEEE 802.15.7 architecture is defined in terms of a number of layers, each of which is responsible for one part of the standard and offers services to the higher layers. The overall architecture is illustrated in Figure 5.2. The network layer provides network configuration and message routing while the application layer provides the intended function of the device. These upper layers are not defined in the standard and are vendor specific.

The logical link control (LLC) layer provides access from the upper layers to the MAC layer through the service-specific convergence sublayer (SSCS). The tasks of MAC layer include beacon management, channel access, guaranteed time slot management, frame validation, acknowledged frame delivery, association, and disassociation of the device. It also provides color function, visibility, color stabilization, and dimming support. MAC data and MAC management information are accessed through the MAC common-part sublayer service access point (SAP) (MCPS-SAP) and the MAC layer management entity SAP (MLME-SAP), respectively. Further details on MAC layers are provided in Section 5.3.

The physical (PHY) layer defines the transceiver functionalities including line coding, modulation, error correction coding, and synchronization. It supports three PHY types, namely PHY I, PHY II, and PHY III. PHY I is designed for outdoor usage with low data rates in the tens to hundreds of Kbps. PHY II is intended for indoor usage with moderate data rates in the tens of Mbps. PHY III is designed for VLC systems with multiple light sources and detectors. It uses color-shift keying (CSK) and supports data rates in the tens of Mbps.

The device management entity (DME) has access to certain dimmer-related attributes in order to provide dimming information to the MAC and PHY layers. The DME also controls the PHY switch for selection of the optical sources and photodetectors for devices in which multiple transmitters/receivers

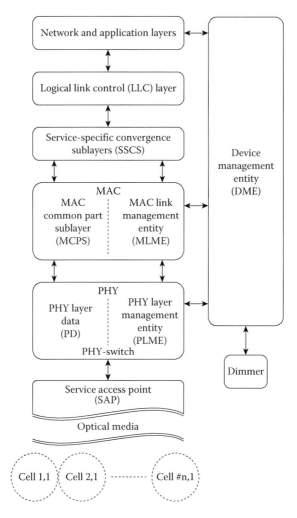

FIGURE 5.2
Network architecture.

are supported. The PHY switch serves as an interface to the optical SAP and connects to the optical media, which consist of a single (in the case of PHY layer types I and II) or multiple optical sources/photodetectors (in the case of PHY layer type III).

Similar to a wireless cellular network, multiple cells can be defined in a VPAN to improve coverage and/or support applications such as location-based services. The size and the position of the cell are variable and can be programmed by the DME, and the physical layer management entity (PLME) controls the PHY switch in order to select a specific cell.

The following sections explain MAC and PHY layer functions in detail.

5.3 MAC Layer

The MAC layer provides two services accessed through two SAPs. MAC management is accessed through the MAC link management entity SAP (MLME-SAP), while MAC data are accessed through the MAC common-part sublayer SAP (MCPS-SAP). The MAC layer handles all access to the physical layer and is responsible for the following tasks:

1. Generating network beacons if the device is a coordinator
2. Synchronizing to network beacons
3. Supporting device association and disassociation
4. Supporting color function (i.e., a function that provides information, such as device status and channel quality to the human eye via color)
5. Supporting visibility to maintain illumination and mitigate flicker
6. Supporting dimming (i.e., reducing the radiant power of a transmitter while preserving the color of the transmitted light)
7. Supporting device security
8. Providing a reliable link between two peer MAC entities
9. Supporting mobility

The standard defines mechanisms to start and maintain a VPAN. The device uses channel scanning to assess the current state of a channel, locate all beacons within its operation environment, or locate a particular beacon with which it has lost synchronization. Beacons are used in networks that either require synchronization or support for low-latency devices. If the network does not need synchronization or support for low-latency devices, it can elect to turn off the beacon for normal transfers. However, the beacon is still required for network discovery.

Following a channel scan and selection of a suitable VPAN identifier (that is not being used by any other PAN in the same area), operation as a coordinator starts. The association/disassociation mechanisms to allow devices to join or leave a VPAN are further defined in the standard.

In Section 5.3.1, we first describe the MAC frame structure and then provide further details on each random access mechanism in Section 5.3.2. Procedures for starting and maintaining a PAN along with device association/ disassociation procedures are described in Section 5.3.3. Transmission, reception, and acknowledgment mechanisms are explained in Section 5.3.4.

5.3.1 MAC Frame Structure

The MAC frame consists of three parts: the MAC header (MHR), the MAC service data unit (MSDU), and the MAC footer (MFR). The overview structure

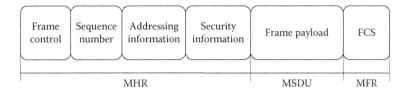

FIGURE 5.3
MAC frame format.

is shown in Figure 5.3. The MHR contains the frame control field, the sequence number field to specify the beacon sequence number (BSN), the address information field to provide source and destination addresses, and the security-related information field required for security processing. An MSDU part can comprise specific information based on the selected frame type (beacon, data, acknowledgment, etc.). An MFR, which contains a frame check sequence (FCS), is an error correction–related footer and located at the end part of a MAC frame.

There are five types of MAC frames:

1. A *beacon frame*, to transmit beacons by a coordinator in any topology
2. A *data frame*, to transmit all data
3. An *acknowledgment frame*, to verify successfully reception of a frame
4. A *MAC command frame*, to manage and transmit all MAC control signal transfer
5. A *color visibility dimming* (CVD) frame, to use visibility and dimming support providing side information such as communication status channel quality among nodes (device-to-device and device-to-coordinator)

The IEEE 802.15.7 standard allows the following two types of channel access mechanisms: beacon-enabled and nonbeacon-enabled. In nonbeacon-enabled mode, the coordinator does not transmit beacons, which means the devices in VPAN cannot be synchronized with each other. Each superframe has two different parts, namely active and inactive portions as detailed in Figure 5.4. The duration of the superframe depends on the value of macBeaconOrder (BO) and macSuperframeOrder (SO) both ranging from 0 to 14 as in the following equations: BI = aBaseSuperframeDuration × 2BO and SD = aBaseSuperframeDuration × 2SO optical clocks for $0 \leq SO \leq BO \leq 14$, where BI and SD denote the beacon interval and active portion length, respectively. Since BO = 15, aBaseSuperframeDuration is discarded and the coordinator will not transmit a beacon frame until it receives a new beacon frame request command. Similarly, if SO = 15, the active portion will be discarded and the superframe

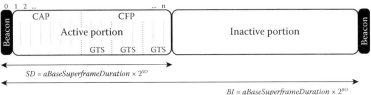

FIGURE 5.4
Detailed superframe structure.

will not remain active after the beacon. The inactive part is used for saving energy (low power mode) by the coordinator.

In the IEEE 802.15.7 standard, with the active region divided into equally spaced slots, a superframe includes three main parts: a beacon, a contention-access period (CAP), and a contention-free period (CFP). Both CAP and CFP are defined by this standard. The beacon starts at slot 0 and is followed by CAP immediately. Since CFP is zero length, CAP might be the end of the active portion of a superframe. The contention slots are used in the CAP. A CAP period should be a different length but also has a minimum length determined by *aMinCAPLength* optical clocks. The *aMinCAPLength* parameter can be reduced by the coordinator due to the inclusion of the new guaranteed time slots (GTSs) increasing temporarily the beacon frame or maintenance. The other issue to determine the period of beacon and CAP is the clock rate. In the IEEE 802.15.7 standard, all devices in same VPAN can support different clock rate ranges. Especially in star topology, the coordinator determines the periods of beacon and CAP at the lowest clock rate to guarantee that all optical receivers can receive beacons and CAP. In star topology, the CAP is used for association requests and the beacon/management frames. A device operating under the IEEE 802.15.7 standard is preferably to use either CAP or CFP, but it may also use both. The CFP should be started immediately following the CAP. The CFP part contains GTSs allocated by the coordinator and could be more than one slot in a sample superframe, illustrated in Figure 5.4.

The IEEE 802.15.7 standard provides two random access methods: slotted and unslotted random access with/without CSMA/CA. With the exception of performing acknowledgment and data frames, all frame types use slotted random access mechanism with/without CSMA/CA to access the channel.

5.3.2 Random Access Mechanisms

Carrier sense multiple access with collision avoidance (CSMA/CA) is used as the random access mechanism by 802.15.7. CSMA/CA was originally

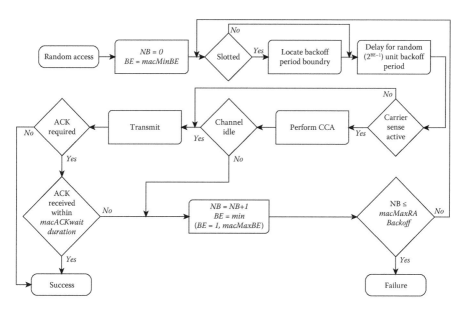

FIGURE 5.5
Flowchart of CSMA/CA method.

designed for the 802.11 standard. In IEEE 802.15.7, each device should ensure that the channel is not used by another device to avoid collision by performing a channel clear assessment (CCA). A CCA is requested by MAC and performed by PHY. Using the CSMA/CA algorithm, each device can sense the transmission channel before transmitting a frame by carefully tuning the timers. There are two types of CSMA/CA: slotted and unslotted. The slotted CSMA/CA method uses superframe structure and relies on a unit backoff period, which is equal to *aUnitBackoffPeriod* optical clocks. Conversely, the unslotted CSMA/CA method is used when there is no superframe structure; consequently, there is no need to use backoff periods. If a device finds the channel free, it starts its transmission immediately. However, if any other devices use the channel, the device will back off for a randomly chosen period of time (backoff time) before trying transmission again. This delay is a function of the backoff exponent (BE) and the unit backoff period for both slotted and unslotted CSMA/CA. It is calculated as backoff = (random integer between 0 and 2^{BE-1}) × (unit backoff period). The initial value of BE is equal to macMinBE. All processes for the CSMA/CA method are summarized in the form of a flowchart in Figure 5.5. In this chart, NB and CW denote the number of backoff attempts and contention window, respectively. In the first stage, NB and BE are set to their initial values: 0 and macMinBE. The value of BE will be incremented in every step, where a CCA is performed and the channel is not busy. The BE can be increased to the maximum value of aMaxBE. If the device waiting for retransmission finds that the channel is still busy, it will then select a new random backoff period for the next retransmission. Similarly, NB will be incremented whenever the channel is still busy after the backoff period. The transmission fails if the NB exceeds the value *macMacCSMABackoff*. The contention window (CW), which is used only in slotted CSMA/CA, is the parameter to determine how many CCAs are needed before starting to transmit.

5.3.3 Starting and Maintaining a VPAN

In IEEE 802.15.7, a coordinator is responsible for starting and maintaining a network. The coordinator also assigns the new devices joining an existing VPAN.

In VPAN, each device or coordinator has a unique address as a 64-bit number. The VPAN identifier called a short address is also a numerical label assigned to each device by a coordinator within the same network. This is the 16-bit address, and it is intended that each node in VPAN has a unique 16-bit network address. There are two types of channel-scanning methods, namely active and passive scanning. A device can perform either active or passive scanning to discover an operating device using the MLME-SCAN. request primitive. The results of the channel scanning are used to choose a suitable PAN. The details of the active and passive scanning procedures are as specified in the following:

- **Active scanning:** This type of scanning is considered to be used in peer-to-peer topology. In peer-to-peer topology, a device can communicate with any other device within its transmission range and can either send an association or active scan command to initiate communication with it. An active channel scanning occurs over a specified set of logical channels. In active scan, the MLME of a device sends a beacon request command and then waits to hear any unique responses from any devices that are within its range of communication. The IEEE 802.15.7 standard specifies how long the client should wait, which is at most [$aBaseSuperframeDuration \times (2n + 1)$] optical clocks, where n is the value of the scan duration parameter. During this waiting time, the device can discard all nonbeacon frames and record all devices with which it wishes to communicate. The number of recorded devices is limited by an implementation-specified maximum number of VPAN descriptors.

- **Passive scanning:** This type of scanning is considered to be used in star and broadcast topologies. Passive scanning is most likely with active scanning with one exception that the device never sends a beacon request command, which is not required for passive scanning. The passive scanning could be used for association based on priority.

Starting a VPAN is only applicable to bidirectional communication modes and not for broadcasting. For a star topology, the coordinator establishes the VPAN by sending beacon frames. For peer-to-peer topology, a device can send either an association or an active scan command to initiate communication with the peer device. The *MLME-SCAN.request* primitive is used to discover the other operating devices using macBeaconOrder (BO) parameter which ranges from 0 to 14. Using *MLME-START.request* primitive, the beacon transmission is started by a coordinator to create a new VPAN or a device to join an existing network. As discussed in Section 5.3.1, the MLME-START.request primitive includes BO and SO parameters which denote the beacon interval and active portion length, respectively. If a device loses its synchronization with the coordinator (e.g., it never received any beacon from the coordinator for a period of aMaxLostBeacons), the MLME of the device informs the higher levels to immediately stop transmitting beacon frames using the *MLMESYNC-LOSS.indication* primitive. Discovering a device is a bidirectional communication and only applicable for star and peer-to-peer topology for PHY I and II. It gives the ability to determine the identity of other devices on the VPAN. In IEEE 802.15.7, a compliant device usually operates in one visible light band. However, the standard also supports use of several visible light bands, and the coordinator needs to indicate these bands.

A device can request to join a VPAN using the results of the channel scanning (active or passive) and configuring the *MLME-ASSOCIATE. request* primitive parameters. The device is allowed to associate with a new VPAN only with permission from the VPAN coordinator. A coordinator can disassociate a device from a VPAN by sending a disassociation notification command to the device. If an associated device wants to leave the VPAN, it sends a disassociation notification command to the coordinator.

5.3.4 Transmission, Reception, and Acknowledgment of MAC Frames

The MAC layer mainly supports two types of service: data and management services. This subsection gives a brief overview of the transmission of data and management services. The data sequence number (DSN) and BSN are the stored numbers in each device (*macDSN*) and coordinator (*macBSN*), respectively. These numbers are copied into the outgoing frames increasing by one whenever a data/MAC frame and a beacon command frame are generated.

All source and destination addresses are stored in the relevant fields. These addresses can be either 16 bit short or 64 bit extended depending on association with a VPAN or not. If the source and destination address field is not present, the device is assumed to be a VPAN coordinator.

The frame transmission can be processed on a beacon- or nonbeacon-enabled VPAN. In a beacon-enabled PAN, the device uses a slotted CSMA/CA random access algorithm except when beacon is not being tracked at most [*aBaseSuperframeDuration* × (2*n* + 1)] optical clocks, where *n* is the value of *macBeaconOrder*. In nonbeacon-enabled PAN, the device uses unslotted CSMA/CA random access algorithm.

In an acknowledged transmission, where its acknowledgment request subfield is set to one, the transmitting device requests the data recipient device to send an acknowledgment frame back containing the same DNS/BNS from the data or MAC command frame, respectively, if the data are received successfully. In unacknowledged transmission, the data recipient does not send an acknowledgment back. The transmission of the acknowledgment frame should start between *aTurnaroundTime-RX-TX* and (*aTurnaroundTime-RX-TX* + *aUnitBackoffPeriod*) optical clocks after the reception of the last symbol of the data or MAC command frame.

5.3.5 Multiple Channel Resource Assignment

The IEEE 802.15.7 standard supports multiple bands when the coordinator needs time slot resources to assign for new users. The coordinator transmits Src_multi_info in the MAC command payload field to the device to use multiple bands. If the coordinator does not use multiple bands due to hardware limitations or interference situation, it sets Src_multi_info in the MAC command payload field with code "0000000" field. Another way to use multiple channels in VLC is mobility. Mobility includes two different types: physical and logical, depicted in Figure 5.6. In this figure, an example of physical and logical mobility is shown. The physical mobility occurs when VLC device D_1 changes its position due to the movement within the coverage area of the infrastructure source I_1. In contrast to physical movement, the logical mobility occurs when VLC device D_1 changes its communication link from a link with infrastructure source I_2 to one with infrastructure source I_3.

Figure 5.7 illustrates a cell configuration for VLC mobility. A VLC coordinator is configured to support the mobility of mobile VLC device 1. The device moves through multiple cells and the VLC coordinator supports

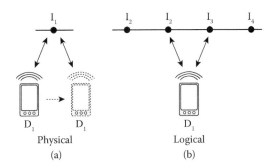

FIGURE 5.6
Physical (a) and logical (b) mobility.

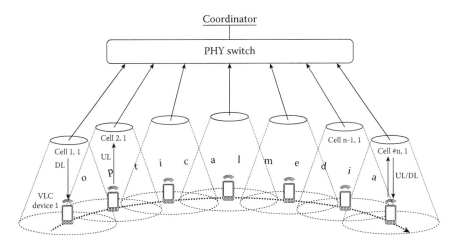

FIGURE 5.7
VLC cell configuration.

mobility using a PHY switch connected to an optical element in each cell and controlled by a DME. The optical element in a cell is denoted by $cell(i, j)$ where j is the index of the optical element in the ith cell. The size and the position of the cells in the optical media may be variable. The determination of the actual size and position of optical elements for the cell by a coordinator's DME is out of scope in the IEEE 802.15.7 standard. A VLC device 1 is assumed to move from $cell(i, j)$ to $cell(i + 1, j)$ and then to $cell(i + 2, j)$. While VLC device 1 is in $cell(i, j)$, it receives data from the coordinator on the downlink. As VLC device 1 moves to the next cell [from $cell(i, j)$ to $cell(i + 1, j)$], VLC device 1 will transmit a response in an acknowledgment frame (or CVD frame) on the uplink from $cell(i + 1, j)$. After movement of VLC device 1, the coordinator will not receive any response from the expected uplink transmission in $cell(i, j)$ and then searches for device 1 through the adjacent cells such as $cell(i + 1, j)$ and $cell(i - 1, j)$ during the same time slots assigned to device 1 in the superframe. This search process may be terminated if VLC device 1 is not found within the link timeout period. This period is defined in MAC PIB attribute *macLinkTimeOut*. VLC device 1 may be considered to be disassociated from the coordinator with an absence of any successful connection in the link timeout period.

While the coordinator may detect the response signal from VLC device 1 in $cell(i + 1, j)$ based on the reception of the uplink signal from VLC device 1, communication with any other devices presented in $cell(i, j)$ will not affect it. If VLC device 1 is assumed to move to $cell(n, j)$ and then stays within the boundaries of $cell(n, j)$, both uplink and downlink communication can occur within the single cell $cell(n, j)$. In this way, the coordinator may detect no further mobility of VLC device 1.

TABLE 5.2

Color Table for Indication

	Color A	Color B	Color C
State	Scan	Association	Disassociation
Color resolution range	0–255	0–255	0–255
Channel quality (current FER [CFER])	$CFER < FER_1$	$FER_1 \leq CFER \leq FER_2$	$CFER > FER_2$
Remaining or transferred file (RTF) size	$RTF < L_{BYTE}$	$L_{BYTE} \leq RTF \leq M_{BYTE}$	$RTF > M_{BYTE}$

5.3.6 Other MAC Functionalities

In the CVD frame, the same color can be assigned to determine the multiple status of a device. Usage of CVD will affect the color of the emitted light. A CVD frame can be used for the MAC state channel quality and file-transfer status indicator. The CVD frames may provide visual information regarding the communication status such as association, scan, and disassociation as summarized in Table 5.2. The CVD frames may also provide information to the users about the quality fluctuation of the communication quality by using frame-error ratio (FER) and remaining/transferred file size. FER is a performance metric and FER_1/FER_2 denotes that threshold value of channel quality, respectively.

5.3.7 Performance Evaluation of MAC Layer

In this section, we present a comprehensive performance evaluation of major MAC layer metrics—throughput, delay, power consumption, collision probability, and packet drop probability based on Markov modeling. We consider a beacon-enabled mode with CSMA/CA in the 802.15.7 MAC layer. Although the IEEE 802.15.7 standard provides the option to use beaconless modes and allows operation without CSMA/CA, such modes become useful for light traffic loads and when a hidden node problem is not observed frequently. Beacon-enabled mode with CSMA/CA allows for accommodating dense traffic and dense node deployment. For this reason, we primarily focus on modeling the beacon-enabled mode with CSMA/CA in saturated environments. The performance of the standard is obtained at different network sizes. In the simulations, N denotes the number of nodes in the network.

The first simulation provides the throughput per node, which corresponds to the fraction of time that a node spends in successful transmission. As shown in Figure 5.8, networks with low numbers of nodes achieve a high per node throughput. Beyond eight nodes, the per node throughput drops below 50%. The average packet drop probability per node is shown in

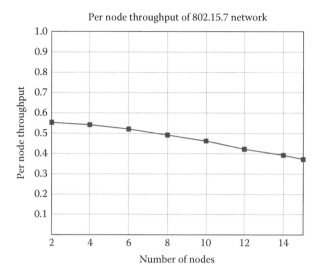

FIGURE 5.8
Per node throughput of 802.15.7 network.

Figure 5.9. In relatively small size of $N = 8$, the drop probability reaches 0.8. The per node collision probability is considered as the probability that during transmission of a given node, at least one other node is transmitting. The simulation results for collision probability are shown in Figure 5.10. Thanks to the CCA algorithm, the maximum value of collision probability is below 0.5 at $N = 15$. The last two simulations are about the average delay and the average power consumption, respectively. Delay for a successfully transmitted packet is considered as the total number of time slots required from the time that packet reaches the head of the line until the acknowledgment is received successfully. The results shown in Figure 5.11 indicate that the delay is between 25 and 30 time slots for moderate to high network sizes. Average power consumption for each node depends on the durations that a node stays in idle, receiving, or transmitting modes. The transceiver is idle when the node is in backoff state. It is in reception mode when the node is either performing CCA or waiting for, or receiving, an acknowledgment, and it is in transmission mode only when the node is transmitting a packet. In order to normalize the power, it is divided by the power consumption for packet transmissions. As shown in Figure 5.12, the average power consumption decreases with increasing network size. The reason for this behavior is the average waiting time increases when collisions occur, because as the offered traffic increases, nodes spend more time in backoff state rather than transmission state. Because transmission incurs more energy consumption than the idle backoff stage, nodes consume less power as the size of the network grows.

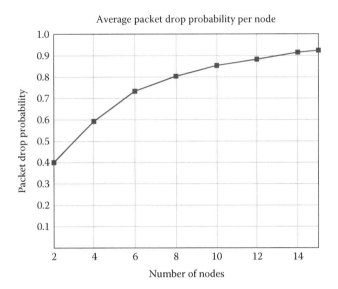

FIGURE 5.9
Average packet drop probability per node.

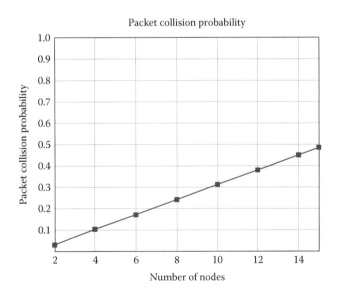

FIGURE 5.10
Packet collision probability.

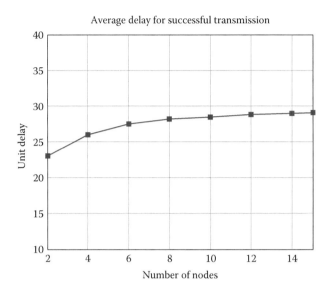

FIGURE 5.11
Average delay for successful transmission.

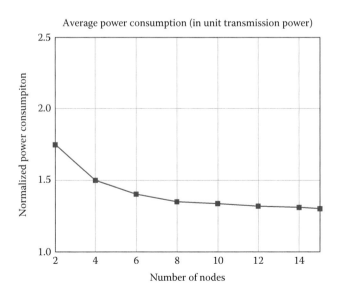

FIGURE 5.12
Average power consumption (in unit transmission power).

5.4 PHY Layer

The functions and services of the physical (PHY) layer are link establishment and termination of a connection to a communications medium. Based on the IEEE 802.15.7 standard for VLC, the PHY layer is responsible for the following tasks:

- Activation and deactivation of the VLC transceiver
- Wavelength quality indication (WQI)
- Clear channel assessment
- Data transmission and reception
- Error correction
- Synchronization
- Supporting dimming

Depending on the intended data rate and usage environment, the IEEE 802.15.7 standard comprises a number of different PHY layer types:

- **PHY I:** This type uses on–off keying (OOK) and variable pulse position modulation (VPPM). It supports concatenated coding with Reed–Solomon (RS) and convolutional coding (CC). This PHY type is intended for outdoor low data-rate applications with rates in the tens to hundreds of Kbps.
- **PHY II:** Similar to PHY I, PHY II uses OOK and VPPM but at higher optical clock rates aiming to achieve data rates in the tens of Mbps. It supports only RS coding. This PHY type is intended for indoor usage with moderate data rate applications. PHY I and PHY II also support a run-length limited (RLL) code to provide DC balance, clock recovery, and flicker mitigation.
- **PHY III:** This type is intended for applications with multiple light sources and detectors. It uses CSK and RS coding. This type aims to achieve data rates in the order of the tens of Mbps.

All layer types with detailed operating modes are summarized in Table 5.3. An IEEE 802.15.7-compliant device must implement at least one of the PHY I and PHY II types. A device implementing the PHY III type should also implement PHY II mode for coexistence. The PHY types may operate in the presence of dimming. OOK under dimming provides constant range and variable data rate by inserting compensation time. On the other hand, VPPM under dimming provides constant data rate and variable range by adjusting the pulse width.

TABLE 5.3

PHY Layers and Operating Modes

	Modulation	RLL Code	Optical Clock r	FEC Outer Code (rs)	FEC Inner Code (cc)	Data Rate
PHY I	OOK	Manchester	200 kHz	(15,7)	1/4	11.67 Kbps
				(15,11)	1/3	24.44 Kbps
				(15,11)	2/3	48.89 Kbps
				(15,11)	None	73.3 Kbps
				None	None	100 Kbps
	VPPM	4B6B	400 kHz	(15,2)	None	3556 Kbps
				(15,4)	None	71.11 Kbps
				(15,7)	None	124.4 Kbps
				None	None	266.6 Kbps
PHY II	VPPM	4B6B	3.75 MHz	(64,32)	None	1.25 Mbps
				(160,128)	None	2 Mbps
			7.5 MHz	(64,32)	None	2.5 Mbps
				(160,128)	None	4 Mbps
				None	None	5 Mbps
	OKK	8B10B	15 MHz	(64,32)	None	6 Mbps
				(160,128)	None	9.6 Mbps
			30 MHz	(64,32)	None	12 Mbps
				(160,128)	None	19.2 Mbps
			60 MHz	(64,32)	None	24 Mbps
				(160,128)	None	38.4 Mbps
			120 MHz	(64,32)	None	48 Mbps
				(160,128)	None	76.8 Mbps
				None	None	96 Mbps
PHY III	4-CSK		12 MHz	(64,32)	None	12 Mbps
	8-CSK			(64,32)	None	18 Mbps
	4-CSK		24 MHz	(64,32)	None	24 Mbps
	8-CSK			(64,32)	None	36 Mbps
	16-CSK			(64,32)	None	48 Mbps
	8-CSK			None	None	72 Mbps
	16-CSK			None	None	96 Mbps

Details on the optical clock rates, data rates, and error correction codes for each PHY type are illustrated in Table 5.3. It is noted from this table that multiple optical rates are provided for all PHY types in order to support a broad class of LEDs for various applications. The choice of optical rate used for communication is decided by the MAC layer during device discovery.

TABLE 5.4

Visible Light Wavelength Band Plan

Band (nm)	Center (nm)	Spectral Width (nm)	Code
380–478	429	98	000
478–540	509	62	001
540–588	564	48	010
588–633	611	45	011
633–679	656	46	100
679–726	703	47	101
726–780	753	54	110
Reserved			111

5.4.1 General Requirements

The visible light spectrum defined in IEEE 802.15.7 covers wavelengths between 380 and 780 nm. A compliant device must operate in one or several visible light wavelength bands as summarized in Table 5.4. LED manufacturers produce LEDs depending on human color perception and not frequency band, so nonlinear widths are needed for a band plan. The VLC standard provides support for seven bands in the visible light spectrum. The bands take into account the nonlinear color sensitivity of the human eye and the corresponding spectral range of LEDs. The standard also supports use of wide bandwidth optical transmitters (such as white LEDs) that can transmit in multiple bands or have leakage in other bands using the concepts of channel aggregation and guard channels.

The two main challenges for communication using visible light spectrum are flicker mitigation and dimming support. Flicker refers to the fluctuation of the brightness of light. Any potential flicker resulting from modulating the light sources for communication must be mitigated because flicker can cause noticeable, negative/harmful physiological changes in humans. To avoid flicker, the changes in brightness must fall within the maximum flickering time period (MFTP). The MFTP is defined as the maximum time period over which the light intensity can change without the human eye perceiving it. While there is no widely accepted optimal flicker frequency number, a frequency greater than 200 Hz (MFTP < 5 ms) is generally considered safe. Therefore, the modulation process in VLC must not introduce any noticeable flicker either during the data frame or between data frames.

Dimming support is another important consideration for VLC for power savings and energy efficiency. It is desirable to maintain communication while a user arbitrarily dims the light source. The human eye responds to low light levels by enlarging the pupil, which allows more light to enter the eye. This response results in a difference between perceived and measured

TABLE 5.5

Definition of Data Mapping for OOK Modulation

Logical Value	Physical Value	
0	High	0 < t < T
1	Low	0 < t < T

levels of light. Hence, communication support needs to be provided when the light source is dimmed over a large range, typically between 0.1% and100% (Rajagopal and Lim, 2012).

5.4.1.1 Modulation

a. **On–Off Keying**

OOK modulation is the simplest modulation scheme for VLC, where the LEDs are turned on or off depending on data bits being 1 or 0. While the modulation is logically OOK, OOK "off" does not necessarily mean the light is completely turned off; rather, the intensity of the light may simply be reduced as long as one can distinguish clearly between the "on" and "off" levels. In Table 5.5, the definition of data mapping for OOK modulation is summarized.

b. **Variable Pulse Position**

VPPM changes the duty cycle of each optical symbol to encode bits. The variable term in VPPM represents the change in the duty cycle (pulse width) in response to the requested dimming level. VPPM optical symbols are distinguished by the pulse position. As shown in Figure 5.13a, VPPM is similar to 2-PPM when the duty cycle is 50%. The logic 0 and logic 1 symbols are pulse-width modulated depending on the dimming duty cycle requirement. As shown Figure 5.13b, the pulse width ratio (b/a) of PPM can be adjusted to produce the required duty cycle for supporting dimming by pulse-width modulation (PWM). Figure 5.14 shows an example waveform of how VPPM can attain a 75% dimming duty cycle requirement, where both logic 0 and logic 1 have a 75% pulse width. In Table 5.6, the definition of data mapping for OOK modulation is summarized.

c. **Color-Shift Keying**

White LED lights are generated by using a mixture of different colors in typically two different methods. White LEDs can be generated using blue LEDs with yellow phosphor. However, yellow phosphor slows down the switching response of the white LEDs. Alternately, faster white LEDs can be generated by simultaneously exciting red, green, and blue LEDs. The use of such multicolor LEDs forms the principle behind CSK modulation; CSK modulation is similar to

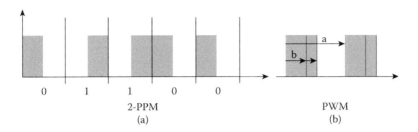

FIGURE 5.13
Basic concept of VPPM: (a) 2-PPM and (b) PWM.

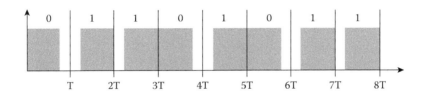

FIGURE 5.14
Waveform of VPPM signal with 75% pulse width.

TABLE 5.6

Definition of Data Mapping for VPPM Modulation

Logical Value	Physical Value	
0	High	$0 < t < dT$
	Low	$dT < t < T$
1	Low	$0 < t < (1 - d)T$
	High	$(1 - d)T < t < T$

Note: d is the VPPM duty cycle ($0.1 < d < 0.9$).

frequency shift keying in that the bit patterns are encoded to color (wavelength) combinations. For example, for 4-CSK (two bits per symbol) the light source is wavelength keyed such that one of four possible wavelengths (colors) is transmitted per bit pair combination.

In order to define various colors for communication, the IEEE 802.15.7 standard breaks the spectrum into seven color bands, according to the CIE 1931 color space standard (CIE, 1931), in order to provide support for multiple LED color choices for communication. Each of these bands has an assigned color code and is mapped into x and y values on the x–y color coordinates. The color codes and x–y coordinate values for each band are shown in Table 5.7.

TABLE 5.7

Color Bands and x–y Color Coordinates

Band (nm)	Code	Center (nm)	(x,y)
380–478	000	429	(0.169, 0.007)
478–540	001	509	(0.011, 0.733)
540–588	010	564	(0.402, 0.597)
588–633	011	611	(0.669, 0.331)
633–679	100	656	(0.729, 0.271)
679–726	101	703	(0.734, 0.265)
726–780	110	753	(0.741, 0.268)

The CSK signal is generated by using three color light sources out of the seven color bands. The three vertices of the CSK constellation triangle are decided by the center wavelength of the three color bands on x–y color coordinates. Certain combinations that cannot make a triangle on the x–y color coordinates are excluded, such as (110–101–100) or (100–011–010). Table 5.7 shows the x–y color coordinates values assuming the optical source is chosen with the spectral peak occurring at the center of each of the seven color bands. It is possible that some of the optical sources would have a spectral peak at a different frequency than the center of the band plan. It is also possible that the spectrum of the optical source would be distributed over multiple frequency bands. Implementers of CSK systems can select the color band based on the center wavelength of the actual optical source.

Table 5.8 shows valid color band combinations that can produce triangles for CSK constellations.

CSK has the following advantages:

- The final output color (e.g., white) is guaranteed by the color coordinates. CSK channels are determined by mixed colors that are allocated in the color coordinates plane.

- The total power of all CSK light sources is constant, although each light source may have a different instantaneous output power. CSK dimming ensures that the average optical power from the light sources is kept constant and maintains the requisite intensity of the center color of the color constellation. Thus, there is no flicker issue associated with CSK due to amplitude variations. CSK dimming employs amplitude dimming and controls the brightness by changing the current driving the light source. However, care needs to be observed during CSK dimming to avoid unexpected color shift in the light source.

- CSK supports amplitude changes with digital-to-analog (D/A) converters (higher complexity), thus allowing higher order modulation

TABLE 5.8

Valid Color Band Combinations for CSK

	Band i	Band j	Band k
1	110	010	000
2	110	001	000
3	101	010	000
4	101	001	000
5	100	010	000
6	100	001	000
7	011	010	000
8	011	001	000
9	010	001	000

support to provide higher data rates at a lower optical clock frequency. PHY I and PHY II allow only OOK modulation, thereby limiting their data rate.

4-CSK constellation design

The 4-CSK symbol points are defined by the design rule in Figure 5.15. In this figure, 4-CSK data mapping is also shown. Two bits are assigned per symbol. Points I, J, and K show the center of the three color bands on x–y color coordinates. S0 to S3 are four symbol points of 4-CSK. S1, S2, and S3 are three vertices of the triangle IJK. S0 is the centroid of the triangle formed by I, J, and K.

8-CSK constellation design

The 8-CSK symbol points are defined by the design rule in Figure 5.16. In this figure, 8-CSK data mapping is also shown. Three bits are assigned per symbol. Points I, J, and K show the center of the three color bands on x–y color coordinates. S0 to S7 are the eight symbol points of 8-CSK.

16-CSK constellation design

The 16-CSK symbol points are defined by the design rule in Figure 5.17. In this figure, 16-CSK data mapping is also shown. Four bits are assigned per symbol. Points I, J, and K show the center of the three color bands on x–y color coordinates. S0 to S15 are the 16 symbol points of 16-CSK.

In CSK modulation, binary data (zeros and ones) are transformed into xy values, according to a mapping rule on the x–y color coordinates by the color coding block. The points on the x–y coordinate are then converted to (R, G, B) values which represent the intensity of the red, green, and blue light emitted

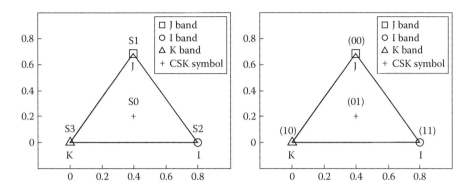

FIGURE 5.15
4-CSK constellation and data mapping.

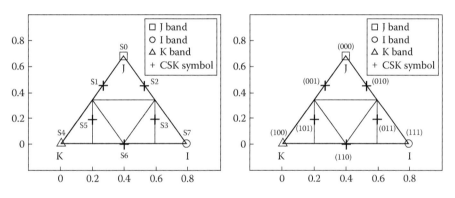

FIGURE 5.16
8-CSK constellation and data mapping.

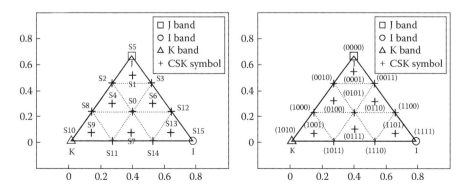

FIGURE 5.17
16-CSK constellation and data mapping.

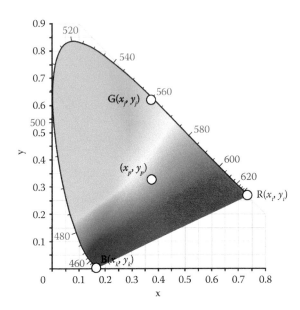

FIGURE 5.18
CIE 1931 x–y color coordinates.

from the RGB LED, respectively. According to Figure 5.18, the points (x_i, y_i), (x_j, y_j), and (x_k, y_k) show the x–y coordinates of three light sources. The point (x_p, y_p) shows one allocated color point in 4-CSK. The color point (x_i, y_i) can be represented by the normalized intensity of the three light sources P_i, P_j, and P_k. The relationship between the coordinates and the intensities is given by the following set of equations:

$$x_p = P_i x_i + P_i x_j + P_k x_k$$
$$y_p = P_i y_i + P_j y_j + P_k y_k$$
$$P_i + P_j + P_k = 1$$

At the receiver side, xy values are calculated from the received intensities of three colors, and then they are decoded into the received data.

Table 5.9 shows color band combination and the x–y coordinate values when color codes (110, 010, 000) are used. Figure 5.18 shows the CIE1931 x–y color coordinates (CIE, 1931) with the color mapping for 4-CSK. In this case, four color points are defined.

5.4.1.2 Forward Error Correction Coding

IEEE 802.15.7 supports various forward error-correcting (FEC) schemes that work reasonably well in the presence of hard decisions that would be

TABLE 5.9

Color Band Combination Example for (110, 010, 000)

Color-Band Combination		x-y Coordinates Values of Symbols		
Color Codes	Center of Band (x,y)	4-CSK[data] – (x_p, y_p)	8-CSK[data] – (x_p, y_p)	16-CSK[data] – (x_p, y_p)
110	(0.730 0.270)	[0 0] – (0.190 0.780)	[0 0 0] – (0.190 0.780)	[0 0 0 0] – (0.190 0.780)
010	(0.190 0.780)	[0 1] – (0.367 0.353)	[0 0 1] – (0.187 0.523)	[0 0 0 1] – (0.249 0.638)
000	(0.180 0.010)	[1 0] – (0.180 0.010)	[0 1 0] – (0.370 0.610)	[0 0 1 0] – (0.187 0.523)
		[1 1] – (0.730 0.270)	[0 1 1] – (0.519 0.383)	[0 0 1 1] – (0.370 0.610)
			[1 0 0] – (0.180 0.010)	[0 1 0 0] – (0.246 0.381)
			[1 0 1] – (0.244 0.253)	[0 1 0 1] – (0.367 0.353)
			[1 1 0] – (0.455 0.140)	[0 1 1 0] – (0.429 0.468)
			[1 1 1] – (0.730 0.270)	[0 1 1 1] – (0.426 0.211)
				[1 0 0 0] – (0.183 0.267)
				[1 0 0 1] – (0.242 0.124)
				[1 0 1 0] – (0.180 0.010)
				[1 0 1 1] – (0.363 0.097)
				[1 1 0 0] – (0.550 0.440)
				[1 1 0 1] – (0.609 0.298)
				[1 1 1 0] – (0.547 0.183)
				[1 1 1 1] – (0.730 0.270)

TABLE 5.10

Generator Polynomials

(n, k)	Polynomials g(x)
(15, 11)	$x^4 + \alpha^{13}x^3 + \alpha^6 x^2 + \alpha^3 x + \alpha^{10}$
(15, 7)	$x^8 + \alpha^{14}x^7 + \alpha^2 x^6 + \alpha^4 x^5 + \alpha^2 x^4 + \alpha^{13}x^3 + \alpha5 x^2 + \alpha^{11}x^1 + \alpha^6$
(15, 4)	$x^{11} + \alpha^9 x^{10} + \alpha^8 x^9 + \alpha^4 x^8 + \alpha^9 x^7 + \alpha^{13}x^6 + \alpha^4 x^5 + \alpha^{12}x^4 + \alpha^4 x^3 + \alpha^5 x^2 + \alpha^3 x + \alpha^6$
(15, 2)	$x^{13} + \alpha^3 x^{12} + \alpha^8 x^{11} + \alpha^9 x^{10} + \alpha^2 x^9 + \alpha^4 x^8 + \alpha^{14}x^7 + \alpha^6 x^6 + \alpha^{10}x^5 + \alpha^7 x^4 + \alpha1^3 x^3 + \alpha^{11}x^2 + \alpha^5 x + \alpha$

generated by the clock and data recovery (CDR). The channel codes support both long and short data frames for high data rate indoor and low data rate outdoor applications. For outdoor applications, stronger codes using concatenated RS and CC are developed to overcome the additional path loss due to longer distances and potential interference introduced by optical noise sources such as daylight and fluorescent lighting. RS and CC are preferred over advanced coding schemes such as low-density parity-check (LDPC) codes in order to support short data frames, hard decision decoding, low complexity, and their ability to interface well with RLL line codes. For indoor applications, where the coding requirements are less stringent for short distances, RS codes are used for FEC since they are better suited to high data rate implementations. RS codes also interface well in conjunction with the RLL line codes, where the errors detected from the RLL line code at the receiver could be marked as erasures to the RS decoder, providing performance improvements of around 1 dB (Rajagopal and Lim, 2011).

1. Reed–Solomon Coding

In coding theory, RS codes are non–binary cyclic error-correcting codes which provide a systematic way of building codes that could detect and correct multiple random symbol errors. In RS coding, source symbols are viewed as coefficients of a polynomial g(x) over a finite field called a Galois field (GF).

For the PHY I outer FEC, systematic RS codes are used with $GF(2^4)$, generated by the polynomial $x^4 + x + 1$. The generators for the RS(n, k) codes for PHY I are given in Table 5.10 where is α a primitive element in $GF(2^4)$.

For PHY II, a systematic RS code operating on $GF(2^8)$ is used to correct errors and increase the system reliability. The RS code is defined over $GF(2^8)$ with a primitive polynomial $x^8 + x^4 + x^3 + x^2 + 1$.

The RS code may be shortened for the last block if it does not meet the block size requirements. No zero padding is required for the RS code. A shortened RS code is used for frame sizes not matching code word boundaries via the following operation to minimize padding overhead. Starting with an RS(n, k) code, one can get an RS(n–s, k–s) shortened code as follows:

   a. Pad the k–s RS symbols with zero RS symbols.
   b. Encode using RS(n, k) encoder.
   c. Delete the padded zeros (do not transmit them).
   d. At the decoder, add the zeros, then decode.

2. Convolutional Coding

Rate 1/3 Code

The inner code is a based on a rate 1/3 mother convolutional code of constraint length seven ($K = 7$) with generator polynomial $g_0 = 1338$; $g_1 = 1718$; $g_2 = 1658$, as shown in Figure 5.19. Six tail bits of zeros need to be added at the end of the encoding in order to terminate the convolutional encoder to an all-zeros state. The tail bit of zeros are applied to both the header and the payload when the inner CC is used.

Rate 1/4 Code

The rate 1/4 code is obtained by puncturing the rate 1/3 mother code to a rate 1/2 code, as shown in Figure 5.20, and then using a simple repetition code as shown in Figure 5.21.

Rate 2/3 Code

The rate 2/3 code is obtained by puncturing the rate 1/3 mother code, as in Figure 5.22.

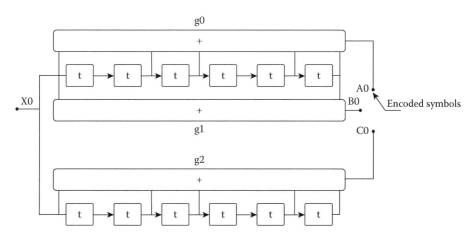

FIGURE 5.19
Rate 1/3 mother CC with constraint length 7.

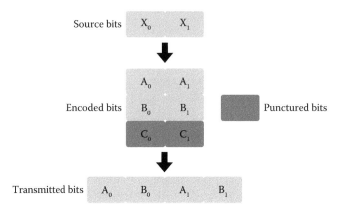

FIGURE 5.20
Puncturing pattern to obtain rate 1/2 code.

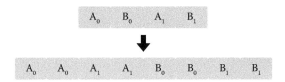

FIGURE 5.21
Repetition pattern used to obtain the effective rate 1/4 code.

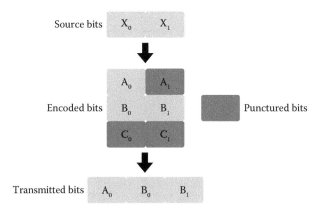

FIGURE 5.22
Puncturing pattern to obtain rate 2/3 code.

5.4.1.3 Interleaving

A block interleaver is used as an interleaver between the inner CC and the outer RS code as shown in Figure 5.23. The interleaver is of a fixed height n but has a flexible depth D, dependent on the frame size. The flexible depth of the interleaver and the puncturing block after it is used to minimize padding overhead.

Table 5.11 introduces the parameters used to describe the interleaver.

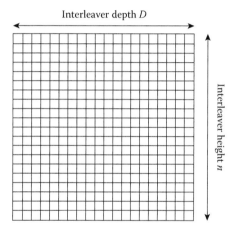

Interleaver depth D

Interleaver height n

FIGURE 5.23
Interleaver for PHY I.

TABLE 5.11

Parameters of Interleaver

Parameter	Description
n	RS codeword length
k	Number of information data symbols in an RS codeword
q	Number of elements in the Galois field: GF(q)
L_{frame}	Input frame size in bytes
S_{frame}	Number of symbols at the input of the RS encoder
S	Number of symbols from the output of the shortened RS encoder
S_{block}	The size of the interleaver used
D	The interleaving depth
i	Ordered indices take the values 0, 1, …, $S_{block}-1$
$l(i)$	Interleaved indices
p	Number of zero RS symbols
t	Ordered indices take the values 0, 1, …, p
$z(t)$	Locations of the bits to be punctured at the output of the interleaver before transmission

TABLE 5.12

Manchester Encoding

Bit	Manchester Symbol
0	01
1	10

The length of the frame is communicated to the receiver in the header so the receiver can adaptively adjust the interleaver based on the frame sizes. When the data rates corresponding to transmissions using the concatenated codes are used, the header is also interleaved according to the procedure shown in above equations. Since the length of the header is fixed, the receiver can deinterleave the header without explicit transmission of the header length.

5.4.1.4 Line Coding

RLL line codes are used to avoid long runs of 1s and 0s that could potentially cause flicker and CDR detection problems. RLL line codes take in random data symbols at input and guarantee DC balance with equal 1s and 0s at the output for every symbol. Various RLL line codes such as Manchester, 4B6B, and 8B10B are defined in the standard and provide trade-offs between coding overhead and ease of implementation.

1. **Manchester Coding**

   According to the IEEE 802.15.7 standard, all OOK PHY I modes should use Manchester DC balancing encoding. The Manchester code expands each bit into an encoded 2-bit symbol as shown in Table 5.12.

2. **RLL Coding**

   4B6B RLL Coding

   According to the IEEE 802.15.7 standard, all VPPM PHY I and II modes will use 4B6B encoding. The 4B6B expands 4-bit to 6-bit encoded symbols with DC balance as illustrated in Figure 5.24. The counts of 1 and 0 in every VPPM encoded symbol is always equal to 3. Table 5.13 defines the 4B6B code.

   The features of the 4B6B code are as follows:

   - Always 50% duty cycle during one encoded symbol
   - DC-balanced RLL code
   - Error detection capability
   - Run length is limited to four
   - Allows reasonable clock recovery

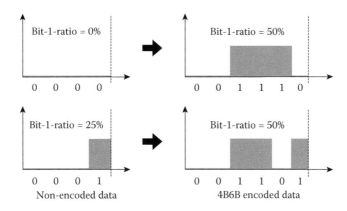

FIGURE 5.24
Comparison between nonencoded and 4B6B encoded symbols.

TABLE 5.13

Mapping Input 4B to Output 6B

Hex	4B (Input)	6B (Output)
0	0000	001110
1	0001	001101
2	0010	010011
3	0011	010110
4	0100	010101
5	0101	100011
6	0110	100110
7	0111	100101
8	1000	011001
9	1001	011010
A	1010	011100
B	1011	110001
C	1100	110010
D	1101	101001
E	1110	101010
F	1111	101100

8B10B RLL Coding

According to the IEEE 802.15.7 standard, all OOK PHY II modes need to use 8B10B encoding as specified in ANSI/INCITS 373. The 8B10B line code converts 8-bit to 10-bit as illustrated in Figure 5.25. To construct an 8B10B code, we can compose the code from compatible

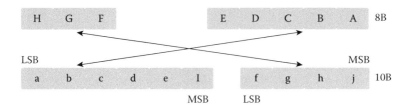

FIGURE 5.25
8B10B encoding structure.

TABLE 5.14

Scrambler Seed Selection

TDP	Seed Value $x_{int} = x[x_i[-1]x_i[-2]...x_i[-14]x_i[-15]]$	PRBS Output First 16 Bits $x[0]x[1]...x[15]$
P1	0011 1111 1111 111	0000 0000 0000 1000
P2	0111 1111 1111 111	0000 0000 0000 0100
P3	1011 1111 1111 111	0000 0000 0000 1110
P4	1111 1111 1111 111	0000 0000 0000 0010

but separate 5B6B and 3B4B codes. The original 8-bit data are separated into two parts: first three bits and last five bits. The first three bits are encoded using the 3B4B RLL encoding scheme and the last five bits are encoded by 5B6B RLL encoding scheme to the output bits.

5.4.1.5 Scrambling

A scrambler should be used to ensure pseudorandom data for the PHY II. The scrambler is applied to the entire PHY service data unit (PSDU). In addition, the scrambler is initialized to a seed value dependent on the topology dependent pattern (TDP) at the beginning of the PSDU.

The 15-bit initialization vector or seed value should correspond to the seed identifier as defined in Table 5.14 and illustrated in Figure 5.26, corresponding to the TDP. The seed values need to be incremented in a rollover fashion for each frame sent by the PHY. For example, if the seed value used is the seed corresponding to P3 in the first frame, the seed value corresponding to P4 is used in the second frame; seed value corresponding to P1 is used in the third frame and so on. All consecutive frames, including retransmissions, are sent with a different initial seed value.

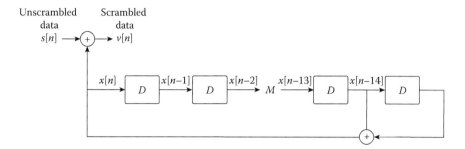

FIGURE 5.26
Scrambler block diagram.

5.4.2 System Models

The IEEE 802.15.7 standard supports three PHY types, namely PHY I, PHY II, and PHY III. PHY I and PHY II are defined for a single light source, and support OOK and VPPM. PHY III is able to support multiple optical sources using CSK. An IEEE 802.15.7-compliant device must implement either PHY I or PHY II types. A device implementing the PHY III type also needs to implement the PHY II mode for coexistence. Details on the optical clock rates, data rates, and error correction codes for each PHY type are summarized in Table 5.3. It is noted from these tables that multiple optical rates are provided for all PHY types in order to support a broad class of LEDs for various applications. The choice of optical rate used for communication is decided by the MAC layer during device discovery. In the following, each PHY type is described.

5.4.2.1 System Model for PHY I

The block diagram of a VLC system with PHY I type is illustrated in Figure 5.27. The input bits are first fed to an (n, k) RS encoder which encodes k-symbol codewords to messages of having n symbols each. The encoding is based on a generator polynomial in $GF(2^m)$ where m denotes the number of bits per symbol. The encoder output is padded with zeros to form an interleaver boundary. The padded zeros are then punctured and fed to a convolutional encoder. Next, through an RLL encoder, data are encoded either with Manchester or 4B6B codes. The former expands each bit into an encoded 2-bit symbol and the later expands 4-bit to 6-bit encoded symbols, both with DC balance. The encoded bits are finally modulated with OOK or VPPM.

At the receiver side, after demodulating the received signal using a threshold detector, the resulting bits are first fed to the RLL decoder, then the Viterbi decoder. The output symbols of the Viterbi decoder are sent to a deinterleaver and the added zeros are removed. At the final stage, an RS decoder generates the final bitstream. Different operating modes of PHY I are summarized in Table 5.3.

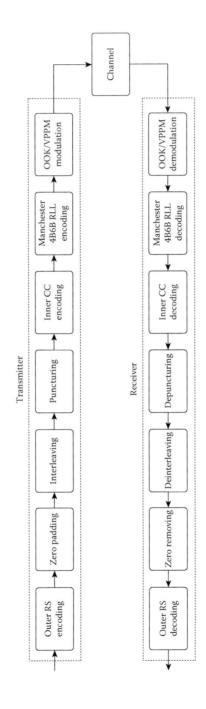

FIGURE 5.27
Block diagram of a PHY I-type VLC system.

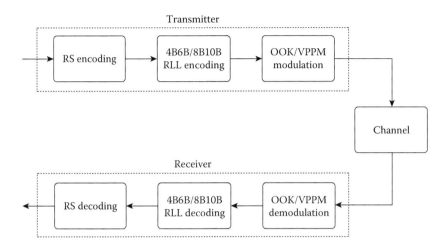

FIGURE 5.28
Block diagram of a PHY II type VLC system.

5.4.2.2 System Model for PHY II

The block diagram of a VLC system with PHY II type is provided in Figure 5.28. The input bits are first fed to an (n,k) RS encoder. The RS encoded output is then sent to an RLL encoder which uses 4B6B or 8B10B encoding for OOK and VPPM modes, respectively. At the receiver side, after demodulating the received signal, symbols are decoded first by RLL and then by the RS decoder. Different operating modes in PHY II are summarized in Table 5.3.

5.4.2.3 System Model for PHY III

The block diagram of a VLC system with PHY III type is provided in Figure 5.29. At first, the input bitstream is sent to a scrambler and converted into a random bitstream avoiding long sequences of the same value in the stream. Then it passes through an RS encoder. After scrambling and channel coding, binary data are modulated using CSK.

In CSK modulation, binary data are first parsed into groups of $\log M$ where M is the modulation size. Each modulation symbol is mapped into x and y values. Three of the modulation symbols are the three vertices of the constellation triangle. Other symbols are then placed within this triangle to form 4-CSK, 8-CSK, or 16-CSK constellations using different constellation design methods. The points on the x–y coordinate are then converted to (R, G, B) values which represent the intensity of the red, green, and blue light emitted from the RGB LED, respectively. In PHY III, the information is transmitted via these three normalized intensities. At the receiver side, three photodetectors with different wavelength ranges are used to detect the intensity of

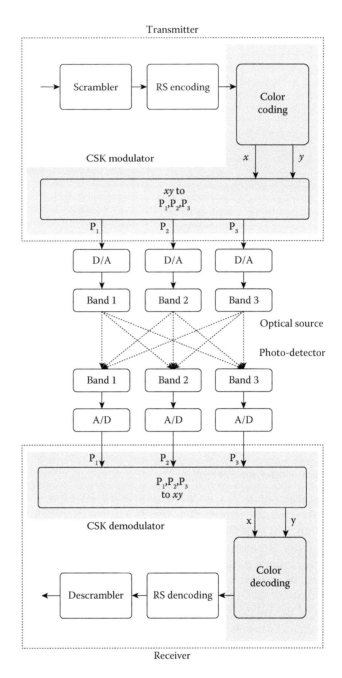

FIGURE 5.29
Block diagram of a PHY III type VLC system.

each color band (intensities of red, blue, and green components of the received light signal). The received intensities are then inversely mapped to points in the x–y color coordinate. A minimum distance detector can then be used to find the corresponding constellation symbols and the original binary data. Different operating modes in PHY III are summarized in Table 5.3.

5.4.3 Performance Evaluation of PHY Layer

In this section, we present results for the BER performance evaluation of PHY I, PHY II, and PHY III types through Monte Carlo simulation. The multipath channel model used in the simulations can be found in Sarbazi et al. (2014).

In Figure 5.30, the BER of PHY I modes with OOK modulation and f_{clk} = 200 kHz (modes a, b, c, d, and e) are provided. At a targeted BER of 10^{-3}, signal-to-noise ratio (SNR) of 2.86 dB is required for the uncoded system (mode PHY I.e). For a coded system with RS(15,11) and rate 2/3 CC, the required SNR decreases to –0.3739 dB. Comparing the modes b and c, as the CC rate increases, the required SNR for fulfilling the BER of 10^{-3} also increases. In Figure 5.31, the BER of PHY I modes with VPPM modulation and f_{clk} = 400 kHz (modes f, g, h, and i) are compared. Comparing these modes, as the RS code rate increases, the required SNR for fulfilling the BER of 10^{-3} also increases. In other words, RS(15,2) provides more performance improvement than RS (15,4). It can also be concluded that VPPM is more robust than OOK.

In Figure 5.32, the BER of PHY II modes with VPPM modulation and f_{clk} = 3.75 MHz (modes a and b) are compared. Comparing these modes, as the RS code rate increases, the required SNR for fulfilling the BER of 10^{-3} also increases. In other words, RS(64,32) provides more performance improvement than RS(160,128). In Figure 5.33, the BER of PHY II modes with VPPM modulation and f_{clk} = 7.5 MHz (modes c, d, and e) are compared. At a targeted BER of 10^{-3}, SNR = 3.25 dB is required for the uncoded PHY II (mode PHY II.n), which is greater than the required SNR when using RS(64,32) in mode c (SNR = –0.21 dB) and RS(160,128) in mode d (SNR = 0.96 dB). In Figures 5.34 through 5.37, the BER of PHY II modes with OOK modulation are provided. Similar to the previous case, it can be generally stated that OOK is less robust than VPPM.

Figure 5.38 presents the BER performance of PHY III. Similar to PHY II, this type deploys only RS coding. For example, with an RS(64,32) and 16-CSK modulation, an SNR of 32.81 dB is required to obtain BER = 10^{-3}. In Figure 5.39, the BER of PHY III modes with f_{clk} = 12 MHz (modes a and b) are compared. Comparing these modes, as the modulation size increases, the required SNR for fulfilling the BER of 10^{-3} also increases. In other words, 4-CSK provides more performance gains than 8-CSK. In Figure 5.40, the BER of PHY III modes with f_{clk} = 24 MHz (modes c, d, e, f, and g) are compared. At a targeted BER of 10^{-3}, SNR = 38.38 dB and SNR = 39.79 dB are required for the uncoded PHY III modes f and g,

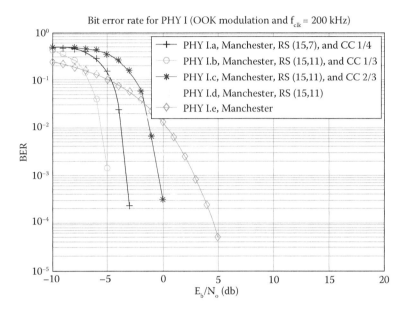

FIGURE 5.30
PHY I modes with OOK modulation and f_{clk} = 200 kHz.

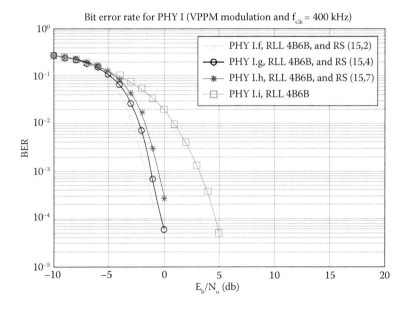

FIGURE 5.31
PHY I modes with VPPM modulation and f_{clk} = 400 kHz.

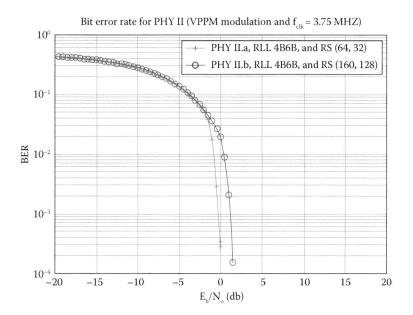

FIGURE 5.32
PHY II modes with VPPM modulation and f_{clk} = 3.75 MHz.

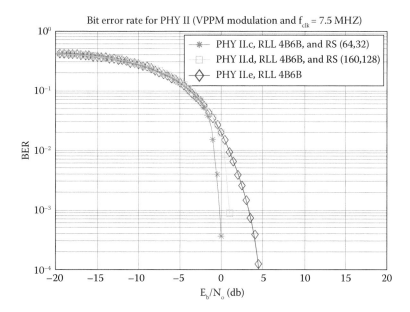

FIGURE 5.33
PHY II modes with VPPM modulation and f_{clk} = 7.5 MHz.

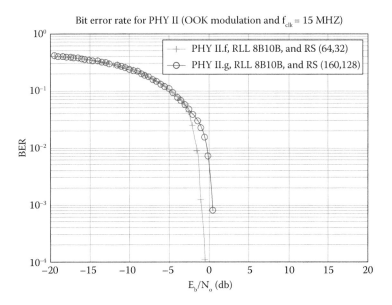

FIGURE 5.34
PHY II modes with OOK modulation and f_{clk} = 15 MHz.

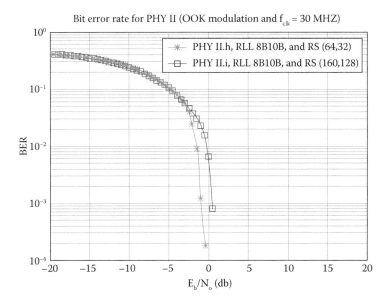

FIGURE 5.35
PHY II modes with OOK modulation and f_{clk} = 30 MHz.

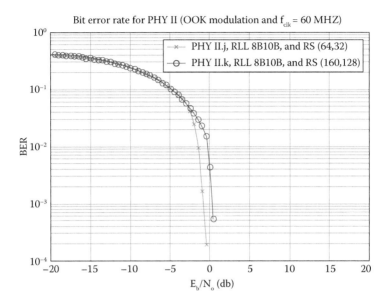

FIGURE 5.36
PHY II modes with OOK modulation and f_{clk} = 60 MHz.

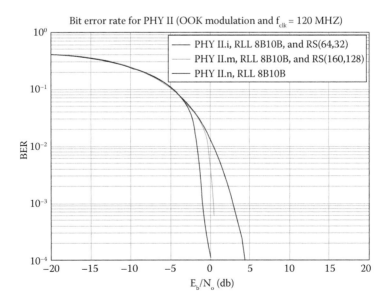

FIGURE 5.37
PHY II modes with OOK modulation and f_{clk} = 120 MHz.

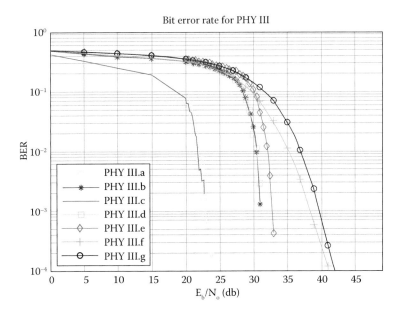

FIGURE 5.38
Bit error rate for PHY III.

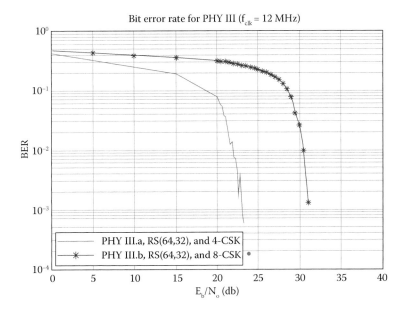

FIGURE 5.39
Bit error rate for PHY III (f_{clk} = 12 MHz).

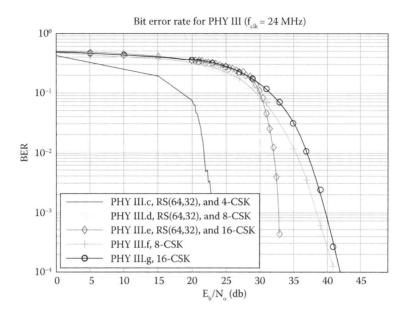

FIGURE 5.40
Bit error rate for PHY III (f_{clk} = 24 MHz).

respectively, which are greater than required SNR values for coded systems with the same modulation (modes d and e, respectively). It is also concluded that, as the alphabet size of the modulation increases, performance decreases. In general, it can be observed that modes a and c with 4-CSK modulation outperform the modes b and d with 8-CSK modulation in terms of BER. Therefore, the cases with 4-CSK modulation have a better performance in terms of BER but there is a trade-off between this BER and the data rate.

5.5 Recent Activities in IEEE Standardization

In 2013, the IEEE 802.15 working group (WG) decided to investigate if an amendment to the standard is necessary by establishing a study group. The discussions within the group and suggestions from industry and academia led to a project authorization request which states:

This amendment defines a physical layer (PHY) ... using light frequencies over the spectral range of 10,000 nm (infrared [IR]) to 190 nm (near ultraviolet [UV]) and any MAC changes specifically required to support this PHY. Transmitting devices include such sources as displays, typically found on cameras and mobile devices, and other LED based sources such as flashes, flashlights, LED tags, LED/laser sources. (IEEE P802.15.7r1, 2016)

The acceptance of the project authorization request by the IEEE Standards Association enabled the IEEE 802.15 WG to work on a new standard which is open to almost any type of VLC communication. A technical requirements document has been prepared to guide prospective standard proposals (Jang et al, 2015). The document uses the term optical wireless communication (OWC) and classifies OWC into:

- Image sensor communications
- Low-rate photodiode communications
- High-rate photodiode communications

In regard to the definition of low speed and high speed, the throughput threshold data rate is 1 Mbps as measured at the PHY layer output of the receiver. Throughput of less than 1 Mbps rate is considered low rate and higher than 1 Mbps is considered high rate. The group determined possible applications that can be served by each communication type. Image sensor communications enable OWCs using an image sensor as a receiver. Main applications of image sensor communications are listed as:

- Offline to online marketing/public information system/digital signage
- Internet of Things (device-to-device/Internet of light [IoL])
- Location-based services/indoor positioning
- Vehicular communication/vehicular positioning
- Underwater communication
- Point-to-(multi)point/relay communication

Low-speed photodiode receiver communications, which is a wireless light ID system using various LEDs with a low-speed photodiode receiver, can be used in the below applications:

- Underwater/seaside communication
- Point-to-(multi)point/communication
- Digital signage
- Internet of Things (device-to-device/Internet of light [IoL])
- LOS authentication
- Identification based services

The high-speed photodiode receiver communications is high-speed, bidirectional, networked, and mobile wireless communications using light with a high-speed photodiode receiver. Main applications for high-speed photo-diode receiver communications are:

- Indoor office/home applications (conference rooms, general offices, shopping centers, airports, railways, hospitals, museums, aircraft cabins, libraries, etc.)
- Data center/industrial establishments, secure wireless (manufacturing cells, factories, hangers, etc.)
- Vehicular communications (vehicle-to-vehicle, vehicle-to-infrastructure)
- Wireless backhaul (small cell backhaul, surveillance backhaul, LAN bridging)

Another task of the group was to determine if a channel model is necessary to compare different standard proposals. The group decided that all proposals which include the PHY algorithms for the high-rate PD communications must use the channel impulse responses provided in TG7r1 Channel Model Document for High-rate PRD Communications (Jang et al. 2015) for the specific scenario that they intend to address in their proposal. The exact channel impulse responses are provided in *TG7r1 CIRs Channel Model Document for High-rate PD Communications* (Uysal et al. 2016). The task group aimed to receive proposals for each type of OWC in 2016; the planned finalization date of the standard is 2018.

5.6 Conclusions

In this chapter, we have first provided an overview of the IEEE 802.15.7 Visible Light Communication standard describing the main features of PHY and MAC layers. Then, we have presented simulation results to demonstrate key performance metrics. At the PHY layer, we have investigated the BER performance and compared the performance of different PHY types. At the MAC layer, we have evaluated key network parameters such as throughput, transmission, delay, packet drop, and collision probability. We have concluded the chapter with a brief overview of ongoing IEEE standardization activities.

Acknowledgments

This work is carried out as an activity of the Centre of Excellence in Optical Wireless Communication Technologies (OKATEM) funded by Istanbul Development Agency (ISTKA) under the Innovative Istanbul

Financial Support Program, 2015. The statements made herein are solely the responsibility of the authors and do not reflect the views of ISTKA and/or T.R. Ministry of Development.

References

[1] Acolatse K., Bar-Ness Y. and Wilson S.K. 2011. Novel techniques of single-carrier frequency-domain equalization for optical wireless communications. *EURASIP J. Adv. Signal Process.*, vol. 11, pp. 1–13.

[2] Bykhovsky D. and Arnon S. 2014. Multiple access resource allocation in visible light communication systems. *J. Lightw. Technol.*, vol. 32, no. 8, pp. 1594–1600.

[3] CIE. Commission Internationale de l'E'clairage Proceedings. Cambridge University Press, USA, 1931.

[4] Elgala H. and Little T. 2015. Polar-based OFDM and SC-FDE links toward energy-efficient Gbps transmission under IM-DD optical system constraints. *IEEE/OSA J. Opt. Commun. Network.*, vol. 7, no. 2, pp. 277–284.

[5] Fernando N., Hong Y. and Viterbo E. 2012. Flip-OFDM for unipolar communication systems. *IEEE Trans. Commun.*, vol. 60, no. 12, pp. 3726–3733.

[6] Gancarz J., Elgala H. and Little T.D.C. 2013. Impact of lighting requirements on VLC systems. *IEEE Commun. Mag.*, vol. 51, no. 12, pp. 34–41.

[7] Hong Y., Chen J., Wang Z. and Yu C. 2013. Performance of a precoding MIMO system for decentralized multiuser indoor visible light communications. *IEEE Photon. J.*, vol. 5, no. 4.

[8] Hsu C.W., Chow C.W. and Yeh C.H. 2015. Cost-effective direct-detection all-optical OOK-OFDM system with analysis of modulator bandwidth and driving power. *IEEE Photon. J.*, vol. 7, no. 4, pp. 1–7.

[9] Hussein A.T. and Elmirghani J.M. 2015. Mobile multi-gigabit visible light communication system in realistic indoor environment. *J. Lightw. Technol.*, vol. 33, no. 15, pp. 3293–3307.

[10] IEEE. 2011. IEEE standard for local and metropolitan area networks. Part 15.7: Short Range Wireless Optical Communication using Visible Light, 802.15.7.

[11] IEEE. 2014. The IEEE P802.15.7r1 Short-Range Optical Wireless Communications Task Group Project Authorization Request (PAR). Available at: https://mentor.ieee.org/802.15/dcn/15/15-15-0064-00-0007-p802-15-7-revisionpar-approved-2014-12-10.pdf (accessed January 16, 2016).

[12] Jovicic A., Li J. and Richardson T. 2013. Visible light communication: opportunities, challenges and the path to market. *IEEE Commun. Mag.*, vol. 51, no. 12, pp. 26–32.

[13] Kashef M., Abdallah M., Qaraqe K., Haas H. and Uysal M. 2015. Coordinated interference management for visible light communication systems. *IEEE/OSA J. Opt. Commun. Network*, vol. 7, no. 11, pp. 1098–1107.

[14] Kizilirmak R.C., Narmanlioglu O. and Uysal M. 2015. Relay-assisted OFDM-based visible light communications. *IEEE Trans. Commun.*, vol. 63, no. 10, pp. 3765–3778.

[15] Le N.-T. and Jang Y.M. 2015. Technical Considerations Document. IEEE 802 15-15-0492-03. Available at: https://mentor.ieee.org/802.15/dcn/15/15-15-0492-05-007a-technical-considerations-document.docx (accessed January 16, 2016).

[16] Lee K., Park H. and Bary J. R. 2011. Indoor channel characteristics for visible light communications. *IEEE Commun. Lett.*, vol. 15, no. 2, pp. 217–219.
[17] Li X., Zhang R. and Hanzo L. 2015. Cooperative load balancing in hybrid visible light communications and Wi-Fi. *IEEE Trans. Commun.*, vol. 63, no. 4, pp. 1319–1329.
[18] Mesleh R., Elgala H. and Haas H. 2011. On the performance of different OFDM based optical wireless communication systems. *IEEE/OSA J. Opt. Commun. Network.*, vol. 3, pp. 620–628.
[19] Miramirkhani F. and Uysal M. 2015. Channel modeling and characterization for visible light communications. *IEEE Photon. J.*, vol. 7, no. 6, pp. 1–16.
[20] Nuwanpriya A., Ho S., Zhang J., Grant A. and L. Luo. 2015. PAM-SCFDE for optical wireless communications. *J. Lightw. Technol.*, vol. 33, no. 14, pp. 2938–2949.
[21] Rajagopal R.D.R.S. and Lim S.K. 2011. IEEE 802.15.7 Physical Layer Summary. Globecom Workshops, Houston, 2011, pp. 772–776.
[22] Rajagopal R.D.R.S. and Lim. S.K. 2012. IEEE 802.15.7 visible light communication: Modulation schemes and dimming support. *IEEE Commun. Mag*, vol. 50, pp. 72–82.
[23] Sarbazi E., Uysal M., Abdallah M. and Qaraqe K. 2014. Ray tracing based channel modeling for visible light communications. IEEE 22nd Signal Processing, Communication and Applications Conference (SIU), Trabzon, Turkey.
[24] Uysal M., Baykas T., Miramirkhani F. and Jungnickel V. 2016. TG7r1 Channel Model Document for High-rate PD Communications. Available at: https://mentor. ieee.org/802.15/dcn/15/15-15-0746-01-007a-tg7r1-channel-model-document-for-high-rate-pd-communications.pdf (accessed January 16, 2016).

6

Techniques for Enhancing the Performance of VLC Systems

Hoa Le Minh, Wasiu O. Popoola, and Zhengyuan Xu

CONTENTS

This chapter discusses the techniques for enhancing the performance of visible light communications (VLC) systems. It begins with the introduction to a parallel data transmission technique that helps to increase the system transmission capacity. This technique exploits parallel data transmission using multiple light-emitting diodes (LEDs) that are generally used in homes and offices for lighting. Parallel transmission is based on multiple-input multiple-output (MIMO) communications systems where the VLC channel matrix is predetermined to separate and recover multiple data streams. In this chapter, we will present the background theory for this approach and discuss how to recover the transmitted data at the receiver. Case study, simulation, and practical results will be presented to illustrate the MIMO system performance.

As discussed in Chapter 5, VLC capacity (or throughput) can also be enhanced with the use of orthogonal frequency division multiplexing (OFDM). To realize this however, the problematic high signal peak in OFDM has to be resolved. The second part of this chapter presents a viable technique for addressing this problem.

6.1 Multiple-Input Multiple-Output

Office and home lighting sources are currently evolving from the traditional fluorescent and incandescent sources to modern energy-saving light bulbs, and now to solid-state lighting using LEDs [1]. This trend has been spurred on through global awareness of the urgent need to reduce the size of our carbon footprint. The introduction of solid-state LED lighting has attracted the attention of communications engineers worldwide, enabling the achievement of the dual functionality of room illumination while simultaneously transmitting wireless data via VLC [2–4]. Currently the bulk of reported VLC research relates to increasing data rates and demonstrating that the idea of combining data transmission with illumination is actually viable. As discussed in the previous chapters, white illumination LEDs are based on combining red, green, and blue (RGB) LED chips integrated on a single package or using a single blue LED chip with a yellowish phosphor coating (YB). The latter option is quite popular because of the reduced complexity of the driving circuitry and it does not require any color balancing. However, the modulation bandwidth (BW) of the phosphor-converted white LED is typically 3–4 MHz; this is considerably lower than that of an RGB-based white LED.

In the view of this BW limitation, different approaches to increase data rate have been reported. These mainly focus on two areas: (i) equalization to extend the modulation bandwidth using analog/digital filters and (ii) the use of complex modulation such as quadrature amplitude modulation (QAM) with multicarrier modulation to effectively utilize the limited LED bandwidth. Further work on increasing the BW is reported in [5,6] with the use of pre- and post-emphasis on the data signals through passive resistor–capacitor (RC) equalization. This technique, although low cost and very successful, when used to increase the BW by more than a certain order of magnitude can reduce the possible transmission distance to only a few centimeters [5].

Spectrally efficient modulation schemes that also take advantage of the high signal-to-noise ratio (SNR) available in VLC systems have been employed to increase the data throughput [7–10]. OFDM and discrete multi-tone (DMT) modulation have been shown to produce data rates of up to

1 Gb/s [9] using complex power loading and frequency-domain equalization (FDE) algorithms. Large-scale Fourier transforms (IFFT and FFT) are also required (2048 points for 512 subcarriers) at such high data rates, resulting in the use of large amounts of computer processing. These systems are hence mostly possible only with offline processing; they are not yet demonstrated for real-time applications.

Another approach to increase the data rate is available through the use of the several lighting units typically employed to supply full room illumination. By simultaneously transmitting parallel data streams through the M independent available units, data rates can then be linearly aggregated. MIMO systems for VLC applications have been studied recently [11–13]. The VLC MIMO technique mainly uses two approaches, namely: (i) nonimaging and (ii) imaging systems. In the first approach, each receiver will capture the combined signals from all transmitters whereas in the imaging system each receiver will receive signal from only one transmitter with the aid of optical lenses.

In [11], a comprehensive simulation of a 4×4 nonimaging VLC MIMO was presented. The number of small LEDs in each transmitter unit was set to 3600, which seems an unrealistic amount for existing commercial deployment. Furthermore, Zeng et al. [11] found a forbidden area on the receiving plain where decomposition of the transmitted signals cannot occur due to symmetry in the system geometry.

An experiment with a 4×9 imaging VLC MIMO system and OFDM transmission has been reported in [12] with an aggregate data rate of 1 Gb/s, bit error rate (BER) of 10^{-3}, and transmission distance of 1 m. This also requires complex algorithms and much computing as previously mentioned for OFDM. In [13], an artificial neural network-based VLC MIMO nonimaging system with a data rate of 1.8 Mb/s was demonstrated. The link length in this paper was 10 cm and was based on an organic photodetector with a dynamic ~ 130–180 kHz bandwidth.

An alternative MIMO configuration is the spatial VLC technique described in [14] where just a single transmitter is active at any given time. While this removes the issue of channel matrix inversion, it does not allow for a significantly improved data rate due to the lack of parallel transmission and is therefore not an optimal solution for extremely fast VLC MIMO systems.

In this section, a conventional nonimaging MIMO system will be discussed. Data recovery techniques such as zero forcing (ZF), pseudoinversion, minimum mean square error (MMSE), and Vertical Bell Laboratories Layered Space Time (V-BLAST) algorithms will be presented. A practical MIMO system achieving 50 Mbit/s error-free transmissions within a standard-sized room with ISO standard illumination will be demonstrated.

6.1.1 VLC Nonimaging MIMO Channel Model and Detection Methods

A schematic for the nonimaging VLC MIMO system is shown in Figure 6.1. The incoming serial data X is split into t parallel transmit data streams

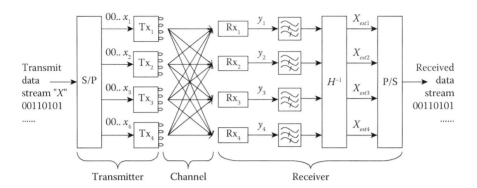

FIGURE 6.1
A typical 4 × 4 nonimaging VLC MIMO schematic.

x_j (j = 1,...,t). Each of the new data streams is used to intensity modulate an LED-based light source, Tx_j (j = 1,...,t). As all data streams are transmitted at the same time, the receiving array picks up cumulative signals composed of x_j. In order to demultiplex the signals and retrieve the transmitted data, the MIMO system has to first estimate the channel coefficients between a transmitter and a paired receiver and constructs a channel matrix. To achieve this, pilot signals are periodically inserted into the transmitted data. During the channel acquisition, a pilot signal is active in one channel while the remaining transmitters are silent, therefore each pilot is transmitted during a unique time slot as with [11,13]. When the pilot signals are received by each of the receivers, a channel matrix **H** detailing the received optical power is logged. The process is repeated until all transmitters send out all their pilots. It is only then that the transmitters can start the simultaneous broadcasting of data.

The general expression for the received cumulative signals y can be expressed by:

$$y = \mathbf{H}x + n \tag{6.1}$$

The elements in Equation 6.1 can be further expanded into:

$$
\begin{bmatrix} y_1 \\ y_2 \\ \vdots \\ y_r \end{bmatrix} =
\begin{bmatrix} h_{11} & h_{12} & \dots & h_{1r} \\ h_{21} & h_{22} & \dots & h_{2r} \\ \vdots & \vdots & \vdots & \vdots \\ h_{t1} & h_{t2} & \dots & h_{tr} \end{bmatrix}
\begin{bmatrix} x_1 \\ x_2 \\ \vdots \\ x_t \end{bmatrix} +
\begin{bmatrix} n_1 \\ n_2 \\ \vdots \\ n_r \end{bmatrix} \tag{6.2}
$$

with y_r being the signal at the rth receiver and n a vector denoting the additive white Gaussian noise. The elements h_{tr} within the matrix are the line of sight (LOS) DC gain between the tth transmitter and the rth receiver.

Assuming that all the LEDs have a Lambertian radiation pattern, the luminous intensity as a function of angle ϕ is therefore given by [15]:

$$I(\emptyset) = \frac{m+1}{2\pi} I \cos^m(\emptyset), \tag{6.3}$$

where I is the total luminous flux of the LED, m is the order of Lambertian radiation determined by the semiangle for half illuminance of an LED, $\Phi_{1/2}$ ($m = -\ln 2/\ln(\cos \Phi_{1/2})$) and ϕ is the angle of irradiance. Thus, the LOS DC channel gain is given by [15]:

$$h_{tr} = \frac{I(\emptyset)A_{det}}{d^2} \cos(\psi)g(\psi), \tag{6.4}$$

with A_{det} representing the active area of the PD, d is the distance between the LED and the detector, ψ is the angle of incidence, and $g(\psi)$ is the gain of the optical concentrator given by:

$$g(\psi) = \begin{cases} \dfrac{n_1^2}{\sin^2\Psi_c}, & 0 \le \psi \le \Psi_c \\ 0, & \psi \ge \Psi_c \end{cases} \tag{6.5}$$

where n_1 denotes the refractive index and Ψ_c denotes the angular field of view (FOV).

Data recovery process can be implemented by a number of approaches including:

6.1.1.1 Zero Forcing

The simplest method to calculate an estimate of the transmitted data would be to invert the **H** matrix and multiply by the received vector y known as a ZF receiver:

$$\mathbf{W}y = x_{est} + n, \tag{6.6}$$

where **W** is the beam former $\mathbf{H}^{-1}$. However, as can be seen from Equation 6.6, if the individual entries in **H** are small, the noise vector n increases leading to noise amplification [16].

6.1.1.2 Pseudoinverse

If **H** is rank deficient, matrix inversion cannot be performed. In such a case, the pseudoinverse of **H** can be used given by [17]:

$$\mathbf{H}^\dagger = (\mathbf{H}^{*H}\mathbf{H})^{-1}\mathbf{H}^{*H} \tag{6.7}$$

where $\mathbf{H}^{*H}$ is the conjugated transpose of **H**. An estimate of x can then be made by substituting Equation 6.7 into Equation 6.6; however, as before this will also result in noise amplification.

6.1.1.3 Minimum Mean Square Error

A regularized inversion of **H** can be performed using the MMSE detection method. This is designed to minimize the error between the received and transmitted vectors [17], and is resilient toward noise enhancement. The pseudoinverse **G** of **H** is chosen so that:

$$\mathbf{G} = \arg\,\min\{E\{||x - \mathbf{G}y||^2\}\}, \tag{6.8}$$

where E{.} denotes expectation and **G**y is the estimate of the transmitted vector. The matrix **G** can therefore be shown as [10]:

$$\mathbf{G} = \rho(\rho\mathbf{H}^{*H}\mathbf{H} + \sigma_n^2\mathbf{I}_N)^{-1}\mathbf{H}^{*H}, \tag{6.9}$$

where σ_n^2 is the receiver noise power variance, $\mathbf{I}_N$ is an identity matrix, and ρ is the average transmission power. An estimate of x can then be recovered, as in Equation 6.6 replacing **W** with **G** before processing.

6.1.1.4 Vertical Bell Laboratories Layered Space Time

The nonlinear detection method V-BLAST employs ordered successive interference cancelation (OSIC), as the impact of each estimated symbol is canceled from the received signal vector y [18,19]. It is a computationally intense iterative process whereby the pseudoinverse of **H** (Equation 6.7 or Equation 6.9), for example, is taken to estimate the symbols from the strongest signal, before canceling that symbol from the received vector. This effectively reduces the size of the **H** matrix from $r \times t$ to $r \times (t - 1)$ matrix. The pseudoinverse of the new **H** is then used to estimate the next symbol, after which the process is repeated until the last symbol has been found. The advantage for using this method is the increase in diversity as the iteration process progresses; however, any errors will propagate through each iteration.

Other methods for decoding y exist such as singular value decomposition (SVD) [20] involving precoding of the transmitted data, sphere decoding with maximum likelihood (ML) detector [21] as well as lattice-reduction-aided (LRA) detection [22]. However these techniques require channel state information available at the transmitter, which is not trivial and usually requires a feedback link, hence are not considered here.

6.1.2 MIMO System Setup

6.1.2.1 System Description

The system setup based on Figure 6.1 consists of four transmitters and four receivers. Without any loss of generality, the MIMO system can be practically configured as 2 × 2 transmitters and 2 × 2 receivers shown in Figure 6.2a and b, respectively. Each transmitter is composed of an array of four high-power white phosphor-converted LEDs acting as a single source, so they all carry

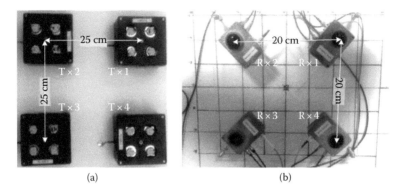

(a)   (b)

FIGURE 6.2
(a) 2 × 2 transmitter array and (b) 2 × 2 receiver array.

the same signal. Each receiver consists of a concentrator lens, PIN photodiode, and a transimpedance amplifier fed into a post amplifier before RC equalization. The transmitters use OOK with NRZ signaling and are fed with cyclic and independent pseudorandom data sequence of length 2^{10}-1 running at 12.5 Mb/s. At the receiver, the received signals are recorded on a real-time oscilloscope and are processed for data recovery. The system parameters are given in Table 6.1.

The root mean square delay spread (DRMS) investigation reported in [23] has indicated that negligible intersymbol interference (ISI) will occur at data rates below 230 Mb/s [23], therefore in the MIMO setup and processing the LOS component of the signal will be mainly considered.

6.1.2.2 MIMO System Performance

The four receivers are uniformly spaced 20 cm apart with the central position denoting the overall location of the receiving array relative to the receiving plane. This is to ensure adequate spatial diversity for the **H** matrix not be singular. At the receiving plane, which is 2 m below the transmitters, a grid of 5 × 5 cm^2 squares was set up for the measurements. The center point of the grid is aligned directly over the center of the four transmitters (with the Cartesian coordinates of [0,0]). For each receiver array position, a training sequence is transmitted to estimate the channel gain coefficients **H**. The noise variance σ_n^2 is taken from the measured received noise when no data signal is being sent.

Figure 6.3 shows the gross BER from all four channels of the MIMO system. Figure 6.4a demonstrates the error-free operation (BER 10^{-6}) within a coverage area of 400 cm^2 (20 × 20 cm^2) using the MMSE algorithm. The contours show the $\log_{10}$(BER) levels against the axes. The MMSE approach gave the best result. Figure 6.4b through d show the BER comparison between the four techniques along the middle row (b), the top row (c), and the far

TABLE 6.1

System Parameters

Parameter	Value
4 × LED Transmitters	
LED device	Luxeon Rebel
Bit rate per channel R_B	12.5 Mb/s
LED pitch	5 cm
Transmitter pitch	25 cm
Optical transmitter power (per LED)	175 mW
Modulation depth	0.45
Modulation bandwidth	4 MHz
Beam angle (full)	120°
Channel	
Test area $w \times l \times h$	$1.4 \times 1.7 \times 2 \text{ m}^3$
4 × Optical receivers	
PIN detector	OSD15-5T
PD reverse bias	50 VDC
Lens *diameter/focal length*	25 mm/25 mm
Receiver field of view (FOV) (full angle)	30°
Receiver pitch	20 cm
Transimpedance amplifier	AD8015
LPF cut-off frequency	$0.75^* R_B$ MHz

right-hand column (d). In all cases, it is shown that there is little or no difference between the methods. For the pseudoinverse receiver, when the inverse of **H** exists (**H** is full rank), the pseudoinverse of Equation 6.7 simply reduces to the ZF. It can also be seen from the MMSE receiver that with a high SNR, Equation 6.9 is also reduced to the ZF. However when the SNR is low, because the algorithm takes into account the noise power, there is a minor improvement in BER. The V-BLAST algorithm also shows no improvement over ZF as this is designed for use in a fading channel environment, which is not the case here but is mentioned as a useful comparison.

Figure 6.5 shows that channel errors have a strong correlation with the spatial positioning of the receiver array in relation to the associated transmitter. Hence the errors from channel 1 only appear along the column that happens to be the greatest distance between transmitter 1 (top right-hand corner) and the receivers (positioned to the far left) where the SNR of the particular channel is at its lowest level. Likewise errors from channels 2–4 occur under similar conditions.

A Q factor analysis of the signal between a single transmitter and single receiver as a function of horizontal displacement beneath the center point of a transmitter was carried out, to investigate at what distance and how fast

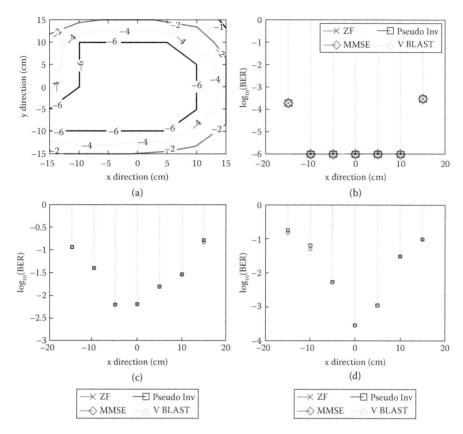

FIGURE 6.3
Overall system BER: (a) $\log_{10}$(BER) scale using MMSE algorithm, (b) BER along middle row in x direction, (c) BER along the top row in the x direction, and (d) the BER down the far right column in the y direction. Results (b) through (d) show a comparison of the four detection methods.

the quality of the signal degrades. Figure 6.5 shows the measured Q factor for channel 1 assuming that the results are symmetrical and follow the same behavior for each of the channels, using OOK NRZ at data rate 12.5 Mb/s.

The BER and Q factor for OOK NRZ have the following relationship:

$$\text{BER} = \frac{1}{2}\text{erfc}\left(\frac{Q}{\sqrt{2}}\right), \tag{6.10}$$

where erfc(.) is the complementary error function; hence the calculated BER at 15, 20, and 25 cm are 10^{-8}, 4.4×10^{-5}, and 1.6×10^{-2}, respectively. Extrapolating this data to estimate the performance of a single-input, single-output (SISO) link, the $\log_{10}$(BER) performance is shown in Figure 6.6.

Each of the four transmitters is located 12.5 cm away from the center of the grid in both the x and y directions. Figure 6.6 shows a single transmitter (Tx)

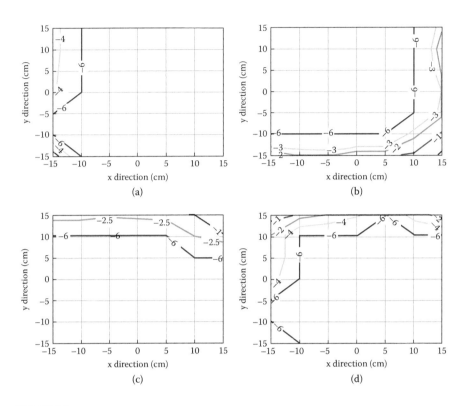

FIGURE 6.4
Aggregated $\log_{10}$(BER) using MMSE method for (a) channel 1, (b) channel 2, (c) channel 3, and (d) channel 4.

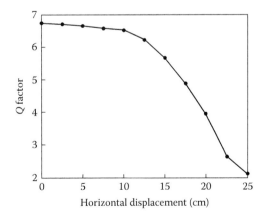

FIGURE 6.5
Q factor with horizontal displacement.

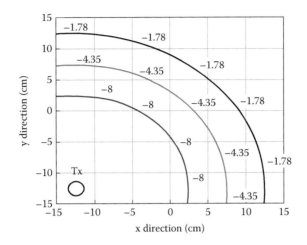

FIGURE 6.6
SISO link performance.

with BER displayed at a radius of 15, 20, and 25 cm. Comparing Figures 6.4 and 6.6 it can be seen that employing the MIMO technique has significantly increased the BER performance and coverage (noted that for Figure 6.4 $\log_{10}$(BER) <-6 has been set to -6). In addition, the MIMO system can increase the data rate M times which greatly enhances the system capacity with a low error rate.

6.2 PAPR Reduction Techniques for Optical OFDM Communications Systems

The OFDM technique is attracting a lot of interest in optical wireless communications because of its resilience to multipath propagation-induced ISI [24–26], high spectral efficiency, and immunity to fluorescent light noise near the DC region [27]. It has also been studied for use in VLCs with data rates in excess of 500 Mbps already demonstrated [28,29]. In spite of the attractions of the OFDM technique, the possibility of the individual subcarrier signals adding up coherently to produce high peaks in the time domain remains one of its main challenges [25]. The presence of these high peaks means that the optical source will have to operate outside its linear region to accommodate the full signal swing. This is very undesirable as it increases, quite considerably, the level of distortion present in the transmitted signal. Consequently, this results in detection error at the receiver.

Moreover in optical OFDM, a DC bias is often added to the electrical OFDM signal in order to make it unipolar and suitable for intensity modulation of an optical source [25,30]. The amount of DC bias required to avoid signal clipping must at least be equal to the most negative peak in the OFDM signal. The transmitted average optical power of the system is also proportional to the DC bias [30]. Therefore, reducing the peaks in the OFDM signal also helps to reduce the transmitted average optical power. Various techniques to address the high peak values and average optical power in optical OFDM literature include block coding between the information bits to be transmitted and the amplitudes modulated onto the subcarriers [30]. In another approach reported in [31], trellis coding was used to reduce the required average optical power (i.e., reduce the negative peaks of the electrical signal). These techniques however have the drawback of increasing the required transmission bandwidth [30]. The addition of out-of-band frequencies, whose amplitudes can be optimized to reduce the average optical power, has equally been investigated in [31]. Alternative concepts of signal transformation, such as selected mapping (SLM) for reducing the signal peak values in an optical OFDM system, have also been studied in [32,33]. A detailed survey and description of other related techniques for reducing the peak values of an OFDM signal is available in [34–40].

This section discusses an optical OFDM wireless communication system in which the signal peak values are reduced by embedding a pilot symbol in the original OFDM signal. In this approach, the phase of each data symbol is rotated by multiplying the data symbol with the pilot. The phase of the pilot symbol is chosen based on the SLM algorithm while the ML criterion will be used at the receiver to estimate the pilot symbol from which the data can then be recovered with improved reliability. The performance of this pilot-assisted technique is then compared with that of peak reduction via signal clipping.

6.2.1 Optical OFDM System Description

The block diagram of the basic OFDM-based optical wireless communication system considered in this section is shown in Figure 6.7. In this system, the signal $X(k)$, with $k = 0, 1, ..., N_{sub}-1$ and N_{sub} being the number of data carrying tones/subcarriers, is the input data stream that is already mapped onto a given M-level QAM constellation.

In an optical OFDM communication system, the time domain signal used to modulate the intensity of the optical carrier must comprise real values only. This condition is ensured by imposing Hermitian symmetry on $X(k)$ prior to the IFFT operation as shown in Figure 6.7. The resulting signal $\mathbf{X_H}$ from this operation is represented as:

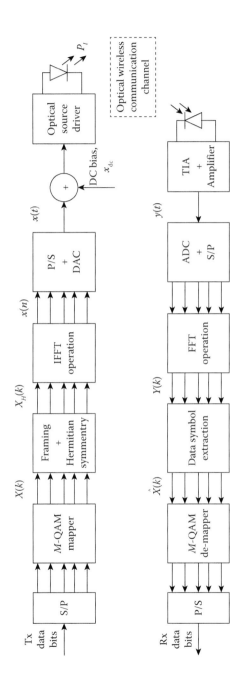

FIGURE 6.7
Block diagram illustration of basic optical OFDM communication system. ADC: analog-to-digital converter, DAC: digital-to-analog converter, P/S: parallel-to-serial converter, S/P: serial-to-parallel converter, TIA: transimpedance amplifier.

$$\mathbf{X_H} = [1, X(k), 0, 0, \ldots, 0, 0, 0, 0, \ldots, 0, X^*(NL - k)], \tag{6.11}$$

where $N = 2(1 + N_{sub})$ and $X^*(.)$ represents the complex conjugate of $X(.)$. The frequency domain signal given by (1) contains $N(L-1)$ padding zeros to account for an L-times oversampling in the time domain signal. The oversampling is necessary to adequately capture all the signal peaks [41]. $\mathbf{X_H}$ is then modulated onto the subcarriers by the NL-point IFFT block. The output of this operation is the L-times oversampled time domain signal $x(n)$. Using the unitary IFFT transform, this L-times oversampled signal $x(n)$ is defined as [25]:

$$x(n) = \frac{1}{\sqrt{NL}} \sum_{m=0}^{NL-1} \mathbf{X_H}(m) \exp\left[\frac{j2\pi nm}{NL}\right]; \quad 0 \le n \le NL - 1. \tag{6.12}$$

A cyclic prefix (CP) is often included in the OFDM signal to combat ISI and intercarrier interference that could be an issue in a dispersive optical wireless communication channel. The CP will however be omitted for simplicity as it has negligible impact on both the required electrical SNR and the spectral efficiency [42].

A continuous time domain signal $x(t)$ is obtained by feeding the discrete signal $x(n)$ into a digital-to-analog converter (DAC); $x(t)$ is then used to modulate the intensity of the optical source. For a linear driver, the radiated optical power can be expressed as:

$$P_t = \mathcal{K}(x_{dc} + \beta x(t)), \tag{6.13}$$

where $\mathcal{K}$ represents the linear modulator's electrical signal to radiated optical power conversion coefficient; x_{dc} is the corresponding DC voltage needed to make the real-valued driving signal $x(t)$ unipolar while β is the optical modulation index. To avoid any lower level signal clipping, the condition $x_{dc} \ge |\min[x(t)]|$ must be met. It is here assumed that the DAC's lower input limit, represented as DAC_{LL}, is less than $\min[x(t)]$. Similarly, to avoid any upper level signal clipping due to device saturation (from either the optical source-modulator combination or the DAC), the condition max $[x(t)] \le$ min $[P_{tmax}, DAC_{UL}]$ must be satisfied. Here, P_{tmax} represents the peak permissible transmit optical power within the dynamic range of the optical source-modulator combination while DAC_{UL} is the upper input limit of the DAC.

At the receiver, the PIN photodetector (PD) converts the incoming optical radiation into an electrical (current) signal. The transimpedance amplifier that follows turns this into a voltage signal. A post-detection amplifier then boosts the signal to an appropriate level $y(t)$ for the analog-to-digital converter (ADC). An FFT operation translates the received signal $y(t)$ into its frequency domain equivalent $Y(k)$ while the conjugate symbols $X^*(NL - K)$ and the padding zeros introduced at the transmitter are removed by the data symbol extraction block as shown in Figure 6.7. An M-QAM demapping operation is then performed to obtain the estimates of the transmitted data.

Using techniques such as predistortion [43], it is possible to reduce the impact of nonlinearities present in an optical source from now on; it will be assumed that the transmitter's response is linear. It is equally very unlikely that the PD introduces additional clipping at the receiver. This is basically due to a combination of high path loss experienced in indoor optical wireless communications, limited transmitted average optical power (due to eye safety regulations), and wide dynamic range of the PD.

As a consequence of the central limit theorem, the OFDM signal $x(n)$, being the sum of a number of independent, identically distributed components, follows the Gaussian distribution. This holds for any practical number of subcarriers; typically $N_{sub} \geq 16$ [44]. Although the Hermitian symmetry introduces some correlation, the distribution of $x(n)$ is still largely Gaussian [44]. This means that occasional high peaks will be present in the time domain OFDM signal. These peaks are often measured with respect to the average value of the signal via the electrical peak-to-average power ratio (PAPR) metric.

The high PAPR implies that, except when some form of transformation is applied, it will be impossible to contain the entire signal within the dynamic range of the transmitter without lower and/or upper level clipping. On the other hand, operating the transmitter into its saturation region to accommodate all the signal swing is highly undesirable. Doing this will result in spectral growth in the form of intermodulation among subcarriers and consequently, error performance degradation. It is possible to keep $x(t)$ within the dynamic range of the transmitter by using a suitably low value of β in Equation 6.13. This is however very unattractive as it results in a 20logβ reduction in electrical SNR at the receiver and eventually, a high error rate. Furthermore, the optical sources, particularly LEDs, experience a significant drop in efficiency when operated outside their dynamic range due to the droop effect [45]. The combination of these factors therefore makes high PAPR a major challenge for an OFDM-based communication system. Finding a suitable technique to reduce the PAPR without a significant loss in error performance is therefore imperative.

6.2.2 The Pilot-Assisted OFDM Technique for PAPR Reduction

The electrical PAPR of a single symbol OFDM signal, oversampled L-times in the time domain, is by definition given as [34]:

$$\text{PAPR} \triangleq \frac{\max\limits_{0 \leq n \leq NL-1} |x(n)|^2}{E[|x(n)|^2]},\tag{6.14}$$

where $E[.]$ denotes the statistical expectation. Any reductions in PAPR are normally illustrated using a PAPR complementary cumulative distribution function (CCDF) diagram. The CCDF of the PAPR is defined as the

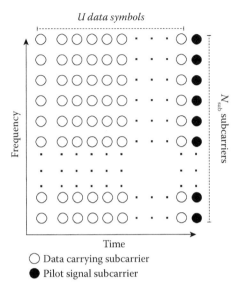

FIGURE 6.8
An illustration of the pilot-assisted OFDM signal frame.

probability that the PAPR of an OFDM frame exceeds a given reference value and it is the most frequently used measure for describing PAPR reduction [34].

An OFDM frame is firstly defined as a cluster of U data carrying symbols and a pilot symbol as illustrated in Figure 6.8. Each of the $(U + 1)$ symbols in the OFDM frame comprises N_{sub} active subcarriers. The data carrying symbols are represented as $X^u(k)$, $u = 1, 2, ..., U$ while the pilot symbol, with amplitude $A_p(k)$ and phase $\theta_p(k)$, is represented as $X_p(k) = X^{U+1}(k) = A_p(k) e^{j\theta p(k)}$ where $k = 0, 1, ..., N_{sub} - 1$.

The pilot-assisted OFDM technique for PAPR reduction begins with modification of the OFDM signal as follows. For a given subcarrier k, the phase of each data entry $X^u(k)$, $u = 1, 2, ..., U; k = 0, 1, ..., N_{sub} - 1$ is rotated by $\theta_p(k)$ while the amplitude is scaled by a factor $A_p(k)$. The resulting signal is denoted as $\tilde{X}^u(k)$. That is:

$$\tilde{X}^u(k) = X^u(k) \times A_p e^{j\theta_p(k)}. \tag{6.15}$$

In principle, $\theta_p(k)$ could take on any value between 0 and 2π.

To make $\tilde{X}^u(k)$ suitable for intensity modulation of the optical source as highlighted above, Hermitian symmetry is invoked on it to obtain $\tilde{X}^u H$. This is then followed by an IFFT operation and the DAC to obtain the discrete and continuous time domain signals $\tilde{X}(n)$ and $\tilde{X}(t)$, respectively.

The reduction in PAPR is achieved by choosing a pilot phase sequence $\theta_p(k)$ that avoids coherent addition of the subcarriers as much as possible. The process of achieving this is summarized as follows:

- Generate R different iterations of the pilot sequence X^r_p, $r = 1, 2, ..., R$
- Each X^r_p comprises a randomly generated phase sequence $\theta_p(k)$, $k = 0, 1, ..., N_{sub} - 1$
- Evaluate the PAPR^r value of each iteration of X^r_p
- Choose the desired pilot as the sequence $\mathbf{X}_p = X^{\tilde{r}}_p$ that gives the minimum PAPR of all the R different iterations

That is:

$$\tilde{r} = \underset{1 \leq r \leq R}{\arg\,\min}(\text{PAPR}^r) = \underset{1 \leq r \leq R}{\arg\,\min} \left[\frac{\underset{0 \leq n \leq (U+1)(NL-1)}{\max} |\tilde{x}^r(n)|^2}{E[|\tilde{x}^r(n)|^2]} \right]. \tag{6.16}$$

In order to preserve the electrical power of the data signals, it is desirable to constrain the pilot signal amplitude A_p to unity. Moreover, to maintain the original constellation of the input symbol sequence and for ease of pilot signal recovery, the phase angle in every pilot phase sequence is constrained to either 0 or π. This results in $X_p(k) \in \{\pm 1\}$, for $k = 0, 1, ..., N_{sub} - 1$.

Although the R pilot sequences here are generated randomly with a uniform distribution, any other phase sequence sets that make the data symbols appear statistically independent as much as possible will suffice. Other viable sequences include [32,46] cyclic Hadamard, Sylvester–Hadamard, Walsh–Hadamard, and Shapiro–Rudin sequences. The random phase sequence set however gives the most PAPR reduction [46].

Also, the pilot sequence used for PAPR reduction could simultaneously be used for channel estimation. It should equally be mentioned that embedding a pilot symbol within an OFDM frame does lead to a reduction of $\frac{1}{U+1}$ in the attainable data rate per OFDM frame. However it is not unusual to insert pilot tones within U-sized OFDM clusters for channel estimation purposes and thus the underlying principle here is not vastly different from well-established embedded pilot techniques.

6.2.3 Pilot Signal Estimation at the Receiver

The received signal, in the frequency domain, is given by:

$$Y^u(k) = H(k) \times \tilde{X}^u(k) + \mathcal{N}(k); \quad u = 1, 2, ..., U + 1, \tag{6.17}$$

where $H(k)$ is the channel's frequency response for the kth subcarrier and $\mathcal{N}(k)$ is the corresponding additive white Gaussian noise with zero mean and variance σ^2. A basic recovery of the transmitted data symbols could be performed by dividing every data element in $Y^u(k)$ by the received pilot signal $Y^{U+1}(k)$ [47]. This will result in an estimate $\hat{X}^u(k)$ of the transmitted data symbol given by Equation 6.18.

$$\hat{X}^u(k) = \frac{Y^u(k)}{Y^{U+1}(k)}; \quad u = 1, 2, \ldots, U. \tag{6.18}$$

The received pilot signal $Y^{U+1}(k) = H(k)\tilde{A}p(k)e^{j\tilde{\theta}_p(k)}$ where $\tilde{A}p(k)$ and $\tilde{\theta}_p(k)$ are the noise-corrupted pilot signal amplitude and phase, respectively. Using (6.18) directly will introduce data recovery errors due to the presence of noise on both the received pilot and data carrying symbols. The effect of this on the received data constellation is obvious when the basic OFDM system with no PAPR reduction shown in Figure 6.9 is compared with the pilot-assisted

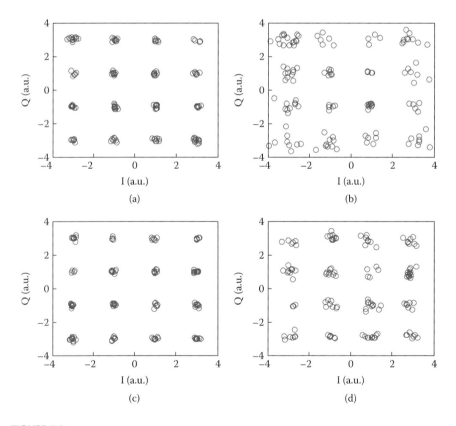

FIGURE 6.9
Constellation diagrams of 16-QAM optical OFDM with $L = 4$, $U = 5$, and $N_{sub} = 127$ active subcarriers at an SNR of 22 dB. (a) Basic OFDM without PAPR reduction, (b) pilot-assisted OFDM technique, symbol estimation based on Equation 6.18, (c) pilot-assisted OFDM technique with ML pilot symbol estimation based on Equation 6.21, and (d) clipped OFDM with $C_{cl} = C_{cu} = 40$.

OFDM technique of Figure 6.9b. The electrical SNR in this case is defined as $\text{SNR} = \frac{(\mathcal{K}\mathcal{R}\beta)^2 E[|\tilde{x}(n)|^2]}{\sigma^2}$ where $\mathcal{R}$ is the PD's responsivity.

To improve on the data recovery, the pilot signal has to be correctly estimated in the presence of noise. To achieve this, we use the ML estimation technique. An estimate $\hat{\theta}_p(k)$ of the pilot signal's phase is taken as the angle $\theta_i, i = 1,2$ (where $\theta 1 = 0$ and $\theta 2 = \pi$) that has the minimum Euclidean distance from $\tilde{\theta}_p(k)$. That is:

$$\hat{\theta}_p(k) = \arg\min_{1 \leq i \leq 2}[(\tilde{\theta}_p(k) - \theta_i)^2]. \tag{6.19}$$

The estimated pilot signal then becomes $\hat{X}_p(k) = e^{j\hat{\theta}_p(k)}$. This ML criterion for estimating the pilot signal's phase is thus equivalent to the condition given by Equation 6.20.

$$\hat{X}_p(k) = \begin{cases} +1 & \text{if } \cos\left(\tilde{\theta}_p(k)\right) \geq 0. \\ -1 & \text{otherwise} \end{cases} \tag{6.20}$$

An estimate of the transmitted data symbol is therefore obtained as:

$$\hat{X}^u(k) = \frac{Y^u(k)}{\hat{X}_p(k)}; \quad u = 1, 2, \ldots, U. \tag{6.21}$$

In addition to maintaining the pilot signal amplitude as unity, the ML condition given by Equation 6.20 will correct for all pilot phase noise that falls within the range $-\pi/2$ to $\pi/2$. A sample received data constellation diagram that is based on Equation 6.21 is shown in Figure 6.9c at an SNR of 22 dB. When this figure is compared with Figure 6.9b, the ML pilot estimation technique's significant improvement in data recovery becomes very obvious. Also the constellation diagrams in Figure 6.9a and c are identical, this implies that the ML pilot estimation technique satisfies a key requirement of not degrading the system error performance. This point will be further highlighted with the BER performance plots in Section 2.6. Moreover, this decoder is quite simple requiring only $2N_{\text{sub}} |.|^2$ operations to solve the ML criterion of Equation 6.19.

6.2.4 PAPR Reduction by Clipping

Signal clipping as a means of PAPR reduction is the most commonly used and the simplest technique to implement [44,48]. The signal clipping takes place at the transmitter in the time domain prior to the DAC stage. The signal could be clipped at lower and/or upper levels [48]. The lower and upper clipping levels, represented as ε_{cl} and ε_{cu}, respectively, will be expressed in terms of the signal variance $\sigma^2_{\mathbf{x}^u}$, where $\mathbf{x}^u = \text{IFFT}(X_H^u); u = 1, 2, \ldots, U.$

$$\varepsilon_{cl} = -\mathcal{C}_{cl}\sigma_{\mathbf{x}^u}^2 \qquad\qquad (6.22)$$

$$\varepsilon_{cu} = \mathcal{C}_{cu}\sigma_{\mathbf{x}^u}^2 \qquad\qquad (6.23)$$

$\mathcal{C}_{cl}$ and $\mathcal{C}_{cu}$ are unit-less coefficients that determine the severity of the clipping at lower and upper levels, respectively. The higher the values of $\mathcal{C}_{cl}$ and $\mathcal{C}_{cu}$, the smaller the amount of clipping and vice versa. The clipping operation thus results in the following clipped OFDM signal $x_c(n)$.

$$x_c(n) = \begin{cases} x^u(n) & \text{if } \varepsilon_{cl} \leq x^u(n) \leq \varepsilon_{cu} \\ \varepsilon_{cu} & \text{if } x^u(n) > \varepsilon_{cu} \\ \varepsilon_{cl} & \text{if } x^u(n) < \varepsilon_{cl} \end{cases}. \qquad (6.24)$$

The required DC bias for the clipped OFDM now becomes $x_{dc} = \min[x_c(n)]$. An illustration of the impact of a very mild signal clipping ($\mathcal{C}_{cl} = \mathcal{C}_{cu} = 40$) on the received data constellation diagram is shown in Figure 6.9d. Comparing this constellation diagram with that of the basic OFDM shown in Figure 6.9a, clearly shows the effect of clipping induced signal distortion.

6.2.5 PAPR Reduction Comparison of Pilot-Assisted and Signal Clipping

The term basic OFDM here refers to the ordinary OFDM with no pilot and no PAPR reduction technique implemented. To illustrate the PAPR reduction capabilities of the pilot-assisted OFDM technique, we show in Figure 6.10 the CCDF as a function of reference $PAPR_0$. It is observed from this figure that for basic OFDM with no pilot and no PAPR reduction method implemented, 1 out of every 10^4 OFDM frames has its PAPR greater than 14.8 dB. But using the pilot-assisted PAPR reduction technique with $R = 5$ iterations, 1 in every 10^4 OFDM frames has its PAPR greater than 12.8 dB. This implies a 2 dB reduction in PAPR at the same value of CCDF. The stated reduction in PAPR increases by a further 0.5 dB with $R = 10$ iterations. Another interpretation of the result is that, for the basic OFDM with no PAPR reduction technique implemented, 3 OFDM frames out of every 100 have a PAPR > 12.8 dB while with the pilot-assisted OFDM method with $R = 5$, the number reduces significantly to 1 in every 10,000.

Another PAPR reduction technique shown in Figure 6.10 is signal clipping. For the 64-QAM under consideration, we have chosen a moderate clipping level of 25 times the signal variance (i.e., $\mathcal{C}_{cl} = \mathcal{C}_{cu} = 25$) for illustrative purpose. Compared with the basic OFDM, clipping at both ends of the signal results in a significant reduction in the PAPR from 14.8 dB to 8.5 dB at a CCDF of 10^{-4}. When the signal is however clipped only at the upper end, there is just a 0.2 dB reduction in PAPR at the same CCDF of 10^{-4}. It can thus be inferred that, clipping at both ends, at $\mathcal{C}_{cl} = \mathcal{C}_{cu} = 25$, offers more PAPR reduction than the pilot-assisted OFDM technique while clipping at the upper end only is not as effective as the proposed

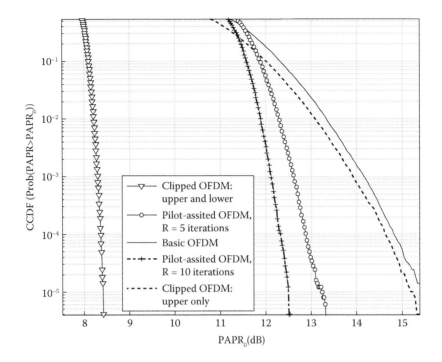

FIGURE 6.10
The PAPR CCDF plot for clipped and pilot-assisted optical OFDM using 64-QAM; $L = 4$, $U = 5$ data carrying symbols and $N_{sub} = 127$ active subcarriers, $C_{cl} = C_{cu} = 25$.

pilot-assisted PAPR reduction technique. In fact, with clipping, it is possible to achieve any desired amount of PAPR reduction but with a severe consequence of signal distortion that subsequently leads to error performance degradation.

In Figure 6.11, we illustrate how the PAPR reduction capability of the pilot-assisted technique is affected by the number of iterations R and data-carrying symbols U contained within an OFDM frame. With $U = 2$ and $R = 4$ iterations, the figure shows that 1 in every 10^4 OFDM frames has its PAPR greater than 13.4 dB while with $R = 10$ iterations, the PAPR is about 0.9 dB lower at the same value of CCDF. With U increased to 10, the PAPR needed at the same CCDF of 10^{-4} with $R = 4$ and 10 iterations reduces to 13.1 and 12.3 dB, respectively. In contrast, for the basic OFDM with no PAPR reduction method implemented, the PAPR increases as U increases. This is because increasing the number of symbols simply increases the probability of high signal peaks due to coherent addition of subcarriers. At a CCDF of 10^{-4}, the basic OFDM system with $U = 2$ has a $PAPR_0$ of 14.5 dB; while with $U = 10$ the $PAPR_0$ increases by a further 0.5 dB.

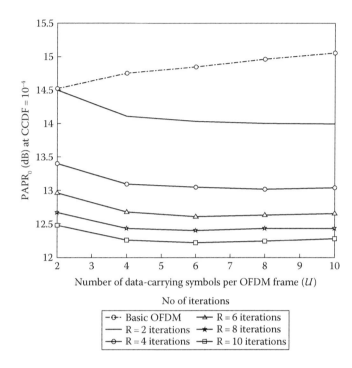

FIGURE 6.11
$PAPR_0$ values required to attain a CCDF of 10^{-4} against U for basic OFDM with no PAPR reduction and pilot-assisted OFDM with $R = [2,4,6,8,10]$ iterations, 64-QAM, $L = 4$, and $N_{sub} = 127$ active subcarriers.

With the pilot-assisted OFDM technique, the amount of reduction in PAPR increases with the number of symbols per OFDM frame. This is due to the fact that, with no PAPR reduction technique in place, the PAPR values increase with the number of symbols per frame. Using higher values of U is also beneficial in improving the payload to frame size ratio $\left(\frac{U}{U+1}\right)$. Doing this does not lead to any significant drop in the PAPR reduction capability of the pilot-assisted OFDM technique as seen in Figure 6.11 for $4 \leq U \leq 10$.

Furthermore, the pilot-assisted optical OFDM technique also leads to a reduction in the required DC bias x_{dc}; this is depicted in Figure 6.12. For the case of 64-QAM and 127 data-carrying subcarriers shown in this figure, it is observed that at a CCDF of 10^{-4}, the DC bias required with the pilot-assisted optical OFDM technique is about 0.41 V and 0.54 V for the basic optical OFDM with no PAPR reduction technique implemented. This amounts to the pilot-assisted technique having a saving of 1.2 dB in transmitted average optical power over the basic OFDM system.

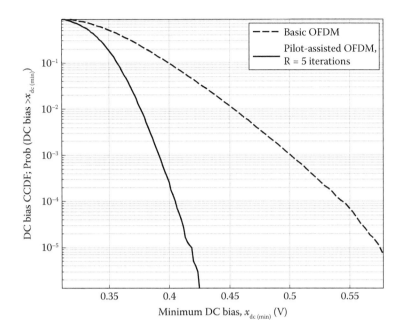

FIGURE 6.12
The DC bias CCDF plot for basic and pilot-assisted optical OFDM using 64-QAM, $L = 4$, $U = 5$ data carrying symbols and $N_{sub} = 127$ active subcarriers.

6.2.6 Effect of PAPR Reduction on Error Performance

In terms of the effect of PAPR reduction on error performance, we show the BER against SNR in Figures 6.13 and 6.14 for 16- and 64-QAM, respectively. These figures show that with the pilot-assisted PAPR reduction technique, the system error performance at a BER of less than 10^{-3} is nearly identical to that of the basic OFDM with no PAPR reduction. Since no reliable communication takes place at BER > 10^{-3} anyway, it can be said that the pilot-assisted PAPR reduction technique does not degrade the system error performance. However, clipping does result in a significant degradation of the BER. For instance at an SNR of 15 dB, very slightly clipping the 16-QAM system at both ends (at $C_{cl} = C_{cu} = 40$) will result in a BER of 10^{-4} while with the pilot-assisted OFDM technique, the BER is less than 10^{-7}. Worst still, with signal clipping, the system reaches an error floor value that depends on the severity of the clipping. For the 64-QAM optical OFDM case with upper and lower level clipping (at $C_{cl} = C_{cu} = 25$), the error floor BER is about 3×10^{-5} and slightly lower at 6×10^{-6} for upper level clipping only. From the foregoing, it can therefore be said that the pilot-assisted optical OFDM technique is a viable method for reducing both the electrical PAPR value and the average transmitted optical power without any noticeable degradation in the system error performance.

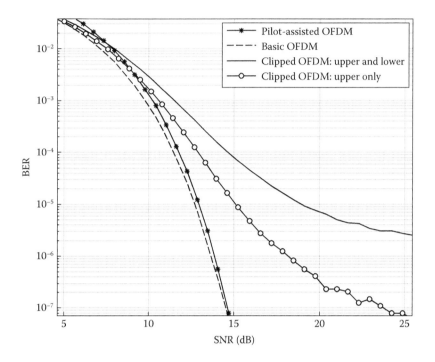

FIGURE 6.13
Comparison of BER performance of clipped optical OFDM ($\mathcal{C}_{cl} = \mathcal{C}_{cu} = 40$) with pilot-assisted optical OFDM; 16-QAM, $R = 5$ iterations, $L = 4$, $U = 5$, and $N_{sub} = 127$ active subcarriers.

6.3 Summary

The first part of this chapter discussed techniques on MIMO VLC system which improved the speed of data communications. Four demultiplexing techniques for data recovery have been presented including ZF, pseudoinverse, MMSE, and Vertical Bell Laboratories Layered Space Time. Here it covered the theoretical development for each approach and provided the system performance evaluation by practical demonstrations. In all cases, parallel transmission using MIMO offers good data rate increment and low error rate.

In the second part of the chapter, the use of a pilot signal in reducing the PAPR of an OFDM-based intensity-modulated optical wireless communication system is discussed and developed. The phase of the pilot signal is chosen based on the SLM algorithm while the ML criterion is used to estimate the pilot signal at the receiver. The pilot insertion technique discussed in this chapter attains PAPR reduction without degrading the systems error performance. That is, the BER performance of the pilot-assisted optical OFDM

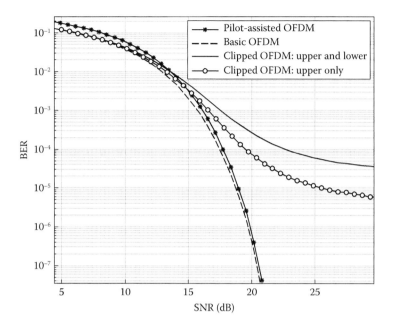

FIGURE 6.14
Comparison of BER performance of clipped optical OFDM ($C_{cl} = C_{cu} = 25$) with pilot-assisted optical OFDM; 64-QAM, $R = 5$ iterations, $L = 4$, $U = 5$, and $N_{sub} = 127$ active subcarriers.

system is identical to that of the conventional optical OFDM with no PAPR reduction technique implemented.

Symbols

A_{det}	Active area of photodetector
δ	Duty cycle
$E\{.\}$	Expectation
G	Pseudoinverse of channel matrix
$g(\psi)$	Gain of the optical concentrator
γ	Dimming factor
H	Channel matrix
h_{tr}	DC gain between the tth transmitter and rth receiver
$\mathbf{I}_N$	Identity matrix
m	Lambertian order
n	Vector of additive white Gaussian noise
ω	Number of optical pulses
ϕ	Angle of irradiance
$\Phi_{1/2}$	Semiangle for half illuminance of an LED

σ_n^2 Receiver noise power variance
ψ Angle of incidence
Ψ_c Angular field of view
τ Pulse duration
T PWM frame duration
v Spectral efficiency
x Transmitted signals
x_{est} Recovered signals
y Received signals

References

[1] B. Heffernan, L. Frater and N. Watson, LED replacement for fluorescent tube lighting, *Australasian Universities Power Engineering Conference (AUPEC)*, Perth, Australia, 9 December, pp. 1–6, 2007.

[2] Visible Light Communications Consortium. Available at: http://www.vlcc.net/e/e_about.html (accessed March 2, 2017).

[3] OMEGA Home Gigabit Access. Available at: http://www.ict-omega.eu (accessed March 2, 2017).

[4] IEEE 802.15 WPAN™ Task Group 7 (TG7) Visible Light Communication. Available at: http://www.ieee802.org/15/pub/TG7.html (accessed March 2, 2017).

[5] H. Le Minh, D. O'Brien, G. Faulkner, L. Zeng, L. Kyungwoo, J. Daekwang, O. YunJe and W. Eun Tae, 100-Mb/s NRZ visible light communications using a postequalized white LED, *IEEE Photon. Technol. Lett.*, vol. 21, pp. 1063–1065, 2009.

[6] H. Le Minh, D. O'Brien, G. Faulkner, L. Zeng, L. Kyungwoo, J. Daekwang and O. YunJe, High-speed visible light communications using multiple-resonant equalization, *IEEE Photon. Technol. Lett.*, vol. 20, pp. 1243–1245, 2008.

[7] J. Vucic, C. Kottke, S. Nerreter, K. Langer and J. W. Walewski, 513 Mbit/s visible light communications link based on DMT-modulation of a white LED, *J. Lightwave Technol.*, vol. 28, pp. 3512–3518, 2010.

[8] W. Fang-Ming, L. Chun-Ting, W. Chia-Chien, C. Cheng-Wei, H. Hou-Tzu and H. Chun-Hung, 1.1-Gb/s white-LED-based visible light communication employing carrier-less amplitude and phase modulation, *IEEE Photon. Technol. Lett.*, vol. 24, pp. 1730–1732, 2012.

[9] A. M. Khalid, G. Cossu, R. Corsini, P. Choudhury, and E. Ciaramella, 1-Gb/s transmission over a phosphorescent white LED by using rate-adaptive discrete multitone modulation, *IEEE Photon. J.*, vol. 4, pp. 1465–1473, 2012.

[10] P. A. Haigh, Z. Ghassemlooy, H. Le Minh, S. Rajbhandari, F. Arca, S. F. Tedde, O. Hayden, and I. Papakonstantinou, Exploiting equalization techniques for improving data rates in organic optoelectronic devices for visible light communications, *J. Lightwave Technol.*, vol. 30, pp. 3081–3088, 2012.

[11] L. Zeng, D. O'Brien, H. Le Minh, G. Faulkner, L. Kyungwoo, J. Daekwang, O. YunJe, and W. Eun Tae, High data rate multiple input multiple output

(MIMO) optical wireless communications using white led lighting, *IEEE J. Sel. Areas Commun.*, vol. 27, pp. 1654–1662, 2009.

[12] A. H. Azhar, T. A. Tran and D. O'Brien, A Gigabit/s indoor wireless transmission using MIMO-OFDM visible-light communications, *IEEE Photon. Technol. Lett.*, vol. 25, pp. 171–174, 2013.

[13] P. A. Haigh, Z. Ghassemlooy, I. Papakonstantinou, F. Arca, S. F. Tedde, O. Hayden and S. Rajbhandari, A MIMO-ANN system for increasing data rates in organic visible light communications systems, *Presented at the IEEE ICC 2013—Wireless Communications Symposium (ICC'13 WCS)*, Budapest, Hungary, 2013.

[14] R. Mesleh, R. Mehmood, H. Elgala and H. Haas, Indoor MIMO optical wireless communication using spatial modulation, *IEEE ICC 2010*, Cape Town, South Africa, pp. 1–5, 2010.

[15] T. Komine and M. Nakagawa, Fundamental analysis for visible-light communication system using LED lights, *IEEE Trans. Consum. Electron.*, vol. 50, pp. 100–107, 2004.

[16] S. Pingping, K. Sooyoung and C. Kwonhue, Soft ZF MIMO detection for turbo codes, *WiMob*, pp. 116–120, 2010.

[17] C. Peel, Q. Spencer, A. L. Swindlehurst and B. Hochwald, Downlink transmit beamforming in multi-user MIMO systems, *Proceedings of the 3rd IEEE Sensor Array and Multichannel Signal Processing Workshop*, 18–21 July, pp. 43–51, Barcelona, Spain, 2004.

[18] S. N. Sur, D. Ghosh, D. Bhaskar and R. Bera, Contemporary MMSE and ZF receiver for V-BLAST MIMO system in Nakagami-m fading channel, *INDICON*, pp. 1–5, 2011.

[19] P. W. Wolniansky, G. J. Foschini, G. D. Golden and R. Valenzuela, V-BLAST: An architecture for realizing very high data rates over the rich-scattering wireless channel, ISSSE, Pisa, Italy, pp. 295–300, 1998.

[20] G. Lebrun, S. Spiteri and M. Faulkner, Channel estimation for an SVD-MIMO System, *IEEE Int. Conf. Commun.*, vol. 5, pp. 3025–3029, 2004.

[21] G. Zhan and P. Nilsson, Algorithm and implementation of the K-best sphere decoding for MIMO detection, *IEEE J. Sel. Areas Commun.*, vol. 24, pp. 491–503, 2006.

[22] Y. Huan and G. W. Wornell, Lattice-reduction-aided detectors for MIMO communication systems, *IEEE GLOBECOM*, vol. 1, pp. 424–428, 2002.

[23] Z. Ghassemlooy, W. Popoola and S. Rajbhandari, *Optical Wireless Communications: System and Channel Modelling*, CRC Press, Boca Raton, FL, 2012.

[24] M. El Tabach, P. Tortelier, R. Pyndiah, and O. Bouchet, Diffuse infrared personal optical wireless based on modified OFDM/OQAM, *Proceeding 6th International Symposium on Communication Systems, Networks and Digital Signal Processing CSNDSP 2008*, Graz, Austria, 23–25 July, pp. 161–164, 2008.

[25] J. Armstrong, OFDM for optical communications, *J. Lightwave Technol.*, vol. 27, no. 3, pp. 189–204, 2009.

[26] O. Gonzalez, R. Perez-Jimenez, S. Rodriguez, J. Rabadan and A. Ayala, OFDM over indoor wireless optical channel, *IEEE Proc. Optoelectronics*, vol. 152, no. 4, pp. 199–204, 2005.

[27] R. Narasimhan, M. D. Audeh and J. M. Kahn, Effect of electronic-ballast fluorescent lighting on wireless infrared links, *Proc. IEEE Optoelectronics*, vol. 143, no. 6, pp. 347–354, 1996.

[28] J. Vucic, C. Kottke, S. Nerreter, K. D. Langer, and J. W. Walewski, 513 Mbit/s visible light communications link based on DMT-modulation of a white LED, *J. Lightwave Technol.*, vol. 28, no. 24, pp. 3512–3518, 2010.

[29] W.-Y. Lin, C.-Y. Chen, H. H. Lu, C.-H. Chang, Y.-P. Lin, H.-C. Lin and H.-W. Wu, 10m/500 Mbps WDM visible light communication systems, *Opt. Express*, vol. 20, no. 9, pp. 9919–9924, 2012.

[30] R. You and J. Kahn, Average power education technique for multiple-subcarrier intensity-modulated optical signals, *IEEE Trans. Commun.*, vol. 49, pp. 2164–2171, 2001.

[31] W. Kang and S. Hranilovic, Power reduction techniques for multiple-subcarrier modulated diffuse wireless optical channels, *IEEE Trans. Commun.*, vol. 56, no. 2, pp. 279–288, 2008.

[32] M. Farooqui, P. Saengudomlert and S. Kaiser, Average transmit power reduction in OFDM-based indoor wireless optical communications using SLM, *International Conference on Electrical and Computer Engineering (ICECE)*, 18–20 December, pp. 602–605, 2010. doi: 10.1109/icelce.2010.5700765.

[33] L. Nadal, M. Svaluto Moreolo, J. Fabrega and G. Junyent, Low complexity bit rate variable transponders based on optical OFDM with PAPR reduction capabilities, *17th European Conference on Networks and Optical Communications (NOC)*, 20–22 June, pp. 1–6, 2012.

[34] S. H. Han and J. H. Lee, An overview of peak-to-average power ratio reduction techniques for multicarrier transmission, *IEEE Wireless Commun.*, vol. 12, no. 2, pp. 56–65, 2005.

[35] T. Jiang and Y. Wu, An overview: Peak-to-average power ratio reduction technique for OFDM signals, *IEEE Trans. Broadcast.*, vol. 54, no. 2, pp. 257–268, 2008.

[36] D.-W. Lim, J.-S. No, C.-W. Lim and H. Chung, A new SLM OFDM scheme with low complexity for PAPR reduction, *IEEE Signal Process. Lett.*, vol. 12, no. 2, pp. 93–96, 2005.

[37] S. Fragiacomo, C. Matrakidis and J. J. OReilly, Multicarrier transmission peak-to-average power reduction using simple block code, *Electron. Lett.*, vol. 34, no. 10, p. 953954, 1998.

[38] R. van Nee and A. de Wild, Reducing the peak-to-average power ratio of OFDM, *Proceeding of 1998 IEEE Conference on Vehicular Technology*, May 1821, p. 20722076, Ottawa, ON, Canada, 1998.

[39] J. Tellado and J. M. Cioffi, Efficient algorithms for reducing PAR in multicarrier systems, *Proceeding of 1998 IEEE International Symposium on Information Theory*, 16–21 August, p. 191, New York, 1998.

[40] D. Wulich and L. Goldfeld, Reduction of peak factor in orthogonal multicarrier modulation by amplitude limiting and coding, *IEEE Trans. Commun.*, vol. 47, no. 1, pp. 18–21, 1999.

[41] C. Tellambura, Computation of the continuous-time PAR of an OFDM signal with BPSK subcarriers, *IEEE Commun. Lett.*, vol. 5, no. 5, pp. 185–187, 2001.

[42] H. Elgala, R. Mesleh and H. Haas, Practical considerations for indoor wireless optical system implementation using OFDM, *Proceeding of the IEEE 10th International Conference on Telecommunications (ConTel)*, 8–10 June, Zagreb, Croatia, 2009.

[43] H. Elgala, R. Mesleh and H. Haas, Non-linearity effects and predistortion in optical OFDM wireless transmission using LEDs, *Indersci. Int. J. Ultra Wideband Commun. Syst. (IJUWBCS)*, vol. 1, no. 2, pp. 143–150, 2009.

[44] L. Chen, B. Krongold and J. Evans, Theoretical characterization of nonlinear clipping effects in IM/DD optical OFDM systems, *IEEE Trans. Commun.*, vol. 60, no. 8, pp. 2304–2312, 2012.

[45] U. Ozgur, H. Liu, X. Li, X. Ni and H. Morko, GaN-based light-emitting diodes: Efficiency at high injection levels, *Proc. IEEE*, vol. 98, no. 7, pp. 1180–1196, 2010.

[46] D.-W. Lim, S.-J. Heo, J.-S. No and H. Chung, On the phase sequence set of SLM OFDM scheme for a crest factor reduction, *IEEE Trans. Signal Process.*, vol. 54, no. 5, pp. 1931–1935, 2006.

[47] B. G. Stewart, Telecommunications Method and System, US Patent 8 126 075, February 28, 2012.

[48] S. Dimitrov, S. Sinanovic and H. Haas, Clipping noise in OFDM-based optical wireless communication systems, *IEEE Trans. Commun.*, vol. 60, no. 4, pp. 1072–1081, 2012.

[49] M. Rea, *Lighting Handbook*, New York: Illuminating Engineering Society of North America (IESNA), 9th ed., 2000.

[50] G. Ntogari, T. Kamalakis, J. Walewski and T. Sphicopoulos, Combining illumination dimming based on pulse-width modulation with visible-light communications based on discrete multitone, *J. Opt. Commun. Networking*, vol. 3, pp. 56–65, 2011.

[51] S. M. Berman, D. S. Greenhouse, I. L. Bailey, R. D. Clear and T. W. Raasch, Human electroretinogram responses to video displays, fluorescent lighting, and other high frequency sources, *Optom. Vis. Sci.*, vol. 68, pp. 645–662, 1991.

[52] M. Dyble, N. Narendran, A. Bierman and T. Klein, Impact of dimming white LEDs: Chromaticity shifts due to different dimming methods, *Proceedings of the SPIE 5491*, Fifth International Conference on Solid State Lighting, pp. 291–299, Bellingham, WA, 2005.

[53] S. Kaur, W. Liu and D. Castor, VLC Dimming Proposal, IEEE 802.15 Working Group for wireless personal area networks (WPANs) 802.15-15-09-0641-00-0007, *Tech. Rep.*, New Work, 2009.

[54] Y. Zhang, Z. Zhang, Z. Huang, H. Cai, L. Xia and J. Zhao, Apparent brightness of LEDs under different dimming methods, in *Proc. SPIE Solid State Lighting and Solar Energy Technologies*, pp. 684109–684109-5, Beijing, China, 2007.

[55] L. Kwonhyung and P. Hyuncheol, Modulations for visible light communications with dimming control, *IEEE Photon. Technol. Lett.*, vol. 23, pp. 1136–1138, 2011.

[56] J. Hyung-Joon, C. Joon-Ho, C. Eun Byeol and L. Chung Ghiu, Simulation of a VLC system with 1 Mb/s NRZOOK data with dimming signal, *International Conference on Advanced Infocom Technology 2011 (ICAIT 2011)*, pp. 1–3, 2011.

[57] C. Eunbyeol, C. Joon-Ho, P. Chulsoo, K. Moonsoo, S. Seokjoo, Z. Ghassemlooy and L. Chung Ghiu, NRZ-OOK signaling with LED dimming for visible light communication link, *16th European Conference on Networks and Optical Communications (NOC)*, pp. 32–35, 2011.

[58] C. Joon-Ho, C. Eun-byeol, K. Tae-gyu and L. Chung Ghiu, Pulse width modulation based signal format for visible light communications, *OptoeElectronics and Communications Conference (OECC)*, pp. 276–277, 2010.

[59] J. Hyung-Joon, C. Joon-Ho, Z. Ghassemlooy and L. Chung Ghiu, PWM-based PPM format for dimming control in visible light communication system, *8th International Symposium on Communication Systems, Networks & Digital Signal Processing (CSNDSP)*, pp. 1–5, 2012.

[60] H. Sugiyama, S. Haruyama and M. Nakagawa, Experimental investigation of modulation method for visible-light communications, *IEICE Trans. Commun.*, vol. 89, pp. 3393–3400, 2006.

[61] H. Sugiyama, S. Haruyama and M. Nakagawa, Brightness control methods for illumination and visible-light communication systems, *3rd International Conference on Wireless and Mobile Communications (ICWMC '07)*, pp. 78–78, 2007.

[62] Z. Lubin, H. Le Minh, D. O'Brien, G. Faulkner, L. Kyungwoo, J. Daekwang and O. Yunje, Equalisation for high-speed visible light communications using white-LEDs, *6th International Symposium on Communication Systems, Networks and Digital Signal Processing (CNSDSP)*, pp. 170–173, 2008.

7

VLC Applications for Visually Impaired People

Rafael Pérez Jiménez, Jose A. Rabadan-Borges, Julio F. Rufo Torres, and Jose M. Luna-Rivera

CONTENTS

7.1 Introduction

Universal design (often inclusive design) refers to broad-spectrum ideas meant to produce buildings, products, and environments that are inherently accessible to older people or people with disabilities. This concept emerged from slightly earlier barrier-free concepts, the broader accessibility movement, adaptive and assistive technology, and also seeks to blend aesthetics into these core considerations. As life expectancy rises and modern medicine increases the survival rate of those with significant injuries, illnesses, and birth defects, there is a growing interest in universal design. There are many industries in which universal design is having strong market penetration but there are many others in which it has not yet been adopted to any great extent.

Universal design is also being applied to the design of technology, instruction, services, and other products and environments.

There are some principles to be taken into account on universal design (or universal accessibility) that can be accomplished by using information and communication technologies (equitable use, flexibility in use, simple and intuitive use, perceptible information, etc.) and many of them are also reachable through the use of the Internet of Things paradigm. One of the technologies to be used is wireless optical communications, and, more specifically, visible light communications (VLC). The introduction of LED lighting creates a new opportunity for a whole new set of technological possibilities in communications systems. This was not previously possible with conventional lighting, but LEDs have a number of key advantages. First, LEDs can be modulated at much higher frequencies than conventional lighting, so the signals required for positioning applications can readily be transmitted at frequencies without the effect of visible flicker. Second, although LED lights are initially more expensive, they have a much longer lifetime—typically several years. This means that the added cost of constructing lights with the extra functionality required for positioning will be relatively smaller and with long-lasting benefits.

In this chapter, we will describe some applications that can be employed in order to allow universal accessibility for disabled people. We study two different scenarios: outdoor systems based on the use of traffic and city lights, and indoor applications using illumination and emergency lamps. The key capability of the VLC systems is that the light "talks" so blind people can "see"—or hear—the information transmitted by regular devices used by sighted people. The expected result is that they can use the signaling and traffic lights by means of a small transceiver that converts light into an understandable message such as by voice. This procedure can be also performed by radio technologies, using network standards like Bluetooth or ZigBee, but it loses the directivity provided by the light itself, requires additional devices on city facilities, and can be affected by EM noise and compatibility issues or the presence of signal jammers. Audio signals are also commonly used as a status indicator in traffic lights for blind people, using some kind of "beep" or birdsong, but they can be masked by ringtones of cell phones or other devices. Furthermore, unless some phone applications provide adapted guidance for disabled people, they cannot be precise enough for the specific situation of a signal, especially when dealing with crossroads. We shall present a solution—the SINAI project—as an experimental implementation for overcoming these problems [1], proposing the use of lights all over the city as a new grid of information to be employed.

On indoor scenarios, one of the main challenges is also positioning and guidance for people with special needs (e.g., the elderly, blind, or those affected by Alzheimer's). Positioning, also known as localization, is the process of determining the spatial position of an object or person. Accurate positioning is critical for numerous applications; the most familiar system of this type for outdoor scenarios is the global positioning system (GPS). Unfortunately, GPS is not suitable in many indoor situations. Despite decades of research into indoor positioning using radio technologies, even that based on wireless local area networks (LANs), there is still no system that is cheap, accurate, and widely available [2,3]. The fundamental problem in radio-based systems is multipath propagation. Radio signals may reach a receiver by both direct line of sight (LOS) and multiple reflected paths. This means there is no simple and reliable way of determining the distance or direction of the transmitter from the received signal.

The widespread introduction of white LEDs for illumination provides an unprecedented opportunity for visible light positioning (VLP) to fill this gap as they are widely available, economical, and easy to use [4–8]. In almost any building or facility you will be able to see multiple light fittings, demonstrating that at most indoor locations, a receiver could be designed to detect LOS signals from multiple light sources and, furthermore, calculate its position with precision [9–12]. Most of the positioning techniques used in radio are also suitable for use with lighting LEDs: beaconing, received signal strength (RSS), angle of arrival (AOA), time of arrival (TOA), and time difference of arrival (TDOA)—even with signals of different nature, as in cricket sensors [13] or fingerprinting.

This chapter is organized as follows. In the next section we shall describe the new possibilities opened by VLC technologies to the problem of increasing mobility in the cities, introducing these possibilities inside the smart city paradigm. We will then describe a specific implementation using streetlights as a resource for the mobility of blind people (SINAI). Section 3 describes the use of VLC systems for indoor positioning and guidance, reviewing the state of the art from some of the many groups that are by now working in this area, and presenting a VLC-ultrasound hybrid solution, with an implementation very similar to the well-known cricket sensor architecture [14,15]. We shall also present some specific applications for safety and emergency management tools. Finally, some conclusions will be presented.

7.2 VLC for Outdoor Mobility

7.2.1 VLC in the "Smart City" and the "Smart Building"

The increased growth in population and mobility are leading many countries to rethink their present and future city planning, especially focusing on integrated socioeconomic infrastructure supported by sustainable development. To support evolving dynamics in modern urban environments, there is a need for the city planners to establish comprehensive information and communication technology (ICT) infrastructure. Establishing this level of integrated ICT infrastructure allows the creation of a "smart city," where people, government, economy, and environment are seamlessly connected. This level of connectivity will benefit diverse stakeholders; immediate beneficial impacts can be felt by the urban population on their quality of life in this utopia. Much of the functionality required for a smart city exists around us due to rapid advancement in ICT during the past decade [16]. However, there still remain major challenges in achieving sustainable connectivity across all the functional layers.

Hard infrastructures such as government institutions, hospitals, airports, and power stations are now connected through distributed networks, where information is distributed and shared across institutions. These networks consist of wired and wireless bearers, where some of these hard infrastructures have preferential bearer technologies to suit their operational requirements. For example, some organizations use wired networks due to their sensitive nature, while others may use mixed networks. To achieve this level of high-speed connectivity and coverage, it has been identified that the existing radio frequency (RF) technologies do not have the adequate bandwidth allocation to fulfill this growing demand. To succeed in carrying out seamless high-speed connectivity and coverage across all functional layers of a modern city, there is a need to evaluate all possible bearer technologies that could support this demand. Following this research guideline, it should be noted that multiple proposals for VLC systems have been recently applied to communication among vehicles and the road infrastructure of the city [17–22]. Some companies and consortia [23–25] have been working on automated vehicles or instrumented roads to develop automation.

The most common research orientation is based on autonomous systems. In this approach, perception solutions consist of radar, lidar, or camera vision systems. The camera is used to detect white lines on the roads since radar or lidar detects other objects on the road. All the perception systems are used to control vehicles' relative position and trajectory to achieve centimetric accuracy. Therefore, the future challenge is to develop communications with high bandwidth to increase the transfer information capabilities. With the development of the new vehicle lighting technology, a competitive way to transmit information with high data rate is made possible. Vehicle lights have been enhanced for position and intensity control, and reliability. Lights based on an LED matrix now appear commonly. The interests in LED lights are numerous; the very high reliability and a long lifetime are the main ones. These advantages lead the automotive industry to replace the classical halogen lamps by all-LED systems in the near future. Another characteristic of solid-state lamps (SSL) is the capability to be current-modulated for optoelectronics transmissions. LED-based power modulations are very common and the technology is well known. For many years, there have been optical communications to enable transmission with high data rates and bandwidth. Transmission of traffic information or mechanical states of vehicles is then fully compatible with this LED capability.

Additionally, many cities (such as Las Palmas de Gran Canaria (LPA), in the Canary Islands) have recently begun to explore a specific concept inside the general smart city frame: smart tourism destination, as a particular view of the smart city paradigm oriented to tourist-oriented services. In this sense, the deploying of cost and energy-saving traffic lights and streetlights with state-of-the-art LEDs, due to their significantly improving efficiency and long lifetime, can be also used to offer services of guidance and additional information for people with diverse capabilities (whether tourists or not).

7.2.2 Electronics for VLC Outdoor Systems

When thinking of VLC systems based on streetlights for outdoor applications, we can roughly classify them in two main categories:

- Broadcasting systems: One-way systems based only on downlink communications from the LED lamp to a receiver, working as an extension of a wired network, or signaling nodes, which send a fixed data frame to receiver devices for positioning, or performing actions that depend on the transmitted information. Broadcast systems need an emission block in the lamp and a reception stage in the terminal, while network nodes need transmission and reception blocks in each network device.

- Network-aware systems: The lamp includes not only the emitter (LED) but also a receiver, working as a network access point for the VLC terminals in its covered area. This kind of system is more commonly used in indoor environments, since the lamps' distribution can assure the coverage and roaming of the terminals.

On outdoor applications, converting a traditional lamp into a dual illumination-communication optical emitter consists of placing a transmission stage—VLC block—between the lamp driver and the LED fixture, as shown in Figure 7.1. This block can be divided in a communication stage to acquire the information to be transmitted, and a transmission control stage which performs the signal encoding and modulation processes that drive the electrical current to the LEDs. For signaling and positioning applications, the communication block is not required as only a fixed, preprogrammed code is needed and can be transmitted periodically or under request.

The complexity and cost of a VLC transmission block depends on the network connection, the data rate, and the modulation scheme. For example, video broadcasting with high efficiency modulations usually requires a complex preprocessing stage, implemented with Field Programmable Gate Array (FPGA) or Digital Signal Processor (DSP) devices. While systems used for positioning, based on lamps transmitting identification codes (beacons) with a simple encoding scheme, can be easily implemented with low cost microcontrollers.

VLC systems for outdoor applications require a huge transmitted optical power, thus, the driver design in VLC systems needs to accomplish this

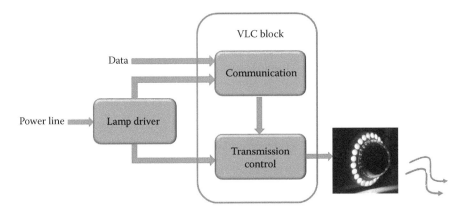

FIGURE 7.1
Emitter scheme.

challenge. Complex modulations such as CSK (Color Shift Keying) or OFDM (Orthogonal Frequency Division Multiplexing), involving multilevel signals, require linear output amplifiers in order to avoid distortion on the transmitted signal (since part of the information is contained in the signal amplitude). These drivers present lower power efficiency than nonlinear ones as they are based on transistors [usually MOSFET (metal–oxide–semiconductor Field Effect Transistor)] working on their ohmic region. This problem becomes more significant for high-power transmissions (as in street lamps or traffic lights), with tens or hundreds of watts of power consumption. On the other hand, modulations based on pulse transmissions [as in OOK (On-Off Keying) or VPPM (Variable Pulse-Position Modulation)], can be nonlinearly amplified without distortion effects, which provides a higher power efficiency, since drivers only switch between ON/OFF stages. This effect makes it more feasible to work with pulsed modulations when dealing with high-power signals. They are baud-rate limited but, for many outdoor applications, hundreds of kb/s are usually more than enough.

Figure 7.2 shows two implementation examples of nonlinear amplifiers for these modulations. Unless for low-power lamps, conventional digital gates can be used for driving the commutation transistors (Figure 7.2a). A MOSFET is required when dealing with higher current values (Figure 7.2b). The low-power device is based on open collector-logical gate chips in a parallel configuration, each of them driving a group of LEDs, for increasing the managed power. This configuration improves the current control and reduces the spurious capacitance values induced by the LEDs that severely affect the available transmission bandwidth. In Figure 7.2b, the scheme includes a high-power MOSFET transistor and a specific integrated driver (e.g., IR2110 from International Rectifier) able to deal even with high current and voltage values.

Another main VLC challenge is the design of the reception stage. It should recover the transmitted data and format the signal to be processed. As in

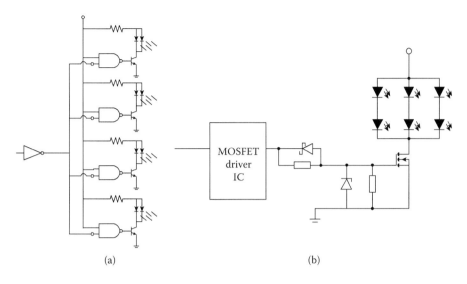

FIGURE 7.2
Transmitter schemes: (a) Parallel configuration with open collector-logical gate chips driving a group of LEDs and (b) transmitter with a MOSFET and an integrated driver.

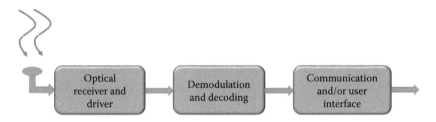

FIGURE 7.3
VLC typical receiver structure that includes the optical to electrical conversion, demodulation/ decoding, and user interface blocks.

the transmission stage, there are two types of elements in the receiver: (1) interfaces between the received data and another device (computer, smartphone, etc.) and (2) systems that process the received data by themselves. The receiver scheme is shown in Figure 7.3. First, there is an optical reception and amplification stage, which performs the optical-electrical conversion and signal conditioning, followed by the demodulation/decoding block. Finally the communication/user interface module transmits the detected data to an external device or, in the case of an embedded system (device working stand-alone), performs some additional processing on the received data. As in transmission, the receiver complexity mainly depends on the network connection, the data rate, and the modulation scheme.

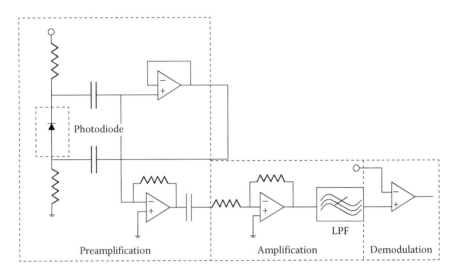

FIGURE 7.4
Optical receiver basic structure.

Receivers for VLC are mostly similar to those used in conventional wireless optical systems, but using a different photodiode spectral responsivity, which includes visible wavelengths for VLC systems instead of the infrared (IR) wavelengths used in optical remote control devices. Figure 7.4 shows an example of a basic structure for an optical receiver with the preamplification block, generally with a transimpedance configuration, a second amplification stage, and a comparison block in the case of pulse-based modulations. As mentioned above, for network-aware systems it is necessary to implement both emitter and receiver stages in the lamp network access point and in the VLC nodes. The general scheme is shown in Figure 7.5.

7.2.3 Using VLC to Improve the Mobility of Blind People: SINAI Project

SINAI [26] was originally conceived as a support system for blind people to provide them with the capability of identifying the state (red, green, caution, and so on) of a traffic light or any other element of urban signage. The focus was placed on sending information by using LOS, high-directivity VLC, avoiding uncertainties in conflicting traffic light scenarios such as busy inter-sections, as illustrated in Figure 7.6. A typical message takes the form "you are in yyyy street, no. zzzz, you can cross" or "you are in avenue yyyy with street xxxx, please wait"; nevertheless, any other required information, not only the state of the traffic light, can be sent (and using any language, depending on the programming of the receiving device). It can be also used to provide guidance through short messages within the city (Figure 7.6).

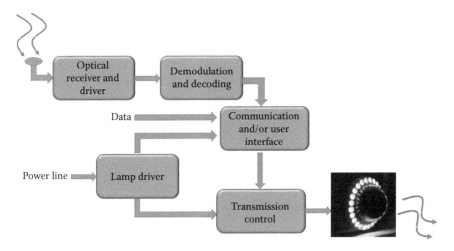

FIGURE 7.5
TX–RX block for VLC network access point and nodes.

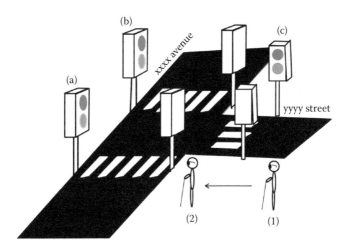

FIGURE 7.6
Basic functionality for the SINAI system: when one user is approaching the streetlight, they will receive one signal with the information about the state of the signaling. As the transmission is highly directive and power can be regulated, they will only receive signal from one emitter (e.g., semaphore ["a"] but not from the others ["b" and "c"]).

Although there are other solutions to aid blind people, this solution offers some advantages that would facilitate their deployment. The main one is that the information is transmitted in a direct way, in contrast to an acoustic system or any RF-based system. Using this, the user only receives light from a source which is in the field of view (FOV) of the detector system. The minimum range of the transmitter is about 10 meters with a half-power angle from the perpendicular focus of 15°. The user's system is autonomous and powered by batteries, and can be activated or deactivated at will. Information is transmitted as a voice advice through a headset that can be connected via Bluetooth. The regulator device installed at the traffic light can also be used to restrict the light output power (and hence power consumption) depending on environmental conditions (day/night, fog, sun, etc.). Figures 7.7 and 7.8 show the practical testing of the SINAI system which is based on commercial and inexpensive devices, so that implementation has

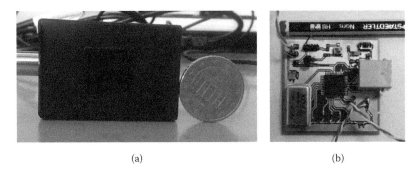

(a)   (b)

FIGURE 7.7
(a) Prototype implementation of the VLC receiver, external view and (b) circuit implementation. In the final, commercial design the dimensions of the receiver can be significantly reduced.

FIGURE 7.8
Laboratory testing of the SINAI system; it has been trialed over typical distances of LPA streets (6 meters for a small road and up to 20 meters for avenues).

a low level of technological risk and the total cost to end users will make it universally accessible. Communication protocol is similar to that proposed in the PHY 1 of IEEE 802.15.7 [27], with a baud rate of 115 kb/s that can be easily scaled. Nevertheless, we consider that high baud rates are not required for the basic funcionality of the device.

7.3 VLC for Indoor Mobility

7.3.1 VLP Fundamentals

In this section, we will describe the use of VLC systems for indoor positioning, also called visible light positioning or VLP. Pedestrian support for visually impaired people involves the use of textured paving blocks, guide dogs, GPS-based voice navigation systems, among others. On the other hand, studies aimed at visually impaired people report that there is a need for voice information inside buildings, and that in the future, we will need adequate indoor pedestrian support systems in large commercial facilities, such as shopping centers and underground shopping malls. However, compared to public spaces and transport facilities, no progress is being made in providing commercial facilities with passive help infrastructure which could be textured paving blocks or audio beacons. VLP can help to provide an adaptive data transmission grid for cost-efficient guiding that could be used not only for visually impaired people in indoor scenarios, but in many other applications. One example could be robot guidance inside an industrial facility (that can be heavily affected by EM noise) or positioning inside a building when security forces are under a bomb threat and jamming devices are active in order to block remote control or cell phone detonators.

There are four main challenges to be kept in mind when dealing with indoor guidance:

- *Accuracy:* Many indoor applications require a position accuracy of a few centimeters and an orientation accuracy of a few degrees. The harshness of indoor environments on signal propagation, caused by obstacles, makes it hard to achieve these accuracies. It is also necessary to provide different types of location information to support diverse indoor applications—user space (which requires accurate boundary detection), position in a coordinate system, and orientation.

- *Scalability:* Indoor environments often contain a large number of physical objects and a high density of people, all requiring location. Hence, an indoor location system needs to scale well with the number and the density of users of the system.

- *User privacy:* The ability to obtain user location without the location system tracking the current location of the user is important to build applications that preserve user privacy.

- *Ease of deployment:* The location system should be easy to deploy, configure, and maintain. The amount of manual configuration and precise placement should be as little as possible.

Additionally, some of the current standards for VLC (including IEEE 802.15.7 or JEITA CP-1221/1222) [28] can be oriented to be used in VLP. Perhaps the most directly relevant to VLP is the JEITA CP-1222 Visible Light ID System, published in 2007, which describes a protocol for transmission of identification signals from LEDs. However, while these early standards show great foresight concerning the importance of VLC and VLP, there has since been a large body of international research on these topics, and the results of this recent research are not incorporated in these proposals. Future standards should build on and extend the earlier standards. For example, IEEE P802.15 Working Group for Wireless Personal Area Networks (WPANs) also considered how positioning can be incorporated in evolving camera communications standards. VLP systems could be designed with different architectures depending on whether there is cooperation among the lights transmitting the signals and/or with the device for which the position is being determined.

Let us consider a room with several illumination SSL lamps (e.g., emergency lights). We can imagine the simplest design where light fittings are transmitting pre-recorded information about the contents of the room (e.g., artwork in a museum, the location of a store in a commercial mall) using headsets as currently provided by museums and other places of interest. In this case, the required number to be entered by the user is automatically provided by the nearest LED, which plays the relevant commentary. We can assume for simplicity that there is no cooperation as the lights simply transmit predetermined signals. While this architecture may not provide the most accurate positioning possible, it is simple to install and cost effective, so it has a great potential for widespread adoption. This approach—beaconing— is based on broadcasting a fixed optical signal (usually, a predefined code or message) through illumination (or emergency) lamps. These codes, when perceived by the user, determine their positions as they are inside a coverage area. Telecommunication standards usually specify the format of the signals to be transmitted, leaving the design of receivers to individual manufacturers. The need for such a global identifier has been anticipated in other fields, and there is already an IEEE-managed 64-bit global identifier, called the EUI-64 [29] which also assures that there are available procedures for generating Ipv6 addresses for each EUI-64 code word. A commercial approach has been recently presented by Carrefour in Lille (France) [30] and provides the user information about its position and the way of finding specific products or promotions. This approach is quite imprecise as it only indicates that you are inside a coverage area that could be 3 or 4 m^2. When more accurate positioning is required, the VLP receiver will use the received signals to determine the relative distance and/or direction of a number of LED transmitters. These measurements will then be combined using classical triangulation (using AOA information) or trilateration (using path length or TOA information) to determine the position of the receiver.

Many positioning techniques and their suitability for use with lighting LEDs have been proposed such as beaconing, RSS, AOA, TDOA, and finger-printing. They are briefly described in the next section.

7.3.2 VLP Proposed Solutions

RSS is the simplest and most common solution to estimate the distance of the receiver(s) from several transmitters in known positions. In this approach, a number of measurements (in the form of a value of optical power transmitted from the LED lamps) are obtained at the receiver, and usually the four (three if we consider only 2-D location) strongest signals are used to obtain the position of the receiver. The power is estimated as a function of the distance from emitter to receiver (denoted as d_i, from $i = 1$ to 4), the location of the targeted receiver by $M = \{M_x, M_y, M_z\}$, and the locations of each LED lamps as $L = \{L_{ix}, L_{iy}, L_{iz}\}$; we can obtain a linear algebraic formula that relates the location of the receiver and three lamps ($H \cdot M = d$). Then the receiver location can be obtained by the LS (Least-Square) method ($M_{est\text{-}LS} = (H^T H)^{-1} H^T d$), where the elements of H and d are composed of the coordinates of the lamps and the estimated distances of the lamps, and T denotes the transpose of the matrix. When the distances are estimated from the measurement data, the LS method finds the point that gives the least sum of differences between the functions of data and the estimated point. Although it provides an easy-to-calculate solution, optimality cannot be guaranteed. An alternative solution [31] is based on a maximum likelihood estimation using an iterative solution such as the Newton–Raphson method, and the $M_{est\text{-}LS}$ solution as the initial value.

Unless many of the papers published so far on VLP have used RSS [32–34], it lacks the necessity of not only a good estimation of the optical channel properties between the transmitter and receiver, but a correct estimation of the optical power transmitted by each LED for an accurate positioning. Unfortunately, these conditions are unlikely to be true in practice. The effect of objects blocking, shadowing, and reflecting the signal mean that the relationship between distance and RSS is almost unpredictable, limiting the accuracy of an RSS approach in LED-based systems. Transmitted optical power itself is also quite unpredictable as it depends on the particular LED and the level of dimming. It will also vary with time, even with factors such as how clean the light fitting is, or whether someone or something is partially blocking the path between light and receiver. Note that although the mechanisms that make RSS potentially unreliable in VLP are very different from those in radio-based systems, the overall result is the same; while rough estimates of position can be made, a number of factors limit the accuracy achievable in practice. Fingerprinting [10,35] is based on recognizing patterns of illumination or even lamp configurations, so we can have a "map" of the illumination parameters over the room, then the mobile device to be located only has to compare the power coming from a lamp (identified by a code) with the stored values to obtain its estimated position. The main problem comes from the narrow relationship between obstacle distribution (especially furniture or moving people) and the received light power distribution.

Another promising method to be considered for VLP systems is AOA [2]. Unless AOA positioning is not often used in radio-based systems because there is typically no LOS between a transmitter and receiver, and also because of the problems caused by multipath transmission. In VLP, the receiver will virtually always have LOS to a number of lights. Although, in addition to the LOS component, the received optical signal will often have a diffuse component due to light reflected from walls and other surfaces, this component is usually very small compared to the LOS component, so any resulting error in AOA estimation will be relatively small. A second factor that makes AOA-based positioning interesting for VLP is that lenses with precise designs are economical to manufacture. This means that relatively simple optical systems can provide accurate AOA information. This is very different from radio systems where determination of accurate AOA requires sophisticated antenna systems. On the other hand, AOA is also heavily affected by shadowing and, when using IM/DD receivers (as it is usually made in VLC), we cannot properly separate light coming from two different lamps. TOA is another technique often used in localization, and is the basis of the GPS system. However, it requires the transmitted signals to be very accurately synchronized. For example, the synchronization of the signals transmitted by GPS satellites is based on very accurate atomic clocks. This is clearly not an option for economical positioning systems based on LED lighting, so it is possible but not optimal. The need for accurate transmitter synchronization can be avoided if TDOA [33] rather than TOA is used. In this case, there must be at least two receivers with an accurately known distance between them. The TDOA of signals reaching the two receivers gives information about the difference in path length from the transmitter. However, as the signals travel at the speed of light and the distance between receivers in indoor applications will necessarily be small, extremely accurate time measurement is required. Localization systems based on phase of arrival (POA) and phase difference of arrival (PDOA) present similar drawbacks as the active area of the receiver is usually much larger than the transmitted wavelength, and many different phases are received simultaneously.

Other solutions can also be addressed. Kim et al. [36] and Jung et al. [37] proposed using carrier allocation methods with different frequencies for each lamp. In this last paper [37], the locations of an object in the room are estimated by using three LED lamps, each one with a unique frequency address identifier (F-ID) modulating its signal. TDOA is estimated through detecting phase differences between the transmitted signals. However, unless accuracy in the measurement of signaling is extremely high, this method introduces an additional complexity to the optical receiver design that would increase implementation costs, and therefore, presents few advantages over the other proposals. An additional drawback to be kept in mind is the additional power consumption required by the lamp driver when using a sinusoidal signal, when compared with simple on/off switching.

As the primary motivation for many users when installing LED-based lamps is saving energy, it is hard to explain why a signal scheme that dramatically reduces this advantage should be used. Specifically for impaired people, Nakajima et al. have tested a solution [8,38] combining LED lighting with the geomagnetic sensor in the already widespread smartphone. However, we can easily imagine several situations where geomagnetic sensors cannot detect the accurate direction due to the presence of large metallic objects or heavy EM noise.

7.3.3 Position Estimation in VLP

If we consider, for simplicity, that each LED has first-order Lambertian radiation pattern, and the receiving angle is always smaller than the FOV, we can neglect the difference of impulse delay (nanoseconds while the system is designed to work at tens of kbps). The received signal becomes Equation 7.1:

$$r(t) = P_0 \sum_{i=1}^{4} s_i(t)h_i + n(t) = \frac{P_0 A_R}{\pi} \sum_{i=1}^{4} s_i(t) \frac{\cos(\phi_i)\cos(\theta_i)}{R_i^2} + n(t) \qquad (7.1)$$

Where n is the mode number of the radiation lobe, ϕ_i is the angle between source orientation vector and the vector pointing from source to receiver; θ_i is the angle between receiver orientation vector and the vector pointing from receiver to source; A_R is the receiver area and FOV is the field of view of the receiver. Each lamp individually emits their location information (a code, a modulated signal, etc.), as positioning references to the mobile device. However, if the signals sent from different sources are simultaneous they will be mixed in the air interface. We will then need to retrieve individual signal and the corresponding channel features, for example, using time division multiplexing (TDM) schemes. Thus, in one frame period, the *i*th LED is assigned a specific time slot between T_{i-1} and T_i, in which it sends its encoded location information.

We can encode X, Y, Z coordinates of the LED into a code, or use a unique code (e.g., an EIT-64 code word) for each lamp. This will be the transmitted *s(t)* signal, OOK modulated (following JEITA or IEEE standards), with an average power of $P_0/2$ on each slot (when the source emits constant high light intensity for only illumination purpose, the average power of these slots is P_0). The channel response h_i for each lamp can be easily obtained as a delayed $\delta(t)$ if we use a unique time slot for each transmitted signal. We can now derive the mobile device location considering that for a received code $s_i(t)$ we have the coordinates X, Y, Z of the *i*th LED$_i$ (L_{ix}; L_{iy}; L_{iz}).

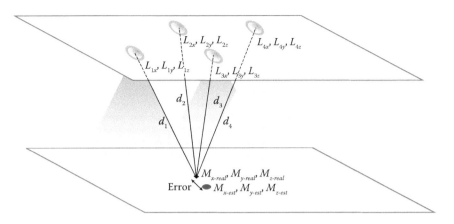

FIGURE 7.9
Setup for the calculation of the position.

The coordinates of the mobile device to be located are unknown and are notated by (M_x; M_y; M_z). Figure 7.9 shows the position estimation setup where we can easily obtain Equation 7.2:

$$\sqrt{\frac{h_i}{K}} = \frac{|L_{iz} - M_z|}{(L_{ix} - M_x)^2 + (L_{iy} - M_y)^2 + (L_{iz} - M_z)^2} = \frac{L_{iz} - M_z}{(L_{ix} - M_x)^2 + (L_{iy} - M_y)^2 + (L_{iz} - M_z)^2}$$

(7.2)

where K is a constant defined as $K = A_R/\pi$. Since L_{iz} is always greater than M_z, we can remove the absolute symbol of $|L_{iz} - M_z|$. When we have four LEDs we obtain an equation group about user locations. Solving the equation group by matrix operation, we have Equation 7.3:

$$2 \cdot \begin{pmatrix} (L_{1x} - L_{2x}), (L_{1y} - L_{2y}), \left(L_{1z} - L_{2z} - \frac{1}{2}\sqrt{\frac{K}{h_1}} + \frac{1}{2}\sqrt{\frac{K}{h_2}}\right) \\ (L_{2x} - L_{3x}), (L_{2y} - L_{3y}), \left(L_{2z} - L_{3z} - \frac{1}{2}\sqrt{\frac{K}{h_2}} + \frac{1}{2}\sqrt{\frac{K}{h_3}}\right) \\ (L_{3x} - L_{4x}), (L_{3y} - L_{4y}), \left(L_{3z} - L_{4z} - \frac{1}{2}\sqrt{\frac{K}{h_3}} + \frac{1}{2}\sqrt{\frac{K}{h_4}}\right) \end{pmatrix} \cdot \begin{pmatrix} M_x \\ M_y \\ M_z \end{pmatrix}$$

$$= \begin{pmatrix} L_{1x}^2 + L_{1y}^2 + L_{1z}^2 - L_{2x}^2 - L_{2y}^2 - L_{2z}^2 + L_{2z}\sqrt{\frac{K}{h_2}} - L_{1z}\sqrt{\frac{K}{h_1}} \\ L_{2x}^2 + L_{2y}^2 + L_{2z}^2 - L_{3x}^2 - L_{3y}^2 - L_{3z}^2 + L_{3z}\sqrt{\frac{K}{h_3}} - L_{2z}\sqrt{\frac{K}{h_2}} \\ L_{3x}^2 + L_{3y}^2 + L_{3z}^2 - L_{4x}^2 - L_{4y}^2 - L_{4z}^2 + L_{4z}\sqrt{\frac{K}{h_4}} - L_{3z}\sqrt{\frac{K}{h_3}} \end{pmatrix}$$

(7.3)

The solution (Mx; My; Mz) is the estimated three-dimension (3-D) user location coordinates. Error is defined by the Euclidean distance between the estimated and real locations of the mobile device:

$$\sqrt{\left(M_{x-est} - M_{x-real}\right)^2 + \left(M_{y-est} - M_{y-real}\right)^2 - \left(M_{z-est} - M_{z-real}\right)^2} \qquad (7.4)$$

Since three coordinates are considered, this error is called 3-D positioning error. In most cases, users are more concerned about their two-dimension (2-D) locations (Mx; My); the 2-D estimation error is Equation 7.5:

$$\sqrt{\left(M_{x-est} - M_{x-real}\right)^2 + \left(M_{y-est} - M_{y-real}\right)^2} \qquad (7.5)$$

7.3.4 A Mixed VLP-Ultrasonic Location System

Trying to overcome some of the limitations of the above-described systems, a mixed VLP-ultrasonic location system ("Firefly") [39], based on the same principle of a cricket system, is presented in Figure 7.10. For better understanding, let us consider the above mentioned room with several illumination LED lamps, each one is associated with an ultrasonic receiver and its position can be considered perfectly known. We also have a mobile device which needs to be positioned. It will send an ultrasonic activation signal; each lamp, after a delay dN (depending on its distance to the emitter), will receive this activation signal for sending its optical code. As the time for light propagation can be neglected (at least, when compared with required time for sound propagation), the delay from each signal relies only on the ultrasonic propagation and therefore in the lamp device, distance and lamps can be considered independent as they are switched asynchronously. This optical code will contain an identification code for the lamp and, assuming that its position is previously known, we can establish a nearly exact position with the obtained distances to several "beacons".

We should also take into consideration that, in an arbitrary position and lamp distribution, two lamps could be at the same (or very similar) distance from the receiver, and therefore, there could be a signal collision. To avoid this, a controlled, random delay can be assigned to each emitter, following a strategy based on slotted time assignment used in TDMA. In this case, we have to add another field to the optical transmitted frames with the random delay added to the signal. The proposed 32-bit and 64-bit frame format are presented in Figure 7.11 where the 64-bit frame presents an ID field that can contain part of the lamp's Ipv6 address and a six-bit field reserved for future applications. In both cases, the slot field number can be compared with a hash of the ID since it is originally obtained that way to determine the delay used to avoid collisions.

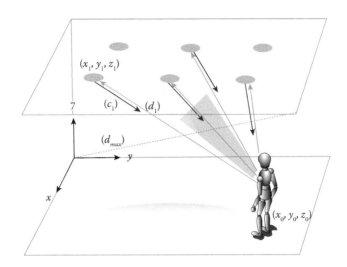

FIGURE 7.10
Schematic representation of the proposed US-VLC system. For a given object on an arbitrary position (X_0, Y_0, Z_0), it will emit an ultrasonic signal (dark gray) to the lamps. After a variable delay (d_i), each lamp, when receiving this signal, will emit an optical code $(C_i$, dark arrow). The optical receiver will calculate the delay for each lamp, assuming this position is known. D_{max} is the maximum distance for the channel, which is used in the Time-Division Multiple Access (TDMA) to prevent collisions among optical transmission codes.

Code frame (32 bits length)					
Syn (8)	Act (4)	ID (12)	Slot (6)	SCS (1)	FCS (1)

Code frame (64 bits length)						
Syn (8)	Act (4)	ID (32)	Slot (8)	SCS (1)	ND (6)	FCS (1)

FIGURE 7.11
Proposed frame format for the optical code, 32 and 64 bits long.

The required time for location can be calculated based on the following assumptions. At T_0, an ultrasonic *ping* is sent by a mobile device. As each lamp is at a different distance, they require a different propagation time, so at $T_0 + T_{1k}$ the ultrasonic ping is received and kth the lamp is switched off. Then, after a variable number of slots (set for each emitter) for avoiding collisions, at instant $T_0 + T_{1k} + i \cdot T_{slot}$ a light code is sent. When the mobile device receives four optical codes, it sends another ultrasonic *ping*. Illumination is set again after $4 \cdot T_{slot}$ or when the second ping is received. So, the total time for location will be less than $T_{1k} + i \cdot T_{slot} + 4 \cdot T_{slot}$ for the farthest lamp, as optical propagation time can be neglected. For a 64-bit frame format and a baud rate of 115 kb/s on a 5 × 5 × 3 m room we have estimated a maximum delay of 20 ms. Nevertheless we can consider substituting the conventional phosphor-blue LED for red, green, and blue (RGB) versions and using alternative energy-efficient modulation schemes such as color-shift keying for decreasing the switched-off time of the lamps to avoid flickering (unless a 20 ms blackout is usually not perceived). In that case we will simply use one color (e.g., red) for the optical code while the remaining color signals are left connected.

Another proposal could rely on preprogramming (e.g., during manufacture) a unique code into each light that has this feature. The need for such a global identifier has been anticipated in other fields, and there is already an IEEE managed 64-bit global identifier, called the EUI-64 [30]. The problem, then, is how to translate this information into a location for location-aware systems. Where are these "directories"? Who performs their maintenance? Who pays for them? One possibility is that, as with telephone directories, organizations could pay to have their codes held in a given directory. Another important advantage of using the EUI-64 code is that standard procedures already exist for generating Ipv6 addresses from them.

One main limitation for the implementation of the proposed system is time resolution for location procedure performing. Time resolution is limited by two sources: the slotted time reserved to avoid collisions, and the time needed for the emission of each optical code (we neglect the time for optical propagation, as it is far smaller than any of these components). As the slotted time is obtained as a function of the time needed by the ultrasonic signal to reach each optical emitter, the main variable is the maximum possible distance among emitters and receivers. This is usually measured as tens of milliseconds, ensuring enough time resolution for people walking in a building or a robot moving through a corridor. The baud rate that can be reached for the lamp does not significantly affect the model even when using either low-speed versions of IEEE 802.15.7 or JEITA CP-1222 [31,32].

Another limitation to be taken into account is the variation of the speed of sound with temperature (0.18% per °C at 25 °C). Since the speed of sound has a relatively large sensitivity to temperature variations, and because indoor temperature can easily vary by 10 °C within the same room, we can even include temperature sensors on the lamp and listeners to compensate for changes in speed of sound due to temperature variations. Each beacon will measure the ambient temperature using an onboard temperature sensor, and include this temperature in its light message. When a listener L computes its distance to a beacon B, the listener measures its temperature T_L, and uses the value $(T_L + T_B)/2$ to represent the room temperature and computes the corresponding speed of sound, where T_B is the temperature at the beacon.

Timing quantization effects should be also considered. In TDOA-based distance estimation, a measured time interval is converted into a corresponding distance. This time interval measurement involves two types of quantization errors. First, measurement of time has a quantization error equal to the period (≈ 1 μs) of the clock used for timing. Second, TDOA-based approaches usually detect the TOA using some kind of reference signal to start the time interval measurement; we will therefore have at least a quantization error equal to a VLC bit duration when detecting the arrival of the incoming code.

Finally, we can also consider the lighting coverage as a limitation for this model, when the mobile device to be located receives less than three source codes (four for 3-D location). This limited source availability can be due to shadowing from obstacles or by the optical properties of the LED light, in particular the HPBW (half-power beam width) that shows the diameter of the illuminated spot for a given distance from the emitter. Considering each lamp as a generalized Lambertian emitter, we can easily obtain the required number of lamps to define the area covered by overlapping the spots from three different emitters. Nevertheless, in real life, a significant part of the incoming optical signal is received after reflections in walls or furniture, so the covered area for each lamp will be larger than the simple LOS component given by the illuminated spot. We have performed simulations using modified Monte Carlo (MMC) and combined deterministic and modified Monte Carlo (CDMMC) algorithms [40,41] in two different room configurations: the first scenario consists of a $8 \times 6 \times 3$ meters room (length, width, and height, x, y, z), six lamps equally distributed, and a maximum distance of 11 meters from any point to the beacon (see Figure 7.12a). The second scenario is a corridor ($3 \times 12 \times 3$) with eight emitters distributed in pairs as shown in Figure 7.13a. In all cases, the LED is profiled as generalized Lambertian with HPBW = 60. As a receiver, a silicon photodiode Hamamatsu S10625-01CT was considered. It has about 1.3×1.3 mm^2 active area, spectral response to visible and IR (340 to 1100 nm), and responsivity of 0.54A/W. This photodiode nominally supports a dynamic range of received power close to 100 dB. In any case, if the dynamic range was at least 10 dB (and is the most general case) in 100% of the surface signal, at least three lamps would be received in both experiments. The effect on the power and the average delay of the received signal is due to reflections. Multipath propagation effects due to walls and obstacles can be neglected since it has been estimated that less than 1% of the total received power comes from the reflected signals (Figures 7.12 through 7.14). Both simulation algorithms offer similar results.

We observed that the measured distance error for each lamp increases with the lamp-to-listener distance d. This increase is to be expected since increasing d causes the received ultrasonic signal strength at the receiver to drop, causing the detection circuits to take a longer time to detect the signal, resulting in an increased positive error. The nonideal ultrasonic transmitter and receiver radiation pattern also explain the increase in error with angular separation between the US emitter direction and the lamp. As the radiation pattern shows, the transmitted power (and receiver sensitivity) drops along directions that are away from the direction facing the ultrasonic transducer, hence the RSS at the listener decreases, again resulting in increased error.

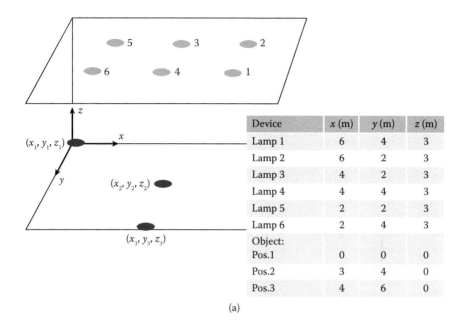

Device	x (m)	y (m)	z (m)
Lamp 1	6	4	3
Lamp 2	6	2	3
Lamp 3	4	2	3
Lamp 4	4	4	3
Lamp 5	2	2	3
Lamp 6	2	4	3
Object: Pos.1	0	0	0
Pos.2	3	4	0
Pos.3	4	6	0

(a)

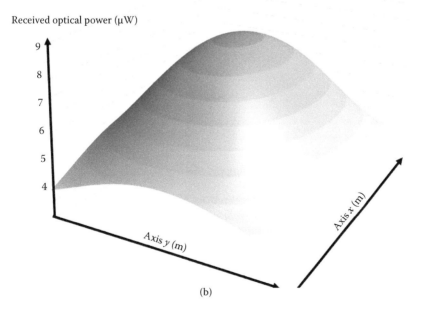

(b)

FIGURE 7.12
(a) Definition of experiment 1, where a rectangular room is illuminated by means of six equally spaced lamps. The location of the lamps and three positions for testing are described in the upper table. (b) The received optical power for experiment 1.

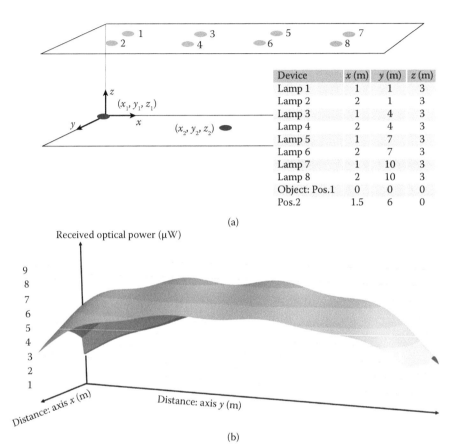

Device	x (m)	y (m)	z (m)
Lamp 1	1	1	3
Lamp 2	2	1	3
Lamp 3	1	4	3
Lamp 4	2	4	3
Lamp 5	1	7	3
Lamp 6	2	7	3
Lamp 7	1	10	3
Lamp 8	2	10	3
Object: Pos.1	0	0	0
Pos.2	1.5	6	0

(a)

(b)

FIGURE 7.13
(a) Definition of experiment 2, where a corridor is illuminated by means of eight equally distributed lamps. The location of the lamps and three positions for testing are described in the upper table. (b) The received optical power for experiment 2.

The ranging accuracy is about 0.5% when the beacon and the listener are 2 m apart and are facing each other; however, the ranging performance degrades as we increase the separation and when they do not face each other. For angles between −40° and 40°, error is less than 5 cm. For large angles (≈ over 80°), the ultrasonic signal at the listener is too weak to be detected and the measure could become erratic.

7.3.5 Experimental Proposal

For the experimental implementation of the whole circuit, in the lamp side, we used its power supply (110 V 60 Hz) for powering not only the SSL lamps but also the ultrasonic receiver and the microcontroller needed for the

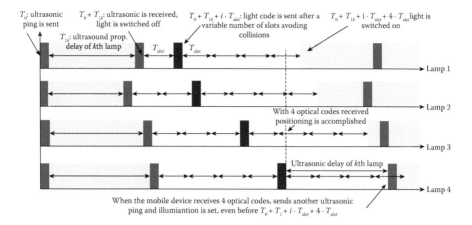

T_0: ultrasonic ping is sent

$T_0 + T_{1k}$: ultrasonic is received, light is switched off

T_{1k}: ultrasound prop. delay of kth lamp

$T_0 + T_{1k} + i \cdot T_{slot}$: light code is sent after a variable number of slots avoding collisions

$T_0 + T_{1k} + i \cdot T_{slot} + 4 \cdot T_{slot}$ light is switched on

T_{slot}

T_{slot}

Lamp 1

Lamp 2

With 4 optical codes received positioning is accomplished

Lamp 3

Ultrasonic delay of kth lamp

Lamp 4

When the mobile device receives 4 optical codes, sends another ultrasonic ping and illumiantion is set, even before $T_0 + T_1 + i \cdot T_{slot} + 4 \cdot T_{slot}$

FIGURE 7.14
Time assignment for the time-slotted structure, dark gray slot is the sonic signal (transmitted and received) while black is the optical one, light gray denotes when the lamp is switched on. (Position 2 of the experiment in Figure 7.12). The time slot can be significantly reduced without loss of generality.

frame setting, therefore voltage conversions are required. First, an AC/DC converter was used to adapt the main 110 V AC to 12 V DC as required by the lamp. An additional DC-DC converter was used to power the microcontroller (e.g., MSP430, fed with 5 V). As the Serial Port Interface (SPI) output from the MSP430 would not provide enough current to switch the lamp, a MOSFET transistor was also used. Due to the switching regime, the transistor dissipates low electrical power, making a heat sink unnecessary. The receiver interface is formed by a transimpedance amplifier, followed by a demodulating and data recovery system. The illumination signal is detected and converted to a current signal by a photodiode. Then, the transimpedance amplifier transforms the current signal to an analog voltage signal, followed by a comparator which eliminates noise and prepares the signal for the demodulation process where another MSP430 could be employed. As it can be seen, all components are low cost and universally available.

As we said before, we have considered until now that there is no network infrastructure connecting the lamps, as it is not required when the object itself desires to know its position. On the other hand, if a network needs to locate an object, we should have an access point (VLC, RF, or any other possibility) to receive the information sent by the user. Lamp interconnection to use a common synchronization is another possibility, but there are many possible drawbacks to be considered. These mainly come from synchronization errors among the lamps due to the physical implementation of the network, and the need of a wired infrastructure that will increase the cost associated to this solution. On the other hand, it will make using commercial lamps possible without the necessity of a pre-recorded code.

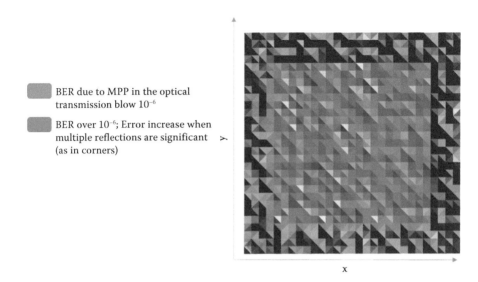

BER due to MPP in the optical transmission blow 10^{-6}

BER over 10^{-6}; Error increase when multiple reflections are significant (as in corners)

FIGURE 7.15
BER estimation due to multipath propagation in a $5 \times 5 \times 3$ square meters room at 115 kb/s.

In order to support visually impaired people who travel indoors, the final implementation will only require positions to be associated inside the building with guidance instructions or services availability at that point. Let us consider as an example one person visiting a hospital, who will have an audio receiver as described for the SINAI system. By means of the US emitter, they will request their position (after a while it will be done automatically). Once the optical codes are received, a mapping application of the building will return their position (usually as a voice message) with additional information which could be "two meters on your left you will find stairs, be careful" or "you are now in front of Dr. House's office." Guide dogs are effective on obstacle-free safe walkways, or avoiding accidents with stairs or furniture, but unfortunately they cannot locate a person's destination. Figure 7.15 shows an example of BER estimation for the VLP-ultrasonic location system in a typical office room.

7.4 Conclusions

In this chapter, we have presented some suitable applications to improve the mobility of visually impaired people. We have taken into account the universal design paradigm regarding buildings, products, and environments that are made inherently accessible to people with disabilities in similar conditions to the majority of the population. We have presented two kind of environments: outdoor and indoor scenarios with different working conditions and necessities. The design goal was optimizing installation by minimizing system cost and complexity as well as power consumption of the receiver. Another design consideration was obtaining an implementation with fully available components ("components off the shelf" or COTS elements). Outdoor devices are simple to build and offer an alternative to classical sound advisement methods in streetlights, opening the possibility of providing more information than the basic red or green state. The same approach is being used to implement information hubs, especially in environments of tourism as an additional application of the smart city (or smart tourism destination) paradigm.

For indoor navigation, there are several available solutions as VLP is becoming one of the hottest topics in VLC research. Technologies such as RSS, AOA, PDoA, TDA, and TDOA have been analyzed, showing some of their strengths and weaknesses, and including references of ongoing experimental implementations. They all share some characteristics: privacy is guaranteed by the inherent security capabilities of the VLC system, and ease of deployment is obtained as it uses a preinstalled facility as the illumination network. On the other hand, one important issue to be taken into account is power consumption and complexity of the portable device, the necessity of maintaining synchronization among the lamps or the energy efficiency related to the signal emitted by the VLC systems. A simple solution, adapting the cricket sensor architecture to be used on VLC environments, has also been proposed. It follows the previous guidelines and the obtained accuracy of the system is similar to regular, RF-based solutions, while avoiding synchronization among the different lamps. The key factor is that the delay measurement is performed over the delay of the ultrasonic signal, so less complexity in the sampling of the received signal is required, compared to when trying to evaluate the delay of an optical transmission, and lower cost devices can be used. Scalability can be easily achieved because, when different users hear the *ping* from one user's ultrasound device, all of them can be on "hearing" state, waiting for the signals from the lamps to locate themselves.

Two different application scenarios were proposed for the indoor guiding system: The first scenario was a device continuously sending beacons to obtain a location identifier, intended for finding paths in airports, hospitals, hotels, and so on. The second scenario was the same device sending beacons but not providing information to the user unless the specific location should produce an alarm or required information. Then, it will send an activation (or "wake up" signal) to the system. This case is well suited not only for visually impaired people but also for certain kinds of illness such as for people affected by Alzheimer's moving through a residence. In this case, the system will signal if the patient is moving out of the attention area and is crossing, for example, the main door to abandon the residence without supervision.

Acknowledgments

Some of the above-described works have been developed in collaboration with multiple companies and administrations. The authors wish to thank the City Government (Ayuntamiento) of Las Palmas de Gran Canaria, Canary Islands, Spain for their support. The SINAI project was funded in part by the Vodafone Foundation. This research is taking place, funded in part by the Spanish Research Administration (MINECO; project ARIES TEC2103-47682-C2-1) and the Mexican Research Administration (CONACYT, 2014 call, project ref. 236188). The authors wish to thank Professor Lopez-Hernandez at the Technical University of Madrid for his advice and kind collaboration.

References

[1] R. Pérez-Jiménez, J. Rabadán, J. M. Luna-Rivera & E. Solana. Visible light communications technologies for smart tourism destinations. Submitted to *First IEEE International Smart Cities Conference (ISC2-2015)*. Guadalajara (México), October 2015.

[2] J. Armstrong, Y. Sekercioglu & A. Neild. Visible light positioning: A roadmap for international standardization. *IEEE Commun. Mag.*, vol. 51, no. 12, pp. 68–73, 2013.

[3] G. Yanying, A. Lo & I. Niemegeers. A survey of indoor positioning systems for wireless personal networks. *IEEE Commun. Surv. Tutorials*, vol. 11, pp. 13–32, 2009.

[4] S. Arnon, J. Barry, G. Karagiannidis, R. Schober & M. Uysal, Eds. *Advanced Optical Wireless Communication Systems*, New York, NY: Cambridge University Press, 2012.

[5] P. Lou, H. Zhang, X. Zhang, M. Yao & Z. Xu. *Fundamental Analysis for Indoor Visible Light Positioning System*, 1st IEEE International Conference on Communications in China Workshops (ICCC), Beijing, pp. 59–63, 2012.

[6] D. O'Brien, R. Turnbull, H. Le Minh, G. Faulkner, O. Bouchet, P. Porcon, M. El Tabach, et al. High-speed optical wireless demonstrators: Conclusions and future directions, *J. Lightwave Technol.*, vol. 30, pp. 2181–2187, 2012.

[7] M. Biagi, A. M. Vegni & T. D. C. Little. LAT indoor MIMO-VLC localize, access and transmit, *IEEE International Workshop on Optical Wireless Communications (IWOW), IEEE*, Pisa, pp. 1–3, 2012.

[8] M. Nakajima & S. Haruyama, Indoor navigation system for visually impaired people using visible light communication and compensated geomagnetic sensing, *2012 1st IEEE International Conference on Communication in China*, 15–17 August, pp. 524–529, 2012.

[9] W. Zhang & M. Kavehrad, Comparison of VLC-based indoor positioning techniques, *Proceeding SPIE 8645, Broadband Access Communication Technologies VII*, pp. 86450M/1–6, 2013.

[10] G. Kail, P. Maechler, N. Preyss & A. Burg. Robust asynchronous indoor localization using LED lighting, *2014 IEEE International Conference on Acoustics, Speech and Signal Processing (ICASSP)*, Florence, Italy, pp. 1866–1870, 2014.

[11] S. Hann, J. H. Kim, S. Y. Jung & C. S. Park. White LED ceiling lights positioning systems for optical wireless indoor applications. *36th European Conference and Exhibition on Optical Communication*, Turin, Italy, pp. 1–3, 2010.

[12] Z. Zhou, M. Kavehrad & P. Deng, Indoor positioning algorithm using light-emitting diode visible light communications, *Opt. Eng.*, vol. 51, no. 8, pp. 085009/1–6, 2012.

[13] N. B. Priyantha, A. Chakraborty & H. Balakrishnan. The cricket location-support system. *ACM Proceedings of the 6th Annual International Conference on Mobile Computing and Networking*, pp. 32–43, August 2000.

[14] A. Smith, H. Balakrishnan, M. Goraczko & N. Priyantha. Tracking moving devices with the cricket location system. *ACM Proceedings of the 2nd International Conference on Mobile Systems, Applications, and Services*, pp. 190–202, June 2004.

[15] D. Iturralde, C. Azurdia, N. Krommenacker, Soto I., Z. Ghassemlooy & N. Becerra. A new location system for an underground mining environment using VLC. *IEEE International Symposium on Communications Systems, Networks & Digital Signals*, pp. 1165–1169, 2014.

[16] S. Ayub, M. Honary, S. Kariyawasam & B. Honary. A practical approach of VLC architecture for smart city, *IEEE Antennas and Propagation Conference (LAPC)*, Loughborough, 2013.

[17] A. Cailean, B. Cagneau, L. Chassagne, S. Topsu, Y. Alayli & J. M. Blosseville. Visible light communications: Application to cooperation between vehicles and road infrastructures. *IEEE Intelligent Vehicles Symposium*, vol. IV, pp. 1055–1059, 2012.

[18] S. Iwasaki, C. Premachandra, T. Endo, T. Fujii, M. Tanimoto & Y. Kimura. Visible light road-to-vehicle communication using high-speed camera. *IEEE Intelligent Vehicles Symposium*, pp. 13–18, 2008.

[19] N. Kumar, D. Terra, N. Lourenço, L.N. Alves & R.L. Aguiar. Visible light communication for intelligent transportation in road safety applications. *7th International Wireless Communications and mobile computing conference (IWCMC)*, pp. 1513–1518, 2011.

[20] M. Akanegawa, Y. Tanaka & M. Nakagawa. Basic study on traffic information system using LED traffic lights. *IEEE Trans. Intell. Transport. Syst.*, vol. 2, no. 4, pp. 197–203, 2001.

[21] N. Kumar, L.N. Alves & R.L. Aguiar design and analysis of the basic parameters for traffic information transmission using VLC. *1st IEEE International Conference on Wireless Communication, Vehicular Technology, Information Theory and Aerospace & Electronic Systems Technology*, 2009.

[22] T. Yendo, M. P. Tehrani, T. Yamazato, H. Okada, T. Fujii & M. Tanimoto. High-speed-camera image processing based LED traffic light detection for road-to-vehicle visible light communication. *IEEE Intelligent Vehicles Symposium* (IV), 2010.

[23] John Markoff. Google Cars Drive Themselves, in Traffic. 2010. Available: http://www.nytimes.com/2010/10/10/science/10google.html?_r=3&partner=rss&emc=rss&pagewanted=all (accessed February 18, 2017).

[24] The SARTRE Project. Available: http://www.sartre-project.eu/en/Sidor/default.aspx (accessed February 18, 2017).

[25] VLCC, Visible Light Communication Consortium. Available: http://www.vlcc.net/?ml_lang=en (accessed February 18, 2017).

[26] Fundación Vodafone España (in Spanish). Available: http://www.fundacionvodafone.es (accessed February 18, 2017).

[27] S. Rajagopal, R. Roberts & S. Lim. IEEE 802.15. 7 visible light communication: Modulation schemes and dimming support. *IEEE Commun. Mag.*, vol. 50, no. 3, 72–82, 2012.

[28] *Visible Light Beacon System, CP-1222*, May 2013. Available: http://home.jeita.or.jp/tsc/std-pdf/CP1222.pdf (accessed February 18, 2017).

[29] *Guidelines for 64-bit Global Identifier (EUI-64TM)*, 2012, Available: http://standards.ieee.org/develop/regauth/tut/eui64.pdf (accessed February 18, 2017).

[30] Philips News Center. Available: http://www.philips.com/a-w/about/news.html (accessed February 18, 2017).

[31] J. Lim. Ubiquitous 3D positioning systems by LED-based visible light communications. *IEEE Wireless Commun.*, vol. 22, no. 2, pp. 80–85, 2015.

[32] G. Del Campo-Jimenez, J. M. Perandones & F. J. Lopez-Hernandez. A VLC-based beacon location system for mobile applications. *IEEE International Conference on Localization and GNSS (ICL-GNSS)*, 2013.

[33] S. Y. Jung, S. Hann, S. Park, & C. S. Park. Optical wireless indoor positioning system using light emitting diode ceiling lights, *Microwave Opt. Technol. Lett.*, vol. 54, pp. 1622–1626, 2012.

[34] T. H. Do & M. Yoo. Potentialities and challenges of VLC based outdoor positioning. *IEEE International Conference on in Information Networking (ICOIN)*, pp. 474–477, January 2015.

[35] Y. Nakazawa, H. Makino, K. Nishimori, D. Wakatsuki & H. Komagata. Indoor positioning using a high-speed, fish-eye lens-equipped camera in visible light communication. *IEEE 2013 International Conference on Indoor Positioning and Indoor Navigation (IPIN)*, pp. 1–8, October 2013.

[36] H. Kim, D. Kim, S. Yang, Y. Son & S. Han. An indoor visible light communication positioning system using a RF carrier allocation technique. *IEEE J. Lightwave Technol.*, vol. 31, no. 1, pp. 134–144, 2013.

[37] S. Y. Jung, S. Hann & C. S. Park, TDOA-based optical wireless indoor localization using LED ceiling lamps. *IEEE Trans. Consum. Electron.*, vol. 57, no. 4, pp. 1592–1597, 2011.

[38] M. Nakajima & S. Haruyama, New indoor navigation system for visually impaired people using visible light communication. *EURASIP J. Wireless Commun. Networking*, no. 1, pp. 1–10, 2013.

[39] I. Marin-Garcia, P. Chavez-Burbano, A. Munoz-Arcentles, V. Calero-Bravo, R. Perez-Jimenez. Indoor location technique based on visible light communications and ultrasound emitters. *IEEE International Conference on Consumer Electronics (ICCE)*, 2015.

[40] F. J. Lopez-Hernandez & R. Perez-Jimenez. Ray-tracing algorithms for fast calculation of the channel impulse response on diffuse IR wireless indoor channels. *Opt. Eng.*, vol. 39, no. 10, pp. 2775–2780, 2000.

[41] M. I. Chowdhury, W. Zhang & M. Kavehrad. Combined deterministic and modified monte carlo method for calculating impulse responses of indoor optical wireless channels. *IEEE J. Lightwave Technol.*, vol. 32, no. 18, pp. 3132–3148, 2014.

8

Car-to-Car Visible Light Communications

Pengfei Luo, Hsin-Mu Tsai, Zabih Ghassemlooy, Wantanee Viriyasitavat, Hoa Le Minh, and Xuan Tang

CONTENTS

8.1 Introduction

The light-emitting diode (LED)-based visible light communications (VLCs) have been gaining attraction in research and applications in recent years, thanks to its huge potential in future energy-saving lighting, display, and wireless data communications. With the ongoing development of white LED devices, the luminous efficiency of commercial white LEDs (WLEDs) has increased to 150 lm/w, which is almost 10 times that of the tungsten incandescent lamp [1]. In addition, WLEDs have an expected lifespan of over 15,000 hours, at least 10 times that of incandescent bulbs [2].

LEDs are much more compact and have higher energy efficiency. Furthermore, LEDs can be switched on and off at the speeds of sub-microseconds [3], thus offering functionalities such as data transmission, sensing, and localization beside illumination [4]. As a result, we are witnessing an explosive growth in the use of LED lamps as replacement for the conventional lamps, which creates huge opportunities for lighting and telecommunications industry, academia, and the way we will use lighting infrastructure in the future. The VLC technology, with its unique characteristics, is an alternative and complementary to the radio frequency (RF) wireless communications, not only for indoor applications but could also be used for outdoor applications such as vehicular communications (vehicle-to-vehicle communications or also known as car-to-car communications [C2C]), as part of the intelligent transportation systems (ITS) in future smart cities.

According to the global status report on road safety 2013: Supporting a decade of action [5] issued by the World Health Organization (WHO), road traffic injuries are the leading cause of death among young people aged 15–29, and the eighth leading cause of death globally; about 1.25 million road traffic deaths occurred on the world's roads in 2013. To address this global problem, urgent actions and concerted efforts are needed to prevent and reduce car accidents as well as improve road safety in the near future. Accordingly, ITS, which involves the application of the advanced information processing, control technologies, sensors, and wireless communications [6], has been proposed to improve road safety, traffic flow, and environmental concerns as well as monitor driving behavior [7].

Current ITS research activities, products, and standardizations mainly focus around the deployment of the RF-based communication technologies for wireless connectivity in vehicular networking. Dedicated short-range communications (DSRC) technology operates at 5.9 GHz [8], and is composed of a set of physical, data link, and higher layer standards for V2V and vehicle-to-infrastructure (V2I) communications. For the V2V communications technology to take root at a global level, a few key aspects that should be considered are: (i) new applications that would be attractive to the driver when purchasing a new vehicle; (ii) simple technology with minimum added cost to the vehicle's price; (iii) reliability and quality of service. The VLC technology based around the wavelength band of 390–750 nm has many inherent advantages over the RF-based DSRC technology, and could be adopted for ITS applications.

- **Low complexity and cost**—LED lamps are already installed in vehicles (e.g., in center high mount stop lamps, 3rd brake lights, brake lights, turn signals, fog lights, and headlamps), traffic lights, and streetlights. Additionally, the VLC transceiver design is much less complex than RF-based systems, because of a much less severe multipath effect.

- **High precision positioning**—Owing to the highly directional line-of-sight (LOS) propagation characteristics, the VLC-based positioning technology is able to reduce the positioning error to a few centimeters, which is more accurate than the RF-based positioning technology [9].
- **Improved link quality**—Important when dealing with traffic congestion, particularly during rush hours, where the traditional RF-based systems would experience undesirable packet collisions and longer delays [10]. Whereas with the VLC-based C2C technology, the vehicles only receive signals from their neighboring vehicles that have the greatest impact on their safety, thus leading to much reduced signal congestion.
- **Scalability**—RF-based V2V communications experience longer delay and lower packet reception rate because of the large number of nodes that participate in channel contention. To overcome this problem, adaptive transmission power schemes could be employed but at the cost of high overhead to precisely estimate the number of vehicles (i.e., nodes) in the locality, thus leading to reduced link reliability and availability. However, with VLC-based V2V communications, only a small number of neighboring vehicles, which are most likely to cause an accident with the host vehicle, could transmit to the host vehicle and participate in channel contention. This selection scheme depends only on the optical propagation property with no requirement for any overhead, and hence is highly scalable [11].
- **Security**—VLC offers high security because of the closer operation range and the LOS-only propagation mechanism. To mount an attack, the attacker has to be in the visual range of the VLC transmission for some time, which is significantly more difficult compared to the RF-based technology. Additionally, the positioning feature of VLC technology can be used to provide an additional layer of security by means of verifying whether the received message is transmitted from a valid spatial location.
- **Weather conditions**—The communications link availability could be reduced due to the high attenuation caused by heavy fog, rain, or snow, thus shortening the transmission range. However, a VLC Rx has better sensitivity than human eyes, which means that under bad weather conditions, VLC will be able to receive a message well before human eyes.
- **Camera-based VLC system**—Also known as optical camera communication (OCC), which employs a camera as a receiver (Rx), this has many unique features compared to the RF-based system. For example, it can spatially separate signals to enable parallel signal transmission, and can utilize the built-in color array to separate signals according to different center wavelengths to establish a wavelength-division multiplexing (WDM) link.

TABLE 8.1

Comparison of VLC and RF (DSRC) Schemes

Type	VLC	RF (DSRC)
Communication mode	LOS—point-to-point	Broadcasting—point-to-multipoint
Target data rate	400 Mbps	27 Mbps
Carrier frequency	400–790 THz	5.85–5.925 GHz
Licensing	Free	Required
Mobility	Low–medium	High
Power efficiency	High	Medium
Coverage area	Short-ranged and narrow	Long-ranged and wide
Security	High	Low

Table 8.1 shows the comparison of VLC and RF (DSRC) schemes [12,13]. The characteristics of a C2C VLC link and channel are summarized as follows:

- Average link duration—determines the time duration during which two cars communicate with each other.
- Link throughput and bit error performance—determines the type of applications that can be put on the C2C VLC links.
- Channel time variation—has implication in both the link throughput and/or the error performance.

Other requirements include standard models or unified mathematical models for:

- Car headlamp optical beam pattern
- Noise sources including sunlight and ambient light
- Influence of road surface (conditions and materials used) on light
- Weather conditions

The latter two will affect the road surface reflection properties, received light intensity, and the maximum data rate R_b that can be transmitted due to the multipath induced interference.

In recent years, we have seen growing research activities both theoretically and experimentally on C2C VLC systems. In [14], the feasibility of a road-to-vehicle communication system using an LED array and a high-speed camera was studied. A hierarchical coding scheme for allocating the data to different spatial frequency components depending on their priorities was adopted. According to the results of both static and driving field trials, there was an improvement in the bit error rate (BER) performance following adaptation of a hierarchical coding scheme. In [15], an outdoor VLC system for the ITS application was investigated, where a direct sequence spread spectrum (DSSS) scheme was used in place of the most commonly and widely used schemes of on-off keying (OOK) and pulse-position modulation (PPM) in order to minimize the effect of ambient noise. The system achieved a low R_b of 20 kpbs over a distance >40 m in the presence of ambient light (e.g., the sun). In [16], a VLC system for vehicle safety applications with R_b of 10 kbps over a distance of 20 m was reported. In [17], the channel characterization of a traffic light to a C2C VLC system was studied, where an analytical LOS path loss model was proposed and validated by experimental measurement. In this work, the background noise interference including solar radiation and artificial lighting are characterized, and the performance of proposed system is evaluated for different modulation schemes. In summary, these existing research and development works on C2C VLC systems were mainly experimental based on much simplified theoretical analysis of the LOS channel. However, in order to expand the study of the C2C VLC system, a more accurate channel model, which considers headlamps and road reflections, is required.

In the C2C VLC system, its capacity can be further increased when the pair of car's headlamps could be simultaneously used for data transmission, for example, establishing a multiple-input multiple-output (MIMO) link. This chapter also outlines the performance analysis and evaluation of a C2C VLC using 2 × 2 MIMO. For channel modeling, a market-weighted headlamp beam pattern model is employed and both LOS and non-line-of-sight (NLOS) configurations are incorporated. The measured light beam from the actual vehicle is also presented. For the BER analysis, a Monte Carlo (MC)-based system-level model is developed, where different communication geometries could be considered. The relationships between the BER and the transmission distance for typical geometries are also given. Finally, network and upper layer performances from the application perspective are outlined and discussed.

The rest of the chapter is organized as follows. In Section 8.2, C2C VLC communication models including both single-input single-output (SISO) and MIMO, noise sources, and road surface are outlined. Characterization of the C2C VLC link duration and channel time variation are presented in Section 8.3, whereas the system performances are outlined in Section 8.4. The application of a C2C VLC system is discussed in Section 8.5, and finally, conclusions and future discussions are given in Section 8.6.

8.2 Car-to-Car VLC Model

In C2C VLC systems, a market-weighted headlamp model, road surface Lambertian reflection model, optical MIMO model, and ambient noise model are adapted. According to [18], a typical C2C VLC scenario is

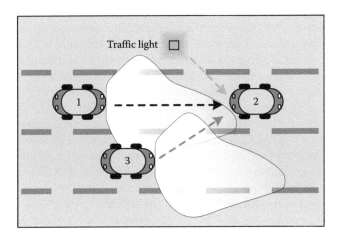

FIGURE 8.1
A typical C2C VLC system with interference from nearby lights.

shown in Figure 8.1, where car 1 sends data to car 2 using its front low-beam headlamps. Unlike the ordinary Lambertian lamp used for illumination, the low-beam headlamps have a special beam pattern, which is suitable for road illumination. Normally, the received light from the front car is a combination of rays from LOS and NLOS (due to reflections from the road surface) paths. According to [19], the road pavement materials (asphalt, concrete), the angle of incidence, and the weather condition (fog, rain, snow, etc.) affect received light intensity (power) power P_{Rx} received by car 2 and result in multipath-induced inter-symbol interference (ISI), which leads to reduced R_b. However, since the typical R_b of a C2C VLC system is not too high (less than ~ Mbps), ISI is not a major concern at all. From Figure 8.1, it can be seen that the major noise sources are lights from nearby cars (e.g., car 3), roadside infrastructures (e.g., traffic lights, streetlights, etc.), and artificial light sources during the nighttime, and the sunlight during the daytime, which affect the link performance.

8.2.1 Car Headlamp Model

In order to make sure that vehicles could provide good road illumination while not to cause glare to other road users, the lamps, reflective devices, and associated equipment must meet the specific requirements [20] outlined by the Economic Commission of Europe (ECE) and the Federal Motor Vehicle Safety Standards (FMVSS) of the US. The high beams are typically used for long-distance visibility with no oncoming cars, and the low beams, with an asymmetrical pattern, provide maximum forward and lateral illumination while minimize glare toward oncoming cars and road users.

The combination of both the high- and low-beam headlamps in vehicles provides a safe and comfortable driving conditions for drivers and other road users during day and night times and in all weather, traffic, and road conditions. The Lambertian model, which has the symmetrical profile, has been widely used in indoor VLC LED modeling and is therefore not appropriate for the modeling of vehicle's headlamp. To increase the reliability of the proposed model for C2C MIMO VLC, a market-weighted headlamp beam pattern model is used as part of the VLC channel model. The headlamps for this market-weighted database [21] were randomly selected from the top 90% of USA vehicle sales for 2010 of which at last 25 samples were used. Following photometric data measurement using a goniophotometer, the data were weighted by the current sales figure for the corresponding vehicle.

Figures 8.2 and 8.3 as reported in [21], demonstrate the isocandela and isoilluminance diagrams of the road surface from a pair of high- and low-beam headlamps (for cars in USA), respectively (luminous intensities

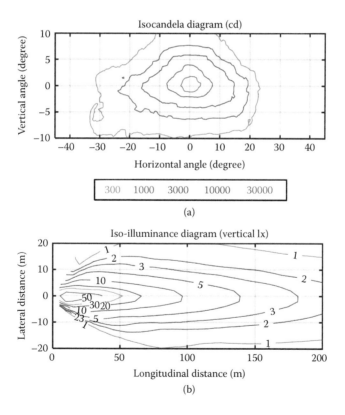

FIGURE 8.2

(a) Isocandela (cd) and (b) iso-illuminance (vertical lx) diagrams of the road surface from a pair of high-beam headlamps, luminous intensities at the 75th percentile (lamp mounting height: 0.62 m; lamp separation: 1.12 m).

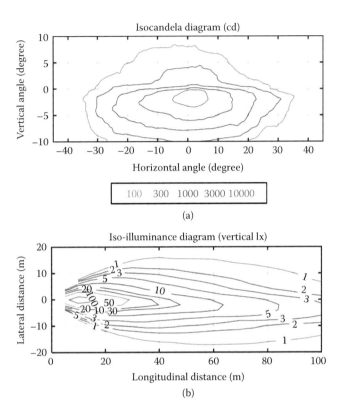

FIGURE 8.3
(a) Isocandela (cd) and (b) iso-illuminance (vertical lx) diagrams of the road surface from a pair of low-beam headlamps, luminous intensities at the 75th percentile (lamp mounting height: 0.66 m; lamp separation: 1.20 m).

at the 75th percentile). It is apparent that for high-beam headlamps, a narrow and flat beam is projected in the horizontal direction of a few degrees to the right, providing a quasi-symmetrical illumination pattern on the road surface, see Figure 8.2b. However, the low-beam headlamps provide an asymmetrical pattern designed to offer adequate forward and lateral illuminations, in addition to controlling the glare by limiting the light being directed toward the other road users, see Figure 8.3b.

According to [22], the illuminance E on the road surface is given by:

$$E = \frac{d\Phi}{dS} = \frac{d\Phi}{d\Omega} \cdot \frac{d\Omega}{dS} = I(\zeta, \xi) \frac{d\Omega}{dS} = \frac{I(\zeta, \xi) \cos \gamma}{d^2}, \tag{8.1}$$

where $d\Phi$ is the luminous flux (lm), S is the area of the road surface (m^2), Ω is the solid angle (sr), $I(\zeta, \xi)$ is the luminous intensity (cd), ζ and ξ are the horizontal and vertical angles (in relation to the headlamp axis), respectively, d is

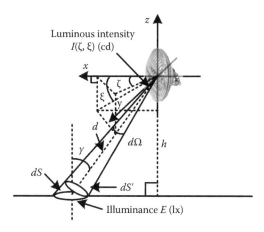

FIGURE 8.4
Illuminance model.

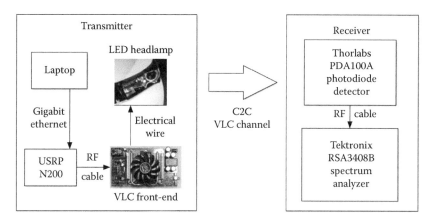

FIGURE 8.5
Block diagram for light intensity measurement for OEM LED lamp.

the distance between the light source and the small area dS, and γ is the angle between the road surface normal and the incident direction, see Figure 8.4.

Figure 8.5 shows the schematic block diagram for measuring light intensity of OEM LED headlamps of a Toyota Corolla Altis (Taiwan model, 2015). The system is composed of a low beam of the left LED headlamp, which is configured to transmit a sinusoidal signal at 1 MHz, and an optical Rx module (Thorlabs PDA100A). A spectrum analyzer (Tektronix RSA3408B) is used to measure the received light intensity (power) within a 500-Hz bandwidth window centered at 1 MHz carrier frequency. The important measurement parameters are summarized in Table 8.2.

TABLE 8.2

Key Parameters for OEM LED Headlamp

Parameter	Value
Transmitter	Headlamp, low-beam, 2015 Toyota Corolla Altis (Taiwan model)
Transmitting current	500 mA
Height of the transmitting lamp	0.7 m
Receiver	Thorlabs PDA100A
Detection area	9.8 mm × 9.8 mm
Gain	750 V/A
Reflectivity	0.2–0.45 A/W (400–700 nm wavelength)
Spectrum analyzer	Tektronix RSA3408B
Received power measurement parameters	Frequency: 1 MHz Window size: 500 Hz

Figure 8.6 shows the measured received illumination intensity (power) patterns of the OEM LED headlamp for a range of Rx heights with respect to the location of the headlamp (note that the difference between the results is in the height of the Rx's location). The trends observed from the measurement results are in-line with results produced using the market-weighted database. For example, both exhibit narrow and flat beam profiles with reduced intensity on the left side to prevent glares to the cars traveling in the opposite direction. It is worth noting that there exists a marginal difference between the two sets of results: the former measures the received power at the carrier frequency, while the latter measures the direct current (DC). Considering the individual headlamp and different Rx heights, the actual received power could be quite different, while this difference could be averaged out in the market-weighted results as they combine results from multiple headlamps.

8.2.2 Modeling of Road Surface Reflection

In general, description of the reflection properties of road surfaces are complex [23]. It can be modeled using luminance coefficients for the range of angles, which has been developed for different road surface classifications based on a large number of photometric measurements [24]. A simplified reflection model with a Lambertian profile is depicted in Figure 8.7. Here, it is assumed that the Lambertian order $m = 1$, which leads to the reflected radiant intensity $R(\phi)$ given as [25]:

$$R(\phi) = \rho \frac{\cos \phi}{\pi}, \tag{8.2}$$

where ρ is the diffuse reflectivity, which varies with different pavement materials, and φ is the polar angle of the scattered light.

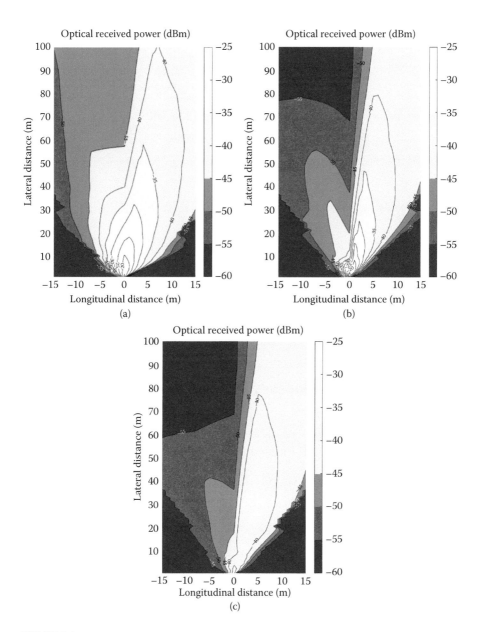

FIGURE 8.6
Measured illumination patterns of the OEM LED headlamp for the receiver height of (a) 55 cm, (b) 70 cm, and (c) 85 cm.

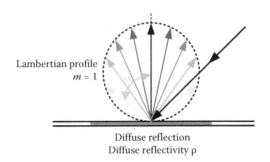

FIGURE 8.7
Road surface reflection with Lambertian profile.

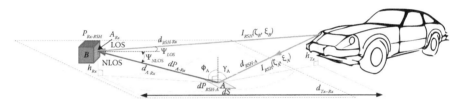

FIGURE 8.8
Configuration of C2C VLC system (only rays from the right headlamp to the Rx are illustrated).

8.2.3 VLC Channel for C2C Communications

The schematic block diagram of the C2C VLC system with the channel model is illustrated in Figure 8.8. Note that the Rx captures light beams from both the right and left headlamps. Assuming that right and left headlamps have almost the same output light distribution [26], then only the right side headlamp (RSH) is considered in the following analysis. As shown in Figure 8.8, both LOS and NLOS paths from a single Tx are calculated.

According to (1) and Figure 8.8, the vertical illuminance E_{RSH-A} at the point A with an area of dS is given by:

$$E_{RSH-A} = \frac{I_{RSH}(\zeta_A, \xi_A)\sin \gamma_A}{d_{RSH-A}^2}, \tag{8.3}$$

where $I_{RSH}(\zeta_A, \xi_A)$ is the luminous intensity of the RSH from the direction (ζ_A, ξ_A), γ_A is the angle between the road surface normal direction of the point A and the incident direction, and d_{RSH-A} is the path length from RSH to the point A, see Figure 8.8.

The luminous efficacy of radiation (LER) of a high-power phosphor-coated WLED of 250.3 lm/W as given in [27] is adopted. Hence, for the RSH, the vertical radiant flux $dP_{RSH\text{-}A}$ at the point A is expressed as:

$$dP_{RSH\text{-}A} = \frac{E_{RSH\text{-}A} \cdot dS}{LER} = \frac{I_{RSH}(\zeta_A, \xi_A)\sin \gamma_A}{LER \cdot d^2_{RSH\text{-}A}} dS. \qquad (8.4)$$

Therefore, P_{Rx} from a single reflected path at the Rx placed at position B is given by:

$$
\begin{aligned}
dP_{Rx-RSH-NLOS} &= \frac{dP_{RSH\text{-}A} \cdot R(\phi_A) \cdot A_{Rx} \cdot \cos \psi_{NLOS}}{d^2_{A-Rx}} \\
&= \frac{I_{RSH}(\zeta_A, \xi_A)\sin \gamma_A A_{Rx}\rho\cos \phi_A \cos \psi_{NLOS}dS'}{LER\pi d^2_{RSH\text{-}A}d^2_{A\text{-}Rx}}
\end{aligned}
\qquad (8.5)
$$

where A_{Rx} and $d_{A\text{-}Rx}$ are the area of the Rx (i.e., the photodetector [PD]) and distance between the point A and the Rx, respectively, ϕ_A is the polar angle of the scattered light from point A to the Rx, and ψ_{NLOS} is the angle of incidence of the NLOS link from the view of the PD.

Therefore, for the RSH, the total received optical power $P_{Rx\text{-}RSH\text{-}NLOS}$ from all reflected paths is expressed as:

$$
P_{Rx-RSH-NLOS} =
\begin{cases}
\iint\limits_{S} dP_{Rx-RSH-NLOS}dS & 0 \leq \psi_{NLOS} \leq \Psi \\
0 & \psi_{NLOS} > \Psi
\end{cases}
, \qquad (8.6)
$$

where Ψ is the half angle of PD's field of view (FOV), and S is the entire area of road surface that has been illuminated. For the RSH, $P_{Rx\text{-}RSH\text{-}LOS}$ of the LOS link is expressed as [28]:

$$
P_{Rx-RSH-LOS} =
\begin{cases}
\dfrac{I_{RSH}(\zeta_B, \xi_B)}{LER \cdot d^2_{RSH-Rx}} \cdot A_r \cdot \cos(\psi_{LOS}) & 0 \leq \psi_{LOS} \leq \Psi \\
0 & \psi_{LOS} > \Psi
\end{cases}
, \qquad (8.7)
$$

where $I_{RSH}(\zeta_B, \xi_B)$ is the luminous intensity and ψ_{LOS} is the angle between the PD surface normal and the incident direction. Therefore, the total $P_{Rx\text{-}RSH\text{-}LOS}$ from the RSH is given by:

$$P_{Rx\text{-}RSH} = P_{Rx\text{-}RSH\text{-}NLOS} + P_{Rx\text{-}RSH\text{-}LOS}. \qquad (8.8)$$

Consequently, the total $P_{Rx\text{-}T}$ is expressed as:

$$P_{Rx\text{-}T} = P_{Rx\text{-}RSH} + P_{Rx\text{-}LSH}, \qquad (8.9)$$

where $P_{Rx\text{-}LSH}$ is P_{Rx} from the left side headlamp (LSH), which is the same as (8) except for the different lateral position.

Hence, the channel DC gain model for C2C VLC system, which includes both LOS and NLOS paths, is derived. Note that this model only considers one Rx.

However, for more Rxs, the channel gains can be determined by using the same model but for different positions of Rxs.

8.2.4 MIMO C2C VLC Channel

Since every car comes with two headlamps and taillamps, the possibility of a 2×2 MIMO [29] configuration to increase the total system throughput by employing two PDs at the Rx is supported. Figure 8.9 illustrates a block diagram of a 2×2 MIMO-based C2C VLC system. Note that $d_{LSH\text{-}RSH}$ and $d_{LSR\text{-}RSR}$ are the distance between two headlamps and two Rxs, respectively. At the Tx, the original serial data are converted into two parallel data streams of x_1 and x_2, which are then used for intensity modulation of 2-LED headlamps (i.e., LSH and RSH). At Rx, the received signals y_1 and y_2 can be represented as:

$$\mathbf{Y} = \mathbf{HX} + \mathbf{N}, \tag{8.10}$$

where $\mathbf{Y}$ and $\mathbf{X}$ are transmitted and received vectors, respectively, $\mathbf{H}$ is the channel matrix, and $\mathbf{N}$ is the noise vector, which are given by:

$$\mathbf{Y} = \begin{bmatrix} y_1 & y_2 \end{bmatrix}^{\mathrm{T}}, \tag{8.11}$$

$$\mathbf{X} = \begin{bmatrix} x_1 & x_2 \end{bmatrix}^{\mathrm{T}}, \tag{8.12}$$

$$\mathbf{H} = \begin{bmatrix} h_{11} & h_{12} \\ h_{21} & h_{22} \end{bmatrix}, \tag{8.13}$$

$$\mathbf{N} = \begin{bmatrix} n_1 & n_2 \end{bmatrix}^{\mathrm{T}}, \tag{8.14}$$

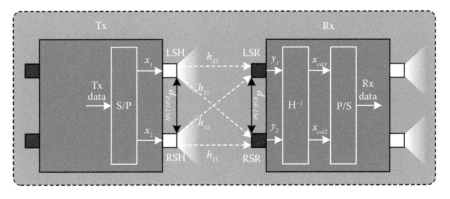

FIGURE 8.9
Block diagram of a 2×2 MIMO C2C VLC system.

where h_{ij} is the channel DC gain from Tx i to Rx j. And i and j equal to 1 for LSH and LSH and equal to 2 for RSH and RSH as illustrated in Figure 8.9. For example, h_{11} can be calculated as:

$$h_{11} = P_{Tx\text{-}LSH\text{-}LSH}/P_{Tx\text{-}LSH}, \tag{8.15}$$

where $P_{Rx\text{-}LSH\text{-}LSH}$ is the received optical power of LSH from LSH, and $P_{Tx\text{-}LSH}$ is the transmitted optical power of LSH.

In order to retrieve the original data from **Y** at the Rx, we have the estimated signal $\mathbf{X_{est}}$, which is given by:

$$X_{est} = \mathbf{H}^{-1} \times Y. \tag{8.16}$$

Note that to successfully determine $\mathbf{X_{est}}$, the channel matrix **H** must be full rank.

8.2.5 Noise in C2C VLC

For optical wireless communication channel, there are two additional light noise sources, which are the background solar radiation during the daytime, and the artificial light (i.e., streetlights, vehicles, static neon signboards, and advertising screens) at nighttime. The background solar radiation is composed of direct and scattered radiations. The former noise is much stronger than latter one and is mostly the dominant noise source. Note that the intensity of the solar radiation received at the earth surface changes with the weather conditions and the position of the sun both during the day and throughout the year [30]. The scattering radiation is not that easy to model due to the surrounding environment. According to [11], the measured electrical power spectrum of the solar radiation is almost constant (i.e., a DC), which can be easily removed by alternating current (AC) coupling. However, the shot noise induced by the solar radiation remains the main source of noise for C2C VLC systems during the daytime [11,17]. Such a noise can be reduced by adopting a combination of optical filter and digital filter (e.g., match filter).

According to [17], artificial light–induced interference has lower intensity than the solar radiation with the frequency spectrum mainly at the low frequency region (i.e., below a few hundreds of kHz). However, the interference introduced by artificial lights dominates during the nighttime. Here, we mainly consider the solar radiation–induced shot noise and the thermal noise, which are considered as additive white Gaussian noise (AWGN). The total noise variance is expressed as:

$$\sigma_{total}^2 = \sigma_{shot}^2 + \sigma_{thermal}^2. \tag{8.17}$$

The shot and thermal noise variances are given by [31]:

$$\sigma_{shot}^2 = 2eRP_{Rx\text{-}Sl}B_s + 2eI_{bg}I_2B_s, \tag{8.18}$$

$$\sigma_{thermal}^2 = \frac{8\pi k T_K}{G} \eta A_{Rx} I_2 B_s^2 + \frac{16\pi^2 k T_K \Gamma}{g_m} \eta^2 A_r^2 I_3 B_s^3, \qquad (8.19)$$

where e is the electron charge $(1.602 \times 10^{-19}$ C$)$, R is the responsivity of the PD, $P_{Rx\text{-}S}$ is the average received optical power of the desired signal, B_s is the system bandwidth, I_{bg} is the received background noise current, k is Boltzmann's constant, T_K is absolute temperature, G is the open-loop voltage gain, η is the fixed capacitance of PD per unit area, Γ is the field-effect transistor (FET) channel noise factor, g_m is the FET transconductance, I_2 is the noise bandwidth factor for the background noise [32], and the noise bandwidth factor $I_3 = 0.0868$ [33].

8.3 Characterization of C2C VLC Channel and Link

In this section, the characterization of C2C VLC channel and links is presented. An approach to obtain these results by utilizing a video footage taken by a dashboard-mounted camera behind the front windshield of a car is utilized. The video footage is processed by computer vision techniques and relevant parameters are extracted. Then, the metrics of interest such as link duration and channel coherence time are estimated. Although this approach would give results which may not be as accurate as using a combination of Tx and Rx, it allows a speedy characterization of the metrics of interest from a relatively large data set without the need for extensive analysis and a time-consuming experimental measurement campaign. The accuracy of the results is sufficient for assessment and evaluation of C2C VLC systems.

8.3.1 C2C VLC Link Duration

Link duration is defined as the time when two nodes can communicate with each other, or in the case of C2C VLC, when two cars can establish a link and maintain communication. Link duration is one of the most crucial parameters to determine the range of applications that can be supported by such communication systems. It is especially of interest in this study since VLC-based C2C is best operated in LOS configuration, and therefore it can help to determine whether the link duration is sufficiently long to support the intended applications. To address this issue, in this subsection, the results from an experimental study will be used to characterize the average transmission duration as a probabilistic distribution in a C2C VLC system. For this purpose, we have used video footage captured using a standard camera mounted behind the front windshield of a taxi, which was driven around in an urban environment. Images extracted from the captured video are post-processed using computer vision techniques to identify the locations of the tail lights. Assuming that cars are equipped with VLC-based tail lights, the time duration that a tail light stays within the FOV of the camera can be readily determined and provides an approximated time duration of the link between the leading car's tail lights and the following car. An empirical probabilistic distribution of the link duration based on 30 hours of video footage was obtained.

Figure 8.10 compares the complementary cumulative distribution functions (CCDF) as a function of the link duration for a range of scenarios (e.g., urban and non-urban areas, normal and red lights, and FOVs). One can observe the following. First, from Figure 8.10a, the link duration in non-urban areas is longer than that in urban areas. This is due to the fact that in urban areas there are more opportunities to establish C2C VLC links, as there are more neighboring cars, which will ultimately result in an increased

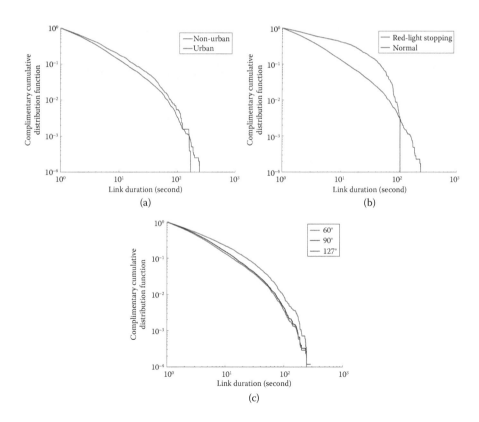

FIGURE 8.10
Empirical link duration distribution: (a) urban versus non-urban, (b) red-light stopping, and (c) field-of-view angle.

number of short-duration links and in terms of a reduced average link duration. Second, from Figure 8.10b, the link duration, when considering only
cars with red lights stopping, drops at a slower rate than the normal case
up to a link duration of <~90 s, beyond which it drops very sharply. This
is because red lights rarely last for more than 100 seconds on average. Our
data show that the average link durations are ~15 s and ~7 s considering
the cases with red-light stopping and with all cases, respectively. Finally,
Figure 8.10c shows the impacts of FOVs on the CCDF, which are obtained
by emulating the effect of having Rxs with different FOVs and cropping
the outermost part of the images. The results for the FOV of 127° and 90°
are mostly the same while reducing from 90° to 60° causes longer link durations, since there are less short-duration links for the FOV of 60°.

8.3.2 C2C VLC Channel Time Variation

Both database-based results and measurement results in Section 2.1 indicate
that the change of relative locations in the lateral direction would cause significant change in the received power. In addition, Equations 8.5 and 8.7,
which include a cosine component, imply that the received power will
change significantly with change of the lateral location of the Rx. These, combined with the fact that the relative location of vehicles changes over time
(i.e., mobility), indicate a fast-changing C2C VLC channel in the time
domain, which will have implications in the effectiveness of channel estimation as well as the achievable system throughput. In this subsection, a video
data stream is utilized to empirically approximate the metric of interest. The
concept is based on identifying and estimating the location of the tail lights of
a neighboring vehicle from the captured image. Based on the relative location
information, values of the relevant parameters such as stand-off distance,
irradiance angle, and incidence angle can be determined. Considering these
parameters in the Lambertian model, the path loss is given by:

$$H(0) = \frac{(n+1)A_R}{2\pi D^\gamma} \cos^n(\phi)\cos\theta, \tag{8.20}$$

where ϕ is the irradiance angle, θ is the incidence angle, D is the stand-off
distance, A_R is the optical detector area, γ denotes the optical path loss exponent, and $n = -\ln 2/\ln(\cos\Phi_{\frac{1}{2}})$ is determined by the half-power angle $\Phi_{\frac{1}{2}}$ of
the Tx.

   The assumption of a Lambertian model for the Tx only applies for a headlamp, but not a tail light. For the tail light, the piecewise Lambertian model
has been adopted as an approximation in [34]. In determining the channel
path loss, it is also assumed that the incidence angle is the same as the irradiance angle, since this is usually the case when the two vehicles travel in the
same direction. The same process can be repeated for images captured at different times, and ultimately the path loss for the C2C VLC link between a

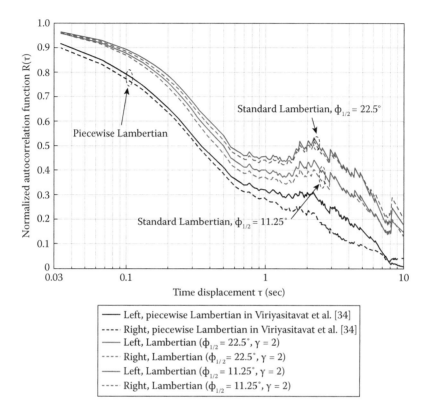

FIGURE 8.11
The normalized autocorrelation function of received power.

neighboring vehicle's tail light and an Rx of the host car can be determined. This information is then used to determine the normalized autocorrelation function of the received power and the channel coherence time. The latter is normally used as a metric to indicate the level of channel time variation, which is a significant parameter in wireless communications.

Figure 8.11 shows the normalized autocorrelation function of received power. The channel coherence time is obtained by finding the first time shift with a correlation value below a threshold, usually 90% or 50%. Here, three different light models are used. The piecewise Lambertian model is directly obtained from real-world measurements, which has the smallest beam width. Table 8.3 summarizes the coherence times for all three models with both left and right tail lights. As mentioned above, the movement of vehicles will result in changes in irradiance and incidence angles. Additionally, when vehicles move around the same angular range, narrower light beam width will result in a larger difference in the path loss and lower values for the coherence time as is the case with piecewise Lambertian model. In contrast,

TABLE 8.3

Empirical Channel Coherence Time of C2C VLC

	Left Tail Light		Right Tail Light	
Light Model	**90% Coherence Time**	**50% Coherence Time**	**90% Coherence Time**	**50% Coherence Time**
Piecewise Lambertian	33 ms	333 ms	33 ms	300 ms
Lambertian, 11.25 degree half angle	67 ms	467 ms	67 ms	433 ms
Lambertian, 22.5 degree half angle	67 ms	533 ms	67 ms	500 ms

the coherence time is greater for the other two models. Note that 50% and 90% of the coherence times of a C2C VLC link are in the order of hundreds of milliseconds and tens of milliseconds, respectively, which are orders of magnitude larger than the RF-based C2C communications (i.e., in the range of 0.45 ms to 5.31 ms in urban areas) [35]. The implication of these results is that C2C VLC links are often more stable and reliable than the RF-based technologies for C2C communications. Therefore, a longer coherence time, which indicates slow variation of the channel, eliminates the need to perform frequent channel estimation by including a training sequence as part of the transmitted frames, thus lowering the level of overhead and improving the data transmission throughput. Additionally, it will also lead to simpler and less complex design for the Tx and Rx, which makes C2C VLC an attractive wireless technology for a wider range of applications with faster market penetration.

8.4 Performance of C2C VLC System

In this section, the BER performance of a C2C VLC system is analyzed. Note that for a C2C VLC system, the channel delay is about 10 ns [36] compared to few MHz bandwidth in a VLC system; therefore, the multipath induced ISI can be considered negligible.

8.4.1 C2C VLC BER Performance

Here we have adopted the mostly widely reported OOK modulation scheme with an AWGN channel. At the Rx, the electrical signal-to-noise ratio (SNR) is given by [25]:

$$SNR = \frac{(\gamma P_r)^2}{\sigma_{total}^2}.$$ (8.21)

Consequently, the BER is given as:

$$BER = Q(\sqrt{SNR}) = Q\left(\frac{\gamma P_r}{\sigma_{total}}\right) = Q\left[\frac{\gamma(P_{Rr} + P_{Lr})}{\sigma_{total}}\right], \quad (8.22)$$

where $Q(x)$ is the Q-function, which is given by:

$$Q(x) = \frac{1}{\sqrt{2\pi}} \int_x^{\infty} e^{-y^2/2} dy. \quad (8.23)$$

For mathematical modeling, we have adopted the following: an optical channel configuration shown in Figure 8.8, low-beam lamps (75% luminous intensity) for the daytime, and a concrete road surface. All the key parameters are listed in Table 8.4. The BER performance of the C2C VLC system at a data rate of 2 Mbps against the distance between two cars for a range of h (the height of the Rx from the ground) is shown in Figure 8.12. As can be observed, for a given BER the coverage distance is higher for lower values of h. For example, for a BER of 10^{-4} and h of 0.2 m, the communication path length is ~20 m, decreasing with increasing h. Note that the BER performance is the worst for h of 0.8 m. The BER distribution at a data rate of 2 Mbps on a vertical plane for a different length between the headlamp and the Rx is depicted in Figure 8.13. It is apparent that as the distance between the Rx and the headlamp increases, the system performance decreases. For a fixed

TABLE 8.4

System Model Parameters

Parameter	Symbol	Value
Diffuse reflectivity	ρ	0.4 [36]
PD area	A_r	1 (cm^2)
Order of Lambertian diffuser	m	1
Luminous efficacy of radiation	LER	250.3 (lm/W)
FOV of the PD	Ψ	30°
Electronic charge	q	1.6×10^{-19} (C)
Responsivity of PD	γ	0.54 (A/W)
Received background noise current	I_{bg}	5100 (µA)
Noise bandwidth factor	I_2	0.562
Boltzmann's constant	k	1.38×10^{-23} (J/K)
Absolute temperature	T_K	298 (K)
Open-loop voltage gain	G	10
Fixed capacitance of PD per unit area	η	112 (pF/cm^2)
FET channel noise factor	Γ	1.5
FET transconductance	g_m	30 (mS)
System bandwidth	B	2 (MHz)

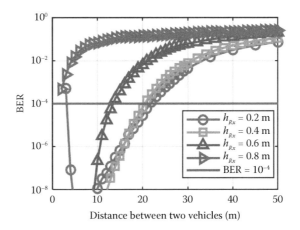

FIGURE 8.12
The BER performance of the C2C VLC system against the different distance between two cars for a range of *h* (height of the Rx from the ground).

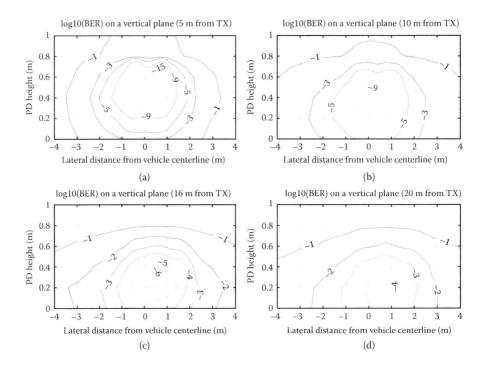

FIGURE 8.13
BER (Log$_{10}$) distribution on a vertical plane for three different distances: (a) log10(BER) on a vertical plane (5 m from TX), (b) log10(BER) on a vertical plane (10 m from TX), (c) log10(BER) on a vertical plane (16 m from TX), and (d) log10(BER) on a vertical plane (20 m from TX).

short distance (e.g., <10 m), the best performance is achieved at $h = 0.3$–0.5 m, and the zones with lowest BER tend to be more skewed to the right; this becomes more apparent when examining lines for the BERs of 10^{-9} and 10^{-4} as in Figure 8.13b and c, respectively. This is because the low-beam head-lamp model adopted is for US vehicles with left-hand drive.

8.4.2 MIMO C2C VLC Performance

In this section, an MC-based system-level model for a C2C VLC system is developed and its BER performance is simulated; the relationship between the transmission distance and BER is also outlined. For better understanding of the headlamp model and the road reflection effects, the received optical power distribution is first calculated on a vertical plane for three different distances of 20 m, 40 m, and 70 m as shown in Figure 8.14. It can be observed that as the distance increases P_{Rx} decreases evidently. The largest contour lines for P_{Rx} degrade from −16 dBm at 20 m to −26 dBm at 70 m, or P_{Rx} is reduced more than 10 times when the distance extends from 20 m to 70 m.

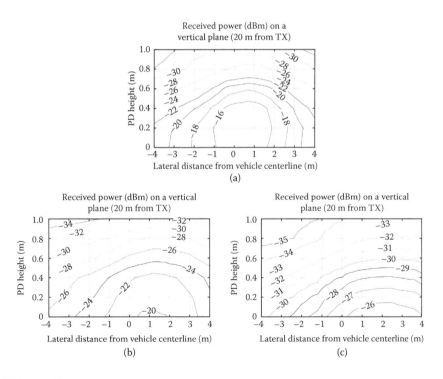

FIGURE 8.14
Received optical power distribution on a vertical plane for three different distances: (a) Received power (dBm) on a vertical plane (20 m from TX), (b) received power (dBm) on a vertical plane (40 m from TX), and (c) received power (dBm) on a vertical plane (70 m from TX).

It can also be observed that the zones with the highest P_{Rx} tend to be smaller and shorter as the distance increases, and their positions are inclined to be more skewed to the right. This becomes more apparent when examining lines for P_{Rx} of −22 dBm in Figure 8.14b and −26 dBm in Figure 8.14c, since the adopted low-beam headlamp model is for US vehicles with left-hand drive.

For MC modeling, we have adopted the channel model shown in Figures 8.8 and 8.9, with low-beam lamps (50% luminous intensity) for the daytime, a concrete road surface, and the key parameters listed in Table 8.4. In order to test the BER performance against d_{Tx-Rx} performance for three difference $d_{LSR-RSR}$, a pair of headlamps and Rxs are always center-aligned facing each other; 1×10^7 random binary bits are generated, transmitted, received, and demultiplexed. Figure 8.15 represents BER performance as a function of d_{Tx-Rx}

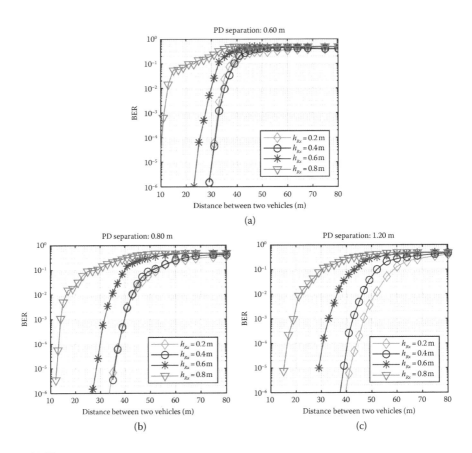

FIGURE 8.15
The BER performance of the C2C VLC MIMO system against the different distance between two cars for a range of h_{Rx} under three different PD separations (0.6 m, 0.8 m, and 1.2 m): (a) PD separation 0.60 m, (b) PD separation 0.80 m, and (c) PD separation: 1.20 m.

for three different $d_{LSR\text{-}RSR}$ of 0.6 m, 0.8 m, and 1.2 m, and four different h_{Rx} of 0.2 m, 0.4 m, 0.6 m, and 0.8 m. The results show that there is an improvement in the BER performance as the Rx separation distance increases. For example, for a pair of Rxs mounted at h_{Rx} of 0.2 m and with $d_{Tx\text{-}Rx}$ of 40 m, the BER is higher than 10^{-1} for $d_{LSR\text{-}RSR} = 0.6$ m; however, the BER drops to less than 10^{-5} for $d_{Tx\text{-}Rx}$ of 1.2 m.

Also observed is that the system is operational over longer distances, provided h_{Rx} is kept low. For instance, for Rxs with a separation distance of 1.2 m (see Figure 8.15c), $d_{Tx\text{-}Rx}$ is less than 20 m for $h_{Rx} = 0.8$ m with a BER of 10^{-4}. However, $d_{Tx\text{-}Rx}$ could increase to more than 40 m when h_{Rx} is reduced to 0.2 m. This is because P_{Rx} increases with a reduced (or shorter) h_{Rx} (see Figure 8.14). Note that h_{Rx} is increased from 0.2 m to 0.4 m, particularly for lower values of $d_{LSR\text{-}RSR}$ (i.e., 0.6 m and 0.8 m (see Figure 8.15a and b). Generally, higher values of h_{Rx} always lead to worse BER performance with a fixed $d_{Tx\text{-}Rx}$. However, as demonstrated in Figure 8.15a and b, the proposed C2C VLC MIMO system has almost the same BER performance for h_{Rx} of 0.2 m and 0.4 m. This is because with MIMO the BER performance does not only rely on the value of received SNR but also on the channel matrix **H**.

8.5 Network and Upper Layers Performance from the Application Perspective

Based on the unique characteristics of the VLC system as well as the link performance results presented in previous sections, it is evident that the VLC technology could be adopted for a number of C2C applications, including collision warning and avoidance, platooning, and cooperative adaptive cruise control (CACC). Based on the link performance shown in previous sections, the following outline whether and how VLC technology could be adopted as part of vehicular communications.

- **Emergency brake lights**—Transmitting a warning message to all vehicles in the vicinity of the vehicle that has made hard braking in an emergency. In addition to an "emergency braking" message, this application can provide critical information such as vehicle's deceleration rate so that drivers can differentiate the level of deceleration and adjust their speeds accordingly. As a result, this application usually assumes that the message will be forwarded to other vehicles that are following. According to the report by National Highway Traffic Safety Administration (NHTSA) [37], emergency electronic brake light application is one of the eight applications that have been identified as high-priority and high potential benefit safety application and is selected as one of the three high-priority applications that is considered for deployment in the near future. Since the emergency braking message content is rather short (< kilobits) and is relevant only to a small number of vehicles, the bandwidth required is very low. In [37], the communication requirements of this application are identified as follows: the maximum communication range is 300 m, less than 100 ms of latency, and the application payload is 36 bytes. These system link requirements can readily be supported using the VLC technology, thus releasing the widely needed RF spectrum for other applications.

- **Cooperative forward collision warning**—This is also one of the eight high-priority safety applications identified by NHTSA. Similar to the emergency brake light information, this application relies on vehicle-to-vehicle communications, whereby vehicles periodically transmit information such as location and speed to nearby vehicles. Exchange of broadcasted information is used to construct a map of the surrounding environment and nearby vehicles, and trajectories. Based on this local map, the drivers can adjust the vehicle's speed and direction to avoid collisions. According to [38], in this type of application, the required maximum communication range is 150 m, the latency is less than 100 ms, and the application payload is about 53 bytes with an update rate of at least 10 Hz.

- **Platooning or CACC applications**—Similar to the cooperative forwarding collision warning application, this application also relies on the information transmitted from nearby vehicles in order to enhance the performance of adaptive cruise controls. This will also improve the traffic flow and capacity by reducing the gap between vehicles without compromising road safety. Even though the applications require more frequent and up-to-date information (at a rate of 10–50 Hz) [37], the host vehicle mainly relies on the information from the preceding vehicles in order to adjust its own speed. The narrow- and short-range coverage required in this application is perfect for the VLC technology. With the VLC technology, the communication interferences (e.g., between vehicles in the platoon or following one another) are kept minimal (compared to other technology such as RF). In addition to the interference and coverage, the VLC system can also provide sufficient data rate and delay performance as well as very accurate positioning capability, thus eliminating the need for other costly positioning systems such as radar [39].

- **Lane change assistance and warning application**—Similar to the collision forwarding warning and avoidance application, this application also relies on a map containing position and speed information of all neighboring vehicles. This application is triggered when the driver uses the turning left/right indicator for switching the lane or overtaking a vehicle, for example, when there is not sufficient distance between the vehicles in the target lane to permit a safe lane change. This application requires a periodic update of information from neighboring vehicles (i.e., an update rate of at least 10 Hz), a maximum communication range of 150 m, and a maximum latency of 100 ms. This level of requirement can also be easily accommodated by the VLC technology.

Finally, for C2C VLC systems to be widely adopted by the vehicles manufacturers, there are a number of challenges that need to be addressed including: (i) design and development of dedicated devices; (ii) comprehensive subsystems and systems modeling; (iii) PD or camera-based receiver; (iv) hybrid VLC–RF; (v) standards.

8.6 Conclusion

This chapter has discussed the C2C VLC as part of the intelligent transportation system. The chapter outlined modeling together with the system elements and the associated specifications. For channel modeling, a market-weighted headlamp beam pattern model was employed together with LOS and NLOS paths. The measured light beam from actual vehicle was also presented and analyzed. Also presented was the road surface reflection modeling. A MIMO C2C VLC system based on the vehicles' headlights and tail lights was introduced, demonstrating the potential for increased data rate. For BER analysis, an MC-based system-level model was developed considering a range of communication geometries.

Network and upper layer performance from the application perspective was discussed, with the focus on how the VLC system can be integrated and utilized in C2C applications. The issues including emergency electronic brake light, cooperative forward collision warning application, adaptive cruise control, and lane change assistance were also discussed. As far as future work is concerned, further theoretical and experimental work is required in order to address many remaining challenges. In particular, different categories of road surfaces, temperature, and weather conditions as well as effects of multipath interference and integration with the backbone network should be considered and incorporated as part of high data rate C2C VLC systems.

References

[1] K. D. Jandt and R. W. Mills, A brief history of LED photopolymerization, *Dent. Mater.*, vol. 29, pp. 605–617, 2013.

[2] W. K. Lin, S. W. Chen, C. Chao, et al., The analysis of the thermal resistance structure of LEDs by measuring its transient temperature variation, *Microsystems, Packaging, 2013 8th International Assembly and Circuits Technology Conference (IMPACT)*, pp. 214–217, 2013.

[3] L. Grobe, A. Paraskevopoulos, J. Hilt, et al., High-speed visible light communication systems, *IEEE Commun. Mag.*, vol. 51, pp. 60–66, 2013.

[4] P. A. Haigh, F. Bausi, Z. Ghassemlooy, et al., Visible light communications: Real time 10 Mb/s link with a low bandwidth polymer light-emitting diode, *Opt. Express*, vol. 22, pp. 2830–2838, 2014.

[5] T. Toroyan, *WHO Global Status Report on Road Safety 2013: Supporting a Decade of Action*, Geneva, Switzerland: World Health Organization, 2013.

[6] K. Rumar, D. Fleury, J. Kildebogaard, et al., *Intelligent Transportation Systems and Road Safety, Brussels*, Belgium: European Transportation Safety Council, 1999.

[7] P. Papadimitratos, A. La Fortelle, K. Evenssen, et al., Vehicular communication systems: Enabling technologies, applications, and future outlook on intelligent transportation, *IEEE Commun. Mag.*, vol. 47, pp. 84–95, 2009.

[8] Y. Morgan, Notes on DSRC and WAVE standards suite: Its architecture, design, and characteristics, *IEEE Commun. Surv. Tutorials*, vol. 12, no. 4, pp. 504–518, 2010.

[9] J. Armstrong, Y. A. Sekercioglu and A. Neild, Visible light positioning: A roadmap for international standardization, *IEEE Commun. Mag.*, vol. 51, pp. 68–73, 2013.

[10] T. D. Little, A. Agarwal, J. Chau, et al., Directional communication system for short-range vehicular communications, *Vehicular Networking Conference*, 2010.

[11] S. H. Yu, O. Shih, H. M. Tsai, et al., Smart automotive lighting for vehicle safety, *IEEE Commun. Mag.*, vol. 51, pp. 50–59, 2013.

[12] Z. Ghassemlooy, W. Popoola and S. Rajbhandari, *Optical Wireless Communications: System and Channel Modelling with MATLAB®*, Boca Raton, FL: CRC Press, 2012.

[13] P. J. F. Ruiz, F. B. Hidalgo, J. Lozano, et al., Deploying ITS scenarios providing security and mobility services based on IEEE 802.11p technology, in *Vehicular Technologies—Deployment and Applications*, eds., Lorenzo Galati Giordano and Luca Reggiani, InTech, 2013. DOI: 10.5772/55285.

[14] S. Arai, S. Mase, T. Yamazato, et al., Feasible study of road-to-vehicle communication system using LED array and high-speed camera, *15th World Congress on Intelligent Transport Systems and ITS America's 2008 Annual Meeting*, New York, 2008.

[15] N. Lourenco, D. Terra, N. Kumar, et al., Visible light communication system for outdoor applications, *8th International Symposium on in Communication Systems, Networks & Digital Signal Processing (CSNDSP)*, pp. 1–6, 2012.

[16] S. H. You, S. H. Chang, H. M. Lin, et al., Visible light communications for scooter safety, *Proceeding of the 11th Annual International Conference on Mobile Systems, Applications, and Services*, pp. 509–510, 2013.

[17] K. Y. Cui, G. Chen, Z. G. Xu, et al., Traffic light to vehicle visible light communication channel characterization, *Appl. Opt.*, vol. 51, pp. 6594–6605, 2012.

[18] P. Luo, Z. Ghassemlooy, H. L. Minh, et al., Performance analysis of a car to car visible light communication system, *Appl. Opt.*, vol. 54, no. 7, pp. 1696–1706, 2015.

[19] A. Ylinen, M. Puolakka and L. Halonen, Road surface reflection properties and applicability of the r-tables for today's pavement materials in Finland, *Light Eng.*, vol. 18, pp. 78–90, 2010.

[20] Wikipedia. Headlamp, 2015. Available: http://en.wikipedia.org/wiki/Headlamp (accessed April 1, 2016).

[21] B. Schoettle and M. J. Flannagan, *A Market-Weighted Description of Low-Beam and High-Beam Headlighting Patterns in the U.S.*, Report No. UMTRI-2011-33, University of Michigan, Ann Arbor, MI, 2011.

[22] J. L. Lindsey, *Applied Illumination Engineering*, The Fairmont Press, Lilburn, GA, 1997.

[23] D. A. Schreuder, *Reflection Properties of Road Surfaces*, Leidschendam, The Netherlands: Institute for Road Safety Research SWOV, 1983.

[24] R. E. Stark, *Road Surface's Reflectance Influences Lighting Design, Lighting Design + Applications*, 1986.

[25] J. M. Kahn and J. R. Barry, Wireless infrared communications, *Proc. IEEE*, vol. 85, pp. 265–298, 1997.

[26] M. Sivak, M. J. Flannagan, S. Kojima, et al., *A Market-Weighted Description of Low-Beam Headlighting Patterns in the U.S.*, University of Michigan, Ann Arbor, MI, 1997.

[27] G. He, L. Zheng and H. Yan, *LED White Lights with High CRI and High Luminous Efficacy*, 2010, p. 78520A.

[28] P. Luo, Z. Ghassemlooy, H. Le Minh, et al., Fundamental analysis of a car to car visible light communication system, *9th International Symposium on Communication Systems, Networks & Digital Signal Processing (CSNDSP)*, 2014.

[29] P. Luo, Z. Ghassemlooy, H. L. Minh, et al., Bit-error-rate performance of a Car-to-Car VLC system using 2×2 MIMO, *Mediterr. J. Comput Networks*, vol. 11, pp. 400–407, 2015.

[30] W. F. Marion, C. J. Riordan and D. S. Renné, *Shining on: A Primer on Solar Radiation Data*, National Renewable Energy Laboratory, 1992.

[31] T. Komine and M. Nakagawa, Fundamental analysis for visible-light communication system using LED lights, *IEEE Trans. Consum. Electron.*, vol. 50, pp. 100–107, 2004.

[32] T. Komine, L. Jun Hwan, S. Haruyama, et al., Adaptive equalization system for visible light wireless communication utilizing multiple white LED lighting equipment, *IEEE Trans. Wireless Commun.*, vol. 8, pp. 2892–2900, 2009.

[33] I. E. Lee, M. L. Sim and F. W. L. Kung, Performance enhancement of outdoor visible-light communication system using selective combining receiver, *IET Optoelectronics*, vol. 3, pp. 30–39, 2009.

[34] W. Viriyasitavat, S.-H. Yu and H.-M. Tsai, Short paper: Channel model for visible light communications using off-the-shelf scooter taillight, *Proceeding IEEE Vehicular Networking Conference (VNC)*, pp. 170–173, December 2013.

[35] C. F. Mecklenbrauker, A. F. Molisch, J. Karedal, et al., Vehicular channel characterization and its implications for wireless system design and performance, *Proc. IEEE*, vol. 99, no. 7, pp. 1189–1212, 2011.

[36] S. Lee, J. K. Kwon, S. Y. Jung, et al., Evaluation of visible light communication channel delay profiles for automotive applications, *EURASIP J. Wireless Commun. Networking*, vol. 370, pp. 1–8, 2012.

[37] *Vehicle Safety Communications Project Task 3 Final Report—Identify Intelligent Vehicle Safety Applications Enabled by DSRC*, U.S. National Highway Traffic Safety Administration, Technical Report, 2005.

[38] A. C. P. Association, Albedo: A measure of pavement surface reflectance, *Res. Technol. Update*, Ann Arbor, MI, 2002. Available: http://1204075.sites.myregisteredsite.com/downloads/RT/RT3.05.pdf (accessed April 1, 2016).

[39] S.-H. Yu, O. Shih, N. Wisitpongphan, et al., Smart automotive lighting for vehicle safety, *IEEE Commun. Mag.*, vol. 51, no. 12, pp. 50–59, 2013.

9

Visible Light Communications Based on Street Lighting

Stanislav Zvánovec, Petr Žák, Petr Chvojka, Ivan Kudláček,
Paul Anthony Haigh, and Zabih Ghassemlooy

CONTENTS

9.1 Introduction

Modern street lighting represents the pervasive utilization of light-emitting diode (LED) lamps to illuminate urban and rural areas. In addition, LED-based lighting sources can also be utilized for outdoor data communications and localization. The replacement of traditional street incandescent and fluorescent-based lights by highly efficient LED lights is taking place on a global level. This chapter highlights the main features of LED public lighting systems that could be used for visible light communications (VLC). First, a description of state-of-the-art street lighting, including its main functions, control systems, and typical parameters, is provided. Then, the main aspects associated with lighting performance and aging are outlined. The remainder of the chapter is devoted to recent studies on public lighting for VLC purposes, focusing on ray tracing simulations, noise parameters, and delay profiles among other topics.

9.2 Modern Public Street Lighting

9.2.1 Introduction

One of the basic characteristics of the earth is the regular rotation of the moon and sun, which creates day and night. In such an environment, light is provided by a natural source—the sun. Nighttime is distinguished by the lack of a light source or sources with little brightness such as the moon and the stars. Given the low intensity of these sources, people, throughout history, transformed the nighttime environment with a means of light. It is likely that the first outdoor lighting served to illuminate settlement centers, ritual ceremonies, and celebrations. In ancient times, citizens gradually started illuminating public outdoor spaces and roads in places such as Pompeii, Veset, and Antioch [1]. In later times, exterior lighting was also used in naval and railway transportation systems. Together with indoor lighting, outdoor lighting grew extensively as a result of the advent of electric light sources at the beginning of the 19th century. Nowadays, outdoor lighting creates conditions, whereby outdoor activities are possible at night and it influences the appearance and the atmosphere of this time of a day. Public lighting is now one of the most commonly applied areas of outdoor lighting and serves to illuminate public spaces in towns and villages including roads for motorized traffic, pedestrian and cycle paths, tunnels, underpasses, bridges, squares, and parks. Their primary function is the provision of personal comfort, transport, security and safety, and improved orientation at night. Public lighting creates an ambience and influences not only the appearance of an area but also its attractiveness to tourists and contributes to how local citizens identify with the areas where they live (see Figure 9.1).

A public lighting system affects the appearance of public areas not only at night but also during the day. From a safety point of view for transport, three basic situations are considered when dealing with public roads and spaces: roads for motorized traffic (see Figure 9.2), pedestrian walkways, and conflict areas [2].

Although night traffic density reaches approximately a quarter of the level of daytime traffic, the likelihood and seriousness of road accidents at night are substantially higher. The share of fatal road accidents at night in 13 OECD (The Organization of Economic Co-operation and Development)

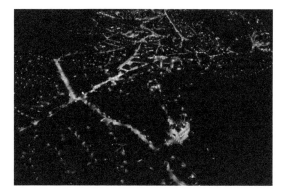

FIGURE 9.1
Aerial photograph of the nocturnal appearance of the historical city center of Kutna Hora, Czech Republic.

FIGURE 9.2
Public lighting of a road for motorized traffic with LED luminaires (Prague, Czech Republic).

countries ranges between 25% and 59%, with an average value of 48% [3]. One of the main reasons is the difference in visual conditions and human sight error reactions. Drivers rely on the car's front lights to help gather visual information. However, at high speed and under worse road and weather conditions, reliance on the headlights in assisting drivers to have a smooth journey decreases significantly. Moreover, a car's lights may not illuminate the entire road surface and also act as a source of glare to oncoming cars or the car at the front.

Of course, street lighting increases safety by making road features such as road alignment, footpaths, furniture, surface condition, other road users, and objects that may be on the road visible to both vehicular and pedestrian traffic as well as lowering the glare effect of cars' front lights. The basic technical parameters adopted in lighting fixtures are the lighting level represented by the road surface luminance L_m (cd/m^2), luminance uniformity U_o (-), U_l (-), glare control TI (%), and the degree of a surrounding's illumination SR (-). At present, dealing with road illumination for motorists, the spectral properties of lighting are not taken into account. Nonetheless, the application of different color tones of street lighting helps drivers navigate through towns or villages.

Illuminating pedestrian walkways is, in many aspects, different from illuminating roads since a significantly lower speed of movement gives pedestrians ample time for their eyes to adjust to road conditions. While drivers need to have a good contour perception of distant objects and people, it is important for pedestrians to recognize the structure of the objects and surfaces in their immediate vicinity. From a public lighting point of view, the aim for pedestrian walkway lighting is to provide both sufficient surface illuminance E_m (lx) (E_{min} (lx) for pedestrians to recognize obstacles and irregularity) and illuminance on vertical planes $E_{v,min}$ (lx) to recognize passers-by. Glare emanating from pedestrians is not as critical as in the case of drivers or cyclists, and, therefore, is not considered in the design. Likewise, there are no restrictions concerning spectral properties. Nonetheless, it is sensible to use light sources with a warm, white color tone. Due to the importance of roads and walkways, it is recommended to use light sources with good color rendition.

Conflict areas include sections of roads where traffic crossings occur (such as junctions, level crossings, pedestrian crossings) or places where the geometry of the road has been changed (such as a reduced number of lanes). These places have a higher probability of accidents, for example, collisions with other vehicles, pedestrians, or with a solid object. The primary purpose of public lighting at a conflict area is to create light conditions which not only draw people's attention to the presence of obstacles in time but also warns them against the location of roadsides, horizontal traffic signs, directions, and the presence of pedestrians and other road users, in addition to vehicles moving in the vicinity of the conflict area. In conflict areas, the level of illuminance E_m is evaluated together with the uniformity of illuminance U_o. To draw attention to the conflict area, it is possible to use a different color tone in contrast to oncoming roads.

9.2.2 Public Lighting System

The following description of the lighting system and its statistical data relate to the Czech Republic, where public lighting systems are operational for about 4,000 hours per year. The situation in other countries may be slightly different depending on the technique used or geographical conditions. No matter which country is discussed, it is important that lighting sources have high electrical energy to light luminous efficacy to reduce both the ever-growing demand for energy, and overall global carbon emissions. For economic reasons, it is also crucial that lights last a long time. While usage intensity of public lighting varies and changes with time, lights are mostly used from the switch-on time around 10:00 p.m. until the switch-off time around 6:00 a.m. when natural light intensity is kept at low levels. Therefore, it makes sense to regulate the lighting level. When designing lighting systems, three key points should be considered: transport safety, the appearance of public areas, and the impact on the environment.

The lighting parameters which must be considered with regard to transport safety are related to motorized traffic, pedestrian walkways, and conflict areas. It also depends on the design speed, the overall layout, traffic volume, traffic composition, and environmental conditions. Based on these features, roads are classified according to what is known as lighting class, which represents the values of lighting parameters. Considering that some of the parameters such as traffic volume or ambient luminosity can change, the lighting class may also change along with lighting requirements. This is what is known as adaptive lighting and the classification of roads is carried out according to corresponding standards [4]. Artificial light generated by the public lighting system may be harmful to the environment; therefore, its level needs controlling and this is why the outdoor environment is divided into environmental zones [5]. Within these zones, the impact of artificial lighting on residential buildings, cities, and a town's appearance is assessed, together with the glare effect on the drivers and astronomical observations. The classification of public areas is carried out according to given standards [6]. There are no prescribed requirements on the esthetic features of public lighting neither in terms of their physical elements nor to their relation to illuminated public space. It is possible to deal with these parameters within the overall concept of public lighting or the lighting master plan [7].

A public lighting system is composed of typical basic elements: a feeder switchboard, power lines, and lighting points consisting of supporting structure and luminaires. The feeder switchboard is used for powering, measuring, circuit protection, and switching; it may also have elements for regulation of the public lighting. In modern lighting systems, the feeder switchboards are equipped by devices for remote communications, and one feeder switchboard typically provides power for about 80 luminaires. Power lines used for supplying the lights with energy may be located above the ground, as is the case of rural areas, or underground, as is the case of the most urban areas. The supporting structure fixes the luminaires in a required position and includes light poles, which are the most commonly found supporting structures, brackets, arms, and suspension wires. The support structure also includes electrical equipment to connect the lighting point to the power lines, a switchboard with terminals for incoming and outgoing wires, and the luminaire fuse. The luminaires themselves incorporate the light sources, the parts necessary for fixing and protecting the lighting sources, the control gears required for the operation of the lighting sources, and those parts which connect them to the power line. The average number of luminaires in towns and villages is one per eight citizens. The average luminaire input power in public lighting systems is 123 W.

9.2.3 Lighting Control System

How lighting systems are controlled and managed affects the efficiency and power consumption of the system. The system for light switching may be done manually or automatically, which relies on information related to time or the amount of daylight. The control signals can be transmitted via power lines, where the signal is modulated to the line voltage, or by using radio frequency (RF) wireless technology. Management systems function for different purposes, such as power consumption optimization, changes in the night time environment, or for lighting system monitoring.

Consumption optimization means that the desired light conditions with defined quantitative and qualitative parameters are reached in the most energetically efficient way. The optimization of energy consumption can be achieved by controlling operation times and/or system power input, or by eliminating lighting system overdesign. In order to optimize the lighting system usage in time, automatic switching systems are employed. These systems may activate the lighting system according to the time of the day by means of astronomical clock, or based on the level of daylight measured using light sensors. The basis for input power optimization from the user point of view may have fixed time plans proceeding from statistical data related to the traffic volume at night. The sensors registering the current traffic situation represent a second method for information gathering. Based on data from the time plans or the sensors, the necessary lighting level is set to correspond to the traffic safety required. Lowering the lighting level is not recommended at any location deemed to be dangerous. For a variety of reasons, it frequently happens that public lighting systems are overdesigned. The first factor is the aging of the system, which gradually lowers the luminous flux. Thus, the required lighting parameters must be maintained at all points of the lighting system's lifetime. Another overdesign may occur as a result of system maintenance when only the luminaires are changed but the original placement is kept the same. Overdesign may also occur when a road is reassigned to a lower lighting class. Lighting system overdesign can be eliminated by installing luminaires capable of gradually regulating luminous flux connected to the central control system.

It is possible to carry out consumption optimization without central control systems. Currently, there are LED luminaires equipped with a control gear, featuring an autonomous regulation which allows a setting-reduced operation regime within given time intervals. Moreover, other luminaires with a constant light output (CLO) function, which eliminates lighting system overdesign owing to the lighting sources aging, are available.

Applying management systems to create specific light atmospheres is now incorporated as a part of lighting in buildings, as well as for the lighting of major urban and public spaces, such as historical sites and pedestrian zones. In this case, the objective of lighting management is to change the atmosphere or appearance of a public space by means of dynamic lighting, which enables the illumination style of the object (by brightness composition), the lighting level, or spectral properties to be changed.

Monitoring of a lighting system is essential to detect operational states and malfunctioning parts. This can be done at the individual luminaires by controlling gear failure or light source problems, power line breakdowns, or failures in the feeder switchboard levels. The management systems also collect information on the total number of hours lights being used and current energy consumption. This data can be used to optimize the maintenance of the system or to assess the entire demand for public lighting. These functions are especially required for lighting systems in larger cities. It is possible to control the operational states of the system from one place using a schematic visualization and to obtain a complete overview of its functions and operations.

9.2.4 Light Sources

Light sources are technical devices used for transforming electric energy into optical radiation and, according to the type of energy transformation, electric light sources are divided into incandescent (incandescent and halogen lamps), discharge (fluorescent tubes, metal-halide, and sodium lamps), and solid-state (LED and organic LED) light sources. Some of the basic technical parameters are:

- luminous flux Φ (lm)
- input power P (W)
- luminous efficacy η (lm/W)
- color temperature T_{cp} (K)
- color rendering index R_a (-)
- lifetime t (hour)

When choosing light sources for public lighting, it is important to pay attention to the main operational characteristics including lengthy operational times and a flexible level of usage during operation. Therefore, relaxed luminous flux regulations, high luminous efficacy (efficiency of transformation of electric energy to the luminous flux), and a long lifetime rank among the most essential factors to consider. Spectral properties, which are significant as they influence the quality of the illuminated space perception, are defined by two parameters: the color rendering index R_a, which renders a faithful surface color perception, and the color temperature T_{cp}, which indicates the color tone of white light (warm, neutral, and cold). When choosing light sources, it is necessary to take into account the dependency of technical parameters on the changing parameters of the surrounding environment; in this case, it consists mainly of the surrounding temperature. The basic parameters of light sources for public lighting are summarized in Table 9.1. Development of the luminous efficacy of electrical light sources is shown in Figure 9.3.

TABLE 9.1

The Basic Technical Parameters of Light Sources for Public Lighting

Source	Lum. Efficacy η (lm/W)	Col. Temp T_{cp} (K)	Col. Render index R_a (-)	Lifetime t (hour)
Sodium lamp	130–170	1,800	0	12,000
Metal-halide lamp	100–125	2,800–4,000	60–90	14,000–24,000
Fluoresc. lamp	70–85	2,700–4,000	80	20,000
Mercury lamp	40–50	3,000–4,000	50	8,000–16,000
Induction lamp	80–100	2,700–6,500	80	60,000
Plasma lamp	80–100	4,000–6,000	70–80	50,000
LED	130–200	2,400–8,000	70–90	100,000

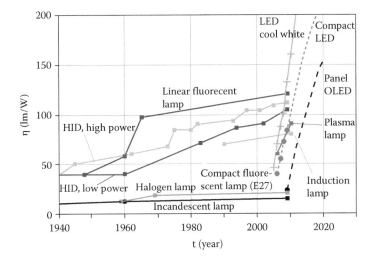

FIGURE 9.3
Evolution of the luminous efficacy of electrical light sources.

Around 2008, LEDs, point light sources suitable for directional lighting, were introduced as a new light source for public lighting. According to the type of material and P–N junction, the emitted spectrum is defined by the dominant wavelength—red (630 nm), yellow (590 nm), green (530 nm), blue (470 nm)—so the LED can be used for colorful illumination with several ways of producing white light. The most commonly used LED sources are blue or near-ultraviolet (nUV) LEDs in combination with a phosphor layer known as a phosphor-converted LEDs (pc-LED). Phosphor is a material applied to the chip which extends the narrow spectral region to the entire region of the visible radiation (~380–780 nm) band, enabling the optical radiation part with higher energy

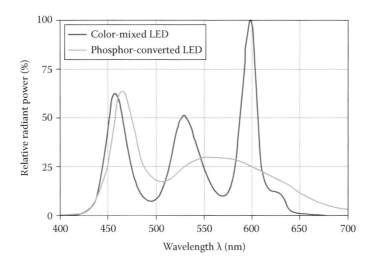

FIGURE 9.4
Spectral power distributions for color-mixed LED and phosphor-converted LED.

to be transformed into lower energy. By the choice of phosphor material, it is possible to influence the color temperature T_{cp} and the color rendering index R_a, the spectral properties of the emitted light. Another way of generating white light is to mix the spectrum of red, green, and blue (RGB) LEDs known as a color-mixed LED (cm-LED): see Figure 9.4.

However, when applying the color-mixed method, it is rather difficult to retain color stability and its tolerance of spectral properties in certain boundaries. Therefore, this type of LED is not widely used for outdoor applications. There are also strategies for white light emission combining both methods and the resulting light source is known as a hybrid LED. LEDs can be divided into four groups based on their power profile: low power (LP), medium power (MP), high power (HP), and super-high power (SHP) and chips on board (COB) LEDs. COB is a packaging technology, and each COB package attaches a number of chips directly to the phosphor layer. The main parameters for describing LED type are the operating current, input power, and luminous flux. In a view of the rapid growth of LED luminous efficacy, in future we may see a recategorization of LEDs. An example of the current state of LED types is presented in Table 9.2.

In the field of public lighting, where it is necessary to keep the values of luminous fluxes within a range of 1,000 lm to 15,000 lm, HP and COB LEDs are used. Applying HP LEDs currently allows us to direct the luminous flux in a desired direction and in a better way. LEDs have gradually superseded all other light sources used in public lighting, thanks to their characteristics. Their luminous efficacy is currently 200 lm/W (HP LED) in mass production and 303 lm/W in the laboratory [8]. The predicted development of HP LED

TABLE 9.2

Example of LED Types Division

Type	Operating Current I_f (mA)	Input Power P_i (W)	Luminous Flux Φ (lm)
LP LED	5–20	0.01–0.1	0.1–6
MP LED	30–150	0.1–0.5	0.2–20
HP LED	350–2,000	0.5–5	50–500
SHP/COB LED	200–3,000	>5	>500

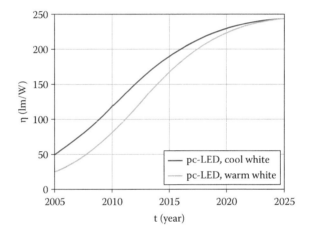

FIGURE 9.5
Predicted luminous efficacy development of pc-LED. (From Bardsley, N., Multi-Year Program Plan: Solid-State Lighting Research and Development, U.S. Department of Energy, 2014. With permission.)

luminous efficacy is shown in Figure 9.5 [9]. It is possible to choose color temperature T_{cp} in the range of 2,400 K to 8,000 K, where an extremely good color rendering index R_a (from 70 to 90) occurs making it easy to dim them. HP LEDs offer a usage lifetime of more than 100,000 hours and a unitary luminous flux of 80,000 lm (COB LED).

9.2.5 Luminaires

The luminaires for public lighting (Figure 9.6) are technical devices which primarily serve to direct a light source's luminous flux in a desirable direction, to limit light source luminance, and to protect light sources and their power supply. In addition, the luminaires consist of a series of devices and parts necessary when attaching and protecting the light sources. Outdoor weather

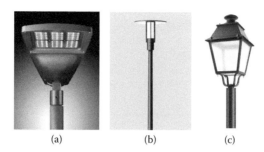

FIGURE 9.6
Examples of luminaires for public lighting (a) for roads and motorized traffic, (b) for pedestrian walkways in a modern design, and (c) for pedestrian walkways in a historical design.

conditions are an important factor which influences the construction design as extended requirements on the degree of ingress protection (IP 54 to IP 66), the resistance to the aggressive environment (such as acid rain), and even the resistance to wind gusts and temperature changes may be necessary. Considering the location of these light sources, they should be well protected against any mechanical damage (IK classification). The basic technical parameters which characterize the luminaires are:

- photometric curve—a description of the light emission
- luminous flux of luminaire Φ (lm)
- input power of luminaire P_i (W)
- luminous efficacy of luminaire η (lm/W)
- electrical protection
- degree of protection (IP code)
- mechanical protection (IK code)

The purpose of the luminaires is to illuminate road surfaces and their immediate surroundings for the sake of motorists and pedestrians. These luminaires exactly and geometrically illuminate defined areas based on their location at the roadside, thereby resulting in a specific photometric shape (character of the light emission) providing us with information about illumination width. The value of the luminous intensity at high angles away from the downward direction provides information on glare limitation. The purpose of luminaires for the illumination of the pedestrian walkways is not only to illuminate the walkway surface, they must also account for vertical surfaces including the space itself, and pedestrians. Unlike luminaires for roads, those for walkways have a different photometric curve shape and they are installed at shorter heights which increases the possibility of glare. This is why it is necessary to control the surface illumination by using, for example, diffusible covers. When choosing the type and appearance of a luminaire, it is recommended to consider the desired light effect, the location, and the atmosphere that will be created.

9.3 Public Lighting Aging and Ecological Aspects

Modern electrotechnical devices are characterized by low power consumption and high output power. Such properties are also expected from light sources and LED light sources fully meet these. The classic 40 W incandescent bulb can be replaced by an LED light source with identical luminous flux, though with a power drain of only 5 W [10]. Aside from their low power drain, LEDs are characterized by long operating lifetimes as the LED lifetime can be 50 times longer than that of the technical lifetime of an incandescent lamp which is close to 1,000 hours. Unfortunately, radiated luminous flux decreases over time and this determines the real operating lifetime of this light source.

9.3.1 LED Source Lifetime

Determination of the operating lifetime is established from the initial luminous flux, and a decrease of the luminous flux below a specified level is considered as the end of a source's operational lifetime. The boundary is usually set at 70% of the initial luminous flux [11] (see Figure 9.7).

A decrease in luminous flux is, according to Figure 9.7, dependent on the operating temperature of the LED. The declared technical lifetime is 150,000 hours at 55 °C and 50,000 hours usually achievable at 75 °C, but every additional 10 °C increase in temperature may lead to at least a 25% reduction in LED lifetime–for example, when the LED chip is assumed to operate at 85 °C, lifetime decrease to 30,000 hours has to be considered. The LED light source consists of a series of diodes whose flux diminishes with usage time. The number of individual diodes with decreased luminous flux (as a percentage) is another parameter that characterizes an LED's lifetime. According to the standard DIN IEC/PAS 62 717, operating lifetime is given by parameter L70B50 which corresponds to 50% of diodes having luminous flux of less than 70%. The real technical life of LED sources is not necessarily expressed by such a parameter. There are, in addition to LED components, other electronic components in the complete light source such as electrolytic capacitors, whose parameters should be taken into consideration since mean time to failure (MTTF) of these components may be much shorter than the MTTF of LEDs.

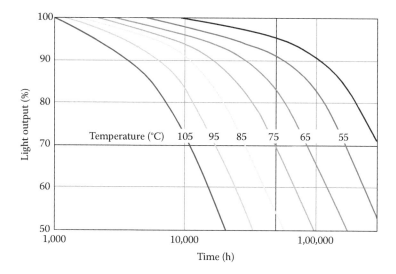

FIGURE 9.7
LED light flux dependence on operating temperature.

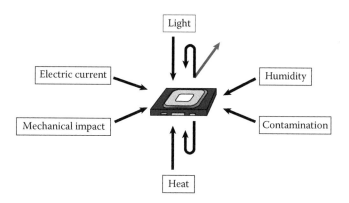

FIGURE 9.8
Factors affecting the operating life of an LED light source.

9.3.2 Factors Affecting the Lifetime and Reliability of an LED Light Source

The operating lifetime parameter L70B50 of the light source with LEDs may be over 50,000 working hours under optimal conditions. The factors affecting the lifetime of an LED are illustrated in Figure 9.8.

- **Light:** Light emitted by a diode and also sunlight may cause the photochemical degradation of plastic parts of the LED light source (chip coating, dispersive reflector). UV radiation (290–400 nm) forms free radicals in the plastic part and this results in cleavage or cross-linking of macromolecular chains. The cover of the chip loses mechanical and, in particular, optical properties which correspond to a reduction of luminous flux in an LED light source. The simultaneous influence of heat, ozone, and oxygen may induce the thermal-oxidative degradation of the plastic, causing the brittleness of the plastic parts of the LED light source and loss of their mechanical properties [12,13].

- **Heat:** Heat generated by an LED and the ambient temperature are key factors that influence the operational reliability of the complete light source. Heat is generated within the PN junction, and the temperature within a diode is given mostly by:

  - The current flowing through a diode
  - The quality of heat dissipation
  - Ambient temperature

The luminous flux of an LED may be increased when operating at a higher electrical current, but the PN junction temperature would also increase. Therefore, this puts increased demand on the heat dissipation efficiency from the PN junction. The optical efficiency of an LED is defined as:

$$\eta = P_l/P_e \tag{9.1}$$

where P_l is the radiation output (W) and P_e is the power drain necessary to achieve the radiation output (W). The output radiation of an LED Φ_e (W) can be expressed as:

$$\Phi_e = \eta P_e \tag{9.2}$$

The heat output P_m (W) of an LED is given by:

$$P_m = P_e - \Phi_e \tag{9.3}$$

It is necessary to dissipate this heat output from the chip. As the PN junction needs to be encapsulated, heat dissipation is mostly realized by heat conduction. The efficiency of heat dissipation is given by the technological implementation of an LED and the dissipation itself is achieved via a diode package, by the leads, or by a special heat-conducting substrate. Further heat dissipation is performed using an outer metallic LED light housing. The aluminum heat sink with proper finning and sufficient surface area is the optimal solution for heat dissipation [14]. The heat dissipation depends significantly on both the technological implementation and ambient temperature. The elevated temperature has a significant negative effect on electroluminescent

LED light sources, where visible light is produced by the incidence of the blue LED light on the phosphor, though an increase in temperature may significantly shorten the lifetime of the phosphor layer [15]. For LEDs to last longer, operating temperatures must be kept below a specific critical temperature given by the diode design. Elevated temperatures lead to a decrease of luminous flux of an LED and a shorter operating lifetime.

*   **Electrical current:** The requested luminous flux is maintained only if the electrical current I is within a specific range. Increasing the electrical current magnitude allows the optimization of emitted light parameters and when a higher electrical current is used, a fraction of the power is dissipated as heat via a serial resistance R within a PN junction. The actual efficiency of LEDs is then given by:

$$\eta = P_l / \left(P_e - (RI^2) \right) \tag{9.4}$$

Higher electrical current, therefore, leads to an undesirably elevated temperature. For this reason, all excess heat must be dissipated. Modern LED light sources are constructed so that the level of output luminous flux throughout the life of the lamp is kept constant which is performed by electronic circuits within an LED light source. Initially, LEDs are powered by a smaller current.

Because the light source is aging, the power drain increases which leads to the greater thermal loading of a light source.

*   **Mechanical influences:** The degradation effect of mechanical strain is tied to the thermal loading of a light source, though reasons for mechanical strain may vary. As a result of the materials used in LED light sources, large temperature fluctuations can produce large mechanical forces. If the light source is exposed to such forces, then the operating lifetime may be adversely affected.

*   **Moisture and chemical contamination of the operating environment:** Exposure to atmospheric moisture and the chemical contamination of the operating environment can lead to irreversible changes in materials, surface protections, and a negative influence on other electronic components of the whole LED light source. The negative effect of moisture on the diode itself is minor, since the diode is encapsulated. Nevertheless, humid atmospheric conditions combined with chemical constituents can be a source of corrosion for the electronic elements and metal parts inside an LED module. Though corrosion can be limited by selecting suitable materials and surface protection, moisture protection is an absolute must, so that an LED module for street lighting achieves the longest operating life possible. The endangering of an LED light source by chemicals depends on the installation site location. It is necessary to take into account all environmental conditions when planning a

lighting system that utilizes LED technology. A maintenance factor of a light source is given for LED lamps for outdoor environments and is calculated by multiplying the aging factor and the pollution factor of the luminaire. The aging factor for quality LEDs reaches 0.8. The pollution factor depends on the environmental pollution rate and the IP code of the lamp. The value of this factor for luminaires with a high-quality protection class (IP 6x) and moderate pollution of the operating environment is 0.89. The resulting maintenance factor is 0.71 in such a case. This factor may be significantly lower for low-quality LED luminaires and poorly maintained lighting systems [16].

- **Evaluation of an LED light source:** The operating lifetime of an LED light source is much longer than other light sources. The main advantage of these light sources is their high lumen output to electric power consumption ratio (luminous efficacy). Superior LEDs have luminous efficiency of up to 200 lm/W. LED light sources also fully comply with ecodesign EU directive 244/2009 [17]. It is necessary to take into account a large number of parameters when the quality of LED lights is evaluated and the most important parameter is the overall efficiency of the lighting module, followed by the type of LED chip, power supply quality, heat dissipation efficiency, the quality of materials used in the luminaire, and finally, the overall protection class. All of this determines the time when the emitted light output is reduced by 30%. Therefore, LED-based lights should be regarded as a complex system and all of the above mentioned aspects should be taken into consideration.

9.4 VLC Communication and Localization by Public Lighting

In recent years, LEDs have been used to replace incandescent lighting for indoor illumination. Car manufacturers use them in headlamps and, in a number of cities around the world, LEDs have been used in street lighting to reduce energy consumption [18,19]. One of the world's largest projects was introduced in Los Angeles, USA, where 140,000 street lights were replaced with energy-efficient LEDs. The result of the project is illustrated in Figure 9.9. It is clear that road illumination is much brighter and less light pollution is observed. Moreover, such a system can be controlled and managed wirelessly. Thus, the energy consumption and carbon emissions were reduced by 63% and 47.583 metric tons a year, respectively [20]. This represents over 8 million USD in annual savings, which is significant.

As the number of LED-based streetlights continues to expand, they can be adopted as part of the VLC network to provide not only efficient illumination, but also data transmission and positioning among other functions. Intelligent street lighting is an interesting solution for the future communications

(a)   (b)

FIGURE 9.9
The replacement of (a) existing streetlights with (b) energy-efficient LEDs in Los Angeles, USA. (City of Los Angelos—The LED Streetlight Replacement Program, 2015, http://bsl.lacity.org/led.html.)

networks which will offer data broadcasting anywhere and anytime. Such smart networks will enable communication among users, vehicles, and existing infrastructure to be established and provide not only internet connections and a local positioning system throughout the city, but also help to prevent car accidents or to transmit traffic and emergency information.

9.4.1 Background Noise

One of the most significant issues that should be taken into account is the effect of ambient light noise which varies during the day and can rapidly reduce VLC system performance. An accurate noise model is needed to combat this issue effectively. In [21], the model based on blackbody radiation is introduced and provides low complexity and accurate results. The background noise power can be given by:

$$Pb = E_{det}T0An2 \tag{9.5}$$

where n is the internal refractive index of the optical concentrator, A is the photodiode area, T_0 is the peak filter transmission coefficient, and E_{det} is given as:

$$E_{det} = \int_{\lambda 1}^{\lambda 2} W_{app}(\lambda)d\lambda \tag{9.6}$$

where W_{app} is the spectral irradiance approximation by the equation at a temperature 6,000 K:

$$W_{app}(\lambda) = S_p \frac{W(\lambda, 6000)}{\max[W(\lambda, 6000)]} \qquad (9.7)$$

where S_p is the peak spectral irradiance. Clearly, spectral irradiance is the only unknown input parameter and it is measured under clear sky conditions. The proposed model showed that background noise power can differ by up to 20 dB during the daytime thus resulting in ~12 dB reduction in signal-to-noise ratio (SNR).

Some methods to combat ambient light impediment are introduced in [22,23]. A special optical filter made from a thin film is performed in [22] to effectively reduce ambient light from incident angles of 30° and higher. This filtering method reduced received light up to 7.2% of the value when no filter was used. This is a promising, simple solution but more research needs to be done on the effect of the filter when the modulated signal comes from covered angles, i.e. > 30°. A more complex method is proposed in [23] featuring a dual receiving, selective combining receiver consisting of two branches, each with a band-pass filter, optical concentrator, and preamplifier. A signal from both branches leads to a decision circuit which selects the branch with higher SNR and reduces background noise power up to 6 dB.

Beside significant background light noise, position changes of the receivers and multipath reflections may cause a rise in intersymbol interference (ISI). Moreover, a signal transmitted over longer distances suffers from indispensable path loss. Both factors result in lower SNR values. Work published in [24] modeled the outdoor channel as a Rician fading channel. With this model, a technique called space-time block-coded orthogonal frequency division multiplexing (STBC-OFDM) with two transmitters was used. Such a modulation scheme outperforms the classic single-input single-output (SISO) system in both low- and high-dispersion channels resulting in higher data rates. Another approach is presented in [25] utilizing a modulation scheme based on a direct sequence spread spectrum (DSSS) technique which shows that SNR values from 8 to 10 dB are required to achieve reliable data reception in medium range and low data-rate applications.

9.4.2 VLC Coverage

The utilization of an LED street lighting network for an outdoor scenario is demonstrated in Figure 9.10(a). A ray tracing mechanism has been used to cover two city roads featuring 10 streetlights and passing traffic.

Figure 9.10(b) illustrates the light distribution of the corresponding outdoor scenario. It is clear that the light covers a significant section of the areas and its distribution can be easily changed or optimized by carefully designing streetlamp placement. The source radiation pattern has a significant influence on the

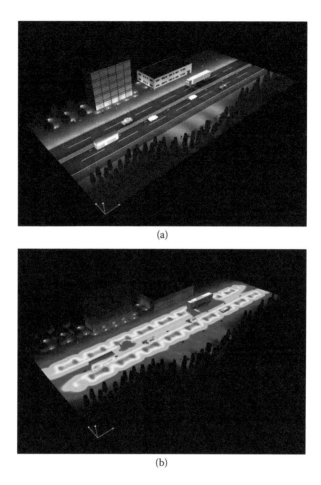

(a)

(b)

FIGURE 9.10
(a) Street lighting configuration and (b) illuminance distribution within a city street.

final distribution. Figure 9.11 shows several radiation patterns [26] formed either by lighting construction or by arrays of LED sources to reach a particular coverage.

Moreover, it has to be considered that such outdoor communication networks will be based mainly on the line-of-sight (LOS) channel in contrast to indoor VLCs where multiple reflections are more significant. High free-space loss typically limits the size of the area covered by a transmitter to tens and hundreds of meters. On the other hand, this feature helps to avoid interference among multiple transmitters deployed. The last significant feature of propagation within optical wavelengths is that the surface roughness has a substantial effect on the impinging wave.

The roughness of common surfaces in the RF domain is negligible, but has to be taken into account when compared with the wavelength of the

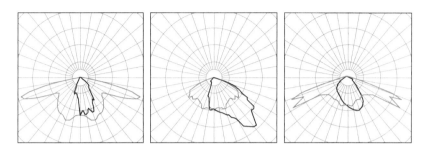

FIGURE 9.11
Example of radiation patterns of public LED lighting sources. (From GE Germany Lighting Outdoor Solutions, http://www.gelighting.com/LightingWeb/na/solutions/industry/roadway/overview/. With permission.)

optical impinging wave, that is, when the wave is scattered significantly more than at lower frequencies. The main motivation for this work arose from the abovementioned propagation issues in association with diffuse reflection and scattering phenomena. Some surfaces are completely irregular and reflect infrared (IR) signals without privileging any particular direction. These surfaces look equally bright when observed from different directions— the reflection patterns of these surfaces are completely diffuse and can be correctly approximated using Lambert's model, which is described by:

$$R(\theta_0) = \rho R_i \frac{1}{\pi} \cos(\theta_0) \tag{9.8}$$

where ρ is the surface reflection coefficient, R_i represents incident optical power, and θ_0 is the observation angle. The expression shows that the shape of the reflection pattern does not depend on the incidence angle, a fact which makes the model simple and easy to implement in software.

The reflection pattern of several rough surfaces is approximated well by Lambert's model except around the specular reflection direction where the pattern presents an intense component. What is more, this model is not able to approximate the mirror-like reflection patterns of smooth surfaces. Within a dense urban area we have to, unlike in suburban and rural areas, enumerate the reflections from smooth and glass building surfaces. Phong developed a model that allows us to approximate those reflection patterns correctly as the model considers the reflection pattern as the sum of two components: one diffuse and the other specular. The percentage of each component depends mainly on surface characteristics and is a parameter of the model. The specular component is modeled by a function that depends on the incidence angle θ_i and on the observation angle θ_0. Phong's model is described as:

$$R(\theta_0) = \rho R_i \frac{1}{\pi} \left(r_d \cos(\theta_o) + (1 - r_d)\cos^m(\theta_0 - \theta_i) \right) \tag{9.9}$$

where r_d represents the percentage of incident signal that is reflected diffusely and assumes values between 0 and 1. Parameter m controls the directivity of the specular component of the reflection. It should be noted that Lambert's model is obtained from Phong's when $r_d = 0$. Thus, implementing this phenomenon onto a further simulation, including the paraxial wave approach, was efficiently treated within the visible or near-IR region (applications of free-space optics) by the Phong model [27]. The reflection pattern of several rough surfaces can be well approximated by this model as the sum of two components: the diffuse component and the specular component. Such patterns were derived by the smoothing of experimental measurements in [27].

A new approach to model diffuse reflections and scattering, based on a fast and simple semideterministic principle, was presented in [28]. This model is based on the combination of precomputed radiation patterns and a fast semideterministic algorithm. An example of a diffuse system with multiple reflections, where scattering patterns were approximated by a simpler 3-D function allowing universal description, is illustrated in Figure 9.12.

The LOS communication channel can be easily blocked by tilting the receiver toward a body or surrounding structures, so there is a need to design the receivers of wireless devices carefully. An intelligent lighting system can adaptively control an individual LED by changing the intensity or by tilting LEDs to provide optimal illumination distribution and energy-saving management as determined by current weather conditions, the time of a day, or the traffic situation. Ray tracing analyses were performed for infrastructure-to-vehicle (I2V) communications in [29,30]. Figure 9.13 shows a similar

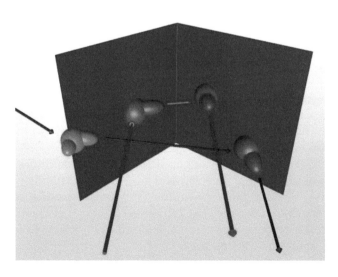

FIGURE 9.12
Example of a simulation of multiple diffuse reflections.

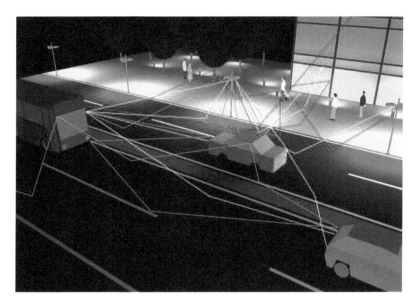

FIGURE 9.13
Example of a 3-D ray-tracing analysis for 12 V communications.

analysis performed via a 3-D ray-tracing model. In such cases, the channel impulse response is composed of multiple LOS and multipath components with a relatively long delay spread.

Several test cases and experimental results have been published for a vehicular VLC network consisting of onboard units, vehicles, and road side units such as traffic lights, street lamps, and digital signage. Cars fitted with LED-based front and back lights can communicate with each other and with roadside units (RSUs) through the VLC technology. Furthermore, LED-based RSUs can be used for both signaling and broadcasting safety-related information to vehicles on the road. Optical wireless communications systems based on the LED transmitter and a camera receiver were proposed for automotive applications in [31]. The signal reception experiment has also been performed for static and moving camera receivers with up to a 15 Mb/s error-free throughput under fixed conditions sustained. In [32], it was shown that the receiver in a driving situation can detect and accurately track an LED transmitter array with error-free communication over distances of 25–80 m. The LED street lighting network can be built in several dense outdoor scenarios as depicted in Figure 9.14a for city centers and Figure 9.14c for rural areas.

Typical metropolitan scenario experiences increased dispersive channel characteristics due to the reflection and diffusion of visible light [30]. When reflectance was set to 40% for buildings, 10% for poles, 20% for cars,

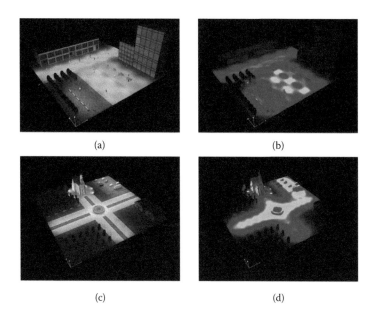

FIGURE 9.14
(a, c) Street lighting and (b, d) illuminance distribution for (a) a city center and (c) a rural area.

TABLE 9.3

Comparison of VLC Channel Profile of the Crossroad
Scenario for Vehicle-to-Vehicle (V2V) and Vehicle-to-
Infrastructure (V2I) cases

	Crossroad–V2V		Crossroad–V2I	
Tap	Relative Delay (ns)	Average Intensity (dB)	Relative Delay (ns)	Average Intensity (dB)
1	5	−1.53	5	−2.25
2	10	−24.25	10	−21.15
3	15	−38.07	15	−33.82
4	–	–	20	−45.83

Source: Lee, S., et al., *EURASIP J. Wireless Commun. Network.*,
2012, 370, 2012.

and 30% for road lines, several incident taps could be determined [30].
Tables 9.3 and 9.4 illustrate results derived in [30] for the VLC channel delay
profile for a crossroad scenario and a metropolitan street scenario, respectively.

As can be seen, the dense metropolitan scenario had more dispersive channel characteristics due to higher orders of reflection and diffusion.

TABLE 9.4

Comparison of the VLC Channel Profile of the Metropolitan Scenario for Vehicle-to-Vehicle (V2V) and Vehicle-to-Infrastructure (V2I) Cases

	Metropolitan—V2V		Metropolitan—V2I	
Tap	Relative Delay (ns)	Average Intensity (dB)	Relative Delay (ns)	Average Intensity (dB)
1	5	−7.52	5	−1.22
2	10	−6.70	10	−31.05
3	30	−100.25	25	−27.81

Source: Lee, S., et al., *EURASIP J. Wireless Commun. Network*, 2012, 370, 2012.

9.5 Conclusion

Contemporary public lighting has the great potential to be used with advances for future VLC and navigation purposes. As this concept has been studied only recently, many challenging research tasks lie ahead. This chapter gave an overview of the main features of LED-based public lighting systems that could be used for VLC. A description of state-of-the-art street lighting, including its main functions, control systems, and typical parameters, were introduced followed by the main aspects associated with lighting performance and aging. Finally, recent studies on public lighting for VLC purposes with the focus on the ray-tracing simulations, noise parameters, and delay profiles were introduced.

References

[1] L. Monzer. *Venkovni osvetleni architektur (Outdoor lighting of architectures)*. SNTL, Prague, 1980.

[2] International Commission on Illumination, CIE 115-2010 (2nd edition). Lighting of Roads for Motor and Pedestrian Traffic, *Commission Internationale de L'Eclairage*, Vienna, Austria, 2010.

[3] International Commision on Illumination, CIE 093-1992, Road Lighting as an Accident Countermeasure, *Commission Internationale de L'Eclairage*, Vienna, Austria, 1992.

[4] British Standard Institution, PD CEN/TR 13201-1:2014 Road Lighting—Part 1: Guidelines on Selection of Lighting Classes. *BSI Standards Limited*, London, 2014.

[5] T. Novak, P. Závada and K. Sokanský, Classification of environmental zones in the Czech Republic. *Lighting Res. Technol.*, 46(1):93–100, 2014.

[6] British Standard Institution, BS EN 12464-2:2014 Light and Lighting—Lighting of Work Places—Part 2: Outdoor Work Places. *BSI Standards Limited*, London, 2014.

[7] U. Brandi and C. Geissmar. *Light for Cities: Lighting Design for Urban Spaces, a Handbook*. Birkhäuser, Basel, 2007.

[8] *Cree First to Break 300 Lumens-per-Watt Barrier*, November 2015. Available: http://www.cree.com/News-and-Events/Cree-News/PressReleases/2014/March/300LPW-LED-barrier (accessed April 19, 2016).

[9] N. Bardsley. *Multi-Year Program Plan: Solid-State Lighting Research and Development*, U.S. Department of Energy, Washington, DC., 2014.

[10] I. Kudlacek and A. Mares. Critical look at eco-design of compact light sources. *Konstrukce*, 20:70–72, 2014.

[11] *Reliability and Lifetime of LEDs, Application Note*, May 2014. Available: www.osram-os.com (accessed May 25, 2016).

[12] H. Neugebauer, et al. Stability studies and degradation analysis of plastic solar cell materials by FTIR spectroscopy. *Proceedings of the International Conference on Science and Technology of Synthetic Metals*, 1999.

[13] B. Singh and N. Sharma. Mechanistic implications of plastic degradation. *Polymer Degradation and Stability*, 93:561–584, 2008.

[14] Thermal Management of Light Sources Based on SMD LEDs: Application Note, July 2013. Available: www.osram-os.com (accesed May 10, 2016).

[15] P. Jeong and C. C. Lee. An electrical model with junction temperature for light-emitting diodes and the impact on conversion efficiency. *IEEE Electron Device Letters*, 26:308–310, 2005.

[16] EN Standard CSN EN 13201-2 Road lighting - Part 2: Performance requirements. Available: https://www.en-standard.eu/csn-en-13201-2-road-lighting-part-2-performance-requirements-3/ (accesed May 25, 2016).

[17] European Union, Official Journal of the European Union, C 022 (Information and Notices), 57:32–33, 2014.

[18] *Liverpool Welcomes Energy-Saving LED Street Lights*, 2014. Available: http://www.liverpoolexpress.co.uk/liverpool-welcomes-energy-savingled-street-lights/ (accessed May 5, 2016).

[19] *Edmonton LED Street Lighting*, 2015. Available: http://www.edmonton.ca/transportation/on your streets/led-lightconversion.aspx (accessed 5 May 2016).

[20] *City of Los Angelos—The LED Streetlight Replacement Program*, 2015. Available: http://bsl.lacity.org/led.html (accessed 5 May 2016).

[21] I. E. Lee, et al. Performance enhancement of outdoor visible-light communication system using selective combining receiver. *IET Optoelectronics*, 3:30–39, 2009.

[22] Y. H. Chung and S. Oh. Efficient optical filtering for outdoor visible light communications in the presence of sunlight or artificial light. *International Symposium on Intelligent Signal Processing and Communications Systems (ISPACS)*, 2013.

[23] I. E. Lee. A dual-receiving visible-light communication system under time-variant non-clear sky channel for intelligent transportation system. *16th European Conference on Networks and Optical Communications (NOC)*, pp. 153–156, 2011.

[24] L. Changping, et al. Outdoor environment LED-identification systems integrate STBC-OFDM. *International Conference on ICT Convergence (ICTC)*, pp. 166–171, 2011.

[25] N. Kumar, et al. Visible light communication for intelligent transportation in road safety applications. *7th International Wireless Communications and Mobile Computing Conference (IWCMC)*, pp. 1513–1518, 2011.

[26] *GE Germany Lighting Outdoor Solutions.* Available: http://www.gelighting. com/LightingWeb/na/solutions/industry/roadway/overview/ (accessed 5 May 2016).
[27] B. T. Phong. Illumination for computer generated pictures. *Commun. ACM,* 18:311–317, 1975.
[28] S. Zvanovec and L. Subrt. Optical approximations for simulations of submillimeter systems. *Proceedings of the Fourth European Conference on Antennas and Propagation (EuCAP),* pp. 1–3, 2010.
[29] S. J. Lee, et al. Simulation modeling of visible light communication channel for automotive applications. *15th International IEEE Conference on Intelligent Transportation Systems (ITSC),* pp. 463–468, 2012.
[30] S. Lee, et al. Evaluation of visible light communication channel delay profiles for automotive applications. *EURASIP J. Wireless Commun. Network.,* 2012:370, 2012.
[31] I. Takai, et al. LED and CMOS image sensor based optical wireless communication system for automotive applications. *IEEE Photon. J.,* 5:6801418–6801418, 2013.
[32] T. Nagura, et al. Tracking an LED array transmitter for visible light communications in the driving situation. *7th International Symposium on Wireless Communication Systems (ISWCS),* pp. 765–769, 2010.

10

Transdermal Optical Communications

Manuel Faria, Luis Nero Alves, and Paulo Sérgio de Brito André

CONTENTS

10.1 Introduction

In the past few decades, we have witnessed an increase of the population life expectancy, as well as the prevalence of illnesses requiring close monitoring by means of implantable medical devices (IMDs), improving the patient's quality of life and contributing to sustaining their lives. Since the development of the first implantable pacemaker in 1958, the field of biomedical engineering has seen phenomenal technological achievements [1]. These achievements have resulted in smaller, safer, more complex, and smarter IMDs.

Nowadays, IMDs offer the possibility to perform real-time monitoring of several functions of the human body, helping in the diagnosis and treatment of illnesses and disorders. However, to perform this function, they require complex electronics systems with the ability to process the collected information and to communicate with an external device.

Currently, millions of people worldwide rely on IMDs, and such devices with external radiofrequency (RF) communications are already being used for a wide variety of applications, including temperature monitors, pacemakers, defibrillators, functional electrical stimulators, blood glucose sensors, and cochlear and retina implants [2]. Thus, the wireless modality for access and remote control of IMD is an increasingly requirement. Many limitations of current IMDs with wireless communication functions come from their RF connections. Three of the most challenging aspects in modern IMDs are electromagnetic interference (EMI), security and privacy, and power considerations [3]. The first two are concerned with the fact that IMDs usually communicate with the external interfaces by means of inductive or RF connections. Therefore, they are subject to interference from other electronic equipment, such as cell phones, or they may be a target of unauthorized access [4,5]. Additionally, patients may not even be allowed to perform some medical examinations, such as MRI (magnetic resonance imaging) [6]. In order to mitigate these problems, optical signals emerged as a viable alternative for wireless data exchange with IMDs [2,7]. Its main advantages are:

- Radiation spectrum not regularized
- High data rates (transdermal optical connections at 50 Mbps were reported [8])
- No radiation hazards
- High EMI immunity
- Security issues
- Maturity of optoelectronic devices

Regarding the power issue, the most used methods employ rechargeable batteries, charged by induction and RF harvesting. Alternatively, optical signals have recently gained attention as an energy harvesting method, suitable for IMDs since it mitigates EMI issues [9,10].

10.2 State of the Art

In the beginning of 1990s, studies started on some applications of optical links through the skin with data rates up to 1 Mbps, such as neuromuscular stimulators [11], artificial hearts and implanted cardiac assist devices [12], stimulating the bladder [13], and laboratory animal monitoring systems [14]. In 1999, Larson [15] studied the benefits of wireless optical communications for transdermal connections aiming at biomedical applications. In this work, a prototype telemeter able to record high-frequency extracellular neuroelectric signals was constructed and implanted in a rabbit. The transmitter used a light-emitting diode (LED) with 880 nm wavelength, in order to improve transmission efficiency through the skin. The receiver consisted of a panel of four GaAlAs photodiodes. The system was designed for an 8-channel connection at 15 kHz/channel. The final integrated circuit consumed 12.5 μA current for signal amplification, encoding, and multiplexing, and used another 7 μA for the optical output.

In 2004, Abita and Schneider [2] reported an important contribution to optical communications in IMDs applications. In this work, the authors conducted transdermal communication tests with samples of porcine skin, establishing connections at 115.2 kbps for several skin samples with an LED transmitter at 860 nm and a PIN photodiode at the receiver.

In 2005, Okamoto et al. [16] showed a development of a bidirectional transcutaneous optical data transmission system that promises adequate performance for monitoring and control of an artificial heart. The developed system used two narrow HPA (half power angle) LEDs in the visible range, with a peak emission wavelength at 590 nm to transmit data from inside the body to an external device. The transmission from the external device used a narrow HPA near-infrared LED with a peak emission wavelength at 940 nm. The system employed an ASK (amplitude-shift keying) modulator with a carrier pulse signal of 50 kHz, to support a maximum data transmission rate of 9600 bps. An in vitro experiment, with porcine skin layers, showed that the maximum tissue thickness of near-infrared optical data transmission without error was 45 mm. Electric power consumption for the data transmission link was 122 mW for near-infrared light and 162 mW for visible light.

In 2007, an optical transdermal connection achieved data communications at 40 Mbps, from an implanted device to a receiver outside the body. The experiment used a sample of porcine skin with 3 mm thickness as the transmission medium [17]. The average power consumption of the transmitter module was 4.3 mW.

In 2008, the innovation proposed for optical transdermal systems already implemented was the use of an LD (laser diode) [18]. There, a transmitter based on a VCSEL (vertical-cavity surface-emitting laser) diode in the infrared region at 850 nm, using Manchester code encoding, was used to test an optical telemetry system for in-body applications. The transmission medium consisted of porcine skin samples with different thicknesses. It demonstrated a system transmitting at data rates up to 16 Mbps, through a skin thickness of 4 mm while achieving a bit error rate (BER) of 10^{-9}, with power consumption of less than 10 mW.

The concern with the energy consumption by the implanted device is evident in [7]. Gil et al. [7] proposed a retroreflector scheme based on micro-electro-mechanical systems (MEMs) principles, to minimize the energy consumption of the implanted device. This work presented a mathematical model and experimental results from measurements of direct and retroreflection links using chicken skin as the transmission medium. The optical window for transdermal communications for both configurations extended from 800 nm to 940 nm. Numerical results showed that it was possible to achieve a BER of 10^{-6}, with transmitter power consumptions of 0.4 μW and 4 mW for the direct and retroreflective links, respectively.

Liu et al. [8] discussed the design of an optical transcutaneous link capable of transmitting data at 50 Mbps through a 4 mm porcine tissue, with a BER less than 10^{-5}, and a maximum power consumption of less than 4.1 mW. The main innovation is the use of a VCSEL driver for the transmitter, using a modified on-off keying for the modulation scheme, which allows less power consumption.

Table 10.1 summarizes the state of the art described above. The main observation to retain is the increasing of the data rates and the decreasing power consumption over time. Some of these papers are also reported in

TABLE 10.1

State-of-the-Art Transdermal Optical Communications

Ref.	Power (mW)	Data Rate	TX	RX	Wavelength (nm)	Skin Thickness (mm)	Skin Type	BER
[15]	–	–	LED	4 PIN GaAlAs	880	–	Rabbit	$<10^{-6}$
[2]	–	115.2 kbps	LED	PIN Si	860	6.9	Porcine	–
[16]	122.0 162.0	9600 bps	LED	PIN Si	940 590	45.0 20.0	Porcine	–
[17]	4.3	40 Mbps	VCSEL	PIN GaAs	850	3.0	Porcine	$<10^{-5}$
[18]	~7.5 ~16.0	16 Mbps	VCSEL	PIN Si	850	2.0 4.0	Porcine	$<10^{-9}$
[7]	$0.4E^{-3}$	–	LED	PIN Si	790	1.0	Chicken	$<10^{-6}$
[8]	2.6 4.1 6.4	50 Mbps	VCSEL	PIN Si	850	2.0 4.0 6.0	Porcine	$<10^{-5}$

one of the most recent review works about telemetry for IMDs, where it addressed media proprieties, standards, power, and data for these types of applications [19–21].

10.3 Transdermal Optical Link Model

This section discusses an approach to model an optical connection through the skin, from a transmitter outside the body to a receiver inside. The model's priority is the simplicity of implementation, based on commonly used components in optical communication systems.

10.3.1 Channel Modeling

In order to model the transdermal channel, three important factors should be taken into consideration: the transmittance of the skin, the misalignment between the transmitter and the receiver, and the background light interference.

10.3.1.1 Skin Transmissivity

Skin is a complex biological structure composed of three essential layers: stratum corneum, epidermis, and dermis (Figure 10.1). All of these layers have different characteristics that make the optical behavior on each one different. Furthermore, human skin is ethnically different, diverse in topology, penetrated by hair and sweat ducts, which makes it a complex, dynamic, and variable optical medium. Thus, a rigorous characterization of skin optical properties is an extremely challenging task, definable only by an approximate approach. There are two main effects to take into account to model the skin optical scattering and absorption.

It is important to find a simple metric that joins all of this complex information and summarizes it into a model parameter—the transmittance of the skin. This is the most challenging factor of skin channel, different from those normally used in OWC (optical wireless communications) [22].

The transmittance of the skin is defined as the ratio between the optical power that passes through the skin and the incident power. This parameter is wavelength dependent and accounts for the effects of absorption, reflection, and scattering. In [23], a dermis transmittance model is presented as a function of wavelength, for a predefined skin thickness. This work considers that the dermis layer is the only one with an important role on the skin transmittance, as previously proposed. In fact, it was demonstrated that most of visible and near-infrared radiation is transmitted through epidermis and stratum corneum layers, with negligible impairments [23].

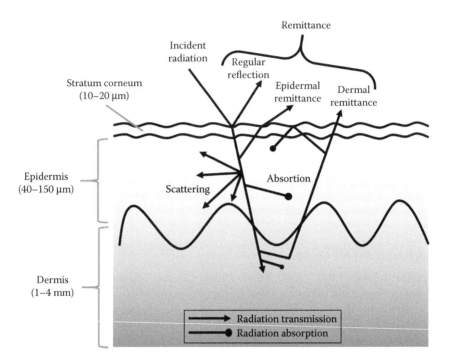

FIGURE 10.1
Schematic diagram of optical pathways in skin.

In order to extend this model to several dermal thicknesses, it is necessary to consider the skin attenuation coefficient, α, in m^{-1} [24]:

$$T(\lambda) = e^{-\alpha(\lambda)\delta}, \tag{10.1}$$

where T is the transmittance of the dermis and δ denotes the total dermis thickness. Hence, from the data reported in [23], it is possible to obtain the total description of the attenuation coefficient as function of wavelength, showed in Figure 10.2 [20], which is consistent with that reported in [25].

10.3.1.2 Misalignment

The directional property of the transmitted beam may be a drawback in terms of additional attenuation, and it is expected that a part of the optical beam power does not reach the photodetector sensitive area. There are three types of misalignment with direct influence on the power level in the receiver: longitudinal, lateral, and angular (Figure 10.3). As in transmittance, it is then necessary to define a single factor that summarizes the three types of misalignment—the misalignment factor, D. This factor

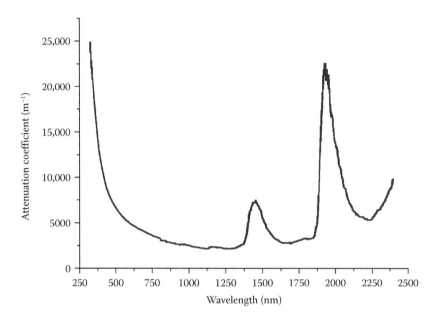

FIGURE 10.2
Attenuation coefficient of human dermis as a function of wavelength. (Computed from Ritter, R., et al., *IEEE Solid State Circuits Mag.*, 6, 47–51, 2014. With permission.)

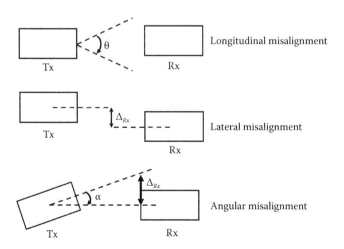

FIGURE 10.3
Different types of misalignment.

varies between 0 and 1 and represents the power loss resulting for the contribution of each type of misalignment.

To model the misalignment loss, it is important to characterize the radial dependence of the transmitted optical beam. Therefore, the optical power distribution in the beam must be known. The model used for the radiation pattern of the transmitter is based on a Gaussian distribution [26]:

$$I(\rho, z) = I_0(z) \, \exp\left[-\frac{2\rho^2}{w^2(z)}\right], \tag{10.2}$$

where I_0 is the maximum optical intensity on the radial direction z, $\rho = x^2 + y^2$ is the radial distance, and $w(z)$ is the radius of the optical beam.

The longitudinal misalignment, also known as beam divergence, arises from the optical beam diffraction from the emitting source. This divergence can significantly reduce the optical power received at the photodiode, since the effective area of the photodiode may be less than the total projection area illuminated by the beam. Following the Gaussian radiation model from Equation 10.2, the total power transmitted by the optical beam, considering a circularly symmetric distribution of the radiation intensity, is given by [27]:

$$P_{tot}(z) = I(0, z) \frac{\pi}{2} w^2(z), \tag{10.3}$$

where the optical beam radius $w(z)$ can be approximated, given the distance between the transmitter and the receiver, d, and the divergence angle, θ_{div}, of the transmitter:

$$w = d \tan\left(\frac{\theta_{div}}{2}\right). \tag{10.4}$$

Thus, considering a perfect alignment between the transmitter and the receiver axes, the power at the photodetector plane is defined as [27]:

$$P_{Rx}(z) = P_{tot}(z) \left\{ 1 - \exp\left[-2\frac{r_{Rx}^2}{w^2(z)}\right] \right\}, \tag{10.5}$$

where r_{Rx} is the radius of the active area of the photodiode. Thus, this factor represents another loss in the emitted power which can be significant, when the illuminated area is considerably larger than the effective area of the photodetector. Note that, to make the divergence angle of the optical transmitter small enough, a great accuracy to align the optical source and the detector is required. So there is a trade-off between the transmitter divergence angle (and its distance to the skin) and the power loss.

Lateral misalignment occurs when the transmission direction axis is not fully aligned with the normal axis of the receiver effective area. The detected optical power depending on the lateral shift, Δ, of the lateral misalignment is given by [28]:

$$P_{REC}(\Delta, z) = \sqrt{\frac{\pi}{2}} w(z) I(0, z) \cdot \int_0^{r_{Rx}} \left\{ \exp\left[\frac{-2x^2}{w^2(z)}\right] \mathrm{erf}\left[-\frac{\sqrt{2}}{w(z)}\left(\Delta + \sqrt{r_{Rx}^2 - x^2}\right)\right] \right.$$

$$\left. - \exp\left[\frac{-2x^2}{w^2(z)}\right] \mathrm{erf}\left[-\frac{\sqrt{2}}{w(z)}\left(\Delta - \sqrt{r_{Rx}^2 - x^2}\right)\right] \right\} dx. \qquad (10.6)$$

In turn, the angular misalignment factor is the power loss due to a difference in angle α between the transmitted beam axis and the axis of the receiving plane normal. If the receiver field of view is 180°, this can be approximated by lateral misalignment, since the misalignment angle α causes a lateral shift in the receiving plane. Thus, the expression of the received power is the same as Equation 10.6, with the lateral shift given by:

$$\Delta_{Rx} = d \, \tan(\alpha). \qquad (10.7)$$

10.3.1.3 Background Noise

Environmental light sources, with emission spectra overlapping the spectrum of data signals, are another disturbing communication factor. The main sources of ambient light are the sunlight and the artificial light sources (e.g., incandescent lamps, fluorescent lamps, and LED based bulbs)—see Figure 10.4. Sunlight is the main source of external noise, since it is the higher intensity source [22]. However, it is also important to study the influence of the typical indoor artificial lighting sources. This work considers that white LEDs are the main source of artificial light. Due to developments in the technology of LEDs, the trend indicates that this will be the main source of lighting in the future, due to its low power consumption, high efficiency, and long lifetime [29–31]. For these reasons, it was decided to simulate an indoor scenario with white LED illumination. The system in the absence of background noise was also studied.

1. Solar light

   To affect the communication link, the solar radiation must penetrate through the skin, reach the photodetector effective area, A_{ef}, and be converted into an electric signal. Thus, the total current produced by solar illumination is given by:

   $$I_{sun} = A_{ef} \int W(\lambda) T(\lambda) R(\lambda) \, d\lambda, \qquad (10.8)$$

   where $W(\lambda)$ is the sun spectral emittance (in $W/m^2.nm$), $T(\lambda)$ is the transmittance of the skin, and $R(\lambda)$ is the photodetector responsivity.

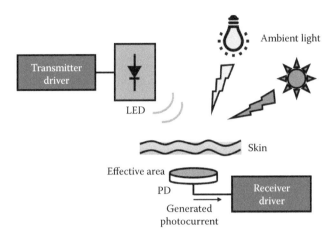

FIGURE 10.4
Illustration of light sources.

The model used for the spectral radiant emittance of the sun was ASTM G173-03, for global tilt conditions.

2. Darkness

The absence of background illumination is considered here as a reference case scenario. Most case scenarios with communications between an external terminal and an IMD will involve lighting. However, from the communications perspective, background illumination traduces into noise generated at the input of the optical receiver. Considering the absence of light, or darkness as previously mentioned, is in this sense an important case study, serving as reference case for low noise conditions.

3. White LED light

A common way to achieve white light employs a scheme similar to fluorescent lamps, performing blue wavelength up conversion with a yellow phosphorous coating. Power LEDs normally employ this method, due to its simplicity and because it translates into cost-effective devices. A simple approach to model the spectral power distribution of white LEDs is to use Gaussian distributions centered on the device response maxima [30]. Following this approach, with two peak wavelengths on blue (~460 nm) and yellow (~550 nm), the white LED's spectral power distribution (SPD) can be approximated by [31]:

$$S(\lambda) = \frac{1}{\sqrt{2\pi}} \left(\alpha \frac{1}{\sigma_1} exp \left[-\left(\frac{\lambda - \lambda_1}{\sqrt{2}\sigma_1} \right)^2 \right] + (1 - \alpha) \frac{1}{\sigma_2} exp \left[-\left(\frac{\lambda - \lambda_2}{\sqrt{2}\sigma_2} \right)^2 \right] \right),$$

$$(10.9)$$

where λ_1 and λ_2 correspond to blue and yellow wavelength peaks, respectively, while α is a weighting factor describing the additive proportions of each peak wavelength. Variables σ_1 and σ_2 represent the power spreading around each respective peak wavelength. The simulated SPD resembles that of the real white LED, where its power level was calibrated to obtain a corresponding total typical illuminance of a representative room, which is around 500 lux [32]. Afterwards, the procedure to acquire the value of the electric current generated by the background optical signal was similar to the solar light method.

10.3.2 Transmitter

The model of the transmitter is composed by a 1 Mbps non-return-to-zero (NRZ) random bit generator. The transmitter model also includes a Bessel filter to reproduce the bandwidth limitation of the optical source, and a gain block representing the conversion of the electrical signal to the optical domain. The transmitter also includes a low-frequency noise source with Gaussian distribution and variance given in [33]. The limitation due to optical signal extinction ratio was also taken into account (>8.2 dB).

10.3.3 Receiver

The model of the receiver considers the responsivity and all the typical impairment sources: thermal noise, electric shot noise, dark current, as well as bandwidth limitations. The thermal noise is caused by thermal fluctuations of the electric carriers in the receiver circuit, with an equivalent resistance, R_L, at temperature, T. This type of noise can be modeled by a white Gaussian noise, with a variance given by [34]:

$$\sigma_{th}^2 = \frac{4k_B T}{R_L} F_n, \tag{10.10}$$

k_B is the Boltzmann constant and F_n is the noise figure. On the other hand, the effect of thermal noise is measured by the NEP (noise equivalent power) [34]:

$$NEP = \left(\frac{4k_B T F_n}{R_L}\right)^{\frac{1}{2}} \frac{1}{R}, \tag{10.11}$$

It is possible to use the photodetector parameters to describe the noise source, combining Equations 10.10 and 10.11, to find the variance:

$$\sigma_{th}^2 = NEP^2 R^2 B. \tag{10.12}$$

The shot noise is a manifestation of charge carrier production statistics in semiconductor junctions. For the present case, the dominant source of shot noise is due to the photodetector. This current fluctuation is mathematically described by a stationary Poisson random process, which can be approximated by a Gaussian process. The shot noise variance is given by [34]:

$$\sigma_s^2 = 2qB(I + I_{dark}), \quad (10.13)$$

where I_{dark} is dark current, which is the current generated by the photodetector in the absence of background light, and comes from electron–hole pairs thermally generated. B is the bandwidth of the filter and q is the electron charge.

The responsivity is modeled by a gain, wavelength dependent, in A/W. The receiver bandwidth limitations were modeled with a Bessel filter, that in addition to simulating the bandwidth limit of the receiver, cut part of the noise present in the signal.

10.3.4 MATLAB Implementation

The model was implemented using the Simulink® toolbox of MATLAB®. The simulator that was built aims to model the behavior of a transdermal communication channel in which the transmitter is outside the body and the receiver inside, immediately after the skin barrier. Figure 10.5 depicts the main blocks of the simulator.

10.3.4.1 Analysis Tools

The purpose of the simulation was to determine the data signal quality at the reception and its current level. Thus, the tools used aim to generate quality indicators such as: eye diagram, Q factor of the eye diagram, and average current value. The Q factor of the eye diagram, as also known as eye signal-to-noise ratio (Eye SNR), is defined as the ratio of the eye amplitude to the sum of the standard deviations of the two binary levels:

$$Q = Eye\ SNR = \frac{\mu_1 - \mu_0}{\sigma_1 + \sigma_0}, \quad (10.14)$$

where μ_1 and μ_0 represent eye level 1 and 0 average amplitudes, respectively, and σ_1 and σ_0 are the standard deviations of the eye level 1 and 0 average amplitudes, respectively. This quality indicator is directly related with BER by [34]:

$$BER = \frac{1}{2}\text{erfc}\left(\frac{Q}{\sqrt{2}}\right), \quad (10.15)$$

which is defined as the average probability of incorrect bit identification. For both indicators (eye diagram and Eye SNR), the eyediagram.commscope tool, from MATLAB's Communications System Toolbox, was used.

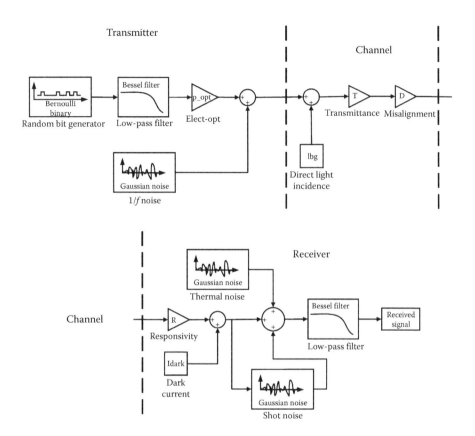

FIGURE 10.5
Simulink® model.

Finally, to measure the average current amplitude of the output signal a simple mean function of all the signal samples was considered.

10.3.4.2 Simulation Parameters

The model was simulated for a spectral range from 400 to 1700 nm through a range of skin thicknesses from 0 to 4 mm.

For the transmitter, an LED was selected due to its low energy consumption and its low cost, which could be consideration factors for commercial IMD design. The considered average emission optical power was 3 mW with a beam divergence angle of 60°. This high value for the divergence angle allows us to mitigate the alignment precision difficulties with the receiver. A distance of 1 cm between the receiver and the transmitter was also considered. The beam divergence was considered constant, since the distance between the emitter and the skin was invariant (1 cm) and the skin thickness

TABLE 10.2

Simulation Parameters

Component	Parameter	Symbol	Value
Transmitter:	Bit rate	D_b	1 Mbps
LED	Emitted optical power	P_{emi}	3 mW
	Wavelength	λ	400–1550 nm
	Beam divergence angle	θ_{div}	60°
Channel:	Skin thickness	δ	0–4 mm
Skin	Distance transmitter–skin	d	1 cm
Receiver 1:	Bandwidth	B	30 MHz
Si PIN	Effective area	A_{ef}	1.1 mm^2
	Noise equivalent power	NEP	6.7 x 10^{-15} W/Hz$^{1/2}$
	Dark current	I_{dark}	0.05 nA
Receiver 2:	Bandwidth	B	18 MHz
InGaAs PIN	Effective area	A_{ef}	0.92 mm^2
	Noise equivalent power	NEP	5 x 10^{-15} W/Hz$^{1/2}$
	Dark current	I_{dark}	0.07 A

has negligible impact in the beam divergence. Lateral and angular misalignments were not considered for this simulation. Due to the wide spectral range of analysis, two types of PIN photodiodes were selected—Si and InGaAs, for 400 nm to 1000 nm, and 1050 nm to 1700 nm, respectively. Table 10.2 shows the main parameters used in the simulation based on the selected components. The skin transmittance used the model previously described in 10.3.1.1. The simulation was performed for a transmission time of 0.1 s in order to generate 100,000 bits, with a total 100 samples per bit.

10.4 Simulation Results

10.4.1 Signal Quality

As already mentioned, the main factors affecting the signal are skin thickness, wavelength, background noise, and misalignment. The eye diagram analysis was one of the indicators chosen to study the signal degradation. The received signal after being converted into the electrical domain, is decoupled into two components—AC (information component) and DC (energy component). Figure 10.6 depicts examples of two different eye diagrams with normalized amplitudes, for each degradation effect, considering the AC component of the signal. As it is possible to observe in Figure 10.6a, the eye diagram quality is consistent with the attenuation coefficient evolution (Figure 10.2), where there

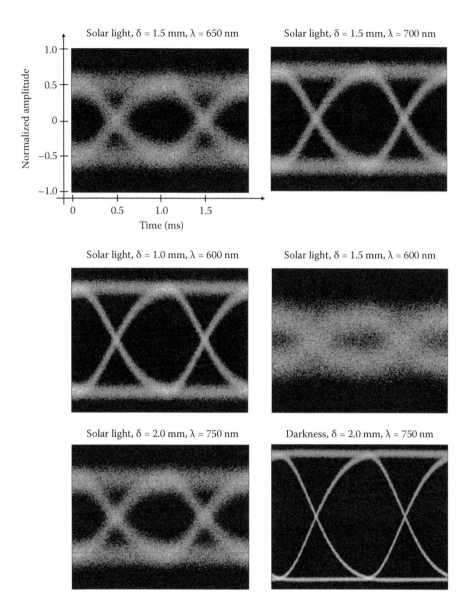

FIGURE 10.6
Received signal eye diagrams, for different values of the signal wavelength, skin thickness, and lighting conditions.

is higher quality for higher wavelength in the visible spectral range, assuming a constant skin thickness. Regarding skin thickness for the same wavelength, the signal degradation increases with higher skin thicknesses, since more tissue corresponds to higher signal attenuation (Figure 10.6b). Finally, in Figure 10.6c, the extreme environments of solar light and total darkness are compared for the same skin thickness and emission wavelength. This figure shows that in presence of sunlight the signal undergoes a much higher degradation effect than in a place without any illumination. This behavior is explained by the background current generated by solar light substantially superimposed on the amplitude of the data signal stream that arrives at the receiver. Consequently, it increases the shot noise since its variance is current dependent, as previously demonstrated. These results confirm that communication is favorable in a scenario without any external illumination source, where it is possible to achieve higher skin depths limits for a certain degradation level of the optical signal.

Since an eye diagram only provides a visual indication of degradation, it is important to have a quantitative metric of signal quality. For that purpose the quality factor, Q, was considered. Figure 10.7 depicts the variation of the Q factor along a range of wavelengths, from visible to IR (400–1800 nm, with minimum simulation step of 50 nm) through a range of skin thicknesses (0–4 mm, with a simulation step of 0.1 mm). As can be seen, the quality factor varies on the spectrum depending on the attenuation coefficient of the skin (Figure 10.2). This demonstrates that the quality factor varies according to the spectral transmittance of the skin for each emission wavelength. It was also confirmed that the quality factor of the signal decreases with the skin thickness, irrespective of the emission wavelength or the background illumination.

Regarding the results in different illumination environments, the highest gap is registered for the solar illumination scenario, in which there is a general decrease of the quality factor compared with the other two. Therefore, it is confirmed that the current produced by the solar lighting will cause a decrease in the quality of the data signal. This is a direct consequence of higher noise levels, which are present for high background illumination, as is the case under normal daylight conditions. Moreover, the illumination obtained by the white LED(s) (500 lux) can be compared to the total darkness environment, where there are no significant differences in the data quality factor for these two scenarios. This can mean an advantage in terms of communication, taking into account that a transdermal optical indoor link, with a typical lighting (500 lux) will not significantly affect communication performance.

From these results, it is concluded that the optimum wavelengths lie in region between 1100 nm and 1300 nm. Data from the simulation indicate wavelengths of 1250 nm and 1300 nm as being the best for communication, once they allow higher skin depths for the same required quality. These results are consistent with the literature presented in the state of the art (Section 10.2), which indicates spectral optical windows that maximize skin penetration, between 600 nm and 1300 nm. Note that, despite the optimal

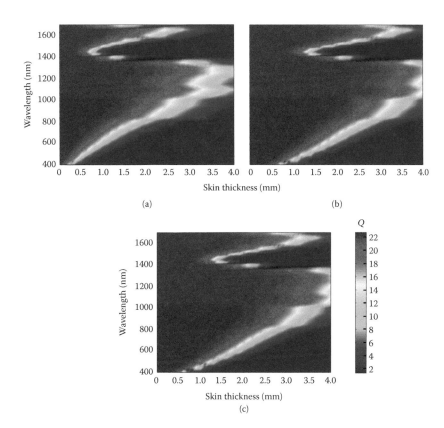

FIGURE 10.7
Q factor in the three lighting scenarios: (a) sunlight, (b) total darkness, and (c) white LED light at 500 lux.

wavelengths, the laboratory tests had to be compatible with the emitters commercially available. Therefore, the wavelengths used in the work are those presented in Table 10.1.

10.4.2 Average Current Level

After evaluating the data component of the optical signal (AC component), the next metric used is intended to compute the energy component of the optical signal (DC component). The aim is to correlate the influence of the received energy level in signal degradation. If there is greater penetration of optical radiation to the skin for certain wavelengths, there is also an increased level of energy received, which will consequently increase noise level. Thus, as for the simulation performed for the quality factor, values of the average current level at reception were extracted for the same wavelengths and skin thicknesses. The average normalized current values are presented in Figure 10.8 in a logarithmic scale.

Figure 10.8 demonstrates once again the similarity of indoor white LED lighting at 500 lux and total darkness environments.

It was also observed that the DC component of the electric current decreases with skin thickness in the three illumination environments, because of the related attenuation increase. However, the current levels are significantly higher in the case of an environment exposed to sunlight (this can be up to two orders of magnitude higher for the same wavelength and skin thickness), which makes the amplitude of the current less dependent on the emission wavelength when compared with the other two cases. These results further justify the degradation of eye diagrams, and thus the overall decrease of the quality factor for the sunlight illumination environment, since the variance of the receiver shot noise is dependent on the current amplitude.

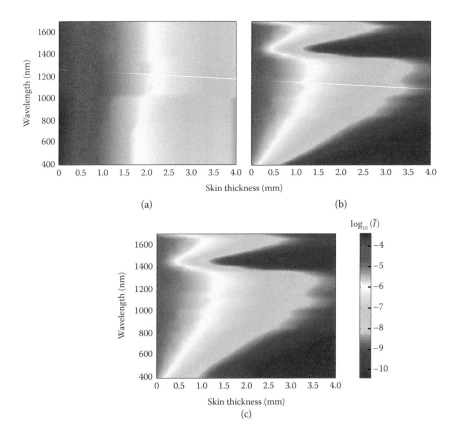

FIGURE 10.8
Average current level in the three lighting scenarios: (a) solar light, (b) total darkness, and (c) white LED light at 500 lux.

Moreover, the energy produced in an environment with white LED(s) with a typical illuminance (500 lux) is not significant, which explains the results obtained for the quality factor in Figure 10.7. In fact, the gap between the solar and the white LED lighting environments is understood, since the intensity of solar radiation can reach 1000 W/m^2 and the white LED, in the case of an illuminance of 500 lux, is limited to 1.5 W/m^2.

10.4.3 Energy Harvesting

The evaluation of the signal energy component (DC level) presented in the previous section can be also used to evaluate the energy harvesting capabilities of the implanted receiver, besides being used to study optical signal degradation. On the one hand, the received electric current can cause signal degradation. On the other hand, it may be used to collect energy for IMD battery charging purposes. Then, in order to find the spectral window that maximizes the signal quality and the energy level received, the multiplication of Q factor and average current level values (normalized at their maximum values) was computed. The results show that the region that maximizes energy harvesting is the same as the one that maximizes the quality of communication, specifically, the region between 1100 nm and 1300 nm, in all illumination environments mentioned. Hence, the ideal photodiode suitable to a transdermal optical system must have a sensitive detection region in the mentioned spectral region, thus it is advisable to use InGaAs PIN photodiodes.

Still regarding the receiver characteristics, in order to increase the collected energy it is possible to increase the effective area of the photodiode. However, receiving more energy can impact the quality of communication, since the level of electric current in the receiver may cause signal degradation, as demonstrated in the previous section. For this reason, the quality factor and the average electric current received as function of the photodiode effective area were computed, for an ideal emission wavelength of 1100 nm and a skin thickness of 4 mm. Figure 10.9 shows that the quality of communication only starts to decrease from 200 mm^2, which is a realistic value for an effective area of a photodetector to implement in an IMD. This effective area corresponds to an average current level of 3.3 µA. Furthermore, for an effective area of 10 cm^2, it is possible to achieve about 15.0 µA of current. A typical nominal supply current for commercial pacemakers is 20 µA [9]. Thus, these results can be relevant to enhance the durability of IMDs with low power consumption.

10.4.4 Misalignment Effect

Another factor described in the transdermal channel model with direct impact on the transmitted power loss is the misalignment between the transmitter and the receiver. To overcome alignment problems between the

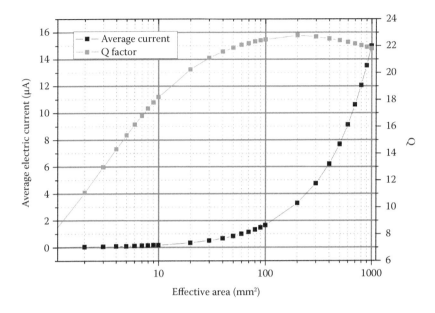

FIGURE 10.9
Average current level and Q factor of optical signal that reaches the receiver, as a function of the photodiode effective area, for a 1100 nm wavelength emission and 4 mm of skin thickness.

emitter and the receiver, the optical beam should be increased through the divergence angle of the transmitter. Moreover, the divergence of the optical beam also causes power losses (longitudinal misalignment). Therefore, this represents a trade-off between the divergence angle of the emitter and the longitudinal misalignment between emitter and receiver. In order to study the relation between the emitter's angle and the lateral misalignment between the axes of the emitter and the receiver, the results of the quality factor of a transdermal link for different divergence angles (10°–170°, with a step of 10°) and lateral shifts (0–10 mm, with a step of 1 mm) were obtained. The results achieved are depicted in Figure 10.10. Applying the Gaussian power distribution of the emitted beam model (Section 10.3.1.2), the simulation was performed for the same parameters shown in Section 10.3.4.2 (Table 10.2) for a typical emission wavelength of an infrared LED—950 nm—and an intermediate skin thickness of 2 mm, in an environment exposed to sunlight. Figure 10.10 shows that the received signal quality decreases with the increase of the emitter's divergence angle for the same lateral shift, since the increase of the area projected by the optical beam means lower received power in the receiver's effective area. It is also confirmed that the quality of the received signal decreases with the increase of the lateral shift, since the total effective area of the receiver is not covered; besides that, there is a

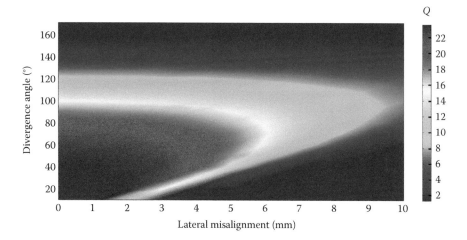

FIGURE 10.10
Quality factor as a function of the emitter's divergence angle and lateral misalignment for a wavelength of 950 nm and a skin thickness of 2 mm in an environment exposed to sunlight.

greater power concentration at the center of the optical beam. From these observations, it follows that it is possible to increase the divergence angle and maintain the quality of the communication under the presence of lateral misalignments. However, there is a limit to this process, where the related power losses compromise the quality.

10.5 Experimental Implementation

An experimental setup was created in order to complement the discussed model and to study optical signal attenuation through a skin sample. Due to ethical, technical, and regulatory barriers relative to the use of human skin, this work was conducted with three animal specimens—pork ham, chicken skin, and porcine skin. Although animal skin has different properties than human skin, it arises as a reasonable alternative since it has similar optical windows [2]. The work conducted consisted of the measurement of the optical signal attenuation measurement and the channel bandwidth. For that purpose, a test system was constructed. This system consisted of an LED as a transmitter (different LEDs with different wavelengths were considered), an interface medium serving as a channel (these included a fixture to hold the skin samples under test), and a photodetector able to recover the optical signal. The following sections present the setup and discusses the measured results.

10.5.1 Experimental Description

10.5.1.1 Spectral Attenuation

Spectral attenuation analysis was employed to estimate the attenuation coefficient of the skin sample. The measurement process consisted of the comparison of the spectral response of a known source acquired with a spectrometer with and without a skin sample. From this comparison, it is possible to disclose the spectral characteristics of the sample. The experimental setup is conceptually depicted in Figure 10.11. Instead of skin, we used a pork ham sample 0.68 mm thick—the results are depicted in Figure 10.14, together with the results of other methods.

10.5.1.2 Frequency Response Analysis

Another method for the measurement of the skin attenuation coefficient employs frequency response analysis, using a vector network analyzer (VNA). The response is also inferred through comparison between the frequency response of the source without a skin sample and the response having the skin sample. The procedure can be repeated for different LED sources with different wavelengths and different skin thicknesses. This method is better suited for the characterization of skin tissues with monochromatic sources, thus providing good results for a single wavelength. The previous method can be employed for wide spectral response sources, such as a white LED. The specimens studied were pork ham, chicken skin, and porcine skin, and the respective thicknesses are indicated in Table 10.3.

The experimental setup consisted of an LED driver driven by the VNA transmitting port and a photoreceiver (achieved with a PIN photodetector

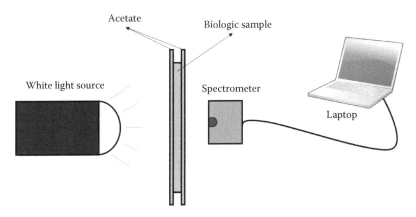

FIGURE 10.11
Illustration of the spectral attenuation analysis setup.

TABLE 10.3

Biological Specimens Used for Experimental Purposes

	Type	Thickness
Specimen 1	Pork ham	0.28 mm
Specimen 2	Chicken skin	1.29 mm
Specimen 3	Porcine skin	2.50 mm

FIGURE 10.12
Picture of the transdermal system setup.

(Thorlabs FDS100) connected to an amplifier (MiniCircuits ZFL 1000LN+) using a low impedance configuration), connected to the VNA receiving port. The LED and photodiode were separated by a 5 mm acrylic support, in which the biological sample was placed—Figure 10.12. Different LEDs were used, in order to register the frequency response of the optical signal received for different emission peak wavelengths, through the different specimens. The technical specifications of the LEDs used are presented in Table 10.4. Figure 10.13 presents the frequency response plots observed for three skin specimens using a blue LED (472 nm).

The analysis in Figure 10.13 confirms that it is possible to transmit an optical signal through a skin layer. The attenuation for different skin samples changes with frequency following the same trend as for the direct incidence case (no skin sample between emitter and receiver). Signal recovery implies amplification, which for the considered conditions is not too high (~25 dB for the worst case). This shows that greater skin depths can be achieved at the expense of more gain. These results also show that there are no significant effects up to 10 MHz, apart from attenuation. Moreover, there is an emission bandwidth of about 3 MHz for direct incidence, which did not change

TABLE 10.4

LEDs Specifications

Denomination	Reference	Wavelength (nm)	Radiant Intensity	Divergence Angle (°)
B	Multicomp MCL034SBLC	472	1.45 cd	36
Y	Optek Technology OVLFY3C7	595	4.00 cd	30
W	Lumex SLX-LX3054UWC	550 (typ.)	3.30 cd	30
IR 1	Kingbright L-53SF4C	880	15 mW/Sr	20
IR 2	Kingbright L-53F3C	940	30 mW/Sr	20

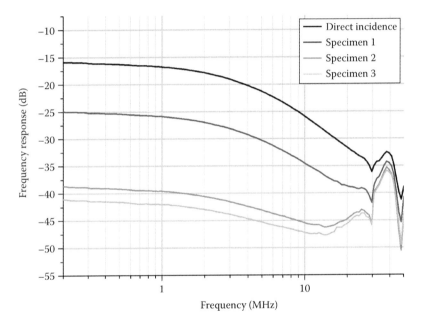

FIGURE 10.13
Frequency response of optical radiation incidence of a blue monochromatic LED (472 nm) to direct incidence and through the specimens 1, 2, and 3.

significantly for different specimens. Therefore, dispersive effects were not detected in the transdermal channel.

10.5.2 Data Analysis

Figure 10.14 depicts the attenuation coefficients inferred from the experimental measurement protocols previously described. For comparison purposes,

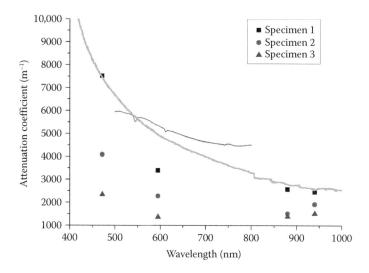

FIGURE 10.14

Attenuation coefficients experimentally obtained were green line–human skin model (Figure 10.2), red line–attenuation coefficient for specimen 1 using spectral analysis, and color points–attenuation coefficient using frequency response analysis for different wavelengths.

these results are depicted together with the attenuation coefficient used for modeling (presented before in Figure 10.2). None of the skin specimens used for this study correspond to human skin (as in Figure 10.2). Nevertheless, it is interesting to correlate results and disclose how the skin sample optical behavior differs from case to case.

Figure 10.14 shows that the closest results to the reported attenuation coefficient for human skin were those obtained from the spectral attenuation analysis data. The discrepancy between the two attenuation curves (red line for pork ham and green line for human skin) increases with wavelength. This discrepancy could be attributed to the biological differences between specimen 1 (pork ham) and human skin.

Single wavelength measurements using the frequency response approach reveal different values of the attenuation coefficient for different specimens. The results that were closer to the human skin (green curve) attenuation coefficient were again those obtained with specimen 1 (pork ham). The difference between the two methods, spectral analysis and frequency response, is noticeable. This can be attributed to the light source used in each case, which differs significantly and may influence the overall measurement. The attenuation coefficient for other specimens exhibited distinct behaviors as expected, but with a general trend to decrease with wavelength increase.

10.6 Conclusions

This chapter discussed the application of OWC means using the visible part of the spectrum to establish communication links through the human skin. As it has been ascertained in the first part of the chapter, optical means may present advantages for the development of IMDs concerning:

- Higher security links
- Low power consumption compared to wireless radio solutions
- Higher immunity to EMI
- The possibility to combine communications and energy harvesting methodologies based on optics

The main contribution of the chapter consisted of the presentation of a channel model suitable for the communications performance analysis. The model considered the characterization and optical transmittance properties of human skin. The main parameter for analysis in this respect is the skin attenuation coefficient. This coefficient varies for different wavelengths. The performance analysis concentrated on the quality factor extracted from the eye diagram of the receiver signal. The conducted analysis revealed how the communication link can be influenced by background illumination, due to either incident sunlight or artificial light produce by white LEDs. It was also investigated how the incident optical power could be explored for energy harvesting considerations. This investigation revealed that the low current levels generated by the incident optical power demand large-area devices in order to achieve the required energy levels. In this sense, optical energy scavenging is not an efficient way of gathering energy for IMDs. The final section of the chapter explored two methodologies for the characterization of skin samples, focusing on how to extract the attenuation coefficient. Different skin specimens were used, providing guidance for further developments in this research direction.

References

[1] F. Nebeker, 50 years of the IEEE Engineering in Medicine and Biology Society and the Emergence of a New Discipline, *IEEE Eng. Med. Biol. Mag.*, vol. 21, no. 3, pp. 17–47, 2002.

[2] J. Abita and W. Schneider, Transdermal optical communications, *Johns Hopkins APL Tech. Dig.*, vol. 3, pp. 261–268, 2004.

[3] W. H. Ko, Early history and challenges of implantable electronics, *ACM J. Emerg. Technol. Comput. Syst.*, vol. 8, no. 2, pp. 1–9, 2012.

[4] U. Lakshmanadoss, P. Chinnachamy, and J. P. Daubert, Electromagnetic interference on pacemakers, *Indian Pacing Electrophysiol. J.*, vol. 2, no. 3, pp. 74–8, 2011.

[5] D. B. Kramer, M. Baker, B. Ransford, A. Molina-Markham, Q. Stewart, K. Fu, and M. R. Reynolds, Security and privacy qualities of medical devices: An analysis of FDA postmarket surveillance, *PLoS One*, vol. 7, no. 7, pp. 1–7, 2012.

[6] S. L. Pinski and R. G. Trohman, Interference in implanted cardiac devices, Part I, *Pacing Clin. Electrophysiol.*, vol. 25, no. 9, pp. 1367–1381, 2002.

[7] Y. Gil, N. Rotter, and S. Arnon, Feasibility of retroreflective transdermal optical wireless communication, *Appl. Opt.*, vol. 51, no. 18, p. 4232, 2012.

[8] T. Liu, U. Bihr, S. M. Anis, and M. Ortmanns, Optical transcutaneous link for low power, high data rate telemetry, *Conf. Proc. IEEE Eng. Med. Biol. Soc.*, vol. 2012, pp. 3535–3538, 2012.

[9] N. K. Pagidimarry and V. C. Konijeti, A high efficiency optical power transmitting system to a rechargeable lithium battery for all implantable biomedical devices, *IFMBE Proc.*, vol. 15, pp. 533–537, 2007.

[10] S. Ayazian and A. Hassibi, Delivering optical power to subcutaneous implanted devices, *Proceedings of Annual International Conference IEEE Engineering in Medicine and Biology Society EMBS*, pp. 2874–2877, 2011.

[11] J. C. Jarvis and S. Salmons, A family of neuromuscular stimulators with optical transcutaneous control, *J. Med. Eng. Technol.*, vol. 15, no. 2, pp. 53–57, 1991.

[12] J. A. Miller, G. Belanger, I. Song, and F. Johnson, Transcutaneous optical telemetry system for an implantable electrical ventricular heart assist device, *Med. Biol. Eng. Comput.*, vol. 30, no. 3, pp. 370–372, 1992.

[13] M. Sawan, K. Arabi, and B. Provost, Implantable volume monitor and miniaturized stimulator dedicated to bladder control, *Artif. Organs*, vol. 21, no. 3, pp. 219–222, 1997.

[14] N. Kudo, K. Shimizu, and G. Matsumoto, Fundamental study on transcutaneous biotelemetry using diffused light, *Front. Med. Biol. Eng.*, vol. 1, no. 1, pp. 19–28, 1988.

[15] B. C. Larson, *An optical telemetry system for wireless transmission of biomedical signals across the skin*, Ph.D. dissertation, Dept. Elect. Eng. Cambridge, MA: Mass. Inst. Technol., 1999.

[16] E. Okamoto, Y. Yamamoto, Y. Inoue, T. Makino, and Y. Mitamura, Development of a bidirectional transcutaneous optical data transmission system for artificial hearts allowing long-distance data communication with low electric power consumption, *J. Artif. Organs*, vol. 8, no. 3, pp. 149–153, 2005.

[17] D. M. Ackermann, *High Speed Transcutaneous Optical Telemetry Link, Master's Dissertation*, Department of Biomedical Engineering, Case Western Reserve University, Cleveland, OH, 2007.

[18] S. Parmentier, R. Fontaine, and Y. Roy, Laser diode used in 16 Mb/s, 10 mW optical transcutaneous telemetry system, *2008 IEEE Biomed. Circuits Syst. Conf.*, pp. 377–380, Baltimore, MA, 2008.

[19] R. Ritter, J. Handwerker, T. Liu, and M. Ortmanns, Telemetry for implantable medical devices—Part 1, *IEEE Solid-State Circuits Mag.*, vol. 6, pp. 47–51, 2014.

[20] R. Ritter, J. Handwerker, T. Liu, and M. Ortmanns, Telemetry for implantable medical devices—Part 2, *IEEE Solid-State Circuits Mag.*, vol. 6, pp. 47–51, 2014.

[21] R. Ritter, J. Handwerker, T. Liu, and M. Ortmanns, Telemetry for implantable medical devices—Part 3, *IEEE Solid-State Circuits Mag.*, vol. 6, pp. 47–51, 2014.

[22] Z. Ghassemlooy, W. Popoola, and S. Rajabhandari, *Optical Wireless Communications: System and Channel Modelling with MATLAB®*, CRC Press, 2013.

[23] R. Anderson and J. Parrish, The optics of human skin, *J. Invest. Dermatol.*, vol. 77, no. 1, pp. 13–19, 1981.

[24] F. A. Duck, *Physical Properties of Tissues: A Comprehensive Reference Book*. Cambridge, Great Britain: Academic Press, 1990.

[25] A. N. Bashkatov, E. A. Genina, V. I. Kochubey, and V. V. Tuchin, Optical properties of human skin, subcutaneous and mucous tissues in the wavelength range from 400 to 2000 nm, *J. Phys. D. Appl. Phys.*, vol. 38, no. 15, pp. 2543–2555, 2005.

[26] J. Vitasek, E. Leitgeb, T. David, J. Latal, and S. Hejduk, Misalignment loss of free space optic link, *16th International Conference on Transparent Optical Networks (ICTON)*, pp. 1–5, Graz, Austria, 2014.

[27] J. Poliak, P. Pezzei, E. Leitgeb, and O. Wilfert, Analytical expression of FSO link misalignments considering Gaussian beam, in *Proceeding of the 2013 18th European Conference on Network and Optical Communications NOC 2013, 2013 8th Conferencece on Optical Cabling Infrastructure, OC I 2013*, pp. 99–104, Graz, Austria, 2013.

[28] J. Poliak, P. Pezzei, E. Leitgeb, and O. Wilfert, Link budget for high-speed short-distance wireless optical link, *Proceedings of the 2012 8th International Symposium on Communication Systems, Networks and Digital Signal Processing, CSNDSP 2012*, pp. 1–6, Poznan, Poland, 2012.

[29] T. Komine and M. Nakagawa, Fundamental analysis for visible-light communication system using LED lights, *IEEE Trans. Consum. Electron.*, vol. 50, no. 1, pp. 100–107, 2004.

[30] H. Chun and S. Rajbhandari, Effectiveness of blue-filtering in WLED based indoor visible light communication, *3rd International Workshop in Optical Wireless Communications*, pp. 60–64, Funchal, Madeira, Portugal, 2014.

[31] P. Butala, H. Elgala, P. Zarkesh-ha, and T. D. C. Little, Multi-wavelength visible light communication system design, *Globecom 2014 Workshop on Optical Wireless Communications*, pp. 1–10, Austin, TX, December 2014.

[32] *European Lighting Standard EN 12464-1*, 2nd Ed., ETAP, 2012.

[33] S. L. Rumyantsev, S. Sawyer, N. Pala, M. S. Shur, Y. Bilenko, J. P. Zhang, X. Hu, A. Lunev, J. Deng, and R. Gaska, Low frequency noise of light emitting diodes, *Noise Devices Circuits III*, vol. 5844, pp. 75–85, 2005.

[34] G. P. Agrawal, *Fiber-optic communication systems*, 4th edition, Rochester, NY: Wiley, 2010.

11

Underwater Visible Light Communications, Channel Modeling and System Design

Mohammad-Ali Khalighi, Chadi J. Gabriel, Luís M. Pessoa, and Bernardo Silva

CONTENTS

11.1 Introduction

Demands for underwater communication systems are increasing due to the ongoing expansion of human activities in underwater environments such as environmental monitoring, underwater exploration, offshore oil field exploration and monitoring, port security, and tactical surveillance. As such, there is a serious requirement to improve the performance of underwater communication systems in order to effectively use the equipment and the resources. The high cost, lack of flexibility, and operational disadvantages of wireline (particularly optical fiber) systems to provide real-time communication in underwater applications become restrictive for many cases. This triggers the growing demand for underwater wireless links. Acoustic communications suffer from a very small available bandwidth, very low celerity, and large latencies due to the low propagation speed. Underwater wireless optical communications (UWOC) which are able to achieve data rates of hundreds of Mbps (even up to Gbps) for short ranges, typically several tens of meters, appear as an attractive alternative or complementary solution to long-range acoustic communications. In fact, water is relatively transparent to light in the visible band of the spectrum and absorption takes its minimum value in the blue–green spectral range (450 nm–550 nm) [1,2]. Thanks to the ability of providing unprecedentedly high-rate data transmission, the UWOC technology enables the establishment of high-speed and reliable links for underwater missions employing robotics or autonomous underwater vehicles (AUVs), for instance. In addition, it is highly energy efficient, compared to the traditional technique of acoustic communication, and also has much less impact on marine animal life (see Figure 11.1) [3,4]. In particular, it is harmless to the cetaceans and coral.

UWOC has been recently the subject of intensive research. A few UWOC units of limited application have been commercialized very recently. For instance, Ambalux [5] has introduced a commercial UWOC system with a maximum data rate of 10 Mbps over ranges up to 40 m. Also, Sonardyne [6] has commercialized the BlueComm 200 UWOC system, claimed to operate over distances of up to 150 m with a maximum rate of 12.5 Mbps.

Our aim in this chapter is to present the fundamentals of UWOC, with a focus on aquatic channel properties, modeling, and characterization. We start by discussing the light propagation in water in Section 11.2, where we describe the different processes that can affect beam propagation in aquatic media, and explain how these phenomena are modeled. Channel characterization using analytical and numerical methods is discussed in Section 11.3. Transmitter (Tx) and receiver (Rx) design is then considered in Section 11.4, where signal modulation, error correction coding, and beam misalignment issues are discussed. The design and implementation of a prototype UWOC system, including experimental evaluation results is presented in Section 11.5. Lastly, some concluding remarks are presented in Section 11.6.

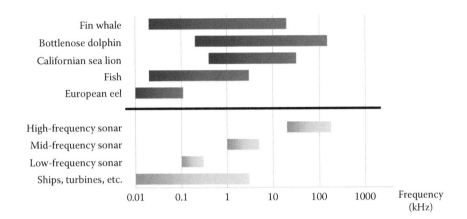

FIGURE 11.1
Hearing ranges of selected fish and mammal species (gray), reflecting some of the typical variety in these taxonomic groups, and frequency range of typical anthropogenic noises (light gray). Vertical lines show the human hearing range (in air). (Adapted from Slabbekoorn, H., et al., *Trends Ecol. Evol.*, 25, 419–427, 2010.)

11.2 Light Beam Propagation in Water

Upon interaction with the particles in suspension or solution within seawater, a propagating light beam is deviated from its initial direction through the scattering process, and a part of its intensity is absorbed and converted into other forms of energy. The scattering and absorption characteristics of a natural water body are called its inherent optical properties (IOPs) and are wavelength dependent [7–9].

11.2.1 Absorption, Scattering, and Turbulence

Absorption is the irreversible loss of power as light propagates in the medium. It is due to the interaction of photons with the water molecules and particles, and depends on the fluctuation of the index of refraction of the medium n and the light wavelength λ. In fact, water is relatively transparent to light in the visible band of the spectrum. Outside this range, light is subject to high absorption rates due to the electron transitions in the far ultraviolet and to different intra- or intermolecular motions in the infrared band. Scattering, on the other hand, refers to the deflection of light from its original path. On the microscopic level, scattering corresponds to the interaction between a photon and a molecule or an atom. Practically, one can divide light scattering in natural waters into pure seawater scattering (of size much smaller than λ) and particles scattering (of size larger than λ). Taking a pure seawater sample, n varies due to the random fluctuations in the concentrations of various ions (Cl^-, Na^+, etc.) [10]. These fluctuations determine the minimum values of the scattering properties. Moreover, particles with different shapes, types, and concentrations effectively determine the scattering properties of the medium. In fact, even in small concentrations, these particles make the scattering highly peaked in the forward direction, which is one of the major characteristics of the visible light propagation in natural waters [11].

The performance of a UWOC system can also be affected by channel fading as a result of oceanic turbulence. This is similar to the atmospheric turbulence in free-space optical communication [12,13]. Blobs of turbulent waters of different sizes can slightly and continuously change the propagation direction of photons due to the variation of n. They are mainly caused by temperature, salinity, and pressure variations in water [14]. In practice, however, the pressure effect on the water refractive index can be neglected [14]. Furthermore, deep seas generally have an approximately constant level of salinity and temperature variations are usually limited. It is shown in [15] that whereas temperature fluctuations have the major impact on turbulence in relatively shallow waters, salinity variations dominate as the water depth increases. Lastly, in deep waters, there is no probable beam blockage caused by bubbles, biologics, or large suspended particles except if the underwater link is implemented near hydrothermal sources, for example. For these reasons, water turbulence has been ignored in most previous works related to UWOC in relatively deep waters [16,17].

11.2.2 IOPs of Sea Water

The spectral beam absorption coefficient $a(\lambda)$ and the spectral volume scattering function (VSF) $\beta(\theta, \lambda)$ are the main IOPs used to model light absorption and scattering in water, respectively. The VSF, in units of $sr^{-1}m^{-1}$, is defined as the fraction of incident power scattered out of the beam through an angle θ. Integrating the VSF over all directions, gives us the beam spectral scattering coefficient $b(\lambda)$ [11]:

$$b(\lambda) = 2\pi \int_0^\pi \beta(\theta, \lambda) \sin \theta \, d\theta. \qquad (11.1)$$

The volume scattering phase function $\tilde{\beta}(\theta, \lambda)$ is defined as:

$$\tilde{\beta}(\theta, \lambda) = \beta(\theta, \lambda)/b(\lambda). \qquad (11.2)$$

Therefore, from the VSF, we can obtain a factor giving the scattering coefficient b with units of m^{-1}, and another one giving the angular distribution of the scattered photons $\tilde{\beta}$ with units of sr^{-1} [11].

Finally, adding $a(\lambda)$ and $b(\lambda)$ gives the spectral beam attenuation coefficient $c(\lambda)$, also called the extinction coefficient:

$$c(\lambda) = a(\lambda) + b(\lambda). \tag{11.3}$$

Another useful parameter is the backscattering coefficient $b_b(\lambda)$ that is obtained by integrating the VSF in the range $[\pi/2, \pi]$. Note that a, b, b_b, and c are in units of m^{-1}.

11.2.3 Seawaters in Suspension and Dissolved Particles

In addition to the wavelength, light absorption and scattering in seawater largely depend on the level of turbidity and the numerous particles that can be found in water [11]. These particles are present in an extraordinary variety of species, sizes, shapes, and concentrations. In fact, natural waters contain a continuous size distribution of particles ranging from water molecules, to organic and inorganic matter, fish, and whales. The constituents of natural waters are traditionally divided into dissolved and suspended particles, of organic or inorganic origins. Note that, dissolved matter corresponds to particles of size less than 0.4 μm corresponding to the shortest wavelength of visible light. Each component of natural waters, regardless of its classification, contributes in some way to the optical properties of a given water body. We discuss the cases of dissolved salts and inorganic and organic matter in the following.

Various dissolved salts constitute about 35 parts per thousand of the weight of sea waters. These salts increase the scattering above that of pure water by about 30%. They have a negligible effect on the absorption at visible wavelengths, but they increase absorption for ultraviolet and higher wavelengths: $\lambda > 800$ nm. Inorganic particles are created primarily by weathering of terrestrial rocks and soils. They consist of finely ground quartz sands, clay minerals, or metal oxides whose sizes range from much less than 1 μm to several tens of μm. Inorganic particles contribute to both scattering and absorption, mainly in turbid waters. On the other hand, organic particles may be in various forms as specified below.

- **Viruses**—in spite of their large numbers, virus particles have a very small impact on both absorption and scattering but can be efficient back scatterers at least at blue wavelengths in very clear waters.
- **Colloids**—part of the absorption and backscattering attributed to dissolved matter is probably due to colloids [18,19].
- **Bacteria**—can contribute to absorption, scattering, and backscattering, especially at blue spectral ranges in clear oceanic waters [20].
- **Organic detritus**—of various sizes, shapes, and origins, they are considered as the major back scatterers in the ocean but are poor scatterers and absorbers except at blue wavelengths [20].

- **Large particles including zooplanktons**—these particles strongly diffuse the light beam at small scattering angles.
- **Phytoplanktons**—widely present in most oceanic waters, they play an important role in defining the optical properties of these waters [11]. In fact, their chlorophyll and related pigments strongly absorb the light in the blue and red spectral ranges. Thus, in high concentrations, they effectively determine the spectral absorption of sea waters. In addition, these particles are generally much larger than the wavelength of the visible light and, hence, are efficient scatterers especially at small scattering angles. However, they scatter weakly at large angles and therefore are not considered as back scatterers.

When the concentration of phytoplankton and other organic matter is high compared to the other particulates, water is considered as Case 1 water [10]. Inorganic particles from land drainage dominate in Case 2 waters. However, about 98% of the world's open ocean and coastal waters fall into the Case 1 category where organic and especially phytoplanktonic particles effectively determine the water absorbance and strongly contribute to the scattering coefficient [11,21,22].

In addition to the above-specified organic matter, both fresh and saline waters contain varying concentrations of dissolved organic components. These components are produced during the decay of plant matter and consist mostly of various humic and fulvic acids [23]. In sufficient concentrations, these particles can color the water with a yellowish brown tint. For this reason, they are commonly referred to as colored dissolved organic matter (CDOMs). CDOMs increase the absorption rate with a decrease in wavelength; this absorption is more pronounced in the blue and ultraviolet spectral ranges. CDOMs are generally more concentrated in lakes, rivers, and coastal waters.

11.2.4 Spectral Beam Coefficients

The spectral beam absorption and scattering coefficients are calculated by adding the contribution of each class of particles to the corresponding coefficients of pure seawater, $a_w(\lambda)$ and $b_w(\lambda)$. Several attempts have been made to measure these coefficients [24,25]; however, determining the exact contribution of the various particulates and dissolved substances to $a(\lambda)$ and $b(\lambda)$ remains a very difficult issue. This difficulty is mainly due to the low concentrations of these constituents, the limitations and uncertainties in the measuring instruments, and the aliasing of the absorption measurements by scattering effects [11].

As mentioned previously, almost all open natural waters and deep seas can be considered as Case 1 waters. Therefore, one can use the chlorophyll concentration, C (in $mg \cdot m^{-3}$), as the free parameter to calculate a and b

based on predictive models, called bio optical models, such as that provided in [21,22]. Based on this model, the absorption coefficient is [22]:

$$a(\lambda) = a_w(\lambda) + a_c^0(\lambda)(C/C_c^0)^{0.602} + a_f^0 C_f \exp(-0.0189\lambda) + a_h^0 C_h \exp(-0.01105\lambda), \tag{11.4}$$

where $C_C^0 = 1$ mg/m^3, $a_w = 0.051$ m^{-1} at $\lambda = 532$ nm, a_c^0 is the specific absorption coefficient of chlorophyll [26], and $a_f^0 = 35.959$ m^2/mg and $a_h^0 = 18.828$ m^2/mg are the specific absorption coefficients for fulvic and humic acids, respectively. Also, C_f and C_h are the concentrations of fulvic and humic acids, in units of mg/m^3. Additional relationships between C and C_f and C_h are presented in [22,27,28] and reproduced below.

$$C_f = 1.74098\ C \exp[0.12327\ (C/C_C^0)] \tag{11.5}$$

$$C_h = 0.19334\ C \exp[0.12343\ (C/C_C^0)] \tag{11.6}$$

On the other hand, b and b_b can be determined by adding the contribution of small and large particles to b_w:

$$b(\lambda) = b_w(\lambda) + b_s^0(\lambda)C_s + b_l^0(\lambda)C_l \tag{11.7}$$

$$b_b(\lambda) = 0.5\ b_w(\lambda) + 0.039\ b_s^0(\lambda)C_s + 6.4 \times 10^{-4}\ b_l^0(\lambda)C_l, \tag{11.8}$$

where $b_s^0(\lambda)$ and $b_l^0(\lambda)$ are the specific scattering coefficients for small and large particles, respectively, in units of m^2/g. Also, C_s and C_l are the corresponding concentrations, in g/m^3, of small and large particulate matter and are given by [22,28]:

$$C_s = 0.01739\ C \exp[0.11631\ (C/C_C^0)], \tag{11.9}$$

$$C_l = 0.76284\ C \exp[0.03092\ (C/C_C^0)]. \tag{11.10}$$

Furthermore, in [22] simple analytical expressions are provided for determining b_w, b_s, and b_l, as a function of the wavelength λ:

$$b_w(\lambda) = 0.005826\ (400/\lambda)^{4.322}, \tag{11.11}$$

$$b_s^0(\lambda) = 1.151302\ (400/\lambda)^{1.7}, \tag{11.12}$$

$$b_l^0(\lambda) = 0.341074\ (400/\lambda)^{0.3}. \tag{11.13}$$

The chlorophyll concentration follows the following relationship in the upper ocean layer [27,28]:

$$C_C = 1.92 \; I_c^{1.8},\tag{11.14}$$

where I_c denotes the color index defined as $I_c = R(550)/R(440)$, with $R(\lambda)$ being the diffuse reflectance at wavelength λ (in nm).

We have shown in Figure 11.2 curves of a, b, and c as a function of λ for two chlorophyll concentrations, C, of 0.31 and 0.83 mg.m^{-3}. As it could be predicted, light absorption in water is at its minimum in the blue–green spectral range, regardless of the water turbidity. In fact, for both chlorophyll concentrations, the minima are around 480 nm. On the other hand, blue light is slightly more scattered than red light due to Rayleigh scattering [29]. However, an increase in C considerably affects b but has a negligible impact on a. Finally, the attenuation coefficient c, is at its minimum in the blue–green range. Note that, in turbid waters, the minimum value of

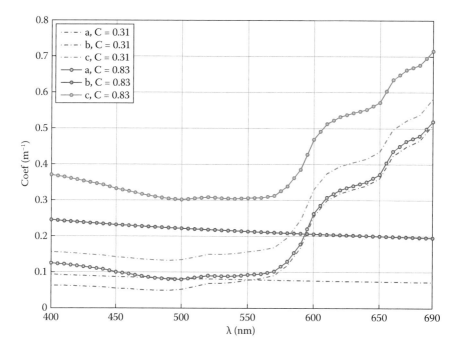

FIGURE 11.2
Absorption a, scattering b, and attenuation c coefficients as a function of the wavelength λ for two chlorophyll concentrations C (in mg.m^{-3}) corresponding to clear ocean and coastal waters using the model. (Adapted from Haltrin, V.I., & Kattawar, G.W., Appl. Opt., 32(27), 5356–5367, 1993; Haltrin, V.I., Appl. Opt., 38(33), 6826–6832, 1999.)

c tends to shift slightly toward the green wavelengths due to the increasing role of scattering.

11.2.5 Water Types

Considering Case 1 waters and knowing that underwater matter and water quality vary from one region to another, four major water types are usually studied in the literature [2,30]:

- Pure sea waters—absorption is the main limiting factor. The low b makes the beam propagate approximately in a straight line.
- Clear ocean waters—have a higher concentration of dissolved particles that affect scattering.
- Coastal waters—have a much higher concentration of planktonic matter, detritus, and mineral components that affect absorption and scattering.
- Turbid harbor and estuary waters—have a very high concentration of dissolved and suspended matter.

We have indicated in Table 11.1 typical values for the parameters a, b, b_b, and c, associated with these water types that we will consider hereafter. For this, we have set the chlorophyll concentration C to obtain close values to the attenuation coefficient values provided in [30,31]. The parameters are calculated using Haltrin's bio optical model presented in the previous subsection for $\lambda = 532$ nm [22].

In practice, water turbidity can be determined using onboard sensors such as a transmissometer, or by using colocated Tx and Rx based on the estimation of backscattered light, as shown in Figure 11.3 [32]. Having an estimate of b_b, we can estimate the chlorophyll concentration C and the attenuation coefficient c.

TABLE 11.1

Absorption, Scattering, BackScattering, and Attenuation Coefficients for the Four Water Types

Water Type	C (mg/m³)	a (m⁻¹)	b (m⁻¹)	b_b (m⁻¹)	c (m⁻¹)
Pure sea	0.005	0.053	0.003	0.0006	0.056
Clear ocean	0.31	0.069	0.08	0.0010	0.15
Coastal	0.83	0.088	0.216	0.0014	0.305
Turbid harbor	5.9	0.295	1.875	0.0076	2.17

Note: Considering typical chlorophyll concentrations [1,11] for $\lambda = 532$ nm.

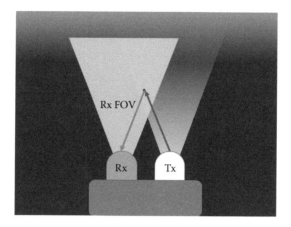

FIGURE 11.3
Using colocated Tx and Rx for estimating optical backscatter coefficient. (Adapted from Simpson, J.A., et al., *IEEE J. Sel. Areas Commun.*, 30(5), 964–974, 2012.)

11.2.6 Phase Function

As explained before, the VSF β is the main IOP that characterizes light scattering in water. It depends on both the spectral coefficient b and the phase function $\tilde{\beta}$. Knowing b, the study of the VSF turns to evaluate the phase function distribution as a function of the scattering angle. To study $\tilde{\beta}$, *in situ* measurements require highly accurate instruments, especially because scattering is highly peaked in the forward direction due to the presence of dissolved particulate matter within natural waters. The generally used technique consists of transmitting a collimated beam of well-known characteristics through the studied water volume and then measuring the scattered intensity as a function of the scattering angle [33]. Unfortunately, implementing such instruments for field applications remains quite difficult. In fact, in addition to the attenuation effect, scattering measurements at the critically small angles (less than 1°) and large angles (about 179°) is extremely difficult and requires precise alignment of the optical elements that can be very hard to achieve in harsh environments such as open seas [11].

The most carefully made and widely cited experimental scattering study is that conducted by Petzold [31] from which all other measurements and approximations were made and compared with. Among other VSF measurements we can mention those of Kirk [23] and Jerlov [34].

Several bio optical models can be used to derive the shape of the VSF. As an example, we can mention the model suggested by Kopelevich based on the addition of the contribution of small and large particles to the scattering [35]. Unfortunately, when compared to Petzold measurements, this model is unable to reproduce exactly the VSF shape, especially at very small angles [11]. Mie scattering calculations can derive β except at very small angles, given the precise optical properties, size, and distribution of the particles. On the other hand, Henyey and Greenstein have proposed a model in [36] for analytically deriving β̃. This popular model is simple and efficient, and can be easily implemented to retrieve the general distribution of the phase function. Originally proposed for galactic scattering in astrophysics [37], the Henyey–Greenstein (HG) phase function is defined by [11]:

$$\tilde{\beta}_{HG}(\theta, g) = \frac{1 - g^2}{2(1 + g^2 - 2g\cos\theta)^{3/2}} \tag{11.15}$$

Here, g is the HG asymmetry parameter that depends on the medium characteristics and is equal to the average cosine of the scattering angle θ over all scattering directions, denoted by $\cos\theta$. It is proposed in [38] to set $g = 0.924$ as a good approximation for most practical situations. In fact, based on Petzold's measurements [31], g is calculated in [39] for clean ocean, coastal, and turbid harbor waters. For these three water types, g is equal to 0.8708, 0.9470, and 0.9199, respectively [1]. In fact, the small difference between these g values has a negligible effect on the optical channel characteristics. This is because the HG model is not accurate at small θ since its shape is broader than most real-phase functions. For collimated beams, the phase function does affect the channel characteristics, as shown in [40] but this is not the case for divergent beams that are typically used in UWOC systems.

A modified phase function, called the two-term Henyey–Greenstein (TTHG), has been proposed later in [11,41]. This model matches the experimental results better, especially those obtained by Petzold [31]. The TTHG phase function is given by:

$$\tilde{\beta}_{TTHG}(\theta, \alpha, g_{FWD}, g_{BKWD}) = \alpha \ \tilde{\beta}_{HG}(\theta, g_{FWD}) + (1 - \alpha)\tilde{\beta}_{HG}(\theta, -g_{BKWD}) \tag{11.16}$$

where α is the weight of the forward-directed HG phase function, and g_{FWD} and g_{BKWD} are the asymmetry factors for forward- and backward-directed HG phase functions, respectively. Relationships between g_{FWD}, g_{BKWD}, α, and $\cos\theta$, are provided in [41,42].

In Figure 11.4, the distribution of the phase function as a function of θ based on the HG and TTHG models are compared with the experimental measurements undertaken by Petzold in [31] corresponding to the average cosine of 0.907. Compared with the HG model, we notice that the TTHG model gives a phase function closer to the Petzold's experimental data, especially at small angles where β̃ is at its maximum (although it is still away from the Petzold's data) [1]. While the TTHG model does not exactly match the Petzold's experimental measurements, it is more accurate than the usually-used HG model.

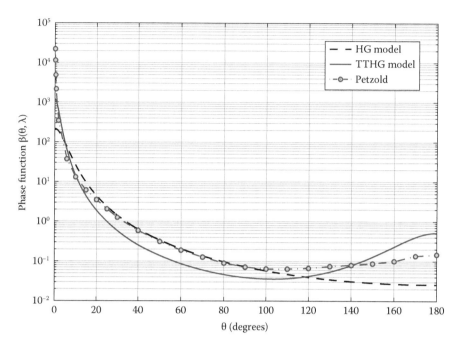

FIGURE 11.4
Contrasting HG and TTHG phase functions with Petzold's experimental measurements [31] for b_b/b=0.038 [1].

11.3 Aquatic Channel Characterization

The study of feasibility and reliability of an underwater optical link necessitates accurate channel modeling by taking into account the seawater optical properties. The propagation of light underwater is typically modeled by the radiative transfer equation (RTE) [38], which involves integrodifferential equations of time and space that characterize a light field traversing a scattering medium. Different analytical and numerical methods can be used to solve the RTE. The main drawback of the analytical methods is their mathematical complexity. Numerical methods based on Monte Carlo simulations are interesting alternative solutions to the RTE and provide a powerful tool that can adequately model light propagation within a scattering medium even if its IOPs vary in space and time [43].

Here, after explaining the principles of the RTE and the Monte Carlo simulation tool, we present some numerical results to show the impact of water channel under various conditions. We focus on the channel impulse response (CIR) that fully characterizes the optical channel.

11.3.1 Radiative Transfer Equation

Once emitted from the Tx, photons initiate a complex chain of scattering and absorption events within the water body. When a photon interacts with a molecule it may be absorbed, leaving this molecule in an excited state with a higher internal energy. In order to return to its stable state, the molecule could emit a photon with the same energy as the absorbed one, in which case the process is called elastic scattering. However, if the released photon has a smaller energy than the original one, the molecule will stay in an excited state. To return to its original state, this molecule could thermally dissipate its residual excess energy, transmit another photon, or wait for another photon to be absorbed and then emit a photon with a higher energy. All these types of photon emission with different wavelengths are called inelastic scattering. On the other hand, all or part of the absorbed photon's energy may be converted into thermal or chemical energy which corresponds to a true absorption process. The reverse process, when the chemical energy is converted into light, is called true emission.

All these processes can be summarized in one mathematical equation—the RTE. This equation describes the light radiance distribution in the propagation medium, given its IOPs and the light beam characteristics. Let us denote the light radiance by $L_r(z, \theta, \phi, \lambda)$, with z being the geometric depth in units of meters and θ and ϕ the polar and azimuthal angles, respectively. The general expression of L_r is given by [11]:

$$\frac{1}{v}\frac{\delta}{\delta t}\frac{L_r}{n^2} + \xi \nabla \left(\frac{L_r}{n^2}\right) = -c\left(\frac{L_r}{n^2}\right) + \ell_*^E + \ell_*^I + \ell_*^S \ (\mathrm{Wm}^{-3}\mathrm{sr}^{-1}\mathrm{nm}^{-1}) \qquad (11.17)$$

where t denotes time, and ξ and v denote the direction and speed of light propagation in the medium, respectively. Also, ℓ_*^E, ℓ_*^I, and ℓ_*^S denote the time-dependent path functions for elastic scattering, inelastic scattering, and true emission source processes respectively, and the factor $-c\left(\frac{L_r}{n^2}\right)$ corresponds to the true absorption process.

As a matter of fact, because of their relatively low contribution to the general solution of the RTE, we can effectively neglect the effects of the inelastic scattering and the true emission processes. In addition, while still neglecting the turbulence effect in water and considering a homogeneous water body with a negligible diffraction impact, the radiative transfer can be considered as time-independent. Considering such conditions, we denote ℓ_*^E by ℓ_E to indicate the time-independency of the path function for elastic scattering [38]. Then, the expression of L_r along a path is reduced to the following:

$$\frac{dL_r}{dr} = -c \ L_r + L_E, \qquad (11.18)$$

where $r = z / \cos\theta$. Integrating this equation with respect to r, we obtain the simplified form of the RTE:

$$L_r(z, \theta, \phi) = L_r(0, \theta, \phi)e^{-cr} + L_\Lambda^E, \tag{11.19}$$

where

$$L_\Lambda^E = \frac{L_E(0, \theta, \phi) \, \exp(-\Lambda \, r \cos\theta)}{c - \Lambda \cos\theta} \, \{1 - \exp[-r(c - \Lambda \cos\theta)]\}. \tag{11.20}$$

Here Λ, which is a function of θ and ϕ, is the diffuse attenuation coefficient for radiance, in units of m^{-1}, and is defined as follows:

$$\Lambda = -\frac{1}{L_r(z, \theta, \phi)} \frac{dL_r(z, \theta, \phi)}{dz}. \tag{11.21}$$

If we neglect beam scattering and consider straight-line propagation, it turns to considering the simple Beer–Lambert's law given below:

$$L_r(z) = L_r(0) \, \exp(-cz). \tag{11.22}$$

Several analytical methods can be used to resolve the RTE [38]. One of them is the invariant embedding solution, which is a computationally efficient and highly accurate method that considers the variation of the IOPs and the boundary conditions at the bottom and the water–air surface. Another method is the discrete ordinates solution and more specifically the eigen matrix solution that can only be applied to homogeneous water bodies [44,45].

On the other hand, a statistical method based on Monte Carlo simulations can be used to resolve the RTE. Although it is not as accurate as the analytical solutions, this method is characterized by its simplicity and flexibility. In addition, a Monte Carlo simulator is a powerful tool that can adequately model light propagation within a scattering media even if its IOPs vary in space and time [43,46,47].

Lastly, note that inverse models can be used to recover the IOPs, given the measured radiometric quantities of an underwater light field. However these methods may encounter several problems such as the non-uniqueness of the solution, the sensitivity to errors in the measured radiometry, and the difficulties of accurately measuring the radiance distribution in the water [38,48].

11.3.2 Numerical Results

Using Monte Carlo simulations, we determine the trajectory of photons launched from the Tx until arriving on the Rx photodetector (PD) active area (if not lost in the meanwhile) [1,49]. We consider the TTHG phase function model for the VSF and the Case 1 waters where organic particles and phytoplankton are dominant. Remember that for this water type, the channel is characterized mainly by the chlorophyll concentration [41], which determines the scattering and absorption properties and depends highly on factors such as the water type, depth, and temperature. Also, we take into account different system parameters such as the Tx beamwidth and beam divergence, wavelength, water type and turbidity, link distance, and the Rx's field of view (FOV) and aperture size. Figure 11.5 shows a simplified schematic of the typical UWOC link we consider here.

Let us investigate the effect of the attenuation coefficient c on the total received intensity that we denote by I_r. We have shown in Figure 11.6 curves of I_r as a function of distance Z for the four water types specified in Table 11.1, a Tx divergence angle of 20°, and a Rx FOV of 180°. Results are presented considering the Beer–Lambert model and Monte Carlo simulation. We notice a difference between the two models that tends to grow with increased turbidity and transmission distance. This difference is mainly due to the fact that the scattering impact is more important for a larger value of cZ that we refer to as the attenuation length. Let us now focus on the results corresponding to the TTHG model and assume a tolerable loss of 50 dB beyond which the signal may not be detectable at the Rx. Note that, in practice, this limit depends on the Tx power, Rx sensitivity, and the safety power margin of the link. With this assumption, we notice from Figure 11.6 that the transmission range is limited to 27 m, 46 m, and 98 m for coastal, clear ocean, and pure sea waters, respectively. When working in turbid or estuary waters, on the other hand, the high signal attenuation limits the communication range to less than 6 m.

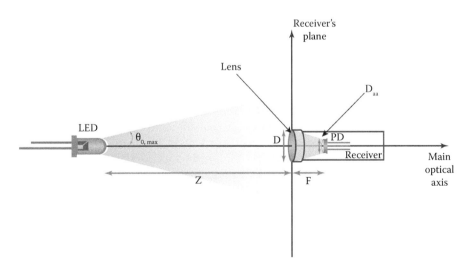

FIGURE 11.5
Schematic of an optical wireless underwater link with distance Z, Tx divergence angle $\theta_{0,max}$, Rx aperture diameter D, and PD active area diameter D_{aa}.

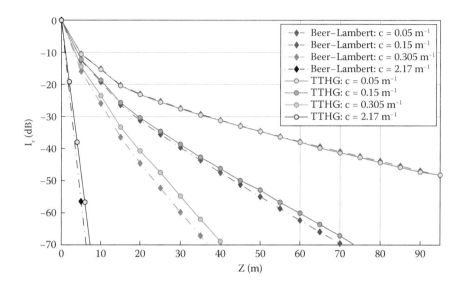

FIGURE 11.6
Received intensity as a function of distance for different water types, $D = 20$ cm.

The channel time dispersion is investigated and quantified in [1] for different system parameters including link distance, Tx beam divergence, and Rx lens aperture size. It is shown that, except for highly turbid waters, the channel time dispersion can be neglected when working over moderate distances. For instance, we have shown in Figure 11.7a the CIR for different water types, where the abscissa represents the absolute propagation time from the Tx to the Rx. Here, the attenuation coefficient c is set to 0.05, 0.15, and 0.305 m^{-1} for the three cases of pure sea, clean ocean, and coastal waters (see Table 11.1), corresponding to attenuation lengths of 1.0, 3.0, and 6.1, respectively.

The abscissa is the absolute propagation time from the Tx to the Rx in (a) and the relative time (with reference to the absolute propagation time) in (b). a, b, and c coefficients correspond to Table 11.1 [1].

The attenuation length, defined as the product cZ, is also indicated in the figure. We notice that the channel dispersion, defined as the duration over which the CIR falls to -20 dB below its peak, is about 0.21 ns, 0.26 ns, and 0.28 ns, for the pure sea, clear ocean, and coastal water cases, respectively. Therefore, for typical data rates (below Gbps), the channel can practically be considered as nondispersive.

We have also shown the CIR for the case of harbor turbid waters with $Z = 6$ and 8 m in Figure 11.7b corresponding to attenuation lengths of 13.02 and 17.36, where the delay spread is about 0.6 and 3 ns, respectively. Note that communication over such distances in highly turbid waters requires very

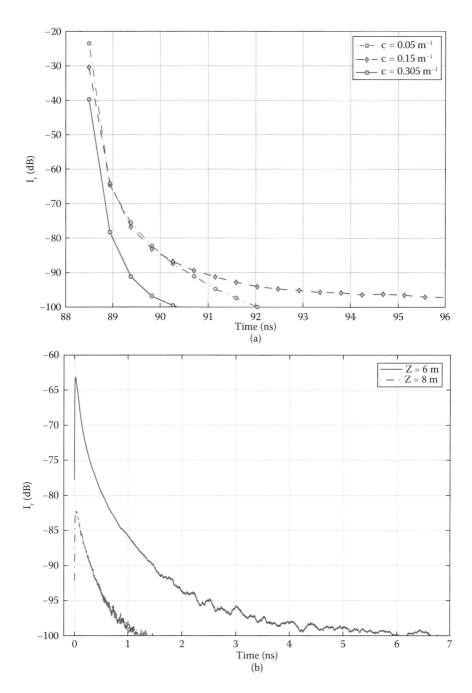

FIGURE 11.7
CIR for (a) pure sea ($c = 0.05$ m^{-1}), clear ocean ($c = 0.15$ m^{-1}), and coastal ($c = 0.305$ m^{-1}) waters; (b) turbid harbor ($c = 2.17$ m^{-1}) waters. Rx FOV = 180°, Tx divergence angle $\theta_{0,max} = 10°$, $D = 20$ cm; $Z = 20$ m in (a).

high power emitters: the intensity loss being about -82.3 dB at $Z = 8$ m. The experimental studies in [30,40,50] confirm these conclusions: it is shown that for attenuation lengths cZ larger than 10, the scattering effect becomes important, and it predominates for $cZ > 15$. The interesting point in Figure 11.7b is that due to the high amount of scattering, the CIR peak occurs slightly after the direct path delay.

11.4 Optical Transmitter and Receiver Design

11.4.1 Signal Modulation

An important design issue is the choice of the modulation scheme that can affect the system performance considerably. Currently, the most widely considered schemes rely on intensity modulation and direct detection (IM/DD), having the advantage of low implementation simplicity. The performances of several IM/DD schemes were compared in [51] by taking into account the energy and bandwidth efficiencies, as well as practical implementation feasibility. Although among IM/DD schemes the pulse-position modulation (PPM) is optimal in the sense of energy efficiency, digital pulse interval modulation (DPIM) [52] makes a good compromise between link performance and complexity, and appears to be a suitable scheme. DPIM outperforms the conventional on-off keying (OOK) modulation in terms of bit error rate (BER) for a given received signal-to-noise ratio (SNR), in particular for large number of bits per symbol [51]. Meanwhile, it has a better bandwidth efficiency.

On the one hand, subcarrier IM [53] schemes allow higher spectral efficiencies at the expense of reduced energy efficiency. This is because of the DC bias added to the signal in order to satisfy the signal positivity constraint (due to IM).

On the other hand, in order to deal with the limited bandwidth of the high-power LEDs that are typically used in UWOC systems, a promising approach is to use discrete multitone (DMT) modulation with the further possibility of bit-loading, as it has been already considered for indoor visible light communication systems [54,55]. An experimental verification of the feasibility of such modulation schemes was presented in [56,57].

We present here some numerical results to compare the performances of OOK, L-ary PPM, and L-ary DPIM modulation schemes. Our criterion for performance comparison is the maximum attainable link distance for an average transmit optical power of $P_t = 0.1$ W and an information bit-rate of $R_b = 100$ Mbps with a target BER of 10^{-4}. We consider uncoded modulation and the use of Si PIN PD at the Rx with the specifications detailed below. Given the limited PD cut-off frequency, we limit L to 8 for PPM and DPIM modulations.

Simulation parameters:

- Tx: beam divergence $\theta_{0max} = 10°$, average transmit optical power $P_t = 0.1$ W.
- Rx: lens diameter $D = 20$ cm, lens focal distance $F = 25$ cm, PD active area diameter $D_{aa} = 3$ mm, corresponding to FOV = 0.069°. Transimpedance resistor $R = 50$ Ω. PD responsivity $R_\lambda = 0.35$ A/W at $\lambda = 532$ nm, corresponding to quantum efficiency $\eta = 0.82$. PD cut-off frequency $f_c = 300$ MHz.

Figure 11.8 shows plots of BER as a function of Z for the case of clear ocean waters and different modulations obtained via Monte Carlo simulations. We notice that, for a target BER of 10^{-4}, the link distance is limited to 38 m when OOK is used. With L-PPM, this distance is about 47.2 and 51.2 m for $L = 4$ and 8, respectively. L-DPIM is less efficient than L-PPM but outperforms OOK for $L = 4$ and 8.

Notice that although for a given P_t and R_b, PPM provides the best BER performance, this advantage comes at the expense of a high peak-to-average optical power ratio (PAPR) and a large bandwidth requirement.

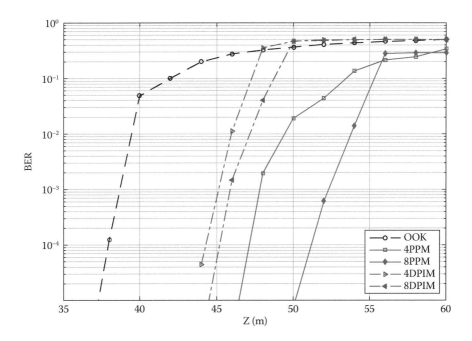

FIGURE 11.8
Contrasting BER performance for OOK, PPM, and DPIM modulations. $P_t = 0.1$ W, $R_b = 100$ Mbps, PIN PD.

Meanwhile, DPIM appears to be an interesting alternative to PPM: although it requires more computationally complex demodulation and may suffer from error propagation in signal demodulation, its BER performance is relatively close to PPM, especially for large L. In addition, DPIM does not require any symbol-level synchronization, has a lower PAPR, and is more bandwidth efficient than PPM.

11.4.2 Error Correction Coding

As in UWOC systems, we should deal with a highly attenuating propagation medium and a weak captured signal at the Rx; the use of channel coding techniques is of particular interest for signal detection under low SNR. The most important noise sources that we are concerned with are thermal noise in the case of using a PIN PD and shot (quantum) noise in the case of using an avalanche photodiode (APD) or a photomultiplier tube (PMT) [58]. Note that background radiations are practically negligible except in relatively shallow waters [59]. In this context, simple block codes such as Reed–Solomon (RS) or more powerful coding schemes such as low-density parity-check (LDPC) codes and turbo codes have been considered so far. These latter schemes necessitate computationally complex decoders and, hence, are suitable for relatively low data-rate links or when data detection and processing can be performed offline. For instance, the performance of RS, LDPC, and turbo codes were compared in [60] using an experimental setup. Higher link reliability can be obtained by using coding at the data link layer, in addition to coding at the physical layer. For instance, in the AquaOptical II modem, an RS inner code is used together with a Luby transform (LT) outer code [61].

11.4.3 Link Misalignment Issues

In a typical UWOC system, angle scattering is highly peaked in the forward direction. Therefore, the optical beam has a high directivity, which turns out to be problematic from the point of view of system implementation. In fact, link misalignments are unavoidable in underwater systems, especially when communicating with an AUV. On the other hand, due to stringent constraints on energy consumption, precise localization and tracking mechanisms may not be employed. Misalignment errors can seriously impact the performance and reliability of the communication link. This is especially the case for small FOV receivers; inserting a lens in front of the PD (which typically has a very small active area) has the advantage of increasing the received optical intensity. However, this seriously limits the Rx FOV and, hence, increases the sensitivity to link misalignments [62]. The problem is slightly alleviated in high-turbidity waters where paradoxically, we can benefit from beam spatial dispersion which helps reduce the sensitivity to link misalignments [33].

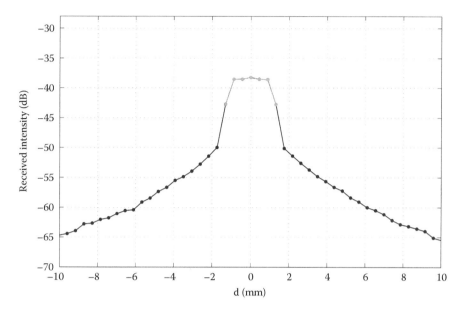

FIGURE 11.9
Intensity distribution over the lens focal plane considering $Z = 20$ m, clean ocean waters. The section in light gray color corresponds to the effective received intensity on the PD active area, assuming $D_{aa} = 3$ mm.

To see the effect of the limited Rx FOV, in Figure 11.9 we have presented the distribution of the received intensity, I_{PD}, as a function of the distance between the impact point on the focal plane and the focal point that we denote by d. Clear ocean waters, a link distance of $Z = 20$ m, and a PIN PD are considered. Other Rx parameters are as specified in Section 11.4.1 for Figure 11.8. The plot is obtained using Monte Carlo simulations [1] based on the TTHG phase function model described in Section 11.2.6. We notice that, given the Rx FOV of 0.69°, around 15% of the captured photons on the lens surface are captured on the PD. Note that the distribution of the received photons has a form of plateau around $d = 0$ which is due to the inaccuracy of the TTHG model at very small angles [1].

Let us investigate the impact of link misalignment when the Rx is displaced from its ideal position on the main optical axis, or when is tilted around it, as illustrated in Figure 11.10. Using Monte Carlo simulations, we have shown in Figure 11.11 plot of I_{PD} as a function of the Rx displacement Δ, considering the same parameters for the Tx and Rx as before. We notice a sharp decrease in I_{PD} just as we move away a little from the main optical axis. In fact, the Rx's tolerable maneuver zone is limited to less than 30 cm outside of which the signal is effectively lost. On the other hand, Figure 11.12 shows I_{PD} as a function of

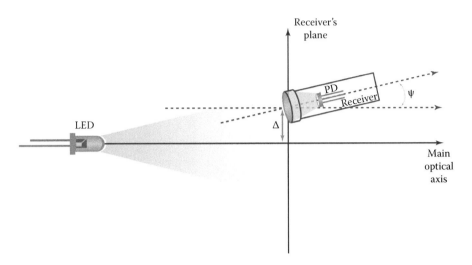

FIGURE 11.10
Illustrating link misalignment due to Rx displacement of Δ from its ideal position and an inclination angle of ψ relative to the main optical axis.

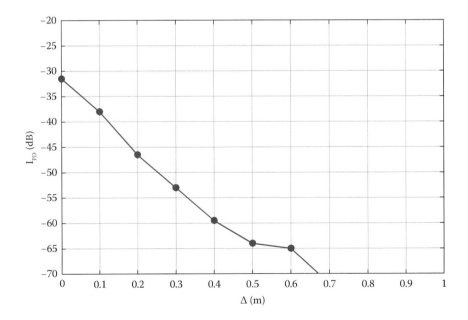

FIGURE 11.11
I_{PD} distribution as a function of Rx displacement Δ.

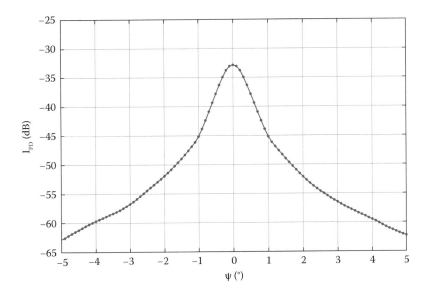

FIGURE 11.12
I_{PD} as a function of the Rx tilting angle ψ, with Δ = 0 m. (From Gabriel, C., et al., *IEEE OCEANS Conference*, June 2013, Bergen, Norway. With permission.)

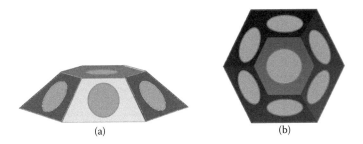

(a)   (b)

FIGURE 11.13
Schematics of lens-photodiode array in the form of truncated hexagonal pyramid structure used at the Rx to increase the FOV. (a) isometric view, and (b) top view. (Adapted from Simpson, J.A., et al., *IEEE J. Sel. Areas Commun.*, 30(5), 964–974, 2012.)

Rx tilting angle ψ assuming Δ = 0. We notice that the received signal is effectively lost for ψ > 2°.

An efficient solution to beam misalignment problems is to use an array of LEDs at the Tx and/or an array of PDs at the Rx. For instance, compact arrays of seven LEDs and seven PDs in the form of truncated hexagonal pyramid structures were used in [32] to achieve quasi-omnidirectional patterns, see Figure 11.13. This allows a large overall Rx FOV and substantial Tx–Rx

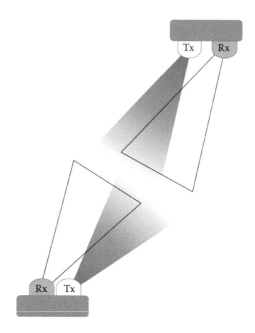

FIGURE 11.14
Electronic beam steering and tracking using arrays of LEDs or PDs. (Adapted from Simpson, J.A., et al., *IEEE J. Sel. Areas Commun.*, 30(5), 964–974, 2012.)

alignment simplification. The further advantage of such arrays is that we can estimate the angle of arrival of the optical signal at the Rx in order to correct the Rx position and direction. Also, we can perform beam steering at the Tx electronically toward the best direction in order to optimize the energy consumption, as schematically illustrated in Figure 11.14. As power consumption is an important issue in underwater missions, the development of energy-efficient solutions for node localization and beam alignment becomes highly interesting.

11.5 Example of UWOC System Prototype

While theoretical investigation and modeling is very important to allow for a rapid evaluation of the link parameters and the main requirements, it is only through prototyping and experimental evaluation that theoretical predictions can be validated. Additionally, certain engineering problems only become apparent in this phase.

In this last part, we present an early UWOC prototype to be integrated into an AUV. This is just a preliminary investigation on the design of UWOC links that can be insightful for a practical design. Here, in addition to the

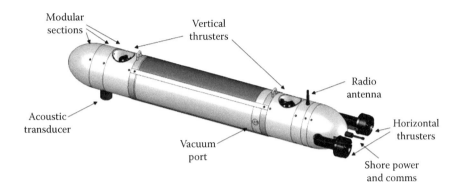

Modular sections   Vertical thrusters   Radio antenna   Acoustic transducer   Horizontal thrusters   Vacuum port   Shore power and comms

FIGURE 11.15
Picture of the MARES AUV, developed at INESC TEC.

limited battery lifetime, which is of special importance and demands a low power transceiver, one should take into consideration the limitations on size and weight of the AUV. Figure 11.15 shows the MARES AUV, which is a highly flexible small-scale AUV developed at INESC TEC. It has a length of 1.6 m and a weight of 32 kg, can operate at a maximum depth of 100 m, and can be configured to carry specific prototypes and logging systems for experimental evaluation [63].

Here, we describe the design of an UWOC for integration into an AUV. The system should be as simple as possible in order to meet the previous size, weight, and power (SWAP) requirements of the AUV, while providing bidirectional communication and an interesting communication range and data rate. We have fixed the target link span and data rate to 5 meters and 1 Mbps, respectively. This would be adequate, for example, to provide real-time video transmission for the purposes of an AUV docking manoeuvre, which requires precise navigation relative to the docking station.

In order to evaluate the system performance, a digital interface was used for Tx/Rx optical hardware with a digital computation platform—a microcontroller. The system was designed with the aim of integrating both Tx and Rx modules into a cylindrical waterproof casing. The prototype concept consists of four layers: light source driver, light source, acquisition and processing of the received signal, and photo-receiver. It allows embedding the Tx and Rx sections into the same module, insuring bidirectional communication. This layered approach allows the replacement of a single layer if necessary, for example, changing the wavelength of the LEDs or the type of photodiode. Two separate prototypes were planned to allow for an experimental evaluation. Figure 11.16 presents a diagram with the interconnection of the different components of the optical Tx and Rx, where each block represents a different layer of the module. The Tx and Rx paths are represented in blue and green, respectively. A cylindrical-shaped waterproof casing of

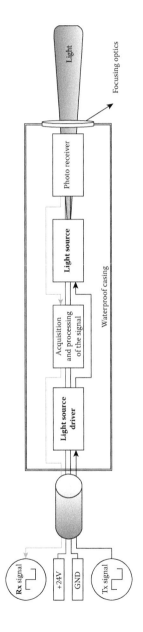

FIGURE 11.16
Concept of the modem prototype implementation in four layers.

40 mm diameter was used as the base platform, which is suitable for deployment on the AUV platform.

11.5.1 Transmitter

For the transmitter, an LED-based light source was considered (laser was not considered due to higher cost and increased pointing requirements). A driver is required to extract the maximum performance from the light source, not only in terms of optical output power but also in terms of operation speed. The driver was designed in order to take TTL signals at its input. A MOS-FET-based solution was considered (reference PMF87EN) that offers a fast switching speed. Additionally, to protect the LEDs from induction charges and linearize the drain voltage a fly-back diode was used—a Schottky diode, reference PMEG4010ETR. The schematic of the driver circuit and a picture of the implemented prototype are shown in Figure 11.17.

The chosen LEDs correspond to the CREE XLamp XP-E2 model in blue wavelength (465 nm, XPEBBL-L1-0000-00201). A standard metal-clad printed circuit board (PCB) was used for assembling seven LEDs in series.

11.5.2 Receiver

The Rx module was designed using a PIN PD in order to avoid high sensitivity to ambient light. The photodetection layer was designed employing

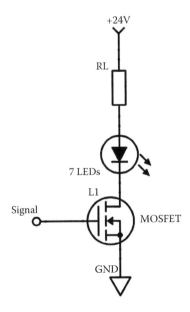

FIGURE 11.17
Transmission light source driver circuit schematic.

six parallel PDs, model BPW34-B with enhanced blue sensitivity. After the PD layer, an acquisition and processing layer was considered, consisting of a transimpedance amplifier (TIA), band-pass filtering and amplification, and a comparator, as shown in Figure 11.18. The TIA was implemented with an operational amplifier, model THS4631, and a feedback resistor of 10 kΩ and a feedback capacitor of 2.2 pF (5.5 MHz cut-off frequency). The active band-pass filter (BPF) has a maximum bandwidth of 160 MHz.

11.5.3 Modem Prototype Implementation

The physical casing was implemented in two parts that screw into each other with an acrylic transparent lens. Two rubber O-rings were used to maintain the waterproofness of the structure. Figure 11.19 shows three pictures of the waterproof casing prototype. The glossy end part of the casing, which we name casing head (CH), can be removed, as shown on the right subfigure.

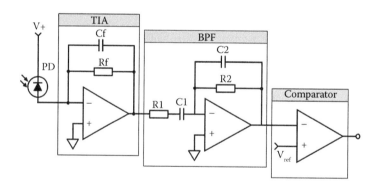

FIGURE 11.18
Receiver circuit schematic.

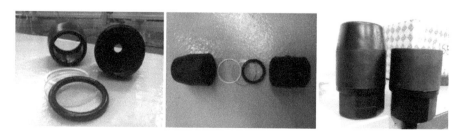

FIGURE 11.19
Waterproof casing prototype.

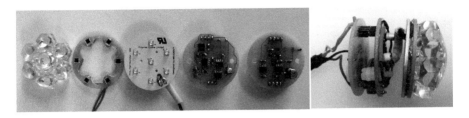

FIGURE 11.20
Overview of the four electronics layers and focusing optics.

We will refer to the casing with and without this section as long and short CH, respectively.

Figure 11.20 shows a global view of the four electronics layers and focusing optics (left), as well as the assembled prototype in side view (right). The PD PCB has been designed to be placed in front of the LED PCB and fill in the six free spaces between the six outermost LEDs. At these points, the considered lens has V-shaped cuts (instead of using circular shaped lenses) to allow each PD to receive the light from the corresponding lens only.

11.5.4 Experimental Results

For the experimental measurements, performance was evaluated by obtaining the minimum LED driving current in order to recover the data at the Rx, using a 1 MHz square wave signal transmitted from a waveform generator. An aluminium profile was employed for testing the misalignment between the modules. Figure 11.21a shows the prototype module holder which screws to the aluminium profile with discrete angle steps of 15° and the prototype system within the test tank.

We have shown in Figure 11.22 the minimum driving current for the link establishment for different link misalignment degrees, for both short and long CHs. We can see that communication can be achieved up to a misalignment of 30°, at the expense of a shorter range, thanks to the relatively large beam divergence. Also, the fact that we notice a better performance for a misalignment of 15° (compared with 0°), is because of the non-optimum positioning of the LEDs relative to the Rx lenses.

Figure 11.23 shows the maximum achievable distance as a function of the misalignment angle. We have tested the system both in the air and underwater. The achieved ranges underwater are logically smaller than those obtained within the air due to beam attenuation in water. The results clearly show that, for small misalignments up to 15°, the short CH is preferable for underwater operation. This is likely due to the water scattering effect, since the short CH allows for more scattered light to be captured.

FIGURE 11.21
Overview of the UWOC prototype during experimental performance assessment.

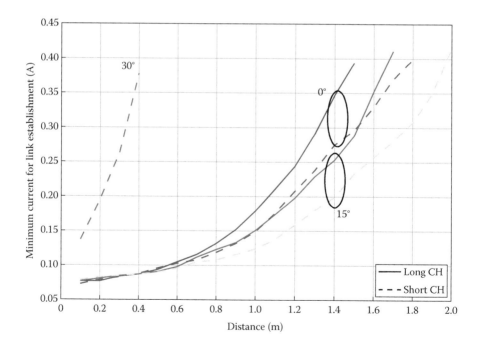

FIGURE 11.22
Performance evaluation within the test tank.

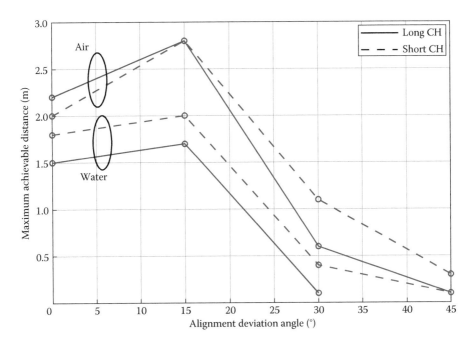

FIGURE 11.23
Performance comparison between air and water media as a function of beam misalignment angle.

11.6 Concluding Remarks

Even though there has been a considerable amount of research work on UWOC during the past few years, this technology requires further research effort to overcome a number of challenges [64]. One remaining aspect is the development of close-to-reality models for vertical links. Indeed, channel characterization has been investigated extensively for the case of horizontal links but in most practical cases the communication is likely to take place rather vertically. For such link configurations, there is a lack of reliable models to take into account, for instance, the variations of water salinity, temperature, and pressure. Taking into consideration accurately the effect of turbulence is another issue that needs further investigation, particularly in the case of relatively shallow waters. Apart from channel modeling, the design of appropriate signaling schemes adapted to the aquatic channel is another issue to explore. Concerning communication in relatively high turbidity waters, energy-efficient modulations and powerful error correcting codes should be developed to provide acceptable link performance and reliability. The development of energy-efficient solutions for Tx/Rx localization and beam alignment through the use of smart Tx and Rx capable of self-adapting to the operational situations is also of crucial importance.

Acknowledgment

Drs. Khalighi and Gabriel would like to acknowledge the support provided by IFREMER Research Center, La-Seyne-sur-Mer, France. They are also grateful to their colleagues of Institut Fresnel, Marseille, France, for the fruitful discussions.

References

[1] C. Gabriel, M.A. Khalighi, S. Bourennane, P. Léon, and V. Rigaud, Monte-Carlo-based channel characterization for underwater optical communication systems, *IEEE/OSA J. Opt. Commun. Networking*, vol. 5, no. 1, pp. 1–12, 2013.

[2] F. Hanson and S. Radic, High bandwidth underwater optical communication, *Appl. Opt.*, vol. 47, no. 2, pp. 277–283, 2008.

[3] A.N. Popper and A. Hawkins, eds., *The Effects of Noise on Aquatic Life*, Springer-Verlag, New York, 2012.

[4] H. Slabbekoorn, N. Bouton, I. van Opzeeland, A. Coers, C. ten Cate, and A. N. Popper, A noisy spring: The impact of globally rising underwater sound levels on fish, *Trends Ecol. Evol.*, vol. 25, pp. 419–427, 2010.

[5] *Ambalux High Bandwidth Underwater Transceivers*, http://www.ambalux.com/underwater-transceivers.html (accessed February 2017).

[6] *Sonardyne Product: BlueComm Underwater Optical Modem*, https://www.sonardyne.com/product/bluecomm-underwater-optical-communication-system/ (accessed February 2017).

[7] F. Pignieri, F. De Rango, F. Veltri, and S. Marano, Markovian approach to model underwater acoustic channel: Techniques comparison, *IEEE Military Communications Conference (MILCOM)*, November 2008, pp. 1–7, San Diego, CA.

[8] C.D. Mobley, B. Gentili, H.R. Gordon, Z. Jin, G.W. Kattawar, A. Morel, P. Reinersman, K. Stamnes, and R.H. Stavn, Comparison of numerical models for computing underwater light fields, *Appl. Opt.*, vol. 32, no. 36, pp. 7484–7504, 1993.

[9] D.J. Bogucki, J. Piskozub, M.E. Carr, and Spiers G.D., Monte Carlo simulation of propagation of a short light beam through turbulent oceanic flow, *Opt. Express*, vol. 15, no. 21, pp. 13988–13996, 2007.

[10] A. Morel and L. Prieur, Analysis of variations in ocean color, *Limnol. Oceanogr.*, vol. 22, no. 4, pp. 709–722, 1977.

[11] C.D. Mobley, *Light and Water: Radiative Transfer in Natural Waters*, Academic Press, San Diego, CA, 1994.

[12] L.C. Andrews and R.L. Phillips, *Laser Beam Propagation through Random Media*, 2nd ed., SPIE, 2005.

[13] M.A. Khalighi and M. Uysal, Survey on free space optical communication: A communication theory perspective, *IEEE Commun. Surv. Tutorials*, vol. 16, no. 8, pp. 2231–2258, 2014.

[14] J.A. Simpson, B.L. Hughes, and J.F. Muth, A spatial diversity system to measure optical fading in an underwater communications channel, *IEEE OCEANS Conference*, October 2009, pp. 1–6, Biloxi, MS.

[15] Y. Ata and Y. Baykal, Structure functions for optical wave propagation in underwater medium, *Waves in Random and Complex Media*, vol. 24, no. 2, pp. 164–173, 2014.

[16] S.Q. Duntley, *Underwater Visibility and Photography, Optical Aspects of Oceanography*, Academic Press, New York, 1974.

[17] F. Hanson and M. Lasher, Effects of underwater turbulence on laser beam propagation and coupling into single-mode optical fiber, *Appl. Opt.*, vol. 49, no. 16, pp. 3224–3230, 2010.

[18] M.L. Wells and E.D. Goldberg, Occurrence of small colloids in sea water, *Nature*, vol. 353, pp. 342–344, 1991.

[19] I. Koike, S. Hara, K. Terauchi, and K. Kogure, Role of submicrometer particles in the ocean, *Nature*, vol. 345, pp. 242–244, 1990.

[20] D. Stramski and D.A. Kiefer, Light scattering by microorganisms in the open ocean, *Progr. Oceanogr.*, vol. 28, no. 4, pp. 343–383, 1991.

[21] V.I. Haltrin and G.W. Kattawar, Self-consistent solutions to the equation of transfer with elastic and inelastic scattering in oceanic optics: I. Model, *Appl. Opt.*, vol. 32, no. 27, pp. 5356–5367, 1993.

[22] V.I. Haltrin, Chlorophyll-based model of seawater optical properties, *Appl. Opt.*, vol. 38, no. 33, pp. 6826–6832, 1999.

[23] J.T.O. Kirk, *Light and Photosynthesis in Aquatic Ecosystems*, 3rd ed., Cambridge University Press, New York, 2011.

[24] R.C. Smith and K.S. Baker, Optical properties of the clearest natural waters (200–800 nm), *Appl. Opt.*, vol. 20, no. 2, pp. 177–184, 1981.

[25] L. Prieur and S. Sathyendranath, An optical classification of coastal and oceanic waters based on the specific spectral absorption curves of phytoplankton pigments, dissolved organic matters, and other particulate materials, *Limnol. Oceanogr.*, vol. 26, no. 4, pp. 671–689, 1981.

[26] R.M. Pope and E.S. Fry, Absorption spectrum (380–700 nm) of pure water. II. Integrating cavity measurements, *Appl. Opt.*, vol. 36, no. 33, pp. 8710–8723, 1997.

[27] H.R. Gordon and A.Y. Morel, *Remote Assessment of Ocean Color for Interpretation of Satellite Visible Imagery: A Review*, Springer-Verlag, New York, 1983.

[28] A. Morel, In-water and remote measurement of ocean color, *Boundary-Layer Meteorol.*, vol. 18, no. 2, pp. 177–201, 1980.

[29] F.A. Jenkins and H.E. White, *Fundamentals of Optics*, 4th ed., McGraw-Hill Education, New York, 2001.

[30] B.M. Cochenour, L.J. Mullen, and A.E. Laux, Characterization of the beam-spread function for underwater wireless optical communications links, *IEEE J. Ocean. Eng.*, vol. 33, no. 4, pp. 513–521, 2008.

[31] T.J. Petzold, *Volume Scattering Functions for Selected Ocean Waters*, Technical Report SIO 72–78, Scripps Institute of Oceanography, San Diego, CA, 1972.

[32] J.A. Simpson, B.L. Hughes, and J.F. Muth, Smart transmitters and receivers for underwater free-space optical communication, *IEEE J. Sel. Areas Commun.*, vol. 30, no. 5, pp. 964–974, 2012.

[33] B. Cochenour, L. Mullen, and J. Muth, Temporal response of the underwater optical channel for high-bandwidth wireless laser communications, *IEEE J. Ocean. Eng.*, vol. 38, no. 4, pp. 730–742, 2013.

[34] N.G. Jerlov, *Marine Optics*, 2nd ed., Elsevier Science, Amsterdam, The Netherlands, 1976.

[35] O.V. Kopelevich, Small-parameter model of optical properties of sea water, in *Physical Ocean Optics*, vol. 1, ed. A.S. Monin, pp. 208–234, Nauka Publishing House, 1983, (in Russian).

[36] L.G. Henyey and J.L. Greenstein, Diffuse radiation in the galaxy, *Astrophys. J.*, vol. 93, pp. 70–83, 1941.

[37] H.C. van de Hulst, *Light Scattering by Small Particles*, Dover, Mineola, NY, 1981.

[38] C.F. Bohren and D.R. Huffman, *Absorption and Scattering of Light by Small Particles*, Wiley-VCH, New York, 2012.

[39] Y.I. Kopilevich, M.E. Kononenko, and E.I. Zadorozhnaya, The effect of the forward scattering index on the characteristics of a light beam in sea water, *J. Opt. Technol.*, vol. 77, no. 10, pp. 598–601, 2010.

[40] L. Mullen, D. Alley, and B. Cochenour, Investigation of the effect of scattering agent and scattering albedo on modulated light propagation in water, *Appl. Opt.*, vol. 50, no. 10, pp. 1396–1404, 2011.

[41] V. Haltrin, One-parameter two-term Henyey-Greenstein phase function for light scattering in seawater, *Appl. Opt.*, vol. 41, no. 6, pp. 1022–1028, 2002.

[42] V.I. Haltrin, Two-term Henyey-Greenstein light scattering phase function for seawater, *IEEE Int. Geosci. Rem. Sens. Symp. (IGARSS)*, June–July 1999, vol. 2, pp. 1423–1425, Hamburg, Germany.

[43] G.N. Plass and G.W. Kattawar, Radiative transfer in an atmosphere-ocean system, *Appl. Opt.*, vol. 8, no. 2, pp. 455–466, 1969.

[44] R.W. Preisendorfer, *Eigenmatrix Representations of Radiance Distributions in Layered Natural Waters with Wind-roughened Surfaces*, Technical Report, Pacific Marine Environmental Laboratory-NOAA, Seattle, WA, 1988.

[45] R. Bellman, R. Kalaba, and G.M. Wing, Invariant imbedding and mathematical physics. I. Particle processes, *J. Math. Phys.*, vol. 1, no. 4, pp. 280–308, 1960.

[46] G.N. Plass and G.W. Kattawar, Monte Carlo Calculations of Light Scattering from Clouds, *Appl. Opt.*, vol. 7, no. 3, pp. 415–419, 1968.

[47] L. Wang, S.L. Jacques, and L. Zheng, *MCML, Monte Carlo Modeling of Light Transport in Multi-layered Tissues*, Technical Report, Laser Biology Research Laboratory, 1995, University of Texas, M.D. Anderson Cancer Center, Houston.

[48] N.J. McCormick, Inverse radiative transfer problems: A review, *Nucl. Sci. Eng.*, vol. 112, pp. 185–198, 1992.

[49] C. Gabriel, M.A. Khalighi, S. Bourennane, P. Léon, and V. Rigaud, Channel modeling for underwater optical communication, *IEEE Workshop on Optical Wireless Communications, Global Communication Conference*, pp. 833–837, December 2011, Houston, TX.

[50] B. Cochenour, L. Mullen, and J. Muth, Effect of scattering albedo on attenuation and polarization of light underwater, *Opt. Lett.*, vol. 35, no. 12, pp. 2088–2090, 2010.

[51] C. Gabriel, M.A. Khalighi, S. Bourennane, P. Léon, and V. Rigaud, Investigation of suitable modulation techniques for underwater wireless optical communications, *International Workshop on Optical Wireless communications (IWOW)*, pp. 1–3, October 2012, Pisa, Italy.

[52] Z. Ghassemlooy, A. Hayes, N. Seed, and E. Kaluarachchi, Digital pulse interval modulation for optical communications, *IEEE Commun. Mag.*, vol. 36, no. 12, pp. 95–99, 1998.

[53] T. Ohtsuki, Multiple-subcarrier modulation in optical wireless communications, *IEEE Commun. Mag.*, vol. 41, no. 3, pp. 74–79, 2003.

[54] J. Armstrong, OFDM for optical communications, *J. Lightwave Technol.*, vol. 27, no. 3, pp. 189–204, 2009.

[55] D.K. Borah, A.C. Boucouvalas, C.C. Davis, S. Hranilovic, and K. Yiannopoulos, A review of communication-oriented optical wireless systems, *EURASIP J. Wireless Commun. Networking*, vol. 91, pp. 1–28, 2012.

[56] B. Cochenour, L. Mullen, and A. Laux, Phase coherent digital communications for wireless optical links in turbid underwater environment, *IEEE OCEANS Conference*, pp. 1–5, September–October 2007, Vancouver, BC.

[57] G. Cossu, R. Corsini, A.M. Khalid, S. Balestrino, A. Coppelli, A. Caiti, and E. Ciaramella, Experimental demonstration of high speed underwater visible light communications, *International Workshop on Optical Wireless Communications (IWOW)*, pp. 11–15, October 2013, Newcastle upon Tyne, UK.

[58] F. Xu, M.A. Khalighi, and S. Bourennane, Impact of different noise sources on the performance of PIN- and APD-based FSO receivers, COST IC0802 Workshop, *IEEE ConTEL Conference*, pp. 211–218, June 2011, Graz, Austria.

[59] T. Hamza, M.A. Khalighi, S. Bourennane, P. Léon, and J. Opderbecke, Investigation of Solar Noise Impact on the Performance of Underwater Wireless Optical Communication Links, *Optics Express*, vol. 24, no. 22, pp. 25832–25845, 2016.

[60] J.A. Simpson, W.C. Cox, J.R. Krier, and B. Cochenour, 5 Mbps optical wireless communication with error correction coding for underwater sensor nodes, *IEEE OCEANS Conference*, September 2010, Seattle, WA.

[61] M. Doniec, M. Angermann, and D. Rus, An end-to-end signal strength model for under- water optical communications, *IEEE J. Ocean. Eng.*, vol. 38, no. 4, pp. 743–757, 2013.

[62] C. Gabriel, M.A. Khalighi, S. Bourennane, P. Léon, and V. Rigaud, Misalignment considerations on point-to-point underwater wireless optical links, *IEEE OCEANS Conference*, June 2013, Bergen, Norway.

[63] N. Cruz and A. Matos, The MARES AUV, a modular autonomous robot for environment sampling, *Proceedings of the MTS-IEEE Conference Oceans'2008*, September 2008, Quebec, Canada.

[64] M.A. Khalighi, C. Gabriel, T. Hamza, S. Bourennane, P. Léon, and V. Rigaud, Underwater wireless optical communication; recent advances and remaining challenges, *Invited Paper, International Conference on Transparent Optical Networks (ICTON)*, July 2014, pp. 1–4, Graz, Austria.

12

VLC for Indoor Positioning: An Industrial View on Applications

Nuno Lourenço and Martin Siegel

CONTENTS

12.1 Introduction

In this chapter, the authors present several considerations on applications of visible light communication (VLC) for indoor positioning, from a professional lighting industry perspective. Rather than discussing vendor-specific technical solutions, the authors aim to provide an overview of applications across the industry's different market segments.

Following this summary, in Section 12.2 the authors provide the general motivation and key enablers that have allowed VLC-based indoor positioning to become technologically feasible from an industry standpoint, as well as market-captivating to both end users and system owners. In this line of thought, Section 12.3 explores the benefits and drawbacks of combining lighting and localization capabilities. A set of basic definitions along with the architectures of typical lighting controls and indoor positioning systems are presented.

Application use cases are introduced in Section 12.4, according to typical professional lighting market segmentation. These aim to provide examples that are closer to real-life scenarios, describing not only supported features but also user interactions and other high-level details. This is followed, in Section 12.5, by an overview of success factors that will influence the adoption of such solutions, throughout the value chain of professional lighting. The chapter ends with a set of considerations and remarks in Section 12.6.

12.2 Motivation and Key Enablers

In a continuously growing and mobile world with limited resources, demands for smarter, more efficient and dynamic systems have become widespread. Building automation, and lighting in particular, have come under the spotlight of the Internet of Things (IoT) momentum. This is mostly due to a large potential for optimization of energy consumption, support of enhanced human assistance systems and general improvement in the usability of spaces. Through deployment of vast networks of interconnected sensors and actuators, which take advantage of the latest communication protocols and technologies, building owners and facility managers ultimately expect this added effort to be translated into energy savings and overall improved user comfort.

With millions of interconnected devices, many of them wireless, being prophesized by the IoT movement, alternatives and complements to the crowded radio frequency (RF) communication spectrum are under the focus of the research community. Although optical wireless communications have been around for several years, the past decade saw VLC becoming accepted as a valid technology for wireless communications, mostly for data downstream.

Widely contributing to this purpose was the industry revolution sparked by the introduction of the white light-emitting diode (LED). Manufacturers are now able to deliver new concepts and designs, providing functionalities enabled by control systems, which were previously out of their domain of expertise. Therefore, VLC presents an attractive proposition to the lighting industry as an enabler for different markets as well as a source of added-value features.

One such feature, which has grown in the interest of both users and service providers, is indoor positioning. This means the ability to determine the position of a person or object inside a building. The familiarity with the generic principle from the domain of global positioning systems (GPS), along with the usability of having a mobile device for interface, makes such solutions highly attractive for end users. Furthermore, combining lighting and localization capabilities in a single infrastructure reduces implementation costs and gives an important added value, which captivates venue owners and managers.

Figure 12.1 represents four main pillars onto which a widespread adoption of VLC for indoor positioning settles. They are split into the two main groups of technology feasibility and market acceptance. In the following subsections, the motivations behind them are discussed.

12.2.1 The LED Revolution

Over the past decade, the lighting industry has undergone one of its most important transformations. The research efforts in the 1990s of later Nobel Prize laureates Shuji Nakamura, Isamu Akasaki, and Hiroshi Amano on blue and white LEDs translated into a stepping stone for the lighting industry into what is known as the LED revolution. However, despite the introduction of the first commercial-grade white LEDs in 1996 by Nichia Corp., it was only in 2006 that the technology reached the critical 100 lumen per watt milestone, allowing LEDs to compete with fluorescent lights for the general lighting market [1]. In the meanwhile several other players have driven the efficiency barrier further, setting new milestones at an impressive rate.

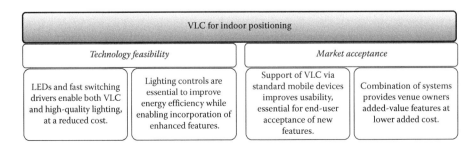

FIGURE 12.1
Key enablers of VLC for indoor positioning.

FIGURE 12.2
Evolution of typical track-mounted luminaire. From left to right: Zumtobel ZX2 (pre-2011); Zumtobel TECTON (2011–2014); Zumtobel TECTON (2015–present).

On a product development perspective, the mechanical and electrical characteristics of LEDs have opened doors to endless possibilities in lighting. The small-form factor enables designers and manufacturers to come up with new solutions in both shape and size that seemed almost impossible a decade ago. Figure 12.2 uses the example of a typical track-mounted luminaire to show how LED lighting has changed luminaire design. From a rather bulky pre-2011 luminaire with two fluorescent tubes in the first step, to an all LED-based luminaire which is already much smaller, to a very thin and miniaturized version launched in 2015. This transformation allowed LED lighting to gain access to diversified market segments reaching from high-end architectural lighting to cost-driven basic performance solutions. Furthermore, the electrical characteristics of these devices, which are also a basis for VLC, have resulted in a simplification of the associated power supply and control logic, also known as gear. For the same equivalent optical output power, LED luminaires can be driven with gears built from switched mode power supplies, which are smaller in shape and lower in cost than those used in standard fluorescent ones.

Enabled by the developments in technology and the potential for new products, the market share of LED light sources has grown tremendously over the last couple of years and by now has reached more than a third of the overall lighting revenue. Estimates of current figures are shown in Table 12.1 [2]. A leading example of LED uptake comes from the Zumtobel Group, a global leader in the professional lighting market, which started LED activities as early as 2001, and reached a quota of 50% LED-driven revenue in the fiscal year of 2014–2015 [3].

Therefore, the massive market uptake sparked by the LED revolution can be seen as a strong motivation toward disseminating VLC on a consumer scale. From a technology point of view, the complexity of adding basic VLC support onto LED luminaires and bulbs is a simple task. Since these are massively deployed devices they can be used to help widespread, what are currently

TABLE 12.1

Global Market Share of LED Lighting Measured as a Percentage of Total Lighting Revenue

Source	Scope	2014	2016	2018	2020	2022
IHS	Lamps	31%	42%	52%	61%	67%
Strategies unlimited	Lamps	41%	56%	68%	76%	80%
Strategies unlimited	Luminaires	33%	44%	53%	61%	69%
LED inside	Lamps and luminaires	26%	34%	54%	–	–

Source: DOE SSL (Department of Energy, Solid-State Lighting) Program, 2015, *R&D Plan*, prepared by Bardsley Consulting, SB Consulting, SSLS, Inc., LED Lighting Advisors, and Navigant Consulting, Inc., DOE Office of Energy Efficiency and Renewable Energy, Washington.

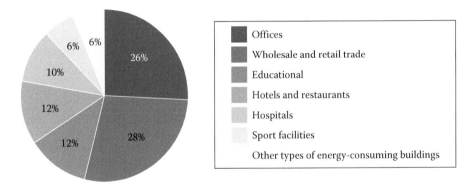

FIGURE 12.3
Share of total energy use per building type in non-residential buildings across Europe. (From Economidou, M., et al., *Europe's Buildings under the Microscope*, Buildings Performance Institute Europe, Brussels, Belgium, 2011.)

scarce, VLC installations. An increase in the installed base can then drive the uptake of such solutions by consumers.

12.2.2 The Need for Controls

Retail and office buildings, which account for over 50% of the energy consumption of non-residential buildings in Europe (see Figure 12.3), are fertile ground for new developments when it comes to smarter use of energy [4,5]. "Green building" legislation has had a strong impact on both building owners and tenants. It has raised awareness toward the impact that choices during construction or refurbishment stages have on the overall sustainability of a building, particularly on its total cost of ownership (TCO). Energy performance assessment schemes such as Leadership in Energy and Environmental Design (LEED), Building Research Establishment Environmental Assessment

Method (BREEAM), and Green Star, analyze buildings and their operation from different angles. They look at long-term sustainability and comfort aspects which include efficient usage of resources, reduction of waste, and levels of occupants' health and productivity. Building automation, including lighting controls, has thus become a standard requirement in order get green building certification [6,7].

The functionality delivered by indoor lighting controls varies significantly according to the demands of the actual installation and intended usage of the building. Features may include little more than on/off and basic grouping of luminaires, all the way to complex functionality. Some of these include reporting mechanisms, scheduling capabilities, support of load-shedding requests, controllable dimming output (often used in combination with daylight harvesting), sensors for detecting motion and monitoring environmental data, actuators to interface windows and blinds, and many others for additional building management systems. In order for these features to work, a backbone network for the control systems needs to be in place. Although several technologies may be used in a given installation, each with a different set of capabilities and limitations, exploiting the lighting control network for added functionality is often feasible within given boundaries. Some examples of what can be achieved and associated limitations will be discussed in the use cases presented in Section 12.4.

Despite the proliferation of lighting controls, the acceptance of enhanced features has not always been trivial. Light is deeply connected to human instinctive reactions, thus lighting systems need to operate seamlessly, providing reliable and repeatable behavior. Users are instinctively accustomed to the reliability provided by the ubiquitous light switch, which means that for users "light is just there and it just works." Over the years several innovations have been taken to the market, only to find that usability and simplicity were a hurdle that was often impossible to overcome. However, the advances in electronics, and particularly from the microcontroller industry, mean that complexity for advanced features could be hidden away from users without significantly increasing final product cost. This paved the way for new ways of thinking about lighting systems and their potential. A clear example comes from the "human-centric lighting" movement that looks not only at the technical aspects of lighting systems, such as the necessary luminous output for a specific task, but also at the non-visual effects of light associated with human health and well-being [8].

Human-centric lighting, backed by the potential of LEDs, pushes for adoption of several features such as artificial lighting that matches the circadian rhythm, color temperature tunable luminaires, task based lighting, among many others [9,10]. These require an effective and enhanced control of the lighting system, and several research efforts are being made related to the quality of light output and its impact on human health, including both intensity and spectral content, and dynamic management of the lighting configuration. Human-centric lighting requires a combination of parameters pertaining not only to the lighting system but also the human user. Under this scope, positioning information is a source of valuable information that can be used to better adapt the lighting system to the needs of the users at the moment and place they need it.

Furthermore, research is also ongoing on advanced modulation techniques of the output optical signal. These often try to fulfill a double goal of providing dimmable illumination capabilities, while reliably supporting data transmissions. Exploring such modulation techniques is of particular importance for scenarios that require coexistence of multiple sources (e.g., a room with multiple luminaires) as well as to improve the data rates [11,12]. Such features become more relevant when they are combined with other digital lighting controls. Lighting controls are, therefore, essential not only to improve energy efficiency, but also to allow introduction of enhanced features, making them a key enabler of VLC for indoor positioning.

12.2.3 Market Acceptance

No matter how good a product idea may be, it can only succeed in the industrial world if the consumer is willing to pay for it. This crucial notion often means that even when all technological hurdles have been overcome, market acceptance is not guaranteed. In the case of indoor positioning, despite several solutions having been available for years, market acceptance was limited. However, with Apple, closely followed by Google, and also Microsoft enabling native support for location-aware context information in new applications, an important milestone was achieved. Supported mostly by Bluetooth Smart technology the mobile industry had found a way to make users interact and learn about their environment at the cost of a fairly inexpensive installation of radio beacons by the venue owner [13,14]. Despite having limited capabilities, such as reduced accuracy and no orientation resolution, early Bluetooth Smart beacons have captivated users due to their incredible simplicity and usability. Familiarity with GPS solutions and integration with mobile platforms means that indoor positioning is a feature which end users instinctively know how to use. This intangible benefit makes this functionality highly desirable.

From a venue owner or manager perspective, there is added motivation to include contextual awareness to either venue-specific or general location applications. This allows them to offer their customers, employees, or visitors a better indoors user experience, increasing visibility and improving user satisfaction. At the same time, added value potential lies in the position information that is received from users. By applying data analytics to movement patterns, user behavior may be better understood. This knowledge may then be applied to improve processes, traffic paths, or planning of usage of spaces. These benefits will initially be realized in large venues, with retail taking an important role in providing this feature to the general public, closely followed by exhibition centers, hospitals, and airports. This represents a broad market worldwide which, according to Market and Markets, is expected to increase from USD 935.05 million from 2014 all the way up to a staggering

USD 4.42 billion by 2019. This translates to a compound annual growth rate of 36.5% [15]. Although these numbers are for the general indoor positioning market, the fast growth rate is a clear sign of demand from both venue owners as well as consumers, which can also be fulfilled by VLC. This makes market acceptance and current demand key enablers of VLC for indoor positioning.

12.3 Combining Lighting and Indoor Positioning

Technological feasibility at an acceptable cost, devices with native capabilities to modulate light, the need for improved controllability of lighting systems, and overall user interest makes VLC and indoor positioning a strong match. With key enablers in place from both technology and market there is a strong potential for large-scale deployments, and there have been already some examples of pilot installations [16] and even product releases [17]. However, in order to prosper, such systems need to keep interoperability in mind. Therefore, a structure or at least common guidelines need to be in place in order to ensure there is consistency in behavior. Eventually, this translates into a common user experience throughout all installations, rather than a sense of one-off solution from a mix of individually constructed systems.

In this section, several definitions necessary for understanding the combination of lighting and positioning systems are provided. This is followed by the benefits of combining both systems and a quick introduction to typical architectures for both the lighting domain and indoor positioning, following the efforts made by the InLocation Alliance [18].

12.3.1 Definitions

Before going deeper into the benefits and drawbacks, some basic definitions around lighting systems must be clarified. For the purpose of this chapter and to avoid misconceptions, a "lighting distribution" corresponds to the pattern generated by the luminous output of light sources, bulbs, or luminaires installed at a given location. A "lighting installation" is then the actual physical infrastructure and network of devices, including luminaires, switches, and other user input methods, cables or wireless connection, sensors, and any other physical or virtual device that is required to perform basic operations. A "lighting management system" (LMS) is the platform for controlling the lighting installation, beyond the basic on/off. The LMS includes all hardware and software required to perform all configuration, control, and monitoring operations. Besides controlling the basic operational state, it also typically allows for grouping, scheduling, sensor reading, data logging, reporting, and interfacing other building systems (e.g., controlling blinds and windows), among others. These capabilities are a combination of hardware and software resources present in the devices of the lighting installation and often centrally controlled. Finally, the generic expression of lighting system, or lighting solution, encompasses both the lighting distribution and associated lighting installation, plus the optional LMS.

Under the topic of indoor positioning, a distinction between location and position also needs to be made. In a broader scope, location is defined as place where an object or person can be found. As for position, it is defined as a condition that the person or object occupies with reference to other people, things, or map. As a practical example, GPS provides the user with a set of coordinates, the position, which they can then use to find their condition relative to a map, thus obtaining their location. For the purpose of this chapter, a location will be the place where the lighting system is installed such as an office space or a retail store, and position will be used to describe the capability to determine the place a person, or object, occupies inside the physical space shared with that installation.

12.3.2 Benefits and Drawbacks

A professional lighting system can vary significantly in dimension and features, according to the size of the installation and the specific application requirements. Figure 12.4 provides a visual mapping with categorization of installations according to the number of nodes (luminaires, sensors, and other control points), type of business, and complexity. As expected, as the numbers in a lighting system increase so does its complexity (x axis). Furthermore, the solutions provided move from a business-to-consumer into a business-to-business market (y axis). Lighting systems can range in numbers from a few units all the way up to thousands of devices [19]. These are spread throughout all areas of a building, from main usage areas such as open spaces in office buildings, to areas with little to no human presence, such as maintenance tunnels. Lighting is a basic necessity and a standard requirement in most forms of indoor spaces, making its infrastructure one of the most ubiquitous available. Therefore, this means that integration of indoor positioning within the lighting system opens the door to virtually all areas inside a building.

Another advantage of the lighting installation is that it includes power distribution. Additional systems deployed on top of this infrastructure should be able to connect to it, thus reducing complexity and cost. Furthermore, many mid- to large-scale professional lighting systems often deploy an LMS which includes a communication backbone, and may in some cases be used for other purposes, such as support to the location platform. However, there is the potential drawback of limited bandwidth that can be inherited from using legacy building automation technologies such as KNX or a local

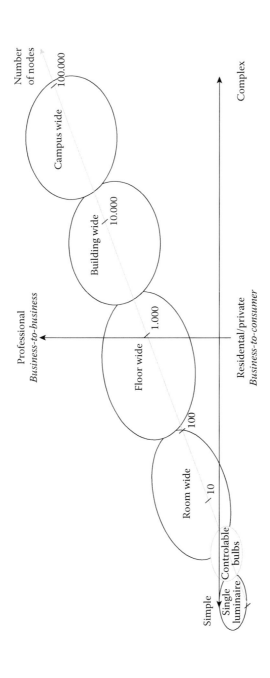

FIGURE 12.4
Classification of lighting systems according to number of nodes, complexity, and business type. (From K. Vamberszky, *Lighting Controls, Zumtobel Group Internal Presentation*, Dornbirn, Austria, 2014. With permission.)

operating network (LON) [20], or other lighting-specific bus technologies like the digital addressable lighting interface (DALI) [21].

In comparison, a stand-alone positioning installation using some form of RF technology will always be faced with problems common to wireless sensor networks. Physical and architectural constrains can limit coverage of the installation inside a building, particularly if new cabling for backbone communications or power distribution is required. Although in some cases batteries may be deployed, as often happens in RF beacon deployments using Bluetooth Smart [22], their short lifetime (about one year) represents an additional maintenance burden on the installation manager. It is also safe to assume that, in most cases, the cost of installing two completely independent systems versus a combined solution is much higher. Finally, although an RF positioning system may also be deployed in combination with a lighting installation, there is a strong benefit from using the output VLC signal from the luminaires for simultaneous information transmission and general lighting purposes [23,24]. Arguments on the benefit of using VLC also for interdevice communication pertaining to LMS operation could be made. However, such consideration carries implications for operation of the lighting system, making it contradictory to the scope of the discussion.

Additional benefits can be found from multiple points of view. Using the perspective of a lighting systems provider, VLC for indoor positioning represents a strong added value to what is often a basic service. Furthermore, position information allows LMS to deploy advanced control and management strategies. As an example, knowing when an office worker enters or leaves a meeting room would allow for intelligent control of the light and dynamic information on usage of the room. From a lighting installation owner or facility manager's perspective, besides the previously stated benefits from having a single installation, tracking people or assets is another valuable capability. As an example, a retail store manager that tracks movement patterns of his customers would then be able to optimize his processes and store layout to improve is profitability. Finally, from an end-user perspective, the main benefit would naturally come from having an assistance tool able to guide or provide them with additional information pertaining to their location [25,26].

12.3.3 The Architecture of the Lighting Management System

The architecture of an LMS is almost always vendor dependent. Despite the proliferation of several standardized interconnection technologies and protocols [20,21] in order to overcome the inherent complexity behind such systems, vendors often include their own particularities in the design. There are, however, a few basic concepts that are widely accepted in the lighting industry. As a reference example for analysis, in Figure 12.5 the architecture of Zumtobel's flagship LMS, Litecom, is depicted [27]. Starting from the bottom up, a lighting infrastructure encompasses a

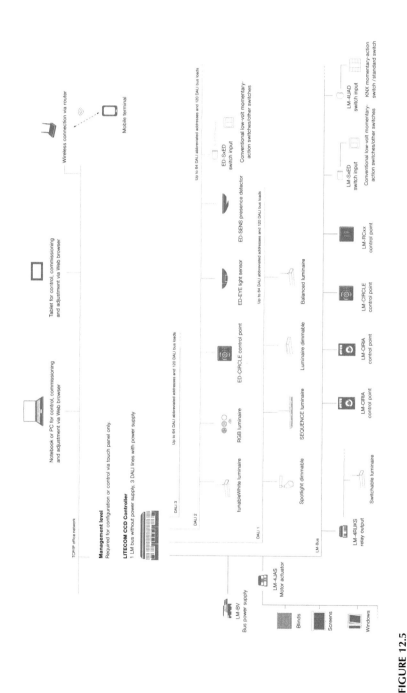

FIGURE 12.5

Architecture of Zumtobel's Litecom lighting management platform. (Zumtobel Lighting GmbH, LITECOM—Next Generation lighting management, Dornbirn, Austria, 2014. With permission.)

field-level network, which can be made out of one or more field bus technologies. It is here that devices such as luminaires, switches, sensors, and other actuators are connected. The DALI bus [21] was designed specifically for lighting applications; currently there are several controllable gears commercially available from multiple vendors, and it is a common choice for interconnecting luminaires. An additional bus technology, such as KNX [20] or a proprietary solution, is also often used to interconnect non-lighting devices such as input switches, blinds, windows, screens, and other controllers.

On top of the field-level devices is the controller, a device that typically sits in an electrical cabinet. It contains an automation engine taking care of all control, reporting, and management operations. Depending on installation dimension, several controller devices may be interconnected in order to cover the necessary number of field devices. This is through a backbone network, similar to standard IT systems but often using separate cabling for security concerns. Remote access control and maintenance features are also possible. The controller devices also contain interfaces for higher level building management systems such as the Building Automation and Control Network (BACnet) [20].

12.3.4 The Architecture of an Indoor Positioning System

Besides some generic concepts, lighting system architectures are quite diversified which makes designing any add-ons, with the expectation of interoperability, an almost impossible task. Therefore, the best bet toward interoperability lies in the design of a high-level architecture for the indoor positioning platform, providing guidelines to which system integrators comply. The InLocation Alliance (ILA) was founded in 2012 by a large consortium of technology driven companies, coming from a wide range of applications. The main aim of the ILA is to drive innovation and accelerate market adoption of indoor positioning by promoting a common understanding of key enablers, components, interfaces, and standards [18]. Its members continuously work together on developing a system architecture which is available in [28]. The document contains several considerations on possible configurations and necessary building blocks to properly design, operate, and manage a location system. Additionally, the ILA also explores use cases to understand market needs and uses them to continuously review the proposed architecture; their work can be found in [29,30].

As seen in Figure 12.6 [28], the first differentiation in the ILA proposed architecture is made based on positioning modes possible; these can either be network centric or mobile device centric. As the name implies, the first mode considers that the positioning calculations, measurements and location network access, and transactions are all initiated and controlled by the network. A possible example would be a network of passive tags that is monitored through an installation of active readers, and where all data are maintained by the location server. The second positioning mode is based

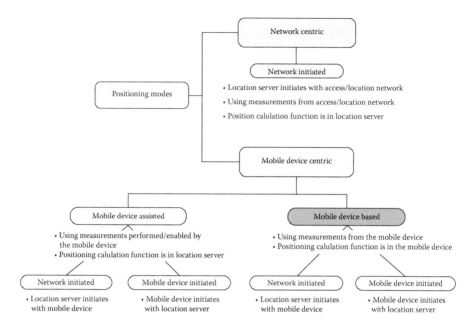

FIGURE 12.6
Possible positioning modes as defined by the ILA system architecture work group. (From ILA System Architecture Working Group, *ILA System Architecture Specifications—Release* 1.0, InLocation Alliance, Piscataway, NJ, 2014. With permission.)

on the mobile device, which may be a smartphone, tablet, or any other dedicated mobile unit. As per the name, the mobile device takes a leading role under this mode which can, depending on configuration, be sub-categorized. This is mobile device assisted, if the measurements are performed or enabled by the mobile device but the calculations are performed in the location server, or mobile device based, if measurements and positioning calculations are made by the mobile device. As an example, most use cases with the objective of providing some sort of guidance system will be mobile device centric. The final distinction is made based on which entity triggers the location process. If the mobile device takes initiative the solution is called mobile device initiated, but if it is the network that triggers the process it designated network initiated [28].

The ILA proposed architecture is quite generic, not being tied down to a specific technology or system topology. The main building blocks are shown in Figure 12.7, where the critical role of the mobile device is easily spotted. As presented in [28], at the core of the system is the access/location network which is only used for the purpose of location; however, simultaneous use for device communication is not prohibited. Associated with this block is the access/location network database, where the access/location network

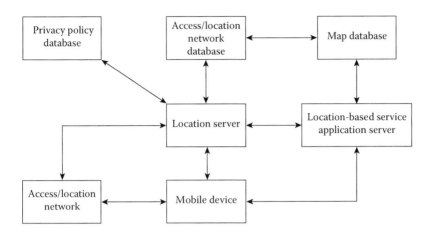

FIGURE 12.7
Building blocks of ILA-proposed system architecture. (From ILA System Architecture Working Group, *ILA System Architecture Specifications—Release 1.0*, InLocation Alliance, Piscataway, NJ, 2014. With permission.)

almanac is stored and to which the location server requests access. The information may be used by the server for different purposes such as calculating the position of a mobile device based on measurements originated in the device or the network. On the right side of the diagram is the location-based service application server, which has the main purpose to provide location-related content to any entity in the system, such as the map content stored in the map database. Furthermore, the architecture also considers a privacy policy database to handle all data privacy and network access issues related to the location platform [28].

The technology-agnostic approach makes it possible to also use this architecture for a VLC-based indoor positioning system, as it already indicates some of the necessary interfaces and logical entities that need to be considered during the design stage.

12.4 Use Cases of VLC for Indoor Positioning

In order to better provide the reader with an understanding of the potential benefits and issues around VLC for indoor positioning when exploring the lighting system, the authors present in this section an introduction to several use cases. Conventional system analysis methodology defines "use case" as a clear way of identifying and organizing system requirements; this section aims to provide added insight into the ideation and design processes that occur in the background of professional lighting systems. Therefore, considerations regarding the impact that VLC-based indoor positioning systems may have on the lighting distribution, installation, and/or LMS, and vice-versa, will be added to the discussion.

Six use cases are presented, reaching across different application areas. The first sub-section, on commissioning and maintenance, is transversal to the different market segments as it addresses the initial system setup and life-time maintenance processes. The following subsections each correspond to a specific application area associated to a market segment, focusing on issues and advantages pertaining to the combination of the lighting and positioning capabilities.

12.4.1 Commissioning and Maintenance: Where is the Light?

A particularly intriguing use case for VLC can be found when looking at the commissioning of a new lighting installation with an LMS. Consider a generic scenario where a specification for a lighting distribution, generated by an architect or lighting designer, was fulfilled by a lighting systems provider. Upon delivery of the devices necessary for the installation an electrician, typically a third party in the process, distributes all the necessary cabling throughout the building and installs the luminaires, according to the plans he was provided. After completing his work, he tests if the luminaires can be turned on/off and signs off on the installation to the commissioning team.

When installed, each luminaire already has a unique identifier and an auto-discover feature. Considering use of DALI protocol, included in the device's control software is the capability to easily detect when devices are connected to the network. Following the detection process, a DALI network address is assigned to a specific luminaire. However, its physical location is usually not known, as the electrician who installs the luminaires does not know their network address. Hence, the main task during commissioning is to identify where each luminaire is located inside a building and to feed this information to the database used for managing the lighting installation, thus finally linking the network address with its physical location. Using today's methods this is typically achieved by having a commissioning engineer at the central controls station switching on and off sections of the lighting installation. In parallel, a second engineer walks through the building to visually identify the position of the selected luminaires. Upon pinpointing the location of the luminaires, the information is sent back to the person on the central controller, who links it to the DALI address and uploads it to a database. As a large-scale commercial building can easily encompass several thousand luminaires, as seen in Figure 12.4, this is typically a time consuming and thus costly task.

Using simple VLC data transmission, as a first step, the commissioning process can already be considerably simplified. As proposed in [31] and shown in Figure 12.8, each luminaire, upon turning on, sends out its unique address via VLC. This in turn can be picked up via smartphone or another handheld device, which then automatically processes the information and

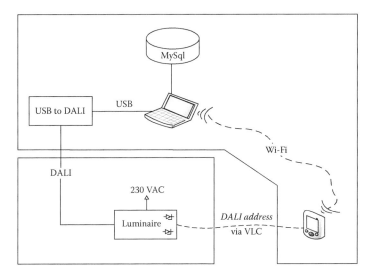

FIGURE 12.8
Commissioning process supported by VLC. (From Studer, M., et al., *VLC mit Handy-Cam*, Bachelor Work Report, 2012. With permission.)

sends it to the central database. By exploring indoor positioning capabilities, in a second step the process could be further optimized as the position information from the user could be estimated from the lighting signals (e.g., by trilateration) and integrated into the database along with the luminaire identification. The main output at this stage would be a reduction in the man hours of expert personnel needed to configure the lighting system and a reduction in the potential for human error.

Picking up on the same concept, VLC can also help in maintenance and servicing tasks. When a luminaire is marked faulty or scheduled for maintenance, the technician could use the VLC indoor positioning system to get directions to the exact location of the device in question, without prior knowledge of the lighting installation. When arriving at the location, if the luminaire is still able to turn on, it would transmit its operational parameters like power settings, operating hours, or even application information, to the receiver device. Via the handheld device, the maintenance engineer would be able to easily compare the information stored in the luminaire with target values, stored in a local file or central server. With additional on-site measurements the technician is then able to determine the proper maintenance operations, either minor adjustments or replacement of the faulty unit, in order to put the system back to working order.

The underlying principle of this use case is the combination of VLC for data transmission and positioning, in order to improve the maintenance process, thus cutting time and costs which helps reduce the building's TCO.

12.4.2 Retail: Guidance and Visitor Tracking

The retail segment is one of the fastest growing when it comes to the demand for indoor positioning. The ability to interact with customers and track their movements and behavior are strong motivators for venue owners. Let us consider a scenario where a new retail store is to be fitted with a VLC indoor positioning platform that ties deeply into the store's mobile app for marketing purposes. A solution that combines lighting and positioning using the same backbone infrastructure is adopted (e.g., through power line communications). With the strong selling point of having reduced cost from a combined system, the venue owner opts in to the solution and puts out a request.

Lighting system specifiers involved in the process now need to take into account the requirement of positioning into the lighting distribution design. Having uniformity and avoiding strong reflections are crucial requirements for a reliable VLC positioning system. Furthermore, a dense installation of light points is also typically necessary to achieve higher resolutions, and this is often contradictory to the designs of some luminaires that try to maximize their spatial coverage.

Once the design is finalized and the installation is complete, the lighting/location system and software application are tested and optimized. Customers and visitors can now come into the store, hold out their own mobile device to access the in-store application and get guided to suggested points of interest. Along the way they may receive information of ongoing promotions on nearby products, and overall have an improved user experience. This solution has the potential to work seamlessly and uniformly across multiple installations. The receiver is the front camera on the customer's smartphone or tablet, and the emitters are the luminaires placed on the ceiling, as depicted in Figure 12.9. There is no need for proprietary hardware, particularly on the customer side.

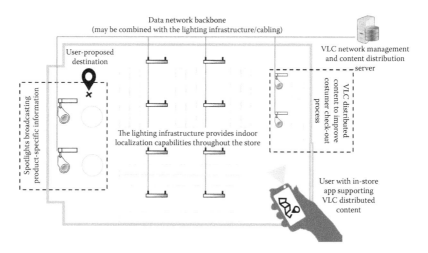

FIGURE 12.9
Representation of a VLC based indoor positioning system for a retail application.

Although dedicated receivers may be fitted into shopping carts, they do not provide the same level of user experience and require a far heavier investment from the venue owner. In any case, usability considerations regarding use of the mobile device's back camera must also be made. Since the user will hold the device in their hands, or attached to some type of fixture in a shopping trolley, the front camera is required to look into the ceiling. Although most high-end devices have such feature, not all mobile devices do, making mass adoption slightly harder. Last, the variety of different smartphones and camera models may also be problematic as the system must work reliably with all of them.

The retail use-case presents another interesting capability that can be explored through data analytics. By storing generic positioning data, even if previously anonymized, the venue owner may use such content and perform data analysis in order to understand movement patterns and behaviors. This would then be used to improve the store layout or check out processes.

Examples of solutions targeting the retail segment using VLC for the purpose of indoor positioning systems have already been presented. ByteLight, now part of the lighting group Acuity Brands, already presented a couple of demonstration installations showcasing a combination of Bluetooth Smart and VLC [17,32]. Phillips Lighting is another strong supporter of VLC for indoor positioning and is continuously promoting their own indoor positioning system supported exclusively by VLC transmission from LED luminaires to a mobile device's camera [16,33].

12.4.3 Office: Light that Follows You

Office buildings offer another interesting opportunity for VLC dissemination. With optimization and productivity highly in the minds of building owners, tenants and facility managers, automation features are often plenty in this market segment. Let us then consider an office space which incorporates a lighting system with LMS, supporting both positioning and basic data downstream from the luminaires. Users are able to find their way to a particular desk or colleague in an open space, find a meeting room, and control the devices within. Furthermore, the system also allows—through the VLC platform—transmission of dedicated information, such as specific drivers or credentials, pertaining to resources available in the vicinity of the user, as depicted in Figure 12.10.

As user position is closely monitored, a basic notion of where all workers are at a given moment enables improved control features. Segments of the lighting installation may be dimmed down or disabled if no one is present. Dynamic scene setting, such as lowering the lights for a presentation when a user gets close to the presentation area, is a capability that can now be enabled. This and other benefits are targeted not only to improve energy efficiency but also human productivity and well-being, by providing the right light when you need it.

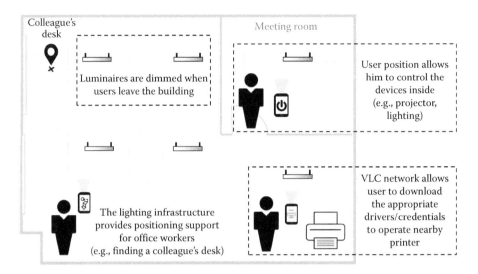

FIGURE 12.10
Representation of a VLC positioning and data transmission solution for an office application.

Furthermore, using VLC would release the burden of transmitting the area-specific information from the often-crowded Wi-Fi network. However, this translates into the precondition that a backbone network for the lighting system, with a bandwidth typically much larger than that of a standard LMS, is required. The proliferous DALI standard which uses a dedicated cabling infrastructure has little more than few bytes per second of effective data throughput, and would not be able to fulfill such requirements. Installations based on power-over-ethernet, power line communication, or other similar network technologies could present a valid alternative. Nevertheless, there is another important drawback as higher data throughput requires advanced modulation schemes, impacting the design and later on cost of lighting gears.

Overall, indoor positioning for office spaces has a strong market potential has it can provide significant benefits for existing automation systems by integrating user position into the control loop. However, when considering VLC applications that require higher data throughput, limitations apply. On the emitter side, a slightly more complex gear design is necessary, which adds to the cost of the device. On the receiver side, the standard design of a mobile device camera may limit the feasible throughputs.

12.4.4 Industry and Warehouse: Self-Driven Vehicles

The industry and warehouse market segments include installations in both manufacturing and storage facilities. Although requirements change frequently to cater for specific physical constraints and application requirements, TCO is the critical argument for any lighting solution in this segment. In order to provide advanced features, beyond typical dimming or even daylight harvesting, the building owner must have a strong requirement for it. Otherwise, features like indoor location must have real-life-proven reliability and benefits before they are even considered by the building owner or tenant. In a factory or warehouse, downtimes are costly and a critical system such as lighting cannot become inoperable, therefore simplicity is chosen over functionality for reliability reasons.

Nevertheless, this scenario allows self-driven robotic units which represent an interesting use case for VLC-enabled indoor positioning. Consider, therefore, a situation as depicted in Figure 12.11, where an industrial building has a lighting system with such capabilities. In this building, there is a set of self-driven vehicles that transport parts and finished goods between the warehouse sections and a loading/unloading bay. A first consideration on such a system is that dedicated receivers will be used for integration into the self-driven unit. This allows for the design of optimized systems as the receiver is dedicated for positioning. However, on the emitter side the complexity still needs to be kept low, as TCO influences the building owner's choice. Also, to reach better resolutions the density of emitters needs to be increased, and in industrial solutions with high-bay luminaires the number of devices is typically low.

In the current use case, the VLC positioning platform coexists with the Wi-Fi data communication platform. The lighting platform only provides the reference signals used by the self-driven vehicles to calculate their

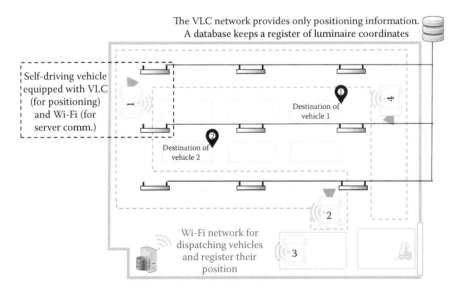

FIGURE 12.11
Representation of a VLC positioning system for an industry or warehouse application.

relative position inside the warehouse. A simple database is used to keep track of the coordinates of the luminaires. All the dispatching requests and data logging is done via the Wi-Fi network.

Indoor positioning/location capability is already provided by several suppliers of robotic transportation units. However, they use a large set of sensors and a dedicated RF network, a fairly complex vision system or a combination of both. Such add-ons are almost always vendor specific, and any kind of maintenance requires trained experts, which are not always readily available. Repairs are time consuming, eventually forcing downtime, and parts are often expensive, making the whole process quite costly. An industrial positioning system for self-driven vehicles based on VLC with the emitter provided by a third party offers the possibility for a multivendor or even a fully standardized ecosystem of products, which ultimately benefits both customers and suppliers.

12.4.5 Health and Care: Tracking Assets

The health and care market segments comprise buildings such as hospitals, medical clinics, rest houses, and others associated with human treatment and recovery. It is a segment that often deals with critical situations, and has extremely diversified and demanding working conditions. Lighting solutions must meet a large set of requirements and cater to the needs of medical professionals, patients, visitors, and other workers. In particular, there is interest in solutions that help the recovery process of patients, but also have an impact on the well-being of all building users. Under these conditions, there is also often the need to track and locate valuable assets, such as staff, patients, or critical equipment, and an indoor positioning system provides great benefits.

The idea of knowing where doctors or nurses are, is not a recent concern. For years, these human assets have been given communication devices in order to make them reachable at any time. Furthermore, and related to liability of the medical institution, tracking of critical patients, such as babies, people with several mental disorders, or elderly people with dementia is a growing concern. Also, given the high costs of medical equipment which limits availability, being able to track critical equipment such as crash carts, portable X-ray machines, and other devices for emergency situations helps improve the overall logistics and readiness for such situations.

Consider a hospital fitted with LED lighting and VLC indoor positioning. Trying to use VLC for real-time tracking of human assets is impractical as the direct line of sight with the receiver unit cannot always be guaranteed. Doctors, nurses, or patients cannot be expected to carry a tag that is always exposed to the lighting system, making the tracking easy to malfunction or override. Furthermore, common video surveillance, and tracking systems based on RF tags, are more reliable and already well established in the market. There is however a strong potential for guidance tools based on VLC. Staff can use mobile devices to instantly know the location of a critical event and use the positioning information to get directions to the location of the event. Applying a similar concept to hospital visitors or check-in patients, the indoor positioning system, through the application on the person's mobile device, would provide location and guidance information. Furthermore, through the hospital's speaker system, personalized audio messages could be played in order to assist them along the way to their destination, thus improving their experience in what is often a stressful location. In practice, this would be a similar approach as presented in the retail use-case, revolving around a mobile application.

Using VLC for tracking people is impractical in such environments. However, there are several advantages in using it for tracking medical equipment and similar assets. In hospital environments, there are strong restrictions with regard to use of RF equipment. As for VLC, some effects are known on the use of PWM modulation and oximeters. However, this does not mean that it is unfeasible, only that the modulation to be used needs to be specified with more care. Hence, VLC receiver tags could be used on the devices, placed in such way that they are always exposed to the lighting system. It is even practical to think of a bi-directional VLC link to relay back the equipment position information. This concept is also applicable to high-risk assets such as certain drugs or organs for transplants. The tag could also be fitted to special containers which would then be monitored in real time.

12.4.6 Museum: Seeing More

The use case for museums, or other exhibition spaces such as galleries, is where the quality of light factor weighs in more than in any other. Although in principle the operation is very similar to the retail use-case, museums are spaces where objects are showcased in a way that is made for them to stand out, and it is quite common to have dedicated lighting systems specially designed for a given exhibit. Besides the possibility to broadcast media content on the nearby artwork with VLC, the interaction with the light regarding its intensity, color, and variation is an important added-value feature. Light plays an important role in perceiving our surroundings, and in interactive environments this becomes even clearer.

Consider therefore a museum of painted artwork with a lighting solution that includes luminaires able to change color and/or color temperature, and is supported by an LMS that supports full reconfigurability as well as dynamic scene setting. This lighting solution also supports VLC for indoor positioning, and the position of users is shared between the mobile device and the LMS software. Users are able to move between different areas of the exhibition and when entering or leaving a particular location, a dynamic light show is used to enhance their experience.

Having the previously described functionality works best if users are truly tracked versus simply using motion sensing; this is, per se, a strong motivation for using VLC for indoor positioning. However, the dynamics of a lighting show means that the brightness of the lights keeps changing, and several channels and sources are being mixed in order to achieve the desired result. This poses one of the greatest challenges for VLC in lighting installations—the need for combining modulation techniques used for controlling the illumination solution and modulation techniques for enabling positioning or even generic data transmission. Furthermore, there is a growing trend to move away from PWM to analog dimming, which may ultimately impact VLC in the near future.

Light quality and output flow control are key topics for lighting, and integration with VLC is still far from being trivial. Although some techniques may be applied, the level of complexity required in the emitter side is still not entirely clear, and the associated costs to manufacture them in volume.

12.5 Success Factors for VLC in Indoor Positioning Along the Value Chain of Lighting

Having looked now at some of the possible use cases for VLC in indoor positioning applications, it makes sense to try and summarize the success factors as well as some potential pitfalls. Toward this goal, it is useful to follow the value chain of the professional lighting industry, as shown in Figure 12.12 [34].

12.5.1 The Value Chain of Professional Lighting

The value chain approach provided in Figure 12.12 follows a line that has been common to the professional lighting industry for years. Although it is mostly focused on the hardware business and does not explore to the fullest the revenue streams that may be generated from added-value functionality or services, it is still a strong model to compare against.

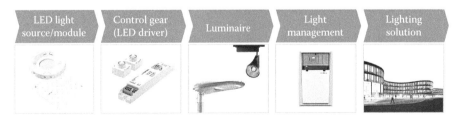

FIGURE 12.12
Typical value chain in professional lighting. (From Zumtobel Group AG, *Passion for Light—Zumtobel Group Corporate Portrait*, Dornbirn, 2015. With permission.)

12.5.1.1 LED Light Sources/Modules

This first stage comprises the base component (the encapsulated LED die) as well as LED boards, which are becoming more and more common. For relatively low-bandwidth VLC applications, like indoor positioning, the requirements of the LED light source are not so strict. In fact, most standard LED light sources should be able to support frequencies in the orders of hundreds of kHz [35]. The exception is some AC-driven luminaires that use phosphor materials with very long relaxation times in order to avoid flicker from the typical AC frequencies. In this case, color filtering on the receiving side might be necessary. Despite having relaxed limitations, an important success factor lies in the selection of appropriate LEDs, and furthermore, when using LED modules, making sure that the board connections allow implementation of the necessary modulation techniques for the final implementation. Although most VLC indoor location systems still rely on basic PWM modulation, this is expected to change in the near future [12].

As for high-bandwidth VLC systems with frequencies in the MHz regime, some care must be taken to pick appropriate LEDs and their configuration. High-power LEDs, in particular, are designed using a large silicon footprint, and therefore have an inherently higher internal capacity which makes them unsuitable for switching at higher frequencies. Finally, it must be noted that VLC is not a key target application for the manufacturers of LED light sources and modules, who very seldom characterize their products for this application. As the focus is always on the optical conversion efficiency, getting manufacturers to design products with VLC applications in mind would also be a strong success factor.

12.5.1.2 Control Gear

The control gear, or LED driver, plays an important role in enabling VLC in general, and indoor positioning applications in particular. It is this device that performs the necessary current modulation to the LED, thus shaping the output-modulated optical signal. One point in favor of VLC is that today, many drivers support some sort of PWM dimming, where essentially the light output of the LED is controlled by switching it on and off very rapidly. Typical frequencies range from several hundred hertz up to several kilohertz. This capability can be exploited to perform basic modulation, thus enabling low debit VLC communications. As the concept is very similar, these LED drivers can—in principle—already be used to provide VLC signals, as long as the frequency modulation lies in the range of what the driver is capable of providing. In fact, there are already commercially available gears supporting VLC deployed by a few manufacturers [16,17].

On the other hand, there are two main pitfalls that may impact VLC for indoor positioning. The first lies with interference between VLC modulation and the dimming process, in particular when using PWM dimming. Since both functionalities are based on pulse modulation of LED current, special care needs to be taken to ensure the two functionalities can coexist in the same device. This is particularly true for low-bandwidth VLC systems that are operating in frequency regions close to those of PWM dimming. The second pitfall comes from a market trend to use more and more analog dimming. Due to cost issues, but also acceptance concerns related to flickering and other potentially harmful health effects, there is a general trend in the lighting industry to move from PWM dimming to analog dimming methods. As a result of this, the percentage of LED drivers capable of supporting fast switching of the LED current, and therefore VLC, is shrinking.

Therefore, within the realm of digitally controllable gears, the main success factor will lie in the ability to perform fast switching. This should preferably be combined with complex modulation schemes, natively from the gear or through a minimum hardware overhead. But, while there is only a small number of manufacturers providing solutions, which are not compatible, a widespread market adoption is almost impossible.

12.5.1.3   Luminaires

The design as well as the type of luminaire affects the potential of VLC for indoor positioning in several ways. Starting with light distribution, depending on application, luminaires are designed to have different patterns. These can range from directed spotlights to completely diffuse area luminaires. As the light distribution directly influences in which area a VLC positioning signal can be received, the light distribution of the luminaires in a given installation needs to be taken into consideration. Otherwise, a spotlight directed at a faraway target or a very diffuse luminaire installed in the ceiling could easily generate false positioning data at the receiver. Direct versus indirect light is also a concern. Many luminaires are designed with two different light distributions, one emitting directly toward the user and another one emitting indirectly toward walls or the ceiling. As these are emitted in opposite directions and the indirect part only reaches the user after one or more reflections, this may also lead to false location signals unless taken into account during the design of the luminaire. As an example, a luminaire with both direct and indirect components should only emit VLC with positioning information from the direct component.

At this stage of the value chain, the most important success factor resides in the ability of luminaire manufacturers to come up with designs that cater for the compromise between lighting uniformity and tailored output for VLC signals. These considerations have a clear relationship with the two previous segments of the value chain. Luminaire manufacturers will need LED modules that can be split or segmented, and control gears that support such segmentation.

12.5.1.4   Light Management

Light management systems that automatically or semi-automatically control the light inside a building need to be taken into account when designing a VLC-based indoor positioning system. Typically, these systems are geared toward minimizing energy consumption and often rely on external sensors to, as an example, turn off the lighting when a daylight sensor recognizes that enough natural light is in the room. To avoid issues with the VLC positioning system, minimum dimming levels need to be set and influences from natural daylight also need to be considered. However, the combination of low dimming levels and high levels of daylight could strongly reduce the VLC signal quality.

Besides these concerns, the main success factor will come from the introduction of positioning systems directly into LMS platforms, as a native feature. With the IoT advent, there is a trend to migrate from bandwidth constrained networks, such as DALI, into solutions with higher throughput, for example, power-over-ethernet. In this final scenario, the VLC for indoor positioning, or even for content distribution, would be able to share the same backbone.

12.5.1.5 Lighting Solutions

Lighting solutions comprise, beyond the classical hardware business, a whole range of features and services through the whole process of designing and implementing a lighting system. A strong benefit lies in having a planned installation, designed with the purpose of VLC and indoor positioning from root. It is far much easier to make sure these features will work properly under new or full refurbishment scenarios, than in an installation that has grown historically over the years. A drawback is that for many applications, light installations are planned in a way that may be less than ideal for VLC systems. One particular example is museum lighting which sometimes is based on completely indirect lighting solutions so as to not draw attention away from the exhibition pieces or to avoid damage from strong light intensities. Ultimately, at this stage of the value chain, the success factor will reside in lighting planners that consider the indoor positioning application upfront, in the design period, and not later, during installation. The lighting distribution, and for this purpose also the luminaire pattern, will have a strong impact on the overall VLC-based indoor positioning system.

12.5.2 The People in the Value Chain

Besides the previous factors stemming from the value chain of professional lighting, there are others that present benefits and drawbacks of VLC and indoor positioning systems. These are related to the different people that somehow interface with the system, or have a decision role in the process of defining it, throughout the value chain.

The sales process is a complex one, starting by finding new projects and customers, and followed by a long negotiation in an overall time-consuming task. Sales people are highly motivated to sign a contract, but will prefer to maintain inside their own, and the customer's, comfort zone. This allows them to swiftly reply to any concern and doubt the customer may have.

The introduction of specialized features, such as VLC or even indoor positioning, is not always seen as a benefit, particularly if it has not been properly validated in the market. Features which customers and users are not trained to handle may cause a bad experience and negative feedback, even if in principle they perform well. Therefore, the sales force of any lighting solutions provider or system integrator needs to be properly trained for understanding and handling VLC systems, so they feel confident in pushing it to market.

For building owners, the main benefit of a VLC-based location system when compared to most competing technologies is that a technology that is based on, and integrated into, the general lighting installation means that no effort for additional infrastructure is required. On the other hand, as already mentioned in the previous sections, special care needs to be taken that the main purpose of general lighting, illumination of the building, and the additional feature of indoor positioning do not collide with each other.

In many cases, the luminaires are installed by untrained personnel who have little understanding for anything that deviates from a standard installation. Consequently, it is necessary to ensure that installation of a lighting system with VLC and indoor positioning is no different from that of a standard lighting system, allowing installers to work seamlessly with it.

For the end user, the main issue with any indoor positioning system is usability. Indoor positioning is often considered as an add-on feature to an existing system, and is not always implemented in a consistent and easily accessible manner. In particular, an indoor positioning system must be compatible with standard electronic equipment, be it smartphones, tablets or smartwatches. Furthermore, the accuracy needs to match the requirements of the use case, allowing the user a reliable and consistent experience. While some use cases like asset tracking or automation may require centimeter accuracy, in others 1–3 meters may be sufficient.

12.6 Final Considerations

Throughout this chapter, the authors presented several viewpoints on VLC for indoor positioning from an application perspective. It is clear that such systems have the basic technology and key market enablers that open up its potential, and there are already several benefits that arise from combining it with a lighting installation. However, there are still a few validation scenarios and, from looking at just a couple of use cases, we realize the complexity that will arise from adding this feature to what is the diversity of lighting systems and installations.

VLC will have to overcome several challenges along the lighting value chain in order to become widespread. Also, it will need to convince several people in the process, besides lighting manufacturers, of its reliability and simplicity. Finally, the research community will also have an important role in this process as they need to continue working not only in enabling enhanced features, such as high data rates, but also in the reliability and simplicity of VLC, validated by practical applications and demonstrations.

From the topics exposed, it seems safe to conclude that VLC in general, and indoor positioning in particular, will require the lighting industry, manufacturers, designers, and planners to look at luminaire and system design in a new way. Components will need to support new modulation techniques, and research efforts should focus on how to achieve this goal with affordable hardware designs. Luminaires will often need to support specific light channels just for VLC, and manufactures should cater for this. As for lighting designers, they will need to weigh in the requirements of these added features, which may sometimes oppose current design rules, into their process.

Finally, and despite all individual efforts, it is crucial for both VLC and indoor positioning to adhere as much as possible to standards. This starts from available modulation techniques to system architecture and even design considerations. A standardized ecosystem of VLC-enabled luminaires with indoor positioning capabilities, which provides users with a consistent and reliable experience, is one of the key factors in making such solutions widespread.

References

[1] E.E. Times, *Product News—Nichia Develops 100-lumens/W efficiency white LED chip*, 2006. Available: http://www.eetimes.com/document.asp?doc_id=1299475 [Accessed June 2015].

[2] DOE SSL (Department of Energy, Solid-State Lighting) Program. 2015. *R&D Plan*, prepared by Bardsley Consulting, SB Consulting, SSLS, Inc., LED Lighting Advisors, and Navigant Consulting, Inc., DOE Office of Energy Efficiency and Renewable Energy, Washington.

[3] Zumtobel Group AG, *Fiscal Year 2014/15 Results*, Dornbirn, Austria, 2015.

[4] M. Economidou, J. Laustsen, P. Ruyssevelt, D. Staniaszek, D. Strong and S. Zinetti, *Europe's Buildings Under the Microscope*, Buildings Performance Institute Europe, Brussels, Belgium, 2011.

[5] A.J. Nelson and O. Rakau, *Green Buildings—A Niche Becomes Mainstream*, Deutsche Bank Research, Frankfurt, Germany, 2010.

[6] Zumtobel Lighting GmbH, *LEED Light Guide*, Dornbirn, Austria, 2014.

[7] Y. Roderick, D. McEwan, C. Wheatley and C. Alonso, Comparison of energy performance assessment between LEED, BREEAM and green star, *Eleventh International IBPSA Conference*, Glasgow, Scotland, July 2009.

[8] A.T. Kearney, *Human Centric Lighting: Going Beyond Energy Efficiency*, LightingEurope, Brussels, Belgium, 2013.

[9] P.R. Boyce, *Human Factors in Lighting*, 3rd ed., CRC Press, Boca Raton, FL, 2014.

[10] L. Schlangen, D. Lang, P. Novotny, H. Plischke, K. Smolders, D. Beersma, K. Wulff, et al., *Lighting for Health and Well-Being in Education, Work Places, Nursing*

Homes Domestic Applications and Smart Cities—Accelerate SSL Innovation for Europe, SSL-erate Consortium, Brussels, Belgium, 2014.

[11] G. Ntogari, T. Kamalakis, J. Walewski, and T. Sphicopoulos, Combining illumination dimming based on pulse-width modulation with visible-light communications based on discrete multitone, *IEEE/OSA J. Opt. Commun. Networking*, vol. 3, no. 1, pp. 56–65, 2011.

[12] X. Ma, K. Lee and K. Lee, Appropriate modulation scheme for visible light communication systems considering illumination, *(IEEE) Electron. Lett.*, vol. 48, no. 18, pp. 1137–1139, 2012.

[13] G. Gruman, *Apple's Next Revolution May be Bluetooth-Powered iBeacons*, InfoWorld, 2013. Available: http://www.infoworld.com/article/2612352/ios/apple-s-next-revolution-may-be-bluetooth-powered-ibeacons.html [Accessed March 2015].

[14] D. Hildebrant, *Tapping into the Potential of Location-Aware Technologies*, White Paper, AT&T, New York, 2015.

[15] Markets and Markets, *Indoor Location Market worth $4,424.1 Million by 2019*. Available: http://www.marketsandmarkets.com/PressReleases/indoor-location.asp [Accessed February 2016].

[16] M. Wright, *LEDs Magazine—Philips Lighting Demonstrates LED-Based Indoor Location Detection Technology*, 2014. Available: http://www.ledsmagazine.com/articles/2014/02/philips-lighting-demonstrates-led-based-indoor-location-detection-technology.html [Accessed March 2015].

[17] M. Halper, Acuity embeds indoor location technology into retail luminaires, *LEDs Magazine*, 2015. Available: http://www.ledsmagazine.com/articles/2015/11/acuity-embeds-indoor-location-technology-into-retail-luminaires.html [Accessed January 2016].

[18] InLocation Alliance, ILA Systems Architecture Specification Release 1.0, http://inlocationalliance.org/about/the-opportunity-for-indoor-positioning/, 2015. Available: http://inlocationalliance.org/ [Accessed February 2016].

[19] K. Vamberszky, *Lighting Controls, Zumtobel Group Internal Presentation*, Dornbirn, Austria, 2014.

[20] H. Merz, T. Hansemann and C. Hübner, *Building Automation—Communication Systems with EIB/KNX, LON und BACnet*, Springer, Berlin, 2009.

[21] DALI AG—ZVEI: Division Luminaires, *Digital Addressable Lighting Interface (DALI) Manual*, DALI AG, Munich, 2001.

[22] C. Iozzio, *Indoor Mapping Lets the Blind Navigate Airports*, Smithsonian.com, 2014. Available: http://www.smithsonianmag.com/innovation/indoor-mapping-lets-blind-navigate-airports-180952292/ [Accessed December 2015].

[23] A.M. Vegni and M. Biagi, An indoor localization algorithm in a small-cell LED-based lighting system, *2012 International Conference on Indoor Positioning and Indoor Navigation (IPIN)*, 2012.

[24] G.D. Campo-Jimenez, J.M. Perandones and F.J. Lopez-Hernandez, A VLC-based beacon location system for mobile applications, *2013 International Conference on Localization and GNSS (ICL-GNSS)*, Turin, 2013.

[25] L. Li, P. Hu, C. Peng, G. Shen and F. Zhao, Epsilon: A visible light based positioning system, *NSDI'14 Proceedings of the 11th USENIX Conference on Networked Systems Design and Implementation*, Berkeley, CA, 2014.

[26] E. Gonendik and S. Gezici, Fundamental limits on RSS based range estimation in visible light positioning systems, *IEEE Commun. Lett.*, vol. 19, no. 12, pp. 2138–2141, 2015.

[27] Zumtobel Lighting GmbH, *LITECOM—Next Generation lighting management*, Dornbirn, Austria, 2015.

[28] ILA System Architecture Working Group, *ILA System Architecture Specifications—Release 1.0*, InLocation Alliance, Piscataway, NJ, 2014.

[29] ILA Use Case Working Group, *Indoor Location Based Services in Retail and Marketing*, InLocation Alliance, Piscataway, NJ, 2015.

[30] ILA Use Case Working Group, *Indoor Location Based Services in Transportation Hubs*, InLocation Alliance, Piscataway, NJ, 2015.

[31] M. Studer, M. Heuberger and M. Zurmühle, *VLC mit Handy-Cam*, Bachelor Work Report, Hochschule Luzern, Lucerne, Switzerland, 2012.

[32] Accuity Brands, *Illuminating the In-Store Experience: Indoor Positioning Services Using LED Lighting Benefit Shoppers and Retailers*, Whitepaper, Steve Lydecker, Acuity Brands, Atlanta, GA, 2015.

[33] Philips Innovation Communications, *Indoor Positioning—Finding your Way Indoors*, Inside Innovation, Eindhoven, The Netherlands, 2014.

[34] Zumtobel Group AG, *Passion for Light—Zumtobel Group Corporate Portrait*, Dornbirn, Austria, 2015.

[35] D. Karunatilaka, F. Zafar, V. Kalavally and R. Parthiban, LED based indoor visible light communications: State of the art, *IEEE Commun. Surv. Tutorials*, vol. 17, no. 3, pp. 1649–1678, 2015.

13

Optical Small Cells, RF/VLC HetNets, and Software Defined VLC

Michael B. Rahaim and Thomas D. C. Little

CONTENTS

13.1 Introduction

As the world has become an increasingly connected ecosystem, data demand has continually pushed the limits of our wireless networks. Each generation of wireless technology has added extensive wireless capacity; however, novel use cases and technological adoption have continually raised the demand at rates previously unseen. This trend continues today as massive data downloads, video streaming, augmented reality, and the Internet of Things push the limits of 4G networks. Furthermore, increasing application complexity and demand per device is paired with an increasing number and density of wireless devices. Hence, the wireless communications industry is challenged to meet an extreme growth in data traffic in the coming years—a demand growth that is unlikely to be met with iterative modifications to the current infrastructure. With this in mind, the world mobile and wireless infrastructure community is looking toward the 5th generation of mobile telecommunications, or 5G, with the expectation that new technologies will be adopted which drastically alter the conventional view of wireless networks [1]. Although 5G has been described as the solution for demands of 2020 and beyond, global standardization efforts have been limited. Given the wide variety of devices, applications, and services expected to be part of next generation networks, the vision for 5G has been defined in several ways [2–4]. Figure 13.1 depicts a set of proposed 5G design principles that relate to the integration of visible light communication (VLC) within the next generation wireless communications landscape. These concepts and their relationship to VLC are described throughout this chapter.

The trend toward *smaller cells and network densification* is expected to be a primary contributor to the aggregate capacity gains. Smaller cells allow for increased spatial reuse, leading to higher bandwidth density ($b/s/m^2$) and area spectral efficiency ($b/s/Hz/m^2$). This has been a common trend

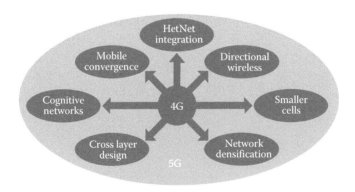

FIGURE 13.1
Next generation wireless communications landscape.

throughout the history of wireless communications, most recently with traffic offloading to RF small cells (RFSCs) such as femtocells or wireless local area networks (WLANs). However, increasing the density of omnidirectional RFSCs is limited by infrastructural constraints including access point (AP) placement and connectivity. Directional small cell (DSC) technologies like VLC allow APs to be distributed at reasonable distances from the mobile terminals (MTs) while generating a small coverage area at the working surface.

In addition, 5G networks are expected to utilize a wide range of access technologies in order to accommodate the variety of use cases. Therefore, *heterogeneous network (HetNet) integration* will play an important role in providing the required aggregate network capacity. HetNets allow for distribution of user traffic among wireless access technologies that are best suited for specific use cases. This provides drastic performance improvements in environments with a diverse set of MTs where the channel and traffic characteristics can vary from one device to another. While VLC DSCs have the potential to increase aggregate capacity due to high bandwidth density, characteristics of the optical medium lead to use cases where it is preferable for an MT to associate with a broader coverage RFSC. As an example, quasi-static MTs (i.e., wireless devices such as laptops or tablets that are typically used in a static location) may associate with a reliable VLC connection if available, whereas MTs with high mobility conditions may associate with the broader RFSC in order to minimize overhead from handover as the device moves through the environment. This is similar to associating with a macrocell when driving down a street as opposed to connecting to every RFSC that is passed.

Finally, the dynamic conditions expected in 5G networks have motivated a great deal of research in adaptive systems. Network cognition provides the ability to predict channel and link characteristics, which can be used in the modulation and coding scheme (MCS) selection for dynamic physical layer or HetNet link selection. Cross layer design improves individual and/or network-wide performance by use of knowledge from various network layers in the configuration process. Implementation of these techniques requires flexibility provided by *software-defined systems*, including software-defined radio (SDR) and software-defined networks (SDNs). Such systems offer the ability to dynamically reconfigure the characteristics of an MT connection in real time. This implies the importance of reconfigurable VLC systems and the need for adaptation of the VLC connection on the fly via software-defined VLC (SDVLC). SDVLC systems can also be used as an educational tool or testbed to show practical implementation of VLC techniques.

While interest in low data-rate VLC applications (e.g., indoor location services) has seen recent growth, VLC is also envisioned as an important component of 5G [4–6]. Integration of VLC within the 5G wireless landscape is viewed as a way to improve aggregate network capacity by offering MTs the ability to opportunistically utilize ultradense optical wireless (OW) DSCs. In this chapter, the application of VLC as a supplemental medium within next generation wireless networks is discussed. In particular, the discussion focuses on VLC in the context of small cells, HetNet integration, and software-defined systems. Section 13.2 presents a motivation for continuation of the small cell evolution and discusses how the directionality of VLC is an ideal medium for network densification beyond RFSCs. Section 13.3 describes the requirements for coexistence of RF and VLC within mixed media environments, and Section 13.4 defines an SDVLC implementation of an RF/VLC HetNet. Section 13.5 concludes the chapter and summarizes the presented material.

13.2 Small Cells

Many techniques have been implemented to meet the growing demand for wireless network capacity. New spectrum allocation has allowed for larger channel bandwidth and increased link capacity (b/s). In addition, novel modulation schemes and signal processing techniques along with methods to increase the signal-to-noise ratio (SNR) or signal-to-interference-plus-noise ratio (SINR) have improved spectral efficiency (b/s/Hz). However, the most significant gains in aggregate wireless network capacity have stemmed from network densification—bringing APs closer to the MTs and reducing cell size in order to increase the number of APs and, accordingly, increase aggregate bandwidth density and area spectral efficiency [7]. Figure 13.2 depicts the evolution of cell size for wireless access and forecasts future needs where ultradense DSCs are implemented with densely distributed VLC APs [8].

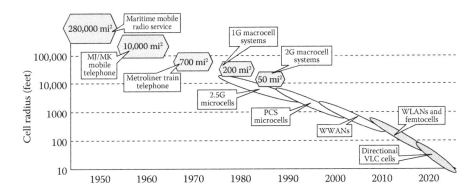

FIGURE 13.2
Evolution of cell size for wireless access. (Adapted from Y. Sheng, A. Kochetkov, and T. Nguyen. *Evolution of Microwave Radio for Modern Communication Networks.* ZTE Grand, 2012. Available: http://wwwen.zte.com.cn/endata/magazine/ztetechnologies/2012/no5/articles/201209/t20120912_343888.html)

The small cell concept has been successfully implemented in recent years with femtocell, picocell, microcell, and other commercially available RFSC devices. Wi-Fi WLANs have also played a major role in offloading traffic from the cellular network. As of 2014, 46% of mobile data traffic was off-loaded to the fixed network via RFSCs, and it is forecasted that more mobile data traffic will be offloaded to Wi-Fi than remains on the cellular networks by 2016 [9].

In addition to the offloaded mobile traffic, the growth in fixed Internet traffic and the increasing percentage of fixed Internet traffic destined for wireless devices are increasing RFSC congestion. While a typical family may have had a single wireless device a decade ago, household WLANs now service multiple smart phones, tablets, and laptops as well as TVs and other networked appliances within the Internet of Things. Offices and commercial environments must also accommodate more MTs assuming every patron or employee requires wireless access and that the infrastructure incorporates wireless sensors and networked devices to meet demands of the "industrial Internet" [10]. Given the increasing demand and density of MTs, a case can be made for a continued densification of wireless cells—specifically in indoor environments; however, infrastructural constraints limit the use of omnidirectional RFSCs when the desired coverage area decreases below that of today's devices. As a resolution, VLC DSCs offer a way to both locate APs away from the MTs and generate a small coverage area. On the other hand, increased AP density and the small coverage area of VLC cells lead to issues with mobility.

This section motivates continued network densification and describes the characteristics of RFSCs and ultradense OW DSCs. In the following, historical trends are presented and a theoretical model is defined in order to analyze performance gains that stem from network densification and motivate the continuation of the small cell trend within 5G networks. Channel characteristics and performance models are then presented for both RF and OW links. The pros and cons of each media as well as ideal use cases are also discussed.

13.2.1 Small Cell Motivation

Motivating the continuation of the small cell trend requires evaluation of potential performance improvements that stem from network densification. Here, performance gains are analyzed in an idealistic scenario and real-world constraints are discussed in relation to how much of these potential gains can be practically achieved. While adding network capacity is the ultimate goal, optimal provisioning should satisfy MT traffic requirements with high probability while avoiding overprovisioning at the expense of additional infrastructure and maintenance costs. Defining system performance in regard to effectiveness (i.e., the ability to meet MT requirements) and efficiency (i.e., the usage of available resources) offers a metric to depict how well a system is provisioned. Weighting the performance toward either of these parameters allows the metric to show preference toward either maximally satisfying user requirements or minimally overprovisioning the system.

The following motivation shows how the current trends of increasing MT traffic requirements and growing numbers of MTs affect system performance. The analysis also shows how increasing cell capacity and cell density can combat the demand growth in order to provide suitable system performance. The example shown here observes ideal operating conditions which are seldom the case in practical implementations; however, the trends relating to the effects of additional MTs, device requirements, cell capacity, and cells can be generalized to more practical dynamic environments. Decisions are assumed to be made in the provisioning process regarding the number of cells, cell distribution, and the capacity of each cell. The number and spatial distribution of MTs as well as their individual requirements are assumed to be dynamic variables where the probability distributions are known.

In order to make the analysis environment agnostic, consider a model where the system consists of M cells each with capacity C_i where i is a cell in the range $1 \leq i \leq M$. In this case, capacity is not the theoretical channel capacity but the maximum throughput a cell can provide based on available MCS. The environment has a variable number of MTs, n, each with rate requirement, r_j, and cell assignment, u_j, where j is a specific MT in the range $1 \leq j \leq n$. The number of MTs in the ith cell, n_i, relates to the total number of MTs, $n = \sum_{i=1}^{M} n_i$. Desired cell capacity, d_i, is the minimum capacity required to satisfy all MTs within the specified cell. These values are formally defined by:

$$n_i = \sum_{j=1}^{n} \delta_{i,u_j} \tag{13.1}$$

$$d_i = \sum_{j=1}^{n} r_j \delta_{i,u_j} \tag{13.2}$$

where δ is the Kronecker delta function. Aggregate capacity, C, and minimum capacity required to fulfill the MT rate requirements, C', are the sum of individual cell capacities and the sum of desired cell capacities (or individual MT rate requirements), respectively. Aggregate throughput, T, is the sum of the throughput from each cell, $T_i = \min(C_i, d_i)$.

$$C = \sum_{i=1}^{M} C_i \tag{13.3}$$

$$C' = \sum_{i=1}^{M} d_i = \sum_{j=1}^{n} r_j \tag{13.4}$$

$$T = \sum_{i=1}^{M} T_i \tag{13.5}$$

Performance for a specific instance is evaluated in terms of network effectiveness, $\tau = T/C'$, and efficiency, $\varepsilon = T/C$. Effectiveness is defined as the ability of the network to satisfy all MT rate requirements, and efficiency is defined as the percentage of available capacity in use by MTs. Instantaneous performance, Ψ, is defined as the weighted sum of τ and ε, where effectiveness and efficiency are given the weights ω_τ and ω_ε, respectively. Note that $0 \le \omega_\tau \le 1, 0 \le \omega_\varepsilon \le 1$, and $\omega_\tau + \omega_\varepsilon = 1$. For scenarios where n, r_j and u_j are not fixed values, the expected performance is $E[\Psi] = \omega_\tau E[\tau] + \omega_\varepsilon E[\varepsilon]$. These parameters are summarized in Table 13.1.

In order to observe idealistic small cell performance, define a system with a fixed $n = N$ and assume cells have the same capacity (i.e., $C_i = C/M \forall i$) and MTs have the same rate requirements (i.e., $r_j = C'/N \forall j$). In this case, the optimal effect of network densification can be shown through a best-case scenario where MTs are laid out according to a structured cell assignment. In a deterministic scenario where MTs are divided among cells such that u_j is incrementally assigned, the first $N\%M$ cells have one MT more than the others where $\%$ is the modulo function (i.e., $a\%b = a - b\lfloor a/b\rfloor$). The per-cell relations and the aggregate system throughput are accordingly:

$$n_i = \begin{cases} \lceil N/M \rceil, & \text{if } i \le N\%M \\ \lfloor N/M \rfloor, & \text{if } i > N\%M \end{cases} \tag{13.6}$$

TABLE 13.1

Parameters for Small Cell Analysis

Parameter	Var	Parameter	Var
Max system capacity	C	# of MTs in the ith cell	n_i
Aggregate throughput	T	Capacity of the ith cell	C_i
# of access points (cells)	M	Desired capacity of ith cell	d_i
# of mobile terminals (MTs)	n	Throughput of the ith cell	T_i
Min capacity requirement	C'	Network effectiveness	τ
Rate requirement of jth MT	r_j	Network efficiency	ε
Cell assignment of jth MT	u_j	Performance	Ψ

$$d_i = \begin{cases} \lceil N/M \rceil r_j, & \text{if } i \le N\%M \\ \lfloor N/M \rfloor r_j, & \text{if } i > N\%M \end{cases} \tag{13.7}$$

$$T_i = \begin{cases} \min(C_i, \lceil N/M \rceil r_j), & \text{if } i \le N\%M \\ \min(C_i, \lfloor N/M \rfloor r_j), & \text{if } i > N\%M \end{cases} \tag{13.8}$$

$$
\begin{aligned}
T &= \left(N - M \left\lfloor \frac{N}{M} \right\rfloor \right) \min \left(C_i, \left\lceil \frac{N}{M} \right\rceil r_j \right) \\
&\quad + \left(M - \left(N - M \left\lfloor \frac{N}{M} \right\rfloor \right) \right) \min \left(C_i, \left\lfloor \frac{N}{M} \right\rfloor r_j \right) \\
&= C_i \left[(N - M \lfloor N/M \rfloor) \min \left(1, \lceil N/M \rceil \frac{r_j}{C_i} \right) \right. \\
&\quad \left. + \left(M - \left(N - M \left\lfloor \frac{N}{M} \right\rfloor \right) \right) \min \left(1, \left\lfloor \frac{N}{M} \right\rfloor \frac{r_j}{C_i} \right) \right]
\end{aligned}
\tag{13.9}
$$

Efficiency and effectiveness can be defined in terms of the ratio of MTs to APs, $\Gamma_M = N/M$, and the percentage of a cell's capacity required to satisfy MT requirements, $\Gamma_C = r_j/C_i$. Equations 13.10 and 13.11 show that efficiency and effectiveness are related by $\Gamma_M \Gamma_C$. This can also be shown as in (Equation 13.12) by observing the definitions of τ and ε along with the fixed capacity and rate requirement definitions for this specific scenario:

$$
\begin{aligned}
\varepsilon &= (\Gamma_M - \lfloor \Gamma_M \rfloor) \min(1, \lceil \Gamma_M \rceil \Gamma_C) \\
&\quad + (1 - (\Gamma_M - \lfloor \Gamma_M \rfloor)) \min(1, \lfloor \Gamma_M \rfloor \Gamma_C)
\end{aligned}
\tag{13.10}
$$

$$
\begin{aligned}
\tau &= \frac{1}{\Gamma_M \Gamma_C} [(\Gamma_M - \lfloor \Gamma_M \rfloor) \min(1, \lceil \Gamma_M \rceil \Gamma_C) \\
&\quad + (1 - (\Gamma M - \lfloor \Gamma M \rfloor)) \min(1, \lceil \Gamma M \rceil \Gamma C)]
\end{aligned}
\tag{13.11}
$$

$$\tau = \frac{T}{C'} = \frac{C}{C'} \varepsilon = \frac{MC_i}{N_{r_j}} \varepsilon = \frac{\varepsilon}{\Gamma_M \Gamma_C} \tag{13.12}$$

Given these equalities, performance can be defined strictly in terms of Γ_M, Γ_C, ω_τ, and ω_ε. In Figure 13.3, the resulting performance is shown across Γ_M and Γ_C for various weights. Lines at $\Gamma_M = 1$ (0 dB) and $\Gamma_C = 1$ (0 dB) indicate $N = M$ and $r_j = C_i$, respectively.

In the top right quadrants, there are more MTs than cells and each MT requires more capacity than a single cell can provide; therefore, requirements cannot be met. In the lower right quadrants, individual cells cannot meet the requirements of a single MT and there are more cells than MTs. Distributed multiple-input multiple-output (MIMO) techniques can be used to satisfy

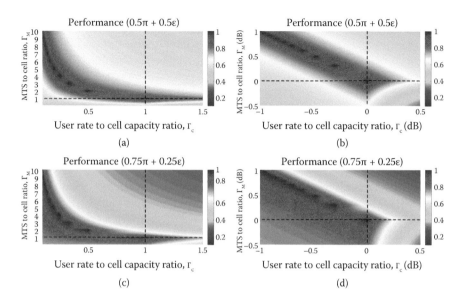

FIGURE 13.3

Small cell performance relating to network effectiveness, τ, and efficiency, ε, given that (a) effectiveness and efficiency are equally weighted and (c) a preference is given to effectiveness. Note that (b) and (d) show equivalent results with Γ_M and Γ_C in dB.

MT requirements in this region [11–13]; however, this is not shown in the results in Figure 13.3. The lower left regions depict performance when per-cell capacity is greater than the MT requirements and there are more cells than MTs. This region relates to overprovisioning where effectiveness is high but efficiency is poor. The top left quadrants depict the region of greatest interest where there are multiple MTs per cell and each cell can accommodate individual requirements. The points where optimal performance is achieved fall on the line $\Gamma_M = 1/\Gamma_C$ when Γ_M is an integer value (i.e., N is a multiple of M). In this idealistic scenario, these are the points where all requirements are met with optimal use of the available capacity. When weights are set to bias performance toward effectiveness (i.e., achieve throughput required to meet demands), there is a preference to be below and to the left of the optimal performance line. With next generation wireless systems trending toward more MTs with higher rate requirements, demands are pushing the performance point up and to the right. In order to move the point back toward the origin, either cell capacity or the number of cells must increase. Current MCS techniques come close to theoretical capacity bounds; hence, densification toward a 1:1 AP to MT ratio is a reasonable method for meeting 5G requirements.

While the results of this analysis indicate that increasingly dense cells provide continued gains in network capacity, there are constraints that limit the benefit of network densification in practical scenarios. First, signal disturbances due to interference from other cells play an important role in the performance of a given link. Hence, increasing the number of cells also requires decreasing cell size in order to mitigate interference. Achievable throughput from an AP to a specific MT is not constant throughout a cell since signal and interference power are each related to the relative distance and orientation of MTs and APs as well as any obstructions in the signal path. Theoretical channel capacity is a function of the received SINR at a specific MT, which is dependent on the location of the MT and the distribution of APs in the environment as well as the allocation of resources among APs and the characteristics of the wireless channel. While the idealistic model defines a fixed cell capacity, practical systems incorporate dynamic rate adaptation techniques such that the optimal MCS and achievable throughput at various locations within a cell can vary with the SINR.

In addition, dynamic traffic patterns and the variety of mobility states lead to a wide range of potential scenarios in a given system. The idealistic scenario where MTs have equivalent requirements and are evenly distributed among cells is unlikely. Traffic requirements can range from low-rate message passing to high-rate video streaming. MT distribution often follows a Poisson point process where the number of MTs in a given region has a Poisson distribution and the number of MTs per cell is not equal across all cells at all times. If provisioning is done such that MT requirements are met with high probability, general operation will often encounter scenarios where resources are unused. This increases effectiveness at the expense of efficiency. When resources are limited, dynamic resource allocation techniques can improve system performance by adapting individual cell capacities according to traffic distribution.

Cell characteristics also change with cell size. Increasing cell density implies additional overhead for coordination and additional infrastructure to connect dense APs to an access network. Price per AP and AP placement also varies greatly from large macrocells to RFSCs. While broad macrocells have traditionally been centrally deployed and controlled, RFSCs are generally owned by local entities and the distribution of APs in an environment tends to be ad hoc rather than planned. Since RFSC interference occurs between cells owned by different entities, configuration should be handled dynamically to account for nearby cells; however, as cell size becomes even smaller and the number of cells increases, the provisioning of cells can be handled at the local level in a manner similar to dense network deployments in universities and office environments [14]. The small coverage area of dense DSC networks also implies that highly mobile devices move between cells at a high frequency—increasing the overhead due to network coordination.

The characteristics of a wireless link also depend on the wireless medium. Sections 13.2.2 and 13.2.3 focus on the traits of RF and OW media that make each attractive in different scenarios. Just as multitier RF networks provide the benefits of macrocell coverage with small cell density, RF/VLC HetNets utilize RFSCs to provide indoor coverage while ultradense VLC DSCs increase network density. The directionality of the VLC channel allows the cells to be positioned in a way to increase network density beyond the capabilities of omnidirectional RFSCs while the RFSCs accommodate high mobility MTs and MTs without a reliable VLC connection.

13.2.2 Radio Frequency Small Cells

Although the evolution of decreased cell size has been occurring for decades, the "small cell" concept began in 3G networks with picocells and femtocells. Wi-Fi WLANs also began to grow in popularity in the early 2000s. It is now common to find a multitude of Wi-Fi WLANs in urban environments; dozens can be found in apartment complexes, and many businesses now provide wireless access. Service providers also deploy Wi-Fi hotspots in order to offload traffic from overcrowded macrocells. When analyzing RFSC networks, the various types of RFSCs as well as the RF channel traits and performance metrics should be evaluated for fair comparison. This evaluation is particularly important in RF/VLC HetNets where fair comparison must be made across connections with different characteristics.

Since the standardization of the first IEEE 802.11 specification in 1999, Wi-Fi has become a nearly ubiquitous means of indoor wireless network access. The unlicensed use of Industrial, Scientific and Medical (ISM) radio bands and contention-based multiple access techniques have made it a favorable technology for the home and office. Due to the availability of Wi-Fi WLANs, techniques to offload cellular data traffic to Wi-Fi became an important component in meeting 4G demands. The available Wi-Fi infrastructure offered a means for network densification without the expensive overhead of a new access network for the wireless APs. New standards have drastically increased the achievable Wi-Fi throughput via increased spectrum usage, high order modulation, and MIMO; however, adoption of the new standards requires backwards compatibility. Accommodating a mix of clients and protocols along with interference from other devices in the ISM bands leads to performance below the specified peak rates. Femtocells are a form of RFSC that utilize licensed frequency bands and coordinated multiple access techniques. While femtocells and Wi-Fi WLANs provide similar coverage area and utilize available infrastructure for connectivity to the access network, femtocells operate in the same licensed band as the broader macrocells and require resource allocation in order to mitigate interference. Since the frequency bands in use by femtocells are licensed, interference can be coordinated by the mobile operator; however, femtocells are purchased by local entities (e.g., home or business owners) and deployed in a way that is not coordinated with the broader cells. Therefore, dynamic resource allocation techniques are used to improve aggregate performance by mitigating interference between femtocells as well as between the femtocells and higher tiers in the cellular network.

The RF link is characterized by the received electrical power of the signal and aggregate disturbance (i.e., noise and interference). Given that AP transmit powers are known, path loss between the APs and the specific MT is used to evaluate the received signal and interference powers. In its simplest form, the RF path is modeled by:

$$\frac{P_r}{P_t} = G_t G_r \left(\frac{\lambda}{4\pi d}\right)^2 = G_t G_r \left(\frac{c}{4\pi f d}\right)^2 \qquad (13.13)$$

where P_t and P_r are the electrical transmit and receive powers, G_t and G_r are the antenna gains, λ and f are the wavelength and frequency of the RF carrier, d is the distance between antennas, and c is the speed of light. Path loss is the inverse of the transmission equation and can be written in dB as:

$$\begin{aligned} L_{dB} &= 20\log_{10}(d) - 20\log_{10}(\lambda) + 21.98\text{dB} - 10\log_{10}(G_t G_r) \\ &= 20\log_{10}(d) + 20\log_{10}(f) - 147.56\text{dB} - 10\log_{10}(G_t G_r) \end{aligned} \qquad (13.14)$$

Note that practical propagation models have an additional environment-specific loss. In indoor environments, this represents attenuation from walls and obstructions. It is dependent on the number of floors and walls that the signal passes through as well as the type of material they are made of. Antenna gain is the gain in the direction of the signal with respect to an isotropic emitter or receiver. While the antenna gain for an ideal isotropic antenna is $G = 1$, practical antennas do not radiate in a uniform pattern in all directions. RF gain is conventionally defined as the ratio of the power produced on the antenna's beam axis to that of the hypothetical isotropic source. Commonly used dipole antennas produce an omnidirectional radiation pattern across the horizontal plane and, accordingly, antenna gain is constant in all directions when the AP is horizontally aligned with the MTs [15].

The signal (i.e., voltage or current) in wireline or RF communications is subject to a power constraint; therefore, SNR is defined in terms of average electrical power in order for fair comparison among various scenarios:

$$SNR_{RF} = \frac{P_{e,SIG}}{P_{e,n}} = \frac{\sigma_{e,SIG}^2}{\sigma_{e,n}^2} \qquad (13.15)$$

In the above definition, $P_{e,SIG}$ [W] is the average received electrical signal power, $P_{e,n}$ [W] is the electrical noise power, and the variances of the signal and noise are $\sigma_{e,SIG}^2$ and $\sigma_{e,n}^2$ [A^2 or V^2], respectively. The second equality holds in this case since the signal and noise are both zero-mean and assumed across an equivalent resistance. This also implies that the signal constraint can be defined as $\sigma_e^2 = \mathrm{E}[x_e^2(t)] \leq C_e$, where x_e (t) is the electrical signal and C_e is proportional to the maximum electrical signal power. In addition, multiple additive white Gaussian noise (AWGN) components such as shot and thermal noise can be included in the SNR definition since total noise variance is the sum of individual noise variances (i.e., $\sigma_{e,n}^2 = \sigma_{shot}^2 + \sigma_{therm}^2$) and the aggregate noise is also Gaussian due to the properties of the Gaussian distribution.

When noise and interference are at the same order of magnitude, SINR is the metric of interest and is defined as:

$$SINR_{RF} = \frac{P_{e,SIG}}{P_{e,I} + P_{e,n}} = \frac{\sigma_{e,SIG}^2}{\sigma_{e,I}^2 + \sigma_{e,n}^2} \quad (13.16)$$

where $P_{e,I}$ [W] is the average electrical power of the aggregate interference and $\sigma_{e,I}^2 >$ [A^2 or V^2] is variance. Assume independent transmitters $P_{e,I}$ and $\sigma_{e,I}^2$ are the respective sums of average electrical interference power and variance across all interferers. When many interferers are present—which is common in RF networks—interference distribution is accurately approximated as Gaussian due to the central limit theorem and the aggregate disturbance is Gaussian with variance $\sigma^2 = \sigma_{e,I}^2 + \sigma_{e,n}^2$. Therefore, $SINR_{RF}$ can be used in theoretical calculations in the same way as SNR_{RF}.

The use of RFSCs in recent years has provided wireless networks with a great degree of additional wireless capacity. RFSCs also offer good coverage and maintain a reasonably reliable connection as MTs move through indoor environments; however, the growing number of RFSCs and the ad hoc nature of their distribution have led to scenarios where the RF spectrum is overly congested and user quality of service (QoS) is below what is desired. Smaller cells can be used to further improve density in these highly congested environments, but omnidirectional RF APs are not well suited for ultradense distribution due to infrastructure constraints.

13.2.3 OW Small Cells

Research in high-speed OW communications originated with infrared (IR) [16,17]; however, the commercial lighting industries' recent trend toward solid-state lighting has led to digitally controlled LED-based luminaires and, accordingly, an interest in dual-use devices offering both illumination and wireless data via VLC [18–20]. Similar to how RFSCs benefit from the available access networks for connecting femtocell and Wi-Fi APs, the distribution of ultradense VLC DSCs has the benefit of utilizing the infrastructure associated with digitally controlled lighting networks that implement either power line communications (PLC) or power over ethernet (PoE) to connect luminaires to the wired network. For the purpose of network densification, directional communications offer the ability to place APs at reasonable distances from the MTs while still providing a small coverage area in the horizontal plane of the working surface. The cone-shape emission of directional communications such as VLC is preferable to omnidirectional emission when distributing dense fixed position APs. In addition, OW signals do not penetrate walls and are relatively contained. Distributed DSCs also add benefit over collocated APs using directional communications to sector the environment since sectors innately become large as the distance from the AP increases, and distributed APs have better coverage due to the higher probability of an available line of sight (LOS) path. Another major difference with DSCs is that the notion of a specific cell's coverage is dependent on device orientation due to the angle-dependent acceptance pattern. Given that multiple distributed VLC APs may be coordinated for use in optical MIMO communications, a cell may also be defined as the area below a set of VLC APs. In this case, cells may be dynamically defined since the set of coordinated APs can vary over time.

Since optical frequencies operate in the terahertz range, OW communications typically implement intensity modulation with direct detection (IM/DD). In this way, signals are transmitted as variations in optical power and the received optical power is directly converted into an electrical current by a photosensor. The OW channel consists of an LOS and multipath component. Assuming a point source emitter and receiver, LOS gain is defined in terms of distance between transmitter and receiver as well as the transmitter and receiver gain functions.

$$\frac{P_{r,o}}{P_{t,o}} = \frac{G_t(\phi_i)G_r(\psi_i)}{d^2} \tag{13.17}$$

Here, gain is angle dependent and ϕ_i and ψ_i are the emittance and acceptance angles. In the simplest form, Lambertian emission with order m is observed at the transmitter and concentrator optics are observed at the receiver:

$$G_t(\phi_i)\frac{m+1}{2\pi}\cos^m(\phi_i) \tag{13.18}$$

$$G_r(\psi_i) = AT_s(\psi_i)g(\psi_i)\cos(\psi_i) \tag{13.19}$$

where $m = -\ln2/\ln(\cos\Phi_{1/2})$ is related to the semiangle at half power, $\Phi_{1/2}$, A is the photodiode area, and $T_s(\psi_i)$ is the filter transmission. Assuming a non-imaging hemispherical concentrator with internal refractive index, n, and field of view (FOV), Ψ_C, the concentrator gain is:

$$g(\psi_i) = \begin{cases} n^2/\sin^2(\Psi_C), & \text{if } 0 \le \psi_i \le \Psi_C \\ 0, & \text{if } \psi_i > \Psi_C \end{cases} \tag{13.20}$$

There are various models and environmental parameters for multipath gain [21,22]; however, the LOS component is typically dominant. Instantaneous received electrical current, $y(t)$ [A], is the summation of the instantaneous received optical power from VLC and ambient light sources passed through a photosensor with responsivity R [A/W] and an AWGN component, $n(t)$ [A], with variance $\sigma_{e,n}^2 [A^2]$. Due to the relationship between current and optical power, the variance of the received electrical signal, $\sigma_{e,rx}^2 [A^2]$ is related to the variance of the transmitted optical signal, $\sigma_{o,tx}^2 [W^2]$, by:

$$\sigma_{e,rx}^2 = R^2 \sigma_{o,rx}^2 = \sigma_{o,tx}^2 \frac{R^2 G_t^2(\phi_i) G_r^2(\psi_i)}{d^4} \tag{13.21}$$

However, the average optical power does not equate to the variance in the same way as average electrical power. Consider a pulse-amplitude-modulated (PAM) signal with M equally weighted and equidistant symbols in the range 0 to $2P_o$ such that P_o is the average received optical power. The variance of the received electrical current is:

$$\sigma_{MPAM}^2 = \frac{M+1}{3(M-1)} (RP_o)^2 \tag{13.22}$$

This shows that the relationship between σ_{MPAM}^2 and average optical power is dependent on the modulation order and proves that the relationship between variance and average is modulation dependent. Note that $\sigma_{MPAM}^2 = (RP_o)^2$ in the case of on-off keying (OOK) where $M = 2$. When analyzing a variable pulse position modulated (VPPM) signal, the idea that signal variance is bounded by $(RP_o)^2$ is disproved. Given a VPPM signal with duty cycle α and peak received optical amplitude $P_{o,max}$, the average optical power is $P_o = \alpha P_{o,max}$ and the received electrical signal has variance:

$$\sigma_{VPPM}^2 = \alpha(1-\alpha)(RP_{o,max})^2 = \frac{1-\alpha}{\alpha}(RP_o)^2 \tag{13.23}$$

Since a specified P_o without a peak constraint can be achieved using any α in the range $0 \le \alpha \le 2$ 1, the relationship $\sigma_{VPPM}^2/(RP_o)^2$ can range from 0 to ∞. Therefore, (RP_o) alone cannot bound the variance of the interfering signal.

Maximum average optical power in IR links is regulated and the lighting requirement in VLC systems specifies average optical power. Therefore, the OW channel is subject to an average optical power constraint, $E[x_o(t)] \le C_o$, where $x_o(t)$ is the optical signal and C_o is the maximum average optical signal power. Given this constraint, OW SNR is defined as [17]:

$$SNR_{OW} = \frac{(RP_{o,SIG})^2}{P_{e,n}} = \frac{E[y(t)]^2}{\sigma_{e,n}^2} \tag{13.24}$$

In this way, fair comparison can be made with a fixed average optical signal power, $P_{o,SIG}$. When defining SINR conventionally, interfering signals are parameterized in the same manner as the true signal. This implies an OW SINR definition where interference is modeled by $(RP_{o,I})^2$, and $P_{o,I}$ is the average optical interference power. This model is occasionally observed in literature [23], but the denominator in this case does not accurately model variance of the aggregate disturbance in all cases due to the dependence on modulation. If the modulation schemes of interferers are known, variance can be directly evaluated and SINR can be defined as:

$$SINR_{OW} = \frac{(RP_{o,SIG})^2}{\sigma_{e,I}^2 + \sigma_{e,n}^2} = \frac{E[y(t)]^2}{\sigma_{e,I}^2 + \sigma_{e,n}^2} \tag{13.25}$$

If both peak and average are known but modulation is ambiguous, interference variance can be upper bounded using the Bhatia–Davis inequality [24,25]. Accordingly, SINR can be bounded from below using the variance upper bound and above by the SNR defined in Equation 13.24.

SNR and SINR of an OW link are sometimes defined in terms of electrical power where the electrical power of the OW signal, $(R\sigma_{o,rx})^2$, relates to the signal's AC component (i.e., variance) [26]. While this does not relate to average optical power constraints, it allows the link performance to be analyzed in terms of the more conventional metric and related to performance of an RF link. When interference is modeled in this manner, SINR is defined as:

$$SINR'_{OW} = \frac{(R\sigma_{o,rx,SIG})^2}{\sum_i (R\sigma_{o,rx,I,i})^2 + P_{e,n}} = \frac{\sigma_{e,SIG}^2}{\sigma_{e,I}^2 + \sigma_{e,n}^2} \tag{13.26}$$

In this definition, $(R\sigma_{o,rx,I,i})^2$ is the received AC electrical power from the ith optical interferer and the denominator is the summation of the variances of electrical current from various interferers and noise components. This definition has been used to model performance of systems implementing optical orthogonal frequency division multiplexing (OFDM) and is related to Shannon capacity as long as signal variance is the constraint applied to the link and interference is Gaussian.

The assumption of Gaussian-distributed interference is not always accurate in the presence of dominant interferers [27]. In OW links, channel directionality leads to scenarios where a small number of transmitters have an LOS path. When the true signal and a single interferer fall within the FOV of the receiver, the interference distribution follows that of the interfering signal. In these scenarios, error rate can be defined in terms of SNR and interference-to-noise ratio (INR) or signal-to-interference ratio (SIR) of the interferer. Here, interference is modeled observing average optical power.

$$INR_{OW} = \frac{(RP_{o,I})^2}{\sigma_{e,n}^2} \tag{13.27}$$

$$SIR_{OW} = \frac{(RP_{o,SIG})^2}{(RP_{o,I})^2} = \frac{P_{o,SIG}^2}{P_{o,I}^2} \tag{13.28}$$

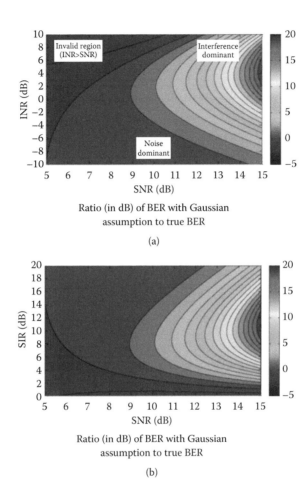

Ratio (in dB) of BER with Gaussian
assumption to true BER

(a)

Ratio (in dB) of BER with Gaussian
assumption to true BER

(b)

FIGURE 13.4
Comparison of bit error rate (BER) calculated with the Gaussian assumption and true BER for an
OOK signal with interference from a single other OOK source. The ratio of the calculations is
shown as a function of (a) SNR and INR as well as (b) SNR and SIR.

As an example, consider an OOK signal in the presence of a single interferer also implementing OOK. The ratio of the theoretical BER calculated
with SINR assuming Gaussian interference to the true BER derived from
modeling four equiprobable received values, 0, $2P_I$, $2P_{SIG}$ and $2P_{SIG} + 2P_I$,
in an AWGN channel is shown in Figure 13.4. In the latter case, the decision
point is placed at the average, $P_{SIG} + P_I$.

$$BER_{Gaus} = Q(\sqrt{SINR}) \qquad (13.29)$$

$$BER_{True} = \frac{1}{2}Q(\sqrt{SNR} + \sqrt{INR}) + \frac{1}{2}Q(\sqrt{SNR} - \sqrt{INR}) \qquad (13.30)$$

Given that $\sigma_{OOK}^2 = (RP_o)^2$ from Equation 13.22, SINR for this scenario can be defined in terms of SNR and either INR or SIR.

$$SINR_{OOK} = SNR\left(\frac{\sigma_{e,n}^2}{\sigma_I^2 + \sigma_{e,n}^2}\right) = SNR\left(\frac{1}{INR+1}\right) = \left(\frac{SNR \cdot SIR}{SNR + SIR}\right) \qquad (13.31)$$

Figure 13.4 shows the ratios of BER calculated with Equations 13.29 and 13.30 for a variety of SNR, INR, and SIR settings. The model for interference becomes irrelevant as noise becomes the dominant source of disturbance (i.e., as INR goes toward 0). When SNR is low, the error rate calculations are also similar since error rates are both very high. The key observation is that the theoretical BER calculated under the assumption of Gaussian interference can be off by multiple orders of magnitude when interference is the dominant source of disturbance. As an example, Equations 13.29 and 13.30 evaluate to $BER_{Gaus} \approx 3 \cdot 10^{-3}$ and $BER_{True} \approx 3 \cdot 10^{-5}$ at $INR = 5$ dB and $SNR = 15$ dB. When interference consists of solely multipath components of the interfering signals, the channel dispersion and combination of more interfering signals can generate an aggregate interference that is near Gaussian and the BER_{Gaus} model becomes more accurate.

The directionality of VLC allows for ultradense distributions of cells and increased aggregate network capacity, but the directionality of VLC links implies that system performance mapping is not dependent solely on location as is typically assumed with RF networks. Accordingly, characteristics of the MTs orientation should also be considered when evaluating performance. In addition, the use of visible light also implies constraints that limit the practicality of stand-alone VLC systems. Besides the fact that RF-based solutions already dominate the wireless communications market, bidirectional VLC links are challenged by issues relating to intrusive uplink and susceptibility to blocking conditions. In addition, the nature of smaller cells that allows for increased network density also implies challenges relating to maintaining connectivity for highly mobile devices. Considering the constraints and limitations described in Sections 13.2.2 and 13.2.3, VLC is ideally suited as a supplemental medium for opportunistically offloading high-rate downlink traffic from the congested RF medium. The different constraints and interference characteristics of the RF and OW networks imply that the individual RF or OW SNR and SINR metrics do not always provide a fair comparison when deciding between available connections or when comparing performance of various scenarios.

13.3 RF/VLC Heterogeneous Networks

In the broad view of 5G, VLC is envisioned within multitier HetNets where MTs are intelligently distributed among cells of various coverage areas and access technologies [6,28,29]. This is an extension of the model used in 4G where traffic is offloaded from macrocells to RFSCs. Multitier HetNets provide reliable coverage from larger cells and utilize densely distributed smaller cells to add wireless capacity in the areas with the highest traffic requirements. The additional offloading to VLC cells is primarily concerned with accommodating high-rate downlink traffic whenever possible. Given the characteristics of RFSCs and VLC DSCs, indoor environments consisting of many MTs with a variety of use cases, traffic patterns, and mobility states are best accommodated by HetNets where highly mobile devices can utilize the broader coverage of the RFSCs while MTs in a static state can improve aggregate performance by offloading their traffic to the ultradense VLC DSCs. In addition, the intrusive uplink and susceptibility to blocking of the VLC link can be mitigated with the available RF channel as an uplink medium in an asymmetric configuration and as an alternative when the VLC link is occluded. The major components in development of the RF/VLC HetNet concept consist of the system layout, design of the VLC connection, and the network implementation where traffic is dynamically distributed among the RF and VLC cells.

13.3.1 System Model

The envisioned RF/VLC HetNet, shown in Figure 13.5, consists of a central RF AP, one or more VLC APs, various MTs, a router, and a gateway to external networks [30]. MTs are ideally assigned to the AP with minimum footprint in order to mitigate interference and maximize aggregate wireless throughput in static scenarios. However, there is typically a trade-off between bandwidth density and coverage area or mobility. As with traffic offloading

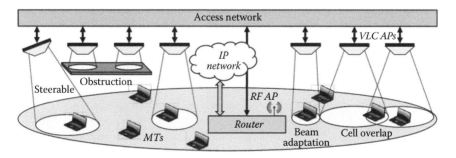

FIGURE 13.5
System model for the RF/VLC HetNet.

to RFSCs, the objective is to increase aggregate throughput via smaller cells while the larger cell provides coverage and reliability. MTs with unreliable VLC signals due to shadowing or movement between cells are better suited for the RFSC. High data rate traffic destined for MTs with a reliable VLC connection can be offloaded to the highly localized VLC DSCs, removing congestion from the RFSC and nearby cells using the same RF band. VLC links at the edge of the RF cell also provide an alternative link where RF SINR and QoS are low.

13.3.1.1   RF Provisioning

RFSC provisioning relates to both the AP placement and the type of RFSC in use. Generally, RFSC coverage is maximized when the AP is centrally located in the environment. The channel quality at any location in the environment is determined by the distance from the AP, the number and type of obstructions between the AP and MT, and the amount of interference from other RF APs. As discussed earlier, the characteristics of the interference depend on the type of cell. In a femtocell, resources are allocated to each MT and interference exists from other femtocells and MTs in the area as well as potential interference from macrocells and MTs associated with the macrocell. Resources are intelligently allocated to mitigate interference, although assigning resources requires overhead and unused assigned resources reduce aggregate system throughput. The carrier sense multiple access (CSMA) technique used by Wi-Fi WLANs allows devices to only reserve the channel when transmission is required, but there is no guarantee on the latency. These factors play an important role when developing handshaking protocols for VLC connections where the RF channel is utilized for uplink.

13.3.1.2   VLC Provisioning

When provisioning VLC cells, interfering APs are typically owned and deployed by the same entity; therefore, AP layout and allocation of resources can be locally planned. Additionally, VLC devices in the same system can be centrally coordinated for dynamic reconfiguration and dynamic resource allocation. In dual-use scenarios, environment-specific lighting levels and uniform illumination is often required to meet lighting specifications. This implies that the provisioning of luminaires should be optimized for communications under the constraints of the lighting system. Parameters relating to VLC provisioning include the AP position and orientation as well as the luminaires' emission pattern and the range of optical signal power.

Modern lighting systems are often deployed in a grid fashion; the distribution of VLC APs in a hexagonal lattice structure has also been explored [31]. The latter relates to the traditional RF macrocell model and provides potential to improve the distribution of SINR in an environment. The emission pattern affects performance since wide-emission luminaires generate better illumination uniformity while narrow emission provides better separation of cells.

Since interference is related to the optical signal, the signal range of VLC APs can also be provisioned such that the output of the luminaire consists of a DC optical power and the VLC signal. Varying the ratio alters the aggregate system performance since increasing signal improves performance of MTs associated with the AP while increasing the non-signal DC component improves performance of other MTs by reducing interference [32].

Design of the receiver's optical front-end also affects performance. In particular, receivers with a very wide FOV have high likelihood of achieving a LOS signal; however, it is also likely that an LOS path exists to interfering APs. On the contrary, narrow FOV receivers have a lower likelihood of achieving a LOS path to multiple APs—mitigating scenarios where interference stems from a dominant source. However, narrow FOV receivers also have a higher likelihood of being positioned such that no LOS path exists and VLC must implement diffuse communications. Spatial diversity techniques such as optical MIMO receivers and diversity selection receivers utilize multiple photosensors to improve the visibility of the receiver while minimizing scenarios where multiple signals land on the same sensor.

Beyond the physical provisioning of APs, resources may also be distributed in order to mitigate interference. This includes various time, frequency, or code division multiple access techniques where overlapping cells utilize different resources. Increasing the reuse factor in an environment adds more separation between interfering APs, but this also decreases the resources available per cell. In order to best accommodate environments where traffic patterns vary over time, dynamic allocation allows resources to be provisioned in relation to the current traffic demands.

13.3.1.3 Access Network

In the RF/VLC HetNet, VLC APs require network connectivity in order to relay data to associated MTs; therefore, data packets must flow between the central RF AP and each of the VLC APs. The access network allows data traffic and additional overhead to flow throughout the system. There are various options for implementing the access network in regards to both the physical channel and the network topology.

As the lighting industry has begun to move toward intelligent systems and dynamic control of individual luminaires, there have been various implementations of the physical channel providing connectivity between devices. Currently, controllable luminaires are often connected with copper wire (e.g., DALI, DMX) or RF mesh networks (e.g., ZigBee). These techniques provide low data-rate throughput—in the order of 100 kbps—which is appropriate for control, although they are not intended for high data rate traffic. Two technologies that provide promise for high throughput are PLC and ethernet, specifically PoE. PLC and PoE provide communication and power, minimizing installation overhead. In-home PLC is capable of operating in the order of 100 Mbps, and PoE can be utilized with gigabit ethernet links.

The access networks topology defines how traffic can flow in the network. A selection of potential topologies is shown in Figure 13.6. PLC is a bus topology since all devices utilize the power line as a transmission medium. PoE topologies are limited by PoE power transmission capability since power constraints limit the number of luminaires that can be powered by a single PoE connection. Therefore, PoE is likely implemented in a star topology, powering a single AP per connection. Other wired connections that are independent of the power requirement may implement tree, line, ring, or mesh topologies. If the channel used for backhaul connectivity is shared between multiple VLC APs, as in the bus topology, it can become a system bottleneck when multiple APs are active. This is also the case when a link is the only path connecting the central AP to a subset of VLC APs, as in the tree topology, since all traffic to the subset will need to be routed through the link. The system does not need to operate under the requirement that all VLC APs are capable of operating at full capacity simultaneously. For example, assume a system where VLC APs are either unused or at maximum capacity of X b/s with $P(\text{max}) = 0.5$. If four VLC APs are connected by a bus with capacity $3X$ b/s, the access network is a bottleneck when all VLC APs are in use. Since this occurs $100 \cdot P(\text{max}) = 6.25\%$ of the time, requirements are satisfied 93.75% of the time. If $P(\text{max}) = 0.2$, the access network satisfies requirements 99.84% of the time. Beyond a certain point, additional access network capacity provides minimal improvement in the probability of being overloaded.

13.3.2 Device Connectivity

In the system shown in Figure 13.5, an MT may either connect to the IP network through the RF connections or with one of the VLC connections. When evaluating the VLC connection, downlink capacity is commonly analyzed in system evaluation, but practical implementation requires that MTs are able to achieve a bidirectional network connection. When defining the VLC connection, it is important to know the logical topology (i.e., flow of network traffic), error correction, and handshaking protocols used to

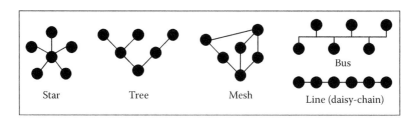

FIGURE 13.6
Potential access network topologies.

mitigate network retransmissions due to packet loss, and dynamics of the VLC channel.

13.3.2.1   Topologies

The potential VLC connection topologies depend on physical capabilities of the VLC AP and MT. Figure 13.7 depicts three topologies broken into categories relating to network traffic flow [33]. A symmetric VLC connection routes all traffic through the VLC AP. An asymmetric VLC connection routes downlink traffic through the VLC AP and uplink traffic through the RF AP. Depending on the uplink medium, symmetric connections can be further divided among scenarios where the uplink utilizes the same channel as the RF connection and scenarios where the uplink is noninterfering. Figure 13.7 depicts these scenarios, each with two MTs accessing individual VLC APs with a VLC connection and a third MT using the RF connection.

13.3.2.1.1   Symmetric Noninterfering

This scenario occurs when RF and VLC connections are independent. It can be further divided into cases where the physical link of the VLC connection is symmetric (i.e., uplink uses VLC) or asymmetric (i.e., uplink is non-VLC and does not interfere with the RF connection). In the prior case, VLC uplink requires an intrusive visible signal from the MT. In the latter, uplink may be implemented with IR or an RF band not used by the RF connection. Interference can occur between the uplink of various VLC connections and resources must be assigned accordingly. As an example, when a femtocell is used for the RF connection, a subset of the resources can be allocated to VLC APs for use in the uplink of VLC connections.

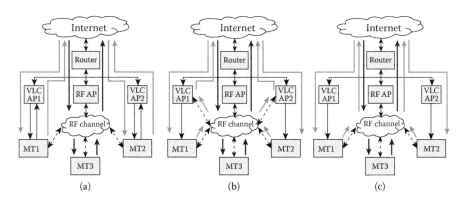

(a)   (b)   (c)

FIGURE 13.7
Traffic flow models for VLC link topologies implementing VLC connectivity through (a) symmetric noninterfering routes, (b) symmetric routes with uplink interference and (c) asymmetric routes.

13.3.2.1.2 Symmetric with Interference

In this scenario, the physical link between a VLC AP and MT is asymmetric and the uplink interferes with RF connections. The RF channel is shared and contention occurs between downlink traffic of all RF connections and uplink traffic from all MTs. The primary example of this scenario is the use of a Wi-Fi WLAN for the RF connection and use of Wi-Fi for uplink between MTs and VLC APs in VLC connections.

13.3.2.1.3 Asymmetric

The asymmetric implementation routes downlink traffic and uplink traffic in different directions. Specifically, VLC APs provide simplex links to MTs with no physical link from MTs to VLC APs. If the RF connection is implemented with a contention-based WLAN, performance gains relative to a strictly RF network come from the offloaded downlink traffic as well as the improved uplink performance from the MTs since they have less contention for the RF channel [34]. In the case of a femtocell RF connection, RF resources can be allocated to the various VLC-enabled MTs for use in the uplink. System gains also come from the offloaded traffic since the femtocell resources required per MT reduce from resource for uplink and downlink to strictly resources for the uplink.

13.3.2.2 Handshaking

Given that wireless links are innately more error-prone than wired links, handshaking is often implemented between two nodes with a wireless connection in order to reduce the network retransmissions required at higher layers. As an example, handshaking protocols are implemented in Wi-Fi systems where packetized data are sent and acknowledged in order to confirm that the destination has received the packet before it is discarded at the transmitting device. If the acknowledgment is not received, the transmitting device can resend the packet so network layer retransmissions are not required. Since the VLC channel is also prone to dynamic variations and potential error, handshaking between the VLC AP and the MT is required to avoid packet loss on the VLC link. In the symmetric noninterfering topology, handshaking at the data link layer can be implemented between VLC APs and MTs in order to minimize network retransmissions. In the symmetric topology utilizing the RF channel for uplink, the RF channel is not reserved during the VLC transmission and the uplink does not have a guaranteed response time; therefore, acknowledgments may be delayed. The asymmetric topology only implements a simplex link between the VLC AP and the MT; therefore, a point-to-point handshaking protocol is not possible, and accurate data delivery relies on handshaking at the higher layers. It is possible to route acknowledgments through the local network but this would imply a cross layer implementation and adds complexity to the system design.

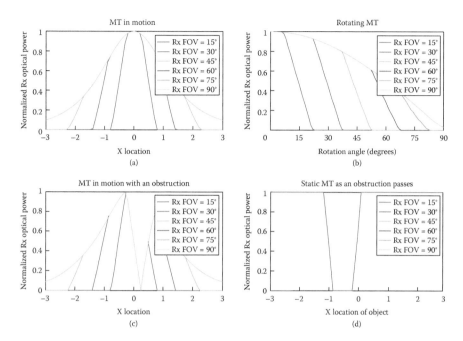

FIGURE 13.8
Comparison of VLC signal loss conditions for various receiver FOVs showing (a) an MT moving below a VLC AP, (b) a fixed position MT rotating away from a VLC AP, (c) an MT in motion with an obstruction in the LOS from the VLC AP to the MT, and (d) a static MT with an obstruction that passes through the LOS path from the VLC AP to the MT.

13.3.2.3 Dynamics

While static environments benefit from VLC cell density, temporal effects in dynamic environments must be understood in order to characterize the true performance of the VLC link. In VLC channels, the primary signal loss conditions stem from physical movement out of range, and occlusions. Assuming static VLC APs, the former relates to movement of the MT—including movement through the environment or device rotation—and the latter may be due to MT movement or movement of other objects. Figure 13.8 depicts the various dynamic changes in received signal at a receiver when the VLC AP is centered at (0,0) and modeled by a 26 × 26 grid of point source emitters each with Lambertian order 1 and a separation of 0.02 m in the X and Y directions.

The LOS path loss models defined in Equations 13.17 through 13.20 are used to calculate received optical power. The top left image shows the normalized received signal for various locations 2 m below the AP. The top right shows signal when a static MT is rotated. The lower left shows signal when an MT moves past a static object, and the lower right shows signal when an object moves past a static MT.

The signal loss due to movement is more gradual than signal loss due to occlusions. This is because the loss of the LOS path occurs abruptly when an object passes between the VLC AP and the MT. When an MT moves out of range, the emission and acceptance angles as well as the distance change gradually until the VLC AP is no longer in the FOV of the MT. In this case, the signal loss can be predicted as the signal begins to degrade. The response to motion has an additional factor of receiver FOV since the concentrator model abruptly loses signal once the acceptance angle becomes greater than the FOV. This loss is similar to the occlusion, and it is preceded by gradual loss. In addition, signal is more likely to return quickly when loss is due to an obstruction than it is when loss is due to movement out of range.

For either of these signal loss conditions, the expected channel quality relates to the frequency and duration of signal outage. Frequency of the outage is defined as the rate at which signal loss conditions occur, and duration is defined as the length of time that the signal is lost. In the case of signal loss due to movement out of range, predictive modeling of the MTs motion can be used to estimate frequency and duration. In the case of occlusions, knowledge of previous occlusions can be used to model expected frequency and duration of future occlusions. Occlusions are often more likely to have a short duration as the occluding object moves through the LOS path between VLC AP and MT. Out-of-range loss conditions are likely to have longer duration since signal loss stems from movement or rotation away from the VLC AP, and the signal from the specific AP will typically require a change in the motion of the MT.

13.3.3 HetNet Implementation

The network implementation of an RF/VLC HetNet implies distribution of network traffic among various wireless connections including distribution between VLC cells as well as between the optical and RF channels. Practical indoor environments consist of a variety of MTs, each with their own time-varying traffic requirements and mobility states. From a user-centric view, the diverse channel options in HetNets provide MTs with the ability to utilize the channel that best fits their current status. From a network-centric view, HetNets provide the diversity to distribute MTs in a way that optimizes aggregate system performance. This typically implies allocation of resources (RF and OW) such that aggregate capacity is maximized under the constraint that individual MTs' minimum requirements are satisfied.

In a dynamic environment, an MT connects to various APs over time. Accordingly, traffic flow must be rerouted when the connection changes. The process of rerouting traffic is called handover. When switching between two APs of the same type, as in transferring from one VLC link to another, a horizontal handover (HHO) occurs. When switching between APs of different types, as in a transfer from Wi-Fi to VLC, a vertical handover (VHO) occurs.

Figure 13.9 shows a general view of an MT handover flow graph for the RF/VLC HetNet, incorporating a single RF link and multiple VLC links. In this model, the MT first accesses the network via the RF connection. It then determines if any VLC links are available by observing the VLC channel while VLC APs send intermittent beacons with unique identifiers. Downlink traffic is rerouted through the appropriate VLC AP if the MT discovers an available link and the assessment determines that handover should be initiated. Once the MT is associated with the VLC AP, channel monitoring continues to observe the quality of the associated VLC link while also discovering other VLC links. Assessment is continually done to determine if the current connection is optimal or whether a switch should be made to another VLC link via HHO or to the RF link via VHO. A subset of this model includes a model where MTs avoid HHO and utilize the RF connection when moving between VLC cells. In the handover process, the MT and network coordinate handover assessment and implementation. The network also analyzes the aggregate system to determine if network performance can be improved via AP reconfiguration. This includes allocation of resources, changes in signal power, or physical orientation.

13.3.3.1 Handover Assessment

Once it has been determined that multiple connections are available, the MT must decide whether a handover should be performed. This assessment can either be done (a) in a distributed user-controlled manner where each MT makes a decision based on the knowledge of the available connections, (b) in a centralized network-controlled manner where the network maintains knowledge of potential connections for the various MTs and distributes MTs according to traffic patterns and assumed connection quality, or (c) in a mobile-assisted manner where MTs relay knowledge of their current status to the network for use in handover assessment such that traffic is distributed in a more optimal way.

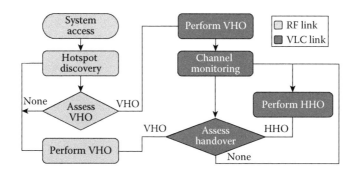

FIGURE 13.9
User-centric handover flow chart.

In each scenario, a utility function evaluates the available connections to decide if an alternative connection is better than the current. The utility function consists of a set of parameters for the network, $p1$ through p_n, and a set of weights for the network or the specific MT, $\omega 1$ through ω_n.

$$U = f\ (\omega_1, p_1, \omega_2, p_2, \omega_n, p_n) \tag{13.32}$$

In the case of user-controlled assessment, MT-centric parameters such as signal strength, channel reliability, and MT power consumption are observed. SINR is an important metric for determining the quality of a specific link; however, it should be noted that the differences in SINR definitions must be accounted for when comparing the SINR of an RF link to that of a VLC link. Temporal properties are also observed by the MT. For the VLC connection, this includes frequency and duration of occlusions as well as device mobility. In the case of network-controlled assessment, network-centric parameters such as channel usage and bandwidth density are considered. In RF/VLC HetNets, network utility has a preference toward VLC links due to the higher bandwidth density. Mobile-assisted assessment observes parameters from both sets.

13.3.3.2 Handover Implementation

Handover implementation defines the method of rerouting traffic once the network or MT makes a decision that another connection is more optimal than the current. Immediately switching to the connection with the highest utility can lead to a ping-pong effect where multiple handovers occur while transitioning between channels; therefore, the implementation can utilize various checks to make sure the new connection is stable. This can include a hysteresis margin or an absolute threshold that the utility of the current channel must drop below such that the utility of a new connection must be greater than the utility of the current connection plus the margin. It can also include a temporal requirement that the new connection must be better than the current for a defined period of time.

The temporal requirement is particularly important when implementing a VHO from a VLC connection to the RF connection since signal loss conditions where an object obstructs the LOS path are short in duration. Due to overhead from handover, it is sometimes preferable to wait for a channel to return rather than switching immediately. Given the conditions shown in Figure 13.8, the optimal type of VHO implementation may differ. An immediate handover occurs as soon as the primary signal is lost, whereas a delayed handover dwells for a specified time to see if the channel returns before initiating the handover. If an out-of-range signal loss occurs, the signal is usually lost for an extended period—implying that the handover should be made immediately. When a blocking condition occurs, it is likely that something is passing through the LOS path and will return soon—implying that the device should delay before handover initiation in the likelihood that the signal will return [35]. As an example, consider a system with an R_vb/s VLC link, R_wb/s Wi-Fi link, X second handover delay, and Y second VLC outage time. After T seconds:

$$D = R_v(T - Y) \tag{13.33}$$

$$I = R_w(Y - X) + R_v(T - Y - X) \tag{13.34}$$

where D is the data transferred to an MT that waited for the VLC signal to return and I is the data transferred to an MT that immediately implemented the handover when the VLC signal was lost and when it returned. The immediate handover performs better when $Y > X((R_v + R_w)/R_w)$, and delayed handover is optimal when $Y < X((R_v + R_w)/R_w)$. Since the system does not know Y a priori, predictive techniques can observe past tendencies, MT motion, or rate of signal loss in order to increase the probability that an appropriate decision is made [36].

Once the network or MT determines that a handover should be initiated, both ends need to coordinate the handover. In a simple case, the router updates where incoming traffic is routed and the MT changes its expectation of where downlink traffic is coming from. If an MT is switching to a resource-allocated channel that is in use by multiple MTs, coordination includes definition of the allocated resources for the MT. For example, if an MT is joining a VLC AP using orthogonal frequency division multiple access, the MT must know which frequency bins to observe.

13.3.3.3 Dynamic Reconfiguration

In order to manage the wide range of traffic scenarios that can occur in hyperdense networks, the network itself can also be adapted to best meet the real-time demands of the environment. Similar to how macrocell networks redistribute resources in order to accommodate varying peaks in traffic during daytime and evening hours, resources in the RF/VLC HetNet can be redistributed to meet time-varying demands within indoor environments. Performance of individual MTs can also be improved via data aggregation techniques where specific MTs are allocated resources from multiple VLC channels—as with spatial MIMO techniques—or by allowing the MT to utilize the RF and VLC connections simultaneously. In addition to the dynamic resource allocation among cells, VLC channels can also be altered through dynamic variations in the network's physical structure. Each of these options requires flexibility at both the higher and lower layers of the communications stack which can be obtained via software defined systems.

13.4 Software Defined VLC

SDR has proven to be an effective and practical tool in RF communications, allowing flexible and rapid exploration of dynamic RF signal processing techniques while accelerating the advancement of configurable RF antennas and front-end hardware. Multiple software toolkits now exist for SDR implementation, including GNU Radio, MATLAB®/Simulink®, and LabVIEW. Low-cost SDR hardware platforms are also broadly available for RF—the most common of which is the Universal Software Radio Peripheral (USRP™) from Ettus Research. The SDR concept can also be adapted to other physical media such as VLC. An SDVLC solution implements an optical front-end to adapt SDR platforms to the constraints of an OW channel using the visible spectrum [37].

SDVLC platforms allow the VLC connection to be dynamically modified in order to meet requirements of a dual-use system, providing both data communications and illumination. The platform also enables concurrent development of signal processing techniques and front-end hardware within an integrated testbed. This modularity, along with the ability to quickly bring up an OW system and implement new test scenarios, makes SDVLC a powerful concept for facilitation of research and experimentation with VLC. The implementation of an SDVLC system requires physical layer development of modulation techniques and the VLC front-end, development of the asymmetric connection using VLC downlink and RF uplink, and network layer development of the dynamic routing techniques for use of the VLC connection as one of many options within a system of multiple MTs and a variety of APs. Figure 13.10 shows an SDVLC connection and network traffic flow. Uplink traffic is routed over Wi-Fi, and downlink traffic is routed through an SDVLC simplex link where the VLC AP uses a USRP to transmit signal through a luminaire and the MT receives signal samples from a USRP connected to a photosensor.

13.4.1 SDVLC Physical Layer

The SDVLC physical layer implementation requires adaptation of the transmitter front-end to produce an intensity-modulated visible light signal. It also requires adaptation of the receiver front-end to detect and convert the received optical power signal, and implementation of VLC modulation and coding schemes that meet the constraints of an OW channel. In the SDVLC link shown in Figure 13.10, the physical layer implementation consists of the path from the VLC AP to the MT. The transmit USRP utilizes a low-frequency transmit (LFTX) daughter card and the output is connected to a voltage controlled current source. A bias T is used to add a DC component to the output of the current source in order to bias the signal into the linear range of the luminaires' LED string. This creates a near-linear

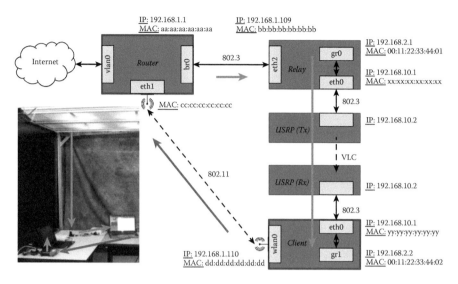

FIGURE 13.10
SDVLC implementation of an asymmetric connection using a Wi-Fi uplink and a simplex VLC downlink.

relationship between the electrical transmit signal of the USRP and the emitted optical signal of the luminaire. At the receiver, an avalanche photodiode (APD) converts the optical signal to an electrical signal which is passed to the receiving USRP utilizing the low-frequency receiver (LFRX) daughter card. At the receiver, the sum of the ambient light and VLC signal should not saturate the APD or else the received signal will be clipped from above.

The modulation and coding of the transmitted signal along with the signal processing at the receiver are implemented in software with one of the various SDR software toolkits discussed above. In the simplest case, available implementations for RF communications can be transmitted over the VLC link using a low-frequency carrier. The benefit that stems from software-defined systems is that the various VLC modulation schemes presented in literature can also be implemented by writing software blocks to fit into the framework of the SDR software toolkit [38]. In addition, the MCS may be dynamically modified based on varying channel conditions or requirements of the lighting system. While this implementation shows the SDVLC hardware and the device running the SDVLC software implemented on separate devices, adapting the concept to platforms where both are handled on the same device is certainly feasible. In addition, tools for implementing SDR signal processing on field programmable gate arrays (FPGAs) are improving, allowing for link testing at higher real-time rates since the FPGA handles much of the intensive processing.

13.4.2 SDVLC Device Connectivity

In order to show practical use of the SDVLC system, a device must be able to receive signal from the VLC AP and also send requests and uplink to the network. In a practical connection, the necessary modifications to connect to a typical network should remain contained within the MT and VLC AP. One technique to implement the asymmetric routing is presented in [39]. If the SDVLC link is defined as a virtual interface, network traffic can be routed through the link. The challenge in the asymmetric connection is to route requests and uplink traffic to the router over the Wi-Fi channel and receive downlink traffic on the virtual interface of the receiver. The key procedures to enable routing to work appropriately are to implement static routing at the router, disable IP packet forwarding and specify the packet relay path at the VLC AP, and implement "operating system spoofing" at the receiver to make sure that higher-layer protocols recognize packets coming into the virtual interface when the request for these packets are sent over the Wi-Fi interface.

In Figure 13.10, this implies that the MT sends requests to the router at 192.168.1.1, and returning downlink traffic into the router that is destined for the MTs virtual interface at 192.168.2.2 should be routed through the VLC AP connected at 192.168.1.109. At the relay, IP packets that arrive at 192.168.1.109 should be sent to the virtual interface at 192.168.2.1. At the MT, traffic is received by the virtual interface at 192.168.2.2. With this asymmetric connectivity, the VLC link can accordingly be used for traditional IP network traffic. This allows for web browsing and network-wide testing in order to see how the addition of VLC can improve performance in practical systems.

13.4.3 SDVLC HetNet Implementation

An SDVLC HetNet implementation utilizes multiple SDVLC links within a system that automates the distribution of network traffic and the selection of traffic routes for multiple MTs. The traffic is distributed among the broad RF connection and the various SDVLC connections. Two options for the physical implementation of such a system are shown in Figure 13.11. The first option scales the single instance from Figure 13.10. This implementation allows each SDVLC link to utilize the processing power of individual relay PCs; however, each channel acts as an independent link. Any coordination between APs in this setup requires attention to the access network's latencies. Network intelligence and control for the VLC APs can be distributed among APs or MTs, centralized within the router, or incorporated within a relay PC.

The second option shown in Figure 13.11 utilizes a single relay PC connected to multiple SDVLC front-ends. This option improves the ability to coordinate signals among various APs—a capability that is particularly important when implementing spatial multiplexing techniques, or spatial modulation schemes where synchronization is needed. In this case, control of the VLC APs can be

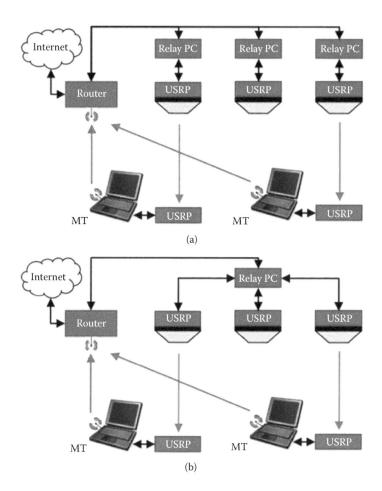

FIGURE 13.11
SDVLC network implementations showing (a) individual APs each with a host processor and (b) multiple VLC luminaires each controlled by a central processor.

centrally coordinated within the relay PC. The difficulty of this option is that the scaling of VLC APs implies a scaling of the signal processing requirements of the relay PC. If its processing is a bottleneck, real-time throughput will be limited accordingly.

13.5 Conclusion

The next generation of wireless communication systems will need to address a rapidly growing demand for capacity and calls for drastic changes in how data are delivered to mobile devices. While peak performance has been the driving force in the standardization of previous wireless networks, 5G demands will motivate an increased focus on the minimum and average performance requirements over a wide range of use cases. In particular, added capacity in indoor environments is needed to satisfy the dense distribution of new devices and traffic requirements of novel applications. Ultradense optical DSCs implementing VLC offer the distribution and directionality to provide additional capacity where it will be needed most.

VLC is still a nascent field in relation to the vast work that has been done in RF communications; however, recent advances show great potential for high data rate communications and densely distributed networks. Use of VLC within the context of multitier and multitechnology HetNets adds the aggregate capacity of VLC DSCs to the wireless infrastructure and provides the reliability of the existing RF network. As a supplemental medium, VLC is able to be utilized in an opportunistic manner where high data rate downlink traffic is offloaded from the congested RF medium. Such an opportunistic approach requires system intelligence and flexibility to recognize and account for the variety of use cases that stem from dynamic environments. The integration of RF and VLC will therefore satisfy the requirements for ubiquitous high-speed network access. Given the novelty of the VLC field, there is a great deal of future research to be done. Research related to VLC systems is an attractive area with many open questions addressing deployment and interaction when multiple VLC connections operate together in a dense environment. Further evaluation of the interaction between these VLC systems and the current and future communications infrastructure is also an important part of practical adoption of VLC.

Acknowledgment

This work was supported primarily by the Engineering Research Centers' Program of the National Science Foundation, under NSF Cooperative Agreement No. EEC-0812056.

References

[1] International Wireless Industry Consortium. *Evolutionary and disruptive visions towards ultra high capacity networks.* White Paper, 2014. Available: http://www.iwpc.org/WhitePapers.aspx

[2] Dahlman, E., Mildh, G., Parkvall, S., Peisa, J., Sachs, J., and Selén, Y. 5G radio access. *Ericsson review,* 6:2–7, 2014.

[3] NGMN Alliance. *Next generation mobile networks*. White Paper, 2015. Available: https://www.ngmn.org/5g-white-paper.html.

[4] C.X. Wang, F. Haider, X. Gao, X.H. You, Y. Yang, D. Yuan, H. Aggoune, H. Haas, S. Fletcher, and E. Hepsaydir. Cellular architecture and key technologies for 5G wireless communication networks. *IEEE Commun. Mag.*, 52(2):122–130, 2014.

[5] M.B. Rahaim and T.D.C. Little. Towards practical integration of VLC within 5G networks. *IEEE Wireless Commun.*, 22(4):97–103, 2015.

[6] S. Wu, H. Wang, and C.-H. Youn. Visible light communications for 5G wireless networking systems: From fixed to mobile communications. *IEEE Network*, 28(6): 41–45, 2014.

[7] V. Chandrasekhar, J.G. Andrews, and A. Gatherer. Femtocell networks: A survey. *IEEE Commun. Mag.*, 46(9):59–67, 2008.

[8] ZTE Grand. *Evolution of microwave radio for modern communication networks*. Available: http://wwwen.zte.com.cn/endata/magazine/ztetechnologies/2012/no5/articles/201209/t20120912_343888.html.

[9] Cisco. *Cisco visual networking index: Forecast and methodology*, 2014–2019. Technical report, Cisco, 2015. Available: http://www.cisco.com/c/en/us/solutions/collateral/service-provider/visual-networking-index-vni/mobile-white-paper-c11-520862.html.

[10] General Electric. *The industrial internet of things*. Available: http://www.ge.com/digital/industrial-internet.

[11] L. Zeng, D. O'brien, H. Minh, G. Faulkner, K. Lee, D. Jung, Y. Oh, and E.T. Won. High data rate multiple input multiple output (MIMO) optical wireless communications using white LED lighting. *IEEE J. Sel. Areas Commun.*, 27(9):1654–1662, 2009.

[12] D. Gesbert, S. Hanly, H. Huang, S. Shitz, O. Simeone, and W. Yu. MultiCell MIMO cooperative networks: A new look at interference. *IEEE J. Sel. Areas Commun.*, 28(9):1380–1408, 2010.

[13] P.M. Butala, H. Elgala, and T.D.C. Little. Performance of optical spatial modulation and spatial multiplexing with imaging receiver. *2014 IEEE Wireless Communications and Networking Conference (WCNC)*, pp. 394–399, April 2014.

[14] J. Florwick, J. Whiteaker, A. Amrod, and J. Woodhams. *Wireless LAN design guide for high density client environments in higher education*. Cisco, White Paper, 2013.

[15] R. Chatterjee. *Antenna theory and practice*. New Age International, New Delhi, India, 1996.

[16] F.R. Gfeller and U. Bapst. Wireless in-house data communication via diffuse infrared radiation. *Proc. IEEE*, 67(11):1474–1486, 1979.

[17] J.M. Kahn and J.R. Barry. Wireless infrared communications. *Proc. IEEE*, 85(2): 265–298, 1997.

[18] T. Komine and M. Nakagawa. Fundamental analysis for visible-light communication system using LED lights. *IEEE Trans. Consum. Electron.*, 50(1):100–107, 2004.

[19] D. O'brien. Visible light communications: Challenges and potential. *2011 IEEE Photonics Conference (PHO)*, pp. 365–366, 2011.

[20] S. Rajagopal, R.D. Roberts, and S.-K. Lim. IEEE 802.15.7 visible light communication: Modulation schemes and dimming support. *IEEE Commun. Mag.*, 50(3): 72–82, 2012.

[21] J.B. Carruthers, S.M. Carroll, and P. Kannan. Propagation modelling for indoor optical wireless communications using fast multi-receiver channel estimation. *IEE Proc. Optoelectronics*, 150(5):473–481, 2003.

[22] M.B. Rahaim, J.B. Carruthers, and T.D.C. Little. Accelerated impulse response calculation for indoor optical communication channels. *2012 IEEE International Conference on Wireless Information Technology and Systems (ICWITS)*, pp. 1–4, 2012.

[23] I. Stefan and H. Haas. Hybrid visible light and radio frequency communication systems. *2014 IEEE Vehicular Technology Conference (VTC Fall)*, pp. 1–5, September 2014.

[24] R. Bhatia and C. Davis. A better bound on the variance. *Am. Math. Mon.*, 107:353–357, 2000.

[25] M. B. Rahaim and T. D. C. Little. Bounding SINR with the constraints of an optical wireless channel. *IEEE International Conference on Communications Workshops (ICC)*, pp. 417–422, Kuala Lumpur, 2016.

[26] B. Ghimire and H. Haas. Self-organising interference coordination in optical wireless networks. *EURASIP J. Wireless Commun. Networking*, 2012:131, 2012.

[27] M.B. Rahaim and T.D.C. Little. Optical interference analysis in visible light communication networks. *IEEE ICC 2015 – First Workshop on Visible Light Communications and Networking (VLCN) (ICC'15 Workshops 24)*, London, UK, June 2015.

[28] M. Ayyash, H. Elgala, A. Khreishah, V. Jungnickel, T.D.C. Little, S. Shao, M.B. Rahaim, D. Schultz, H. Jonas, and F. Ronald. Coexistence of WiFi and LiFi towards 5G: Concepts, opportunities, and challenges. *IEEE Commun. Mag.*, 54(2):64–71, 2016.

[29] Y. Wang and H. Haas. Dynamic load balancing with handover in hybrid Li-Fi and Wi-Fi networks. *J. Lightwave Technol.*, 33(22):4671–4682, 2015.

[30] M.B. Rahaim, A.M. Vegni, and T.D.C. Little. A hybrid radio frequency and broadcast visible light communication system. *2011 IEEE GLOBECOM Workshops (GC Wkshps)*, pp. 792–796, 2011.

[31] J.H. Liu, Q. Li, and X. Zhang. Cellular coverage optimization for indoor visible light communication and illumination networks. *J. Commun.*, 9(11):891–898, 2014.

[32] M.B. Rahaim and T.D.C. Little. SINR analysis and cell zooming with constant illumination for indoor VLC networks. *2013 International Workshop on Optical Wireless Communications (IWOW)*, 2013.

[33] T.D.C. Little and M.B. Rahaim. Network topologies for mixed RF-VLC HetNets. *IEEE Summer Topicals (Invited)*, 2015.

[34] S. Shao, A. Khreishah, M. Ayyash, M.B. Rahaim, H. Elgala, V. Jungnickel, D. Schulz, T.D.C. Little, J. Hilt, and R. Freund. Design and analysis of a visible-light-communication enhanced WiFi system. *J. Opt. Commun. Networking*, 7(10): 960–973, 2015.

[35] J. Hou and D.C. O'Brien. Vertical handover-decision-making algorithm using fuzzy logic for the integrated radio-and-OW system. *IEEE Trans. Wireless Commun.*, 5(1):176–185, 2006.

[36] M.B. Rahaim, G.B. Prince, and T.D.C. Little. State estimation and motion tracking for spatially diverse VLC networks. *2012 IEEE Globecom Workshops (GC Wkshps)*, pp. 1249–1253, December 2012.

[37] M.B. Rahaim, A. Mirvakili, S. Ray, M. Hella, V.J. Koomson, and T.D.C. Little. Software defined visible light communication. *Wireless Innovations Forum Conference on Wireless Communications Technologies and Software Defined Radio (SDR-WInnComm)*, 2014.

[38] A. Mirvakili, V.J. Koomson, M.B. Rahaim, H. Elgala, and T.D.C. Little. Wireless access test-bed through visible light and dimming compatible OFDM. *2015 IEEE Wireless Communications and Networking Conference (WCNC)*, pp. 2268–2272, March 2015.

[39] S. Shao, A. Khreishah, M.B. Rahaim, H. Elgala, M. Ayyash, T.D.C. Little, and J. Wu. An indoor hybrid WiFi-VLC internet access system. *2014 IEEE 11th International Conference on Mobile Ad Hoc and Sensor Systems (MASS)*, pp. 569–574, October 2014.

14

OFDM-Based VLC Systems FPGA Prototyping

Mónica Figueiredo and Carlos Ribeiro

CONTENTS

14.1 Introduction

Visible light communication (VLC) is an emerging field in optical wireless communications, where white light-emitting diodes (LEDs) can be simultaneously used for illumination and data communications. Since 2011, VLC technology has gained momentum supported by the release of the IEEE 802.15.7 draft standard. This defines the physical and medium access control (MAC) layers supporting multiple topologies with data rates up to 96 Mb/s, for indoor and outdoor applications [1]. Also, several live demonstrations of VLC technology and its potential applications have contributed to increase its popularity among the general public [2–6]. It is expected that the popularity of VLC will continue to grow in the future and new applications and services will emerge to help mature this technology. According to [7], the global market for VLC is expected to grow at a compound annual growth rate of 109.2% from 2015 to 2022. This chapter outlines detailed information for VLC system architects who wish to prototype high data-rate systems in real hardware. Section 14.1.1 provides general information about the most popular envisioned applications for both low and high data-rate VLC systems. Section 14.1.2 takes a closer look at high-speed VLC demonstrators, and Section 14.1.3 discusses VLC implementation technologies.

14.1.1 VLC Applications

VLC systems have a myriad of potential applications. Most of these have focused on applications where the LED technology is already used for lighting purposes and thus, communications can support new value-added services. This is the case of applications in the area of intelligent transportation systems (ITS) [8], intelligent lighting [9], indoor networking [10], or indoor positioning [11]. However, VLC also suits other application scenarios where the illumination and data communication synergy is not the key element, such as underwater communications [12,13], electromagnetic interference-sensitive or security-sensitive scenarios [14–16], or the toy industry, where giants like Disney are already aiming to integrate VLC in their products just because it is appealing to their customers [17].

Some of these application environments involve only low data-rate communications, due to the nature of the information to be transmitted. This is the case in indoor VLC-based positioning systems and low data-rate communication applications. VLC positioning has been proposed to support multiple value-added services, guiding users in indoor environments, such as large museums, healthcare facilities, and shopping malls [18]. In ITS, several low data-rate services have also been proposed, such as collision warning and avoidance, lane change assistance or warning, or cooperative adaptive cruise control [8]. Cameras are also being used to capture low data-rate streams transmitted by red, green, and blue (RGB) LEDs [19] or embedded in images, both in vehicles [20] and smartphones [21].

Several high-speed VLC application environments have also recently emerged. VLC is already expected to play a significant role in future 5G wireless access networks, to complement the Radio-Frequency (RF) wireless infrastructure for downloading high data-rate streams in low mobility scenarios [22,23]. Hybrid or aggregated Wi-Fi–VLC systems have been shown to outperform conventional Wi-Fi for crowded environments in terms of throughput and web page loading time [24]. High-speed VLC is also envisioned to provide a highly secure wireless access for aircraft cabins, hospitals, and hazardous environments (where RF-based technologies are prohibited), or security-sensitive offices and laboratories (where RF technologies are prone to snooping).

Currently, low data-rate VLC systems have no significant technical constraint to their commercial deployment, as they can be implemented in low-cost, software-based platforms. When embedded with lighting, they usually resort to low-cost and low-bandwidth phosphor-converted LEDs (PC-LEDs). In these systems, pulse amplitude or pulse-position modulation techniques are generally adopted, with different coding schemes, to guarantee a flicker-free and dimmable ambient light [1]. On the contrary, high data-rate VLC systems have to deal with some technical challenges, the most relevant of which is the high intersymbol interference that results from the low modulation bandwidth of illumination-compatible LEDs.

A number of solutions have been suggested to achieve high data rates in the VLC system. Common solutions include the use of:

- High-bandwidth LEDs—multichip LEDs (typically RGB chips) or micro LED (µ-LED) arrays
- Optical lenses and filters
- Pre- and/or post-equalization [25]
- Multiple access schemes (e.g., wavelength division multiplexing [WDM] with RGB LEDs)
- High-efficiency modulation techniques (such as optical orthogonal frequency division modulation [OFDM] or multiband carrierless amplitude and phase modulation [CAP])
- Parallel communication schemes (i.e., optical multiple-input multiple-output [MIMO]) [26]

However, most of these techniques involve the implementation of complex and computationally intensive signal processing algorithms, which require expensive, high-speed, hardware-based platforms and high nonrecursive engineering costs.

14.1.2 High-Speed VLC Systems

Several research teams have contributed to demonstrate the feasibility of using visible light for transmitting data rates in the range of gigabits per second (Gbps). Most of them have been using arbitrary waveform generators to generate the modulated signal and digital storage oscilloscopes to recover the transmitted bitstream. These will be referred to in this chapter as laboratory VLC demonstrators. Most of the research effort has been concentrated on the signal processing algorithms, required to overcome the limited optical channel bandwidth, and in the visible-light optical front-end. Figure 14.1 shows the reported data rates in these demonstrators over the last few years [26]. Distances between transmitter and receiver vary substantially within these demonstrators. Nevertheless, it is clear that higher rates are obtained with RGB LEDs (resorting to WDM), with a saturation observed around 3.25 Gbps. The attainable data rates for PC-LEDs are also showing a saturation around 1 Gbps, while higher rates have been reported by resorting to micro LEDs and a high-bandwidth 450 nm laser diode. Laser diodes are a promising lighting source for some multigigabit [27] scenarios, but probably not for those where the VLC source is also used for illumination purposes.

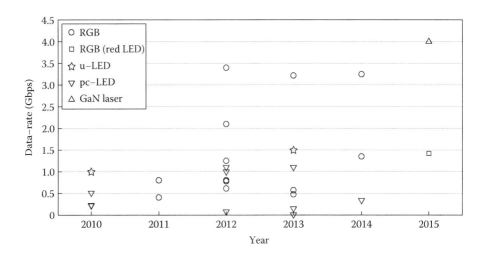

FIGURE 14.1
Data rates reported in VLC laboratory demonstrators. (Adapted from Karunatilaka, D., et al., *IEEE Commun. Surveys Tutorials*, 17(3), 1649–1678, 2015.)

Although these laboratory demonstrators have achieved multigigabit data rates, there are no real-time systems that support such data rates. The first real-time demonstrators were developed under the framework of the European-Community project Home Gigabit Access Network (OMEGA), which resulted in a VLC system running close to 100 Mbps and broadcasting four high-definition video streams from sixteen LED ceiling lamps to a photodetector placed within the lit area [28]. More recently, the same research team developed a bidirectional real-time VLC system with 500 Mbps peak data rate under various lighting conditions [29]. However, this is a commercial system using proprietary transmitter and receiver modules and thus, there is no published information regarding its implementation details. Other research teams are currently focusing on developing real-time demonstrators, with more moderate data rates (around and over 100 Mbps per single LED), as shown in Figure 14.2. This graphic focuses only on high-speed systems, as many other low data-rate, real-time VLC demonstrators have been proposed over the last decade.

To implement these real-time systems, most currently active research teams are resorting to field programmable gate array (FPGA) technology [32–35], which was also the technology of choice for demonstrators developed in the OMEGA project [31]. Digital signal processors have also been used to implement VLC demonstrators, as in [30], but they lack the performance, density, and flexibility offered today by FPGAs. In fact, there has been no high-speed VLC implementation using these devices in the past five years. On the other hand, FPGAs offer a number of ready-to-use intellectual property (IP) modules and system-level development tools, which allow

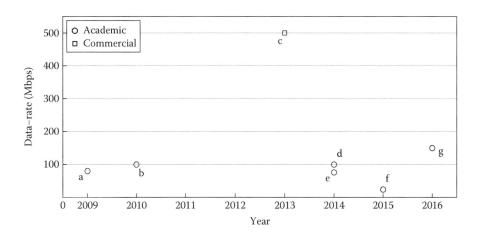

FIGURE 14.2
Data rates in real-time high-speed VLC demonstrators: (a) Elgala et al. [30], (b) Vŭcíc et al. [31], (c) Grobe et al. [29], (d) Shi et al. [32], (e) Yeh et al. [33], (f) Ribeiro et al. [34], (g) Ribeiro et al. [35].

users to easily design complex and highly integrated systems, which is a clear advantage.

Regarding the signal processing algorithms, most high-speed VLC demonstrators resort to OFDM-based modulation schemes with some sort of equalization. Since LEDs must be intensity modulated and the light thus produced must be directly detected, the OFDM signal must be made real and unipolar. To do that, several OFDM flavors have been proposed, with different power and bandwidth efficiencies [36]. One of the major drawbacks of OFDM is its high peak-to-average power ratio (PAPR). Since LEDs are dynamic-range limited, peaks of the OFDM waveform can be clipped causing signal degradation. To avoid this issue, CAP modulation has been proposed as an alternative to OFDM, offering the same spectral efficiency with much lower PAPR [37,38]. In terms of implementation complexity, CAP has the advantage of not requiring the inverse discrete Fourier transform in the transmitter. However, it needs a more complex receiver with either complex time-domain equalization or discrete Fourier transform (DFT) computations in order to perform simple frequency domain equalization. On the other hand, because the VLC channel has limited bandwidth and uneven frequency response, OFDM can maximize its transmission capacity by bit- and power-loading, while CAP requires an additional power pre-emphasis scheme [39]. For these reasons, OFDM-based modulations are currently the most popular solutions in high-speed VLC systems.

In literature, there are several modulation and equalization schemes that promise to enable higher data rates, higher power efficiency, and implementation simplicity. However, most of these schemes have never been tested in real-time environments. Thus, it is not possible to make real performance comparisons between alternative schemes nor evaluate their robustness to real-world hardware and channel imperfections. The first real-time platform available for cooperative research in high-speed VLC has been reported in [34], and will be described in Section 14.4. It enables the comparison between different algorithms in OFDM-based systems, strengthening potential synergies between research teams and helping to mature the VLC technology for high data-rate applications.

14.1.3 FPGAs for Digital Signal Processing

Digital signal processing (DSP) and graphics processing unit (GPU) processors are traditionally used to implement high-speed digital signal processing algorithms. DSP processors have a specially designed architecture to process high-speed streaming digital signals, and provide functionalities that are helpful for DSP applications, such as fast Fourier transform (FFT) computing. However, for applications that require customized DSP function implementations or very high-speed computing, their sequential and fixed hardware architecture limits performance. In certain applications, like image processing and computer vision algorithms, GPU processors have been used to provide extra computing power by exploiting parallelism. Their architecture consists of thousands of highly specialized small cores that can handle multiple tasks simultaneously. However, they also lack the hardware flexibility required to implement generic high-speed communication systems.

FPGAs are completely different from GPUs or DSPs as they offer a totally flexible hardware architecture, with thousands of configurable logic blocks and special embedded silicon features immersed in a sea of programmable interconnects. Some of these features are well-suited for the implementation of DSP functions such as finite impulse response (FIR) filters, FFTs, correlators, equalizers, encoders, decoders, and arithmetic functions. Others allow the designer to create complete systems-on-chip (SoC), such as soft- and hard-core processors, gigabit transceivers, and clock management units. Despite these benefits, their popularity among DSP designers is still limited due to the absence of a transparent C code–based design flow. However, this picture is now changing fast. New and increasingly user-friendly high-level tools and parameterized IP cores (like FFTs and filters) are now being offered by FPGA vendors, making hardware design more accessible to system architects. On the other hand, FPGAs typically run at about a 10 times slower clock rate when compared to CPUs or GPUs, which makes them more power efficient.

For the implementation of high-speed VLC systems, FPGAs stand out as the best candidate. They can efficiently implement the DSP algorithms required to implement the physical layer and simultaneously provide overall system integration and flexibility by allowing the designer to implement higher communication layers in embedded processors. Also, high-density FPGAs are the best solution to implement WDM-VLC or MIMO-VLC, which require very high aggregate data rates.

14.2 Prototyping with FPGAs

When selecting an FPGA development board for VLC prototyping, designers must be aware that there are many options in the market, at many different prices and performances. To help them make an informed and confident choice, Section 14.2.1 describes the major FPGA vendors and their high-performance FPGA families, which are suitable for the implementation of high-speed VLC systems. Then, Section 14.2.2 provides an overview of currently available mezzanine cards that extend general purpose FPGA development board functionality. With regard to OFDM-based VLC systems, these are essentially high-speed digital-to-analog (D/A) and analog-to-digital (A/D) cards. Finally, Section 14.2.3 looks at software development tools currently available to help the designer in the implementation, test, and debug tasks.

14.2.1 Hardware Platforms and Features

The biggest FPGA market share lies in communications and networking applications, which require high-performance and high-density devices. The main FPGA vendors in this high-end market are Xilinx and Altera Corporation. Other players in the FPGA market include Achronix Semiconductor, Microsemi, Lattice Semiconductor, Atmel Corporation, and E2V Technologies. Most of these vendors are mainly focused in providing components for the military and aerospace, with ultra-low power, high-security, and high-reliability applications. The exception is Achronix, which proposes high-end FPGAs with an architecture similar to those of Xilinx and Altera. This vendor has capitalized on the silicon advantage provided by their partnership with Intel, but still has a long way to go (in what relates to tools, services, and IP) before it can really compete with the two main vendors.

For the reasons set out above, and because the implementation of high-speed VLC systems requires the usage of high-density and high-performance FPGAs, only Xilinx and Altera high-end devices will be discussed in this section. Currently, both vendors offer high-density FPGAs built with state-of-the-art technologies with over 4 million logic elements, innovative routing architectures, and an increasing number of hard- and soft-core IP blocks to implement performance-critical functions. These high-end FPGAs are effectively becoming SoCs, with multicore processing subsystems, DSP units, peripherals, memory, and interfaces. These features can be exploited to efficiently implement the demanding real-time transceivers required to implement high data-rate VLC links. In the following paragraphs, FPGA evaluation boards (EVBs) suitable for this application are presented and discussed.

14.2.1.1 FPGA Development Boards—Xilinx

The high-end Xilinx portfolio includes Virtex-6, Virtex-7, and Kintex-7 FPGA families. Within 7-series FPGA families, Ultrascale and Ultrascale+ devices leverage on cutting-edge process technologies to provide increased performance, bandwidth, and reduced latency but are also considerably more expensive. Their usage is justified only in applications such as multigigabit networking and high-performance computing, which is not the case in VLC systems. For this reason, this section will focus only on Virtex and Kintex device families. Table 14.1 shows the currently available FPGA EVBs suitable for this particular application, with prices under $4,000.

All these FPGAs support embedded processing with MicroBlaze, a soft-core 32-bit reduced instruction set computing processor. Also, they are equipped with VITA 57.1 compliant FPGA mezzanine card (FMC) connectors, developed by Xilinx. FMC allows for two sizes of connector, low pin count and high pin count (HPC), each offering different levels of connectivity. FMCs decouple I/O interfaces from the FPGA base board, allowing the user to connect custom or off-the-shelf mezzanine cards. In this particular

TABLE 14.1

EVBs with Xilinx FPGAs (@ 4th Quarter 2015), under US$4000

	ML605	VC707	KC705	AES-K7DSP
FPGA	Virtex-6 XC6VLX240T	Virtex-7 XC7VX485T	Kintex-7 XC7K325T	Kintex-7 XC7K325T
LCs[a]	241 k	486k	326k	326k
DSPS[b]	768	2800	840	840
BRAMs	15 Mb	37 Mb	16 Mb	16 Mb
CMTs[c]	12 MMCMs	14	10	10
SW[d]	ISE LE	Vivado DE	Vivado DE	Vivado SE
HW[e]	–	–	–	4DSP FMC150
FMCs	1HPC, 1LPC	2 HPC	2 HPC	2 HPC
Vendor	Xilinx	Xilinx	Xilinx	Avnet
Cost	US$1,995	US$3,495	US$1,695	US$3,995

[a] Logic Cells.
[b] One DSP slice contains one pre-adder, one multiplier, one adder, and one accumulator.
[c] One CMT (clock management tile) contains one phase-locked loop (PLL) and one mixed-mode clock manager (MMCM).
[d] Software included: (L) Logic, (D) Design or (S) System Edition.
[e] Hardware included.

application, FMC connectors are useful to connect high-speed AD/DA mezzanine cards and/or optical front-ends (such as optical digital-to-analog converters) with guaranteed high throughput and low latency.

14.2.1.2   FPGA Development Boards—Altera

Altera offers two high-performance families, suitable for the current application. Stratix is Altera's high-end FPGA family with very high-density and high-performance devices, especially suited for multigigabit networking and high-performance computing applications. The Arria family is more mid-range, but still offering a rich feature set of memory, logic, and DSP blocks. Table 14.2 provides an overview of the most adequate (for this application), currently available, and reasonably priced (under $4,000) EVBs with Altera's FPGAs.

Sharing the same strategy with Xilinx, Altera boards come equipped with high-speed mezzanine card (HSMC) connectors, which can be used with a variety of application-specific daughtercards. The HSMC was developed by Altera, based on the Samtec mechanical connector and, as for FMC, there is currently a large amount of commercial HSMC mezzanine cards for a wide range of application domains. For compatibility with a broader collection of mezzanine cards, some of the boards presented in Table 14.2 also provide one FMC connector.

TABLE 14.2

EVBs with Altera FPGAs (@ 4th Quarter 2015), under US$4000

	ArriaII GX	ArriaII GX6G	ArriaV GXSt	ArriaV GT
FPGA	EP2AGX125	EP2AGX260	5AGXFB3	5AGTFD7
LEs[a]	124k	256k	362k	2x504k
ALMs[b]	50k	103k	137k	190k Mult[c]
	576	736	2090	2090
MBs[d]	6.6 Mb	8.5 Mb	17.3 Mb	24.1 Mb
PLLs	6	6	12	16
SW[e]	ADS	ADS	Quartus II	Quartus II
MCs	1 HSMC	2 HSMC	1 HSMC	2 HSMC 1 FMC
Vendor	Altera	Altera	Altera	Altera
Cost	$1,495	$3,195	$850	US$3,995

	Arria10 GX	Stratix III	DE3	DE4
FPGA	10AX115	EP3SL150	EP3SL150	EP4SGX230
LEs	1150k	142k	142k	228K
ALMs	427k	57k	57k	91k
Mult	3036[f]	384	384	1288
MBs[d]	53 Mb	5.5 Mb	5.5 Mb	14.3 Mb
PLLs	16	6	6	6
SW[e]	no	Quartus II	no	no
MCs	1 FMC	2 HSMC	2 HSMC	2 HSMC
Vendor	ReFLEX	Altera	Terasic	Terasic
Cost	$3,495	$2,495	$1,795	$2,995

[a] Logic elements.
[b] Adaptive logic modules.
[c] 18 × 18 bit multipliers.
[d] Memory blocks.
[e] Software included: ADS—includes the Quartus II Software Development Kit Edition, Nios II Embedded Design Suite, and MegaCore IP Library.
[f] 18 × 19 bit multipliers.

14.2.1.3   Comparison

It should be noted that Xilinx and Altera FPGAs have significantly different architectures and it is not possible to make direct comparison between their densities, memory availability, nor to infer at first sight which will perform best for a given application. FPGA specifications shown in Tables 14.1 and 14.2 should be understood only as indicative of performance, as actual performance depends greatly on the nature of the circuit to implement, the quality of hardware description language (HDL), implementation tools, and their optimization settings. As register transfer-level (RTL) details are not usually known to the system architect, the choice ends up being about the interfaces available in development boards, the availability of suitable daughterboards, how intuitive the software development tools are (which determines the steepness of the learning curve), and their availability/cost. Regarding hardware, this section has shown that both Altera and Xilinx EVBs can be connected to a broad range of daughterboards. The currently available A/D and D/A boards required to implement the VLC system are discussed in further detail in Section 14.2.2. Software tools will be discussed later in Section 14.2.3.

14.2.2 FPGA Mezzanine Cards

Beyond FPGAs, the successful implementation of a high-speed VLC demonstrator depends on hardware platforms required to make the interface between the digital and analog worlds. High-performance A/D and D/A converters are necessary to minimize signal distortion induced by insufficient time and amplitude resolution, when generating the signal to be transmitted and digitizing the received signal. Another issue is the availability of adequate off-the-shelf A/D and D/A development boards that can be used by research teams to implement these systems. Developing custom boards for these high-end components would be prohibitively time-consuming and unlikely successful without an experienced team of circuit and board designers. This section is intended to help the system architects in the task of selecting the most adequate off-the-shelf A/D and D/A boards for both FMC- and HSMC-based FPGA platforms.

14.2.2.1 FMC A/D and D/A Boards

There are three main vendors of FMC-based high-speed data conversion mezzanine cards: 4DSP, Texas Instruments, and Analog Devices. 4DSP offers the most extensive range of these boards, with A/D and D/A functions available on the same or separate cards. Regarding performance metrics, this vendor is mainly focused on providing high-performance solutions. Reference designs are also available for most high-end Xilinx boards, which is very convenient to speed up the design process. For a bidirectional VLC system implementation, an A/D and D/A board is required in both transmitter and receiver and thus, the most cost-effective solution is to choose an FMC that incorporates both functions. The bad news is cost—these boards are quite expensive with prices ranging from $2,195 (4DSP FMC150) up to $9,015 (4DSP FMC160). Single A/D and D/A FMC boards could be used for broadcast VLC systems, but the price range is equivalent.

Texas Instruments had a different approach. They provide FMC-ADC-ADAPTER and FMC-DAC-ADAPTER boards to enable Xilinx Series 6 and series 7 FPGA EVBs to connect to a broad range of their high-speed A/D and D/A modules. Both the adapter boards and modules are very reasonably priced (around $50 for the adapters and $500 for the modules), but there are no available Xilinx reference designs. This is a significant drawback as it is usually not straightforward to implement all the necessary configuration and control modules.

TABLE 14.3

A/D and D/A Analog Device FMC Boards (@ 4th Quarter 2015)

	AD9434-FMC	AD9467-FMC	AD9739A-FMC
Type	1 ADC	1 ADC	1 ADC
Speed	500 MSPS	250 MSPS	2,500 MSPS
Size	12-bit	16-bit	14-bit
Ref. design	ML605	KC705	ML605
	KC705		KC705
	KC707		KC707
Cost	$399	$399	$349

	FMCOMMS1[a]	FMCADC2[b]	FMCDAQ2
Type	2 ADC + 2 DAC	1 ADC	2 ADC + 4 DAC
Speed	250, 1,200 MSPS	2,500 MSPS	1,000, 2,800 MSPS
Size	14-, 16-bit	12-bit	14-, 16-bit
Ref. design	ML605	VC707	VC705
	KC705		KC707
	KC707		KCU105
Cost	$999	$1,243	$1,380

[a] This board is an RF transceiver, but it can be configured (through solder jumpers) to bypass the RF section.
[b] The reference design for the AD-FMCADC2-EBZ requires a commercial license to use the Xilinx JESD204B core.

Analog Devices followed a similar path, providing low-cost, high-speed analog-to-digital converter FMC interposer boards ($100) that allow certain HSMCs to be used on FMC-based Xilinx EVBs. Nevertheless, they also offer FMC-compatible versions of their A/D and D/A modules, at reasonable prices, which is very convenient. Also, most boards are available with reference designs compatible with high-end Xilinx EVBs, which significantly speeds up the system design process. Table 14.3 shows the most interesting solutions for the current application.

14.2.2.2 HSMC A/D and D/A Boards

As previously mentioned, Altera provides a proprietary HSMC connector in most of their high-end EVBs, which can be used to interface with A/D and D/A mezzanine cards. Table 14.4 provides an overview of currently available HSMC compatible cards, from different vendors. They are based on mid-range speed A/D and D/A converters from Analog Devices and Texas Instruments and only one provides a reference design for a high-end FPGA, which can be a significant disadvantage when compared with available FMC-based cards. The only advantage is price, which is significantly lower.

TABLE 14.4

A/D and D/A HSMC Boards (@ 4th Quarter 2015)

	ADA-HSMC	ADA Card	DAC5682Z	DEV-ADC34J22
Type	2 ADC	2 DAC	2 ADC	4 ADC
	2 DAC	2 DAC		
Speed	65 MSPS	150 MSPS	1,000 MSPS	50 MSPS
	125 MSPS	250 MSPS		
Size	14-, 14-bit	14-, 14-bit	16-bit	12-bit
Ref. design	–	DE3	–	SOCKIT
				Cyclone V
Vendor	Terasic[a]	Terasic[a]	Texas Inst.	Dallas Logic
Cost	US$219	US$390	US$499	US$199

[a] Compatibility with Terasic boards is guaranteed through an adapter available in DE3 and DE4 kits.

14.2.2.3 Comparison

Despite the higher cost, FMC-based cards offer substantially higher sampling rates, which is key in high-speed communication systems. If a low-cost solution is desired (sacrificing data rates), low-cost HSMC-based cards can also be used with FMC-enabled FPGA boards by resorting to an HSMC to FMC adaptor board (FMC2HSMC), available through Kaya Instruments for $395. However, when both costs are considered (HSMC card plus adaptor), FMC-native cards are still more competitive.

14.2.3 Software Tools for FPGA Design

In today's FPGA design world, it is no longer just about the hardware, but also about the design tools and IP resources. Modern FPGA projects require a complete set of electronic design automation (EDA) tools for system-level design, design creation, synthesis, verification, and board-level design. Leading edge customers, used to the application specific integrated circuit (ASIC) implementation flow, usually resort to HDL flows for design entry and verification, and front-end flows provided by companies such as Mentor Graphics or Synopsys. However, for mainstream designers, EDA tools provided by FPGA vendors can be more interesting solutions. They are high-quality tools made available at much lower prices (or even free of charge if they are meant to be used for research only), or included in the price of development boards. Also, they currently include high-level design entry tools that greatly shorten the path from design concept to working hardware. This section describes the development tools provided by Xilinx and Altera.

14.2.3.1 Xilinx Tools

In October 2013, Xilinx released the last version of their Integrated Synthesis Environment (ISE) Design Suite. It has been discontinued in favor of Vivado, a newly re-architected complete FPGA design suite. While ISE is still available and will be supported indefinitely for customers targeting Virtex-6, Spartan-6, and their previous generations, Vivado must be used for 7-series, Ultrascale and future device families. This section will look into both design suites, as many designs may not require the usage of 7-series or Ultrascale FPGAs and thus, must still resort to ISE tools.

Within ISE, Project Navigator is the software tool that manages and processes the FPGA design from design entry to device programming, passing through synthesis, implementation, and verification. The most complete of its editions, ISE System Edition, also includes the embedded development kit (EDK), for designing embedded processing systems:

- PlanAhead, to easily analyze critical logic and improve design performance with floorplanning, constraint modification, synthesis, and implementation settings
- System Generator, for the design and verification of DSP systems
- ChipScope™ Pro, to enable the debug of high-speed FPGA designs
- A large repository of plug and play IP, including MicroBlaze soft processor and peripherals

Concerning the implementation of real-time VCL systems, some of these tools are particularly interesting. System Generator enables the designer to build, simulate, and translate into hardware complete systems within the Simulink® (from MathWorks) environment [40]. It presents a high-level and abstract views of the design that allows the designer to implement DSP algorithms using parameterizable or user-defined (in HDL or MATLAB® programming language) blocks. For system architects, used to MATLAB and Simulink, this is very convenient to quickly translate their algorithms into working hardware. System Generator is also a powerful tool for bit- and cycle-accurate simulation using Simulink environment, source, and sink modules. Designs can also be simulated using ModelSim (from Mentor Graphics) or cosimulated (using hardware in the loop), for considerably faster design verification.

If the designer wants to explore the advantages of designing the VLC system as an embedded system, embedded development kit (EDK) tools offer an integrated development environment for systems with hard or soft-processor cores (e.g., MicroBlaze). It includes Xilinx Platform Studio to configure the embedded system architecture, buses, and peripherals; and the Eclipse-based software development kit (SDK), to support C/C++ software development. EDK also provides real-time and embedded operating system support. Resorting to embedded processors can be particularly useful to implement physical layer control and management functions, as well as implementing higher communication layers, as required in any communication system. Finally, the ChipScope Pro tool offers the possibility to perform in-circuit verification, much as one would do with a logic analyzer. While the design is running, it is possible to trigger when certain events occur and observe internal signals, including embedded hard or soft processors. Captured signals are then displayed and analyzed using the ChipScope Pro Analyzer tool. Although it is limited by the FPGA's available memory and clock frequency, it can be very useful for real-time debug and verification purposes.

The new Vivado Design Suite has been built from the ground up to address productivity bottlenecks in system-level integration and implementation, especially for high-end device families. Vivado is considerably faster than ISE due to its completely new logic synthesis, time analysis, and placement engines, which are now built with a shared scalable data model. On the other hand, the entire design flow of Vivado is now centered on IP-based design and design reuse, which is a significant departure from the traditional approach. Also, it enables designers to create RTL implementation from C level descriptions (C, C++ and SystemC), through the new high-level synthesis (HLS) tool.

Vivado is currently offered in three editions:

- Design Edition, which provides the tools required to support IP integration and the physical implementation flow.
- System Edition, providing software-defined IP generation with Vivado HLS and DSP design integration with System Generator for DSP.
- WebPack (device-limited edition of Design Edition).

While HLS is a completely new tool, System Generator is conceptually the same tool as the one offered in ISE, with some new features to accelerate design verification and IP integration. SDK can be added to both editions as an option.

14.2.3.2 Altera Tools

The Altera Quartus II design suite includes solutions for all phases of FPGA design with a user-friendly graphical user interface (GUI). Over the years, new place-and-route algorithms have been introduced to reduce compilation time and integration with third-party EDA tools. Following the same path as Xilinx, specific tools are also available for IP integration (Qsys), DSP design support (DSP Builder), FPGA floorplanning (ChipPlanner), timing analysis (TimeQuest), embedded system debug (SignalTap), and support for high-level C language descriptions (Altera SDK).

The Qsys system integration tool is the counterpart of Xilinx EDK. It manages the hardware design when resorting to Altera's soft-processor core Nios II. Software development can then be performed using the Eclipse-based Nios II software build tools. DSP Builder is Altera's tool that allows HDL generation of DSP algorithms directly from the Simulink environment, using Altera-specific blocks [41]. Similarly to System Generator, it provides parameterizable IP blocks for most common signal processing functions, which greatly simplifies the implementation and verification of signal processing algorithms. Altera SignalTap embedded logic analyzer shares the same principles with Xilinx ChipScope Pro tool, allowing the designer to debug and verify the design while the device is running in the system. It is possible to select signals, set up triggering events, configure memory, and display waveforms, all within the Quartus II interface. Moreover, this tool is available in the free edition of Quartus II, which can be an advantage when compared to the Xilinx ChipScope Pro tool.

14.2.3.3 Comparison

Traditionally, Xilinx is considered to have better silicon and Altera better software. The Xilinx ISE® design suite is in fact a mixture of tools and technology acquired over the years, which results in a poorly integrated platform. Although the new Vivado suite eliminates these issues and has all the makings of a future winner, it is currently used only by a fraction of customers (as it is limited to high-end devices). On the other hand, Altera Quartus II has a better GUI and provides a more seamless tool integration although it may start showing its age for large and complex designs. For the particular application discussed in this chapter, the implementation of a real-time VLC system, both vendors provide adequate high-level software tools.

14.3 Design and Implementation Issues

FPGAs are an increasingly appealing solution for DSP applications, such as communication systems. However, implementing an algorithm on an FPGA requires much greater design effort compared to a DSP or general purpose processor. Efficient FPGA implementations involve many subtle design choices and complex trade-offs that are unfamiliar to most system-level engineers. This section discusses some of these issues at different design levels.

14.3.1 Architecture-Level Issues

The most important architecture-level design choice is the system's clocking architecture—the synchronous versus asynchronous design paradigm. Traditional design flows and methodologies usually lead the designer to the adoption of a totally synchronous approach, with single or multirate system design. This approach lends itself to a simpler clocking structure, enabling a faster development, and simpler test and validation of the system. Design, debugging, and validation need not be concerned with clock domain-crossing issues and associated synchronization units. However, this may come at the price of lower overall system performance, as will be discussed in the following paragraphs.

The implementation of a high-speed OFDM-based VLC link is not a simple task, as it involves multiple complex operations that build up a significantly complex and high-density system. FPGAs with high resource utilization may start exhibiting excessive routing delays and fail to meet the designer's timing objectives. On the other hand, the need to distribute a low-skew clock to the entire design limits the system's maximum clock frequency and affects system performance. Pipelining can be used to shorten critical paths (at the cost of designer development time), but it adds path latency that must be accounted for in the rest of the design.

To maximize the system's performance by design, a better approach can be to resort to a globally asynchronous, locally synchronous (GALS) design [42]. Performance improvements provided by this approach arise from many different factors, as listed below:

- Shorter critical paths, which are now confined to synchronous domains. This reduces the probability of having critical timing violations precluding the system's implementation on mid-range (and cheaper) FPGAs.
- Reduced complexity and power consumption associated with distributing a low-skew and high frequency clock to the entire system. Each synchronous block can operate at its minimum required speed and draw on the multiple global and regional clock routing resources, currently available in most FPGA families [41,43].
- Better usage of available implementation area, due to reduced routing complexity and simpler floorplanning.
- More scalable and flexible designs, as new modules can be easily introduced or eliminated.
- Lower on-chip power supply noise, as clock current demands are spread in time.

Unfortunately, higher performance comes at the price of higher complexity, as the designer must implement small synchronous functional modules that communicate using an asynchronous handshake protocol. Adding this communication blocks to each functional module means higher resource utilization and higher developing time. Also, verification tasks are more difficult in datapaths that cross clock domains. Nevertheless, performance gains may justify the effort.

TABLE 14.5

Timing Results for Different OFDM Transceiver Implementations

Implementation	Score	Period Requirement	Actual Period
GALS	0ps	4.167 ns	2.5 ns (400 MHz)
Synchronous	1,321,201 ps	4.167 ns	9.107 ns (110 MHz)

For illustration purposes, Table 14.5 shows timing results for two different FPGA implementations of an OFDM transceiver, operating with a 240 MHz clock. The sum of negative slacks for each endpoint is represented by the timing score parameter, showing how much the constraints are failing. This system was designed according to a GALS methodology, with several synchronous modules communicating asynchronously through dual-clock first-in-first-out (FIFO) memories. However, the mode of operation can be settled by defining appropriate timing constraints. If timing-ignore constraints are added to inform the tools that data paths between synchronous modules are nonsynchronous paths, it becomes a GALS system. If not, the tools have to enforce period timing constraints in every data path (even between modules), thus configuring a totally synchronous scheme. Results show that the synchronous implementation clearly fails to meet performance requirements, as the tools fail to distribute the 240 MHz transceiver clock signal. On the contrary, the GALS implementation can achieve a maximum clock frequency of 400 MHz.

14.3.2 Circuit-Level Issues

To boost the performance and implementation efficiency of fundamental DSP functions, hard- and soft-IP cores are currently embedded in most mid-range and high-end FPGAs. However, to take advantage of these resources, the designer must be aware of their characteristics, as bad design choices can significantly impact performance and device utilization [44]. Some of these issues are especially relevant when implementing a complex high-speed system and thus, are discussed below. This discussion is illustrated with implementation data for Xilinx Virtex-6 FPGA.

14.3.2.1 *Multiplications*

OFDM transceivers require the implementation of complex multiplications, especially in the receiver where the data outputted by the FFT are used for channel estimation and equalization. A complex multiplication implemented in rectangular coordinates requires four real multiplications, one addition, one subtraction, and optional buffering of intermediate stages for performance enhancement. In a Virtex-6 FPGA, this can be done by resorting to DSP48E1 slices, which basically hold one 18×18 multiplier followed by an accumulator [45]. Depending on the number of bits used to represent data and the optimization flavor selected (area or performance), resource utilization and latency varies significantly. Table 14.6 shows implementation data for a fully pipelined complex multiplication with different configurations. Although latency has little impact in the system's performance for streaming data, performance may be affected by excessive routing delays resulting from high levels of resource usage (especially for DSP48, which are scarce and location constrained).

To quantify hardware cost, let *CMC* be the complex multiplication cost in rectangular coordinates. The exact formulation of this cost depends on the designer's perception of which are the cheapest resources in the target FPGA, for his particular design. Based on this perception, different weights can be given to each resource type. If a given design requires *N* complex 18-bit multiplications, then the hardware implementation cost (*HIC*) is just given by (14.1).

$$HIC_{rect} = N * CMC \tag{14.1}$$

Alternatively, complex multiplications can be made in polar coordinates, which may lend itself to a more efficient implementation. When operands are expressed in polar coordinates, their product is reduced to a real multiplication and a phase rotation (sum of angles), which significantly reduces the number of DSP slices required. However, there is a cost in the conversion from rectangular to polar coordinates and vice versa. If *RMC* is the real multiplication cost and *AC* is the addition cost, *HIC* is now given by (14.2), where *CC* is the conversion cost from rectangular to polar and back to rectangular.

$$HIC_{polar} = 2 * CC + N * (RMC + AC) \tag{14.2}$$

The most efficient way to make these conversions is to resort to a coordinate rotation digital computer (CORDIC) algorithm [46], provided by most FPGA vendors as a parameterized soft-IP. Not only it is useful to convert rectangular to polar coordinates (and vice versa), it can solve a broader range of equations, including hyperbolic and square root [47]. The algorithm

TABLE 14.6

Complex Multiplication with Rectangular Coordinates in Virtex-6 FPGA

Optimization	Performance	Resources	18-bit Performance	32-bit Resources
Latency	4	6	10	13
DSP slices	4	3	16	12
Slice registers	0	0	286	458
Slice LUTs	0	0	207	451

Note: DSP, digital signal planning; LUT, Look-up Table.

iteratively approaches the solution, performing a sequence of successively smaller rotations up to the desired precision. The latency is therefore associated with output width and precision, and requires no multiplier block. The CORDIC algorithm is limited to the first quadrant, and its output is affected by a scale factor that depends on the number of iterations. If there is a need to extend the algorithm to the full circle and to compensate the CORDIC scaling factor, real multiplications are needed to scale the output, increasing the resource count (either logic slices or DSP48).

Table 14.7 shows implementation data for different CORDIC implementations in a Virtex-6 FPGA, which provides a sense of the CC featured in Equation 14.2. It shows data for parallel architectures (for best performance) with rotation and output scaling, and different choices regarding pipelining. Maximum pipelining has a higher CC but allows the circuit to operate faster when streaming large bursts of data, while no pipelining incurs, in a two-cycle latency, a penalty for each processed sample.

For comparison purposes, Figure 14.3 shows the HIC for an increasing number of multiplications (N), when using rectangular or polar coordinates. Parameters CMC, CC, and RMC are calculated as a weighted sum of DSP count ($DSPC$), LUT count ($LUT\ C$), and register count ($REGC$), as shown in Equation 14.3. To be fair, LUT and REGC values were weighted by the number of LUTs required to implement an 18-bit multiplier in the FPGA fabric (520 LUTs). CORDIC was considered to be implemented with DSP48 scaling (worst-case cost) and the RMC to be only the number of DSP48 blocks (the hardware cost of the adder required for phase rotation is negligible). Also note that while $RMC = 1$ for an 18-bit multiplication, it increases to 4 for a 32-bit operation.

$$CMC, CC, RMC = DSPC + (LUTC + REGC)/520 \qquad (14.3)$$

Graphics show that, for 18-bit operands, the polar format is advantageous for three or more multiplications in a cascade. For 32-bit operands, polar operands are advantageous for $N > 2$ and have similar cost for $N = 1$. The

TABLE 14.7

CORDIC Implementation Details for Virtex-6 FPGA

	18 bit				32 bit			
Pipelining	No		Maximum		No		Maximum	
Scaling	LUT	DSP48	LUT	DSP48	LUT	DSP48	LUT	DSP48
Latency	2	2	25	25	2	2	40	42
DSP slices	0	2	0	2	0	4	0	4
Slice reg.	74	74	355	199	129	129	842	380
Slice LUTs	58	58	286	122	560	100	678	218

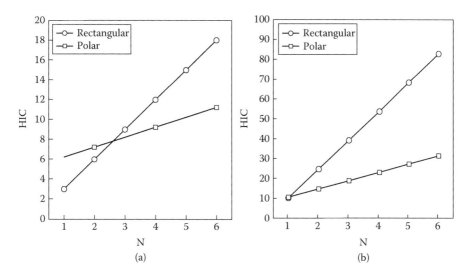

FIGURE 14.3
HIC for a cascade of N complex multiplications, in rectangular and polar coordinates, for (a) 18-bit data operands and (b) 32-bit data operands.

only disadvantage is latency, but as these are fully pipelined architectures; it is not very significant for long data bursts.

14.3.2.2 Memory

When it comes to memory, the designer should also be aware of the features available in the chosen FPGA device. Modern mid- and high-end FPGAs include variable amounts of dedicated RAM, typically organized in large blocks. These block RAMs (BRAMs) have many different aspect ratios and can be spread or located in specific regions inside the FPGA fabric, depending on the manufacturer and device family. Although the width of these blocks can usually be adjusted, not every combination is possible. Thus, if the designer defines a non-supported ratio, he may be compromising the efficient usage of FPGA resources. Memory can also be implemented using distributed RAM (using small RAMs or flip-flops in the FPGA logic elements), which is ideal for small-sized memories. However, when comes to large memories, this may cause extra routing delays and the usage of a significant amount of resources. Thus, to ensure that performance is not compromised by memory design choices, the designer cannot ignore the underlying FPGA architecture.

14.3.2.3 FFTs

The most efficient way to implement an FFT (or inverse—IFFT) is to resort to IP cores. These cores usually come with different implementation architectures,

TABLE 14.8

Implementation Data for a 1,024-point 16-bit FFT, in a Virtex-6 FPGA

	Radix-2 Burst	Lite Radix-2	Radix-4 Burst	Pipelined Streaming[a]
Transform time[b]	12,317	7,324	3,436	3,172
DSP slices	3	6	20	16
Slice reg.	0	0	1,775	2,713
Slice LUTs	0	0	1,068	2,143
18k BRAMs	3	3	7	9

[a] Scaled, Natural Order, 5-Stage BRAM, 4-multiplier structure, CLB Butterfly.
[b] Transform time is measured in clock cycles.

to allow the designer to choose the most efficient (in resource usage or performance) for his application. Xilinx provide four different architectures: Radix-2 Lite, Radix-2, Radix-4, and Pipelined Streaming. The first three require much less resource but have higher latency as data load and unload operations are performed in each transform cycle. These are best suited when data arrive in small bursts and latency can be accommodated without affecting system performance. For long data bursts or continuous data processing, pipelined streaming becomes more efficient at the cost of higher resource usage. Implementation data for a 1,024-point 16-bit FFT from Xilinx core are shown in Table 14.8.

In this table, transform time is shown for a single 1,024-point computation. For that reason, values shown for Radix-4 and Pipelined Streaming architectures are quite similar. For a big burst of streaming data (e.g., M times 1,024 samples), the first would require $3,436 \times M$ clock cycles to complete while the last would take only $3,172 + (M - 1) \times 1,024$ clock cycles, which is a very significant performance difference. For the streaming architecture, there is also the possibility to configure other parameters, allowing the designer to save BRAMs and DSP48s at the cost of higher logic utilization. For this core, resource usage is also dependent on the number of bits used to represent samples, so a good design practice is to keep it below 18.

14.3.3 Data and Control Signals

This section presents some good design practices related to the definition and usage of data and control signals inside the FPGA.

14.3.3.1 Signal Data Types

FPGAs support different data types, both fixed-point and floating-point. High-level DSP design tools, such as System Generator, support both definitions although just for a selection of hardware blocks and IP cores. Floating-point data are especially convenient to represent very large and very small numbers in the same data path (such as with accumulators) or when you need to rapidly develop a functional hardware prototype. However, designers must be aware that the data type selection has a significant impact on FPGA resource usage and ultimately, in performance.

Even when selecting fixed-point data types, the number of bits used to represent data must be carefully chosen. Performance is severely affected by this choice because long data types result in long datapaths. While synthesis and implementation tools are very efficient in implementing small circuits, their efficiency usually diminishes when large datapath widths need to be accommodated, especially at high clock frequencies [48]. For example, large multiplexer structures can require cascading of many instances of smaller multiplexers with a significant routing overhead, which can be significantly spread throughout the FPGA (depending on the upstream functions). Heavy utilization of routing between slices is known to be a key contributor to slower performance. Pipelining or choosing alternative implementation styles can help mitigate this issue. However, it is always more effective (and less time consuming) to take the preventive approach and select, in each data path, the shortest fixed-point representation that can accurately represent data.

14.3.3.2 Control Signals

Another significant issue is the design of the reset network. The designer should check whether resets are natively synchronous or asynchronous, in the FPGA fabric. An optimal reset structure will enhance device utilization, timing, and power consumption in an FPGA [49]. This implies a correct choice of reset type (synchronous or asynchronous) and its polarity. In Xilinx FPGAs, active-high synchronous resets enhance FPGA utilization, performance, and overall power consumption [50]. It is also important to synchronize the reset with the clock to guarantee the correct operation of state machines and avoid metastability in flip-flops. If the reset is synchronous, it is sufficient to use two back-to-back flip-flops in each clock domain to generate the global reset signal. If however an asynchronous reset is necessary, a scheme to assert reset asynchronously and de-assert it synchronously should be employed [51].

14.3.3.3 Asynchronous Inputs

Designers should also follow good design practices when dealing with asynchronous inputs or when a signal transfers between circuitry in unrelated or asynchronous clock domains. A simple synchronizer made up with a sequence of registers should be inserted in the destination clock domain to reduce the probability of metastability. Both Altera and Xilinx provide dual-clock FIFOs that use synchronizers to transmit control signals between two clock domains and dual-port memory blocks to transfer data [52,53].

14.4 An FPGA-Based VLC Prototype

This section describes a real-time high-speed VLC prototype transceiver based on DCO-OFDM, implemented in a Xilinx Virtex-6 FPGA. The transceiver was designed as a GALS system, so it is very easy to add or remove blocks, offering a real-time test bed to evaluate the performance of different modulation schemes and DSP algorithms. The system's architecture is presented in Section 14.4.1 and implementation details discussed in Section 14.4.2. The system transmits over 2 m with a 12 MHz bandwidth using quadrature phase shift keying (QPSK), thus achieving 24 Mbit/s with a bit error rate (BER) lower than the 3.8×10^{-3} forward error correction (FEC) limit [34]. For shorter distances, for example, 50 cm, the system is able to transmit at 150 Mbps, using 64-QAM and 25 MHz bandwidth [35]. Detailed performance results will be presented and further discussed in Section 14.4.3.

14.4.1 System Architecture

The VLC system-level architecture is depicted in Figure 14.4. It comprises the transceiver implemented in a Xilinx FPGA (ML605 development board), a DAC/ADC board from Analog Devices (AD-FMCOMMS1-EBZ, in baseband configuration) and an optical front-end. The optical transmitter uses a single blue LED (Seoul Semiconductor X42180-07), while the receiver is currently based on a Hamamatsu C12702 Avalanche photodetector (APD) receiver module. The VLC transceiver and analog modules are easily configured

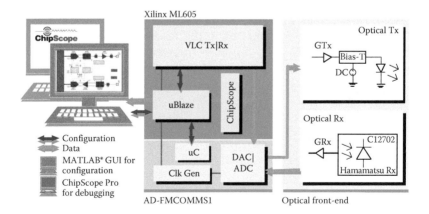

FIGURE 14.4
VLC system's high-level architecture.

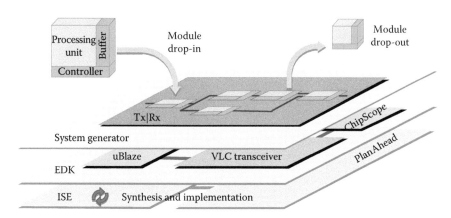

FIGURE 14.5
Software development flow and tools.

via MicroBlaze, using a MATLAB GUI. For debugging and testing purposes, Xilinx ChipScope Integrated Logic Analyzers (ILA) modules are used. Future developments include the migration to a higher density FPGA (KC705 development board) to enable the system to support RGB LED transmission using WDM.

Figure 14.5 depicts the software development flow and Xilinx tools used in this platform. Processing units were developed, tested, and evaluated in System Generator; the transceiver netlists were integrated with MicroBlaze and ChipScope modules in EDK; and finally, synthesis and implementation steps were performed in ISE.

To better illustrate how these tools were integrated in the design flow, Figure 14.6 depicts an implementation flowchart. The transceiver is designed in System Generator and its performance evaluated using the built-in simulator, which is bit and clock cycle accurate. At this point, the design is independent of the target FPGA, although the designer should be aware of its architecture in order to implement efficient modules, as explained in Section 14.3. Once the transceiver is complete, it must be connected to the FMC card drivers, which are available in the FM-COMMS1 embedded reference design. This reference design was adapted according to the system's requirements and modified so that MicroBlaze could also be used for user interface and system configuration. Synthesis and implementation steps were performed in ISE, after defining the necessary pinout, area, and timing constraints. PlanAhead has also been used to identify and correct critical paths when ISE failed to meet timing constraints. Once the system is correctly implemented, the FPGA can be programmed through EDK and performance evaluated in real time with the ChipScope Analyzer.

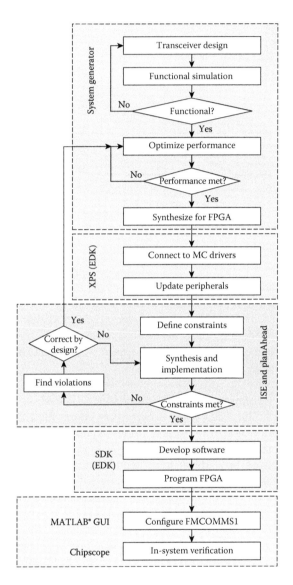

FIGURE 14.6
Implementation flow chart.

14.4.2 Transceiver Implementation

The transceiver is based in a DC-biased optical OFDM (DCO-OFDM) modulation scheme, due to its bandwidth efficiency. The analog signal is constrained to be real and positive by imposing a Hermitian symmetry to the transmitter inverse FFT (IFFT) input vector and by adding a DC bias in the analog domain. Figure 14.7 shows the transceiver's block diagram, which is described next.

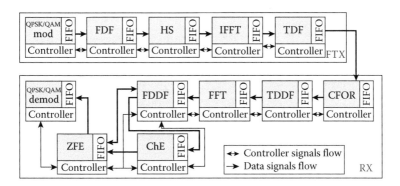

FIGURE 14.7
DCO-OFDM transceiver architecture.

The transmitter (TX) includes:

- A QPSK/QAM modulator
- A frequency-domain framing (FDF) unit
- A Hermitian symmetry (HS) unit
- An IFFT unit
- A time-domain framing (TDF) unit

The modulator takes data from a pseudorandom binary sequence (PRBS) generator and generates QPSK, 16 QAM, 32 QAM, or 64 QAM (configured by the user). The FDF unit is then responsible for making up the frame, placing modulated data, pilots, and null carriers in their proper locations. Each symbol has 1,024 carriers, where only 400 out of the possible 512 are loaded (due to the required HS). The first carrier is located at 2 MHz and their separation is 30 kHz, for a total modulation bandwidth of 12 MHz (for a sampling frequency of 30 MHz). As the LED used in this setup has a 2 MHz 3 dB-bandwidth, this framing arrangement highlights the advantages of using OFDM to explore the LED's out-of-band bandwidth. To track the full channel bandwidth, pilots are inserted in odd symbols, evenly spaced with a separation of four carriers. The HS block generates data for the second half of the frame, according to the data present in the first half. The 1,024-point IFFT block, besides transforming the signal to the time domain, adds a cyclic prefix (CP) with 256 samples to each OFDM symbol. In the time domain, the TDF unit appends a high autocorrelation synchronization symbol.

The receiver (RX), includes a clock frequency offset removal (CFOR) block that implements the required timing synchronization tasks: estimation of start of OFDM symbol, estimation of start of OFDM frame, estimation of the frequency offset, and compensation of the frequency offset returned by the estimation block. The three required estimations are performed by a joint maximum likelihood algorithm [54] that takes advantage of the presence of the higher power synchronization symbol and the presence of CP in each symbol of the frame. The estimation of start of OFDM symbol and the estimation of start of OFDM frame are used to define the time boundaries of each received frame. The estimated frequency offset affecting each received symbol, caused by the mismatch between TX and RX oscillators, is fed to a CORDIC-based phase-rotation block that compensates the frequency offset.

After, the CFOR block follows a time-domain deframing (TDDF) unit, to remove the CP and a FFT module. The FFT is implemented using the Cooley–Tukey algorithm, with unscaled (full-precision) fixed-point arithmetic, Radix 4 decompositions for computing the DFT (the N-point FFT consists of log4(N) stages, where each stage holds N/4 Radix-4 butterflies) and decimation-in-time [55]. This implementation takes advantage of the presence of DSP48 and BRAM blocks to lower the implementation area and keep the maximum path length low enough to enable the required bandwidth. After the FFT, a frequency domain deframing (FDDF) unit is used to separate pilot carriers, data carriers, and null carriers. While the values in the data carriers are feed to the zero forcing equalizer (ZFE) module, the values in the pilot carriers are fed to the channel estimator block (the null carriers are discarded). The first module compensates large channel gain differences without significantly distorting the resulting constellation. The latter estimates the channel in the pilot positions using a least-squares algorithm and performs a linear interpolation to extend the channel estimation to the data carriers. To lower the implementation area, the ZFE is implemented in polar coordinates, built around a single real divider block that compensates the magnitude distortion of each carrier and a CORDIC-based phase rotation block that rotates the phase of received value in each data carrier to the original phase. Finally, the demodulator performs a hard decision to extract the transmitted bitstream. Key transceiver design and implementation parameters are given in Table 14.9. Note however that a higher signal bandwidth can be set if a higher sample frequency is used (up to 54 MHz) and/or more subcarriers are loaded (up to 512).

OFDM systems commonly require the transmission of a synchronization sequence, marking the starting point of each OFDM frame. Usually, a CAZAC (constant amplitude zero autocorrelation) waveform is used, such as the Zadoff–Chu sequence. However, these are complex signals and cannot be used in optical OFDM. In published VLC systems, authors refer only the use of a PN (pseudonoise) sequence, providing no further details, although it is implied that it must have HS in the frequency domain to guarantee that the OFDM transmitted signal is real. In this implementation, it is additionally required that the sequence is bandpass (between 2 MHz and 14 MHz). To fulfill these requirements, a bandpass 64-tap linear feedback shift register PN sequence was designed, with a configurable starting frequency and HS. The sequence was designed to have an average transmitted power 6 dB higher than the OFDM symbols, to improve the system's timing synchronization performance.

TABLE 14.9

Transceiver Design and Implementation Parameters

Design Parameters		Implementation Parameters	
System carriers	1,024	Sample frequency	30 MHz
Loaded carriers	400	Clock frequency	100 MHz
Pilot carriers	100	Signal bandwidth	12 MHz
Pilot separation	4	Carrier separation	30 kHz
OFDM frame	5 symbols	Digital data width	32 bit
Sync. symbol	BPH-PN	DAC resolution	16 bit
CP length	256	ADC resolution	14 bit

Figure 14.8 shows the frequency domain, time domain, and autocorrelation characteristics of the designed Bandpass Hermitian (BPH)-PN sequence. For comparison purposes, a BPH Zadoff–Chu sequence was designed using the same procedure and is shown superimposed in the graphics. The design procedure followed four steps: (i) compute the FFT of a 64 carrier Zadoff–Chu sequence, (ii) shift the signal in the frequency domain to the transmission bandwidth, (iii) force the signal to have HS, and (iv) compute the time-domain sequence by applying an IFFT (inverse FFT) to the manipulated spectrum. Results show that the BPH-PN sequence has slightly inferior frequency and autocorrelation properties, but has a significantly lower PAPR (as shown by the sequence's amplitude levels in the time domain). This means that the BPH-PN synchronization sequence will experience lower

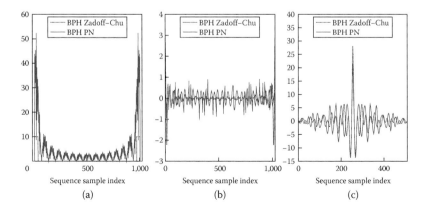

FIGURE 14.8
BPH-PN and BPH Zadoff–Chu sequences compared (a) in the frequency domain, (b) in the time domain, and (c) autocorrelation.

distortion resulting from the LEDs non-linear response, as compared to the BPH Zadoff–Chu (which exhibits a much higher dynamic range).

14.4.3 Performance Results

The implemented DCO-OFDM transceiver was tested at different levels of the implementation flow. First, its performance was evaluated through simulation, using the tools available in System Generator and Simulink. Once the correct operation was verified, functionality was evaluated in real time with the analog transmitted signal connected to the receiver's with a sub-miniature version A (SMA) cable (here referred as back-to-back electric). Results were obtained by resorting to four ChipScope ILAs implemented in key locations in the transceiver. Figure 14.9 shows the signal constellation for QPSK, 16-QAM and 64-QAM. Back-to-back electric results show an excellent transceiver performance, with a negligible signal-to-distortion noise ratio (56 dB for QPSK).

The system's performance was then evaluated using the setup depicted in Figure 14.10. The digital signal generated in the TX FPGA is converted to the analog domain by the FMCOMMS1 DAC. The analog signal is then passed through a variable attenuator, amplified (Mini-Circuits ZFL-500+ followed by a ZHL-6A+), and then fed to a single blue LED via a bias-T. The bias point was set to 3.5V in order to maximize transmitted power and minimize signal distortion. The receiver includes a plano-convex lens coupled and the Hamamatsu receiver, followed by a Mini-Circuits ZFL-500+ amplifier. The signal was then fed to the FMCOMMS1 ADC connected to the RX FPGA board.

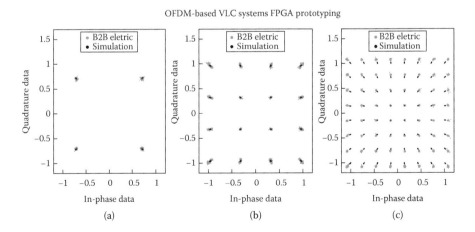

FIGURE 14.9
Signal constellations for (a) QPSK, (b) 16-QAM, and (c) 64-QAM.

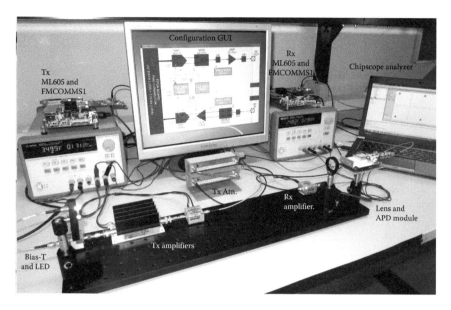

FIGURE 14.10
Experimental setup.

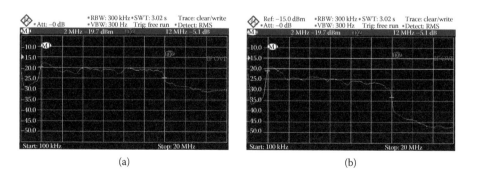

FIGURE 14.11
For a 30 MHz DAC/ADC sample frequency, spectrum of (a) transmitted signal and (b) received signal for TX @ 1.5 m distance.

With the design parameters described in Table 14.9, and for a 1.5 m distance between TX and RX the transmitted and received signal spectrum can be observed in Figure 14.11. The transmitted signal was captured after the bias-T, with 22 dB set in the variable attenuator, and the received signal after the RX amplifier. In the pictures, the synchronization sequence can be easily observed in the lower end of the spectrum, as it is transmitted with 6 dB more power than the OFDM signal.

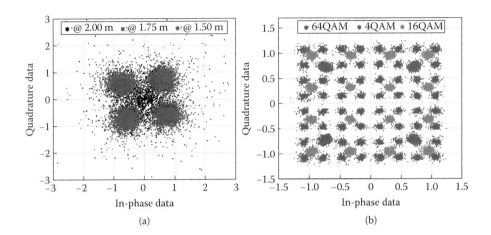

FIGURE 14.12

Received constellations: (a) QPSK for increasing distances and 12 MHz modulation bandwidth and (b) *M*-QAM at a fixed distance (50 cm), for increasing M and 25 MHz modulation bandwidth.

In [34], the system's performance was evaluated for increasing distances. Results were obtained using a PRBS generator in the transmitter and a BER measurement system in the receiver (both implemented in the FPGA). The BER was measured for increasing distances up to 2 m. At distances below 1.5 m, errors were not detected after the transmission of several millions bits. So, it is reasonable to assume that BER is below 1×10^{-6}. At greater distances, BER was measured to be 1.8×10^{-4} @ 1.75 m and 2.2×10^{-3} @ 2 m, both below the FEC limit.

Figure 14.12a presents the signal constellations obtained at the distances of 2 m, 1.75 m, and 1.5 m. The measured signal-to-noise ratio (averaged over the 400 carriers) for the depicted distances was 15 dB, 17.9 dB, and 22.6 dB, respectively. The image confirms the degradation of the constellation as the distance increases. Transmission at higher distances would require higher transmitted optical power, either by increasing the modulation depth or increasing the number of transmitting elements (LEDs). Note that these results are based on the usage of a single LED, which is clearly lower than what would be required for illumination purposes.

To increase data rates, the transceiver's performance was optimized by pipelining some datapaths that were limiting performance. With a higher digital throughput, it was possible to increase the modulation bandwidth to 25 MHz, with 470 OFDM carriers, thus increasing the achievable data rates [35]. Figure 14.12b shows the received constellations for QPSK, 16-QAM, and 64-QAM, achieving a maximum net data rate of 150 Mbps over a distance of 50 cm. To the authors' knowledge, this is the higher data rate achieved in a real-time VLC prototype (i.e., excluding the commercial

VLC system reported in [29]). The experimental setup used to obtain these results was similar to the one depicted in Figure 14.10, except for an additional plano-convex lens in the TX.

14.5 Conclusions

This chapter focused on the implementation of real-time, high data-rate VLC systems in FPGAs. The goal was to provide the VLC system architect some key information about the practical means available to convert simulation-based high-speed prototypes into working hardware. This would help providing clear evidence of the capabilities of this new technology and help mature DSP algorithms and modulation schemes (either already proposed in literature or not). In fact, most of them are based in simulation-based platforms and thus, their performance and robustness cannot undoubtedly be certified without real-time measurements and real conditions of use.

With this goal in mind, the chapter started providing an overview of the envisioned future high-speed VLC applications and the current state of the art in what concerns proof-of-concept demonstrators. To help the designer choose an adequate FPGA platform for system implementation, Section 14.2 provided an overview of currently available development boards, mezzanine cards, and software tools suitable for the implementation of a high-speed VLC prototype. Some design and implementation issues regarding the FPGA implementation of complex DSP systems in FPGAs were discussed in Section 14.3. Finally, Section 14.4 presented an FPGA-based VLC prototype, describing implementation details and providing performance results.

Authors expect to have provided useful information and awakened the interest of the VLC research community to the possibility of using FPGAs to implement real-time demonstrators. FPGAs are increasingly becoming the platform of choice for prototyping (and sometimes also deploying) complex communication systems. With multiple high-speed VLC applications being sought for future niche applications, where custom hardware solutions may not be cost effective, FPGA-based solutions will definitely be around.

Acknowledgments

This work is funded by FCT/MEC through national funds and when applicable cofunded by FEDER PT2020 partnership agreement under the project UID/EEA/50008/2013. The work of Carlos Ribeiro was financially supported by FCT/MEC and its funding program under the postdoctoral grant SFRH/BPD/104212/2014.

References

[1] IEEE Std 802.15.7-2011. *IEEE Standard for Local and Metropolitan Area Networks—Part 15.7: Short-Range Wireless Optical Communication Using Visible Light.*, Pages 1–309, 2011.

[2] H. Haas. *TED Talks, Li-Fi: Technology Light*, 2011. https://www.youtube.com/watch?v=gjqSgsKbagQ (accessed February 2017).

[3] IP Nexus. *Fujitsu Transmits Data Using Video Get Info on your Phone Right from your TV Screen*, 2013. https://www.youtube.com/watch?v=l75zp5qnMcY (accessed February 2017).

[4] Disney Research. *(In)visibleLight Communication: Combining Illumination and Communication*, 2014. https://www.youtube.com/watch?v=vnlv7EjRKqY (accessed February 2017).

[5] Channel Panasonic. *Smartphone Optical Data Communication Technology—Light ID Technology*, 2015. https://www.youtube.com/watch?v=UKI-Tw8sAvM. (accessed February 2017)

[6] Philips Lighting. *Philips LED Indoor Positioning Technology at Carrefour*, 2015. https://www.youtube.com/watch?v=uQw-o6bjrec (accessed February 2017).

[7] Transparency Market Research. *Visible Light Communication Market—Global Industry Analysis, Size, Share, Growth, Trends and Forecast 2015–2022*. Transparency Market Research, Albany, NY, 2014.

[8] S.-H. Yu, O. Shih, H.-M. Tsai, et al. Smart automotive lighting for vehicle safety. *IEEE Commun. Mag.*, Pages 50–59, December 2013.

[9] M. Miki, E. Asayama, and T. Hiroyasu. Intelligent lighting system using visible-light communication technology. In *2006 IEEE Conference on Cybernetics and Intelligent Systems*, Pages 1–6, 2006.

[10] H. Elgala, R. Mesleh, and H. Haas. Indoor optical wireless communication: Potential and state-of-the-art. *IEEE Commun. Mag.*, 49(9):56–62, 2011.

[11] W. Chunyue, W. Lang, C. Xuefen, et al. The research of indoor positioning based on visible light communication. *China Commun.*, 12(8):85–92, 2015.

[12] G. Baiden, Y. Bissiri and A. Masoti. Paving the way for a future underwater omni-directional wireless optical communication systems. *Ocean Eng.*, 36(910): 633–640, 2009.

[13] G. Cossu, R. Corsini, A.M. Khalid, et al. Experimental demonstration of high speed underwater visible light communications. In *2013 2nd International Workshop on Optical Wireless Communications (IWOW)*, Pages 11–15, October 2013.

[14] D.R. Dhatchayeny, A. Sewaiwar, S.V. Tiwari, et al. EEG biomedical signal transmission using visible light communication. In *2015 International Conference on Industrial Instrumentation and Control (ICIC)*, Pages 243–246, May 2015.

[15] R. Prasad, A. Mihovska, E. Cianca, et al. Comparative overview of UWB and VLC for data-intensive and security-sensitive applications. In *2012 IEEE International Conference on Ultra-Wideband (ICUWB)*, Pages 41–45, September 2012.

[16] D. Krichene, M. Sliti, W. Abdallah, et al. An aero-nautical visible light communication system to enable in-flight connectivity. In *2015 17th International Conference on Transparent Optical Networks (ICTON)*, Pages 1–6, July 2015.

[17] G. Corbellini, K. Aksit, S. Schmid, et al. Connecting networks of toys and smartphones with visible light communication. *IEEE Commun. Mag.*, 52(7):72–78, 2014.

[18] M. LaMonica. Philips creates shopping assistant with LEDs and smart phone. *IEEE Spectrum Mag.*, 2014. Available: http://spectrum.ieee.org/techtalk/computing/networks/philips-creates-store-shopping-assistant-with-leds-and-smart-phone (accessed February 2017).

[19] P. Luo, M. Zhang, Z. Ghassemlooy, et al. Experimental demonstration of RGB LED-based optical camera communications. *IEEE Photon. J.*, 7(5):1–12, 2015.

[20] T. Yamazato, I. Takai, H. Okada, et al. Image- sensor-based visible light communication for automotive applications. *IEEE Commun. Mag.*, 52(7):88–97, 2014.

[21] H. Aoyama and M. Oshima. Visible light communication using a conventional image sensor. In *2015 12th Annual IEEE Consumer Communications and Networking Conference (CCNC)*, Pages 103–108, January 2015, Las Vegas, NV, USA.

[22] S. Wu, H. Wang, and C.-H. Youn. Visible light communications for 5G wireless networking systems: From fixed to mobile communications. *IEEE Network*, 28(6):41–45, 2014.

[23] S. Zvanovec, P. Chvojka, P.A. Haigh, et al. Visible light communications towards 5G. *Radioengineering*, 24(1):1–9, 2015.

[24] S. Shao, A. Khreishah, M. Ayyash, et al. Design and analysis of a visible-light-communication enhanced WiFi system. *IEEE/OSA J. Opt. Commun. Networking*, 7(10):960–973, 2015.

[25] P.A. Haigh, Z. Ghassemlooy, S. Rajbhandari, et al. Visible light communications: 170 Mb/s using an artificial neural network equalizer in a low bandwidth white light configuration. *J. Lightwave Technol.*, 32(9):1807–1813, 2014.

[26] D. Karunatilaka, F. Zafar, V. Kalavally, et al. LED based indoor visible light communications: State of the art. *IEEE Commun. Surveys Tutorials*, 17(3): 1649–1678, 2015.

[27] K. A. Denault, M. Cantore, S. Nakamura, et al. Efficient and stable laser-driven white lighting. *AIP Adv.*, 3(7):072107-1–072107-6, 2013.

[28] K.-D. Langer, J. Vucic, C. Kottke, et al. Exploring the potentials of optical-wireless communication using white LEDs. In *2011 13th International Conference on Transparent Optical Networks (ICTON)*, Pages 1–5, June 2011.

[29] L. Grobe, A. Paraskevopoulos, J. Hilt, et al. High-speed visible light communication systems. *IEEE Commun. Mag.*, 51(12):60–66, 2013.

[30] H. Elgala, R. Mesleh, and H. Haas. Indoor broadcasting via white LEDs and OFDM. *IEEE Trans. Consum. Electron.*, 55(3):1127–1134, 2009.

[31] J. Vucic, L. Fernandez, C. Kottke, et al. Implementation of a real-time DMT-based 100 Mbit/s visible-light link. In *2010 36th European Conference and Exhibition on Optical Communication (ECOC)*, Pages 1–5, September 2010.

[32] J. Shi, X. Huang, Y. Wang, et al. Real-time bidirectional visible light communication system utilizing a phosphor-based LED and RGB LED. In *2014 Sixth International Conference on Wireless Communications and Signal Processing (WCSP)*, Pages 1–5, October 2014.

[33] C.H. Yeh, Y.L. Liu, and C.W. Chow. Demonstration of 76 Mbit/s real-time phosphor-LED visible light wireless system. In *2014 OptoElectronics and*

Communication Conference and Australian Conference on Optical Fibre Technology, Pages 757–759, July 2014.

[34] C. Ribeiro, M. Figueiredo, and Alves L.N. A real-time platform for collaborative research on visible light communication. In *2015 International Workshop on Optical Wireless Communications (IWOW)*, September 2015.

[35] M. Figueiredo, C. Ribeiro, and L.N. Alves. Live demonstration: 150Mbps+ DCO-OFDM VLC, *2016 IEEE International Symposium on Circuits and Systems (ISCAS)*, Montreal, QC, Canada, pp. 457–457, 2016.

[36] R. Mesleh, H. Elgala, and H. Haas. On the performance of different OFDM based optical wireless communication systems. *IEEE/OSA J. Opt. Commun. Networking*, 3(8):620–628, 2011.

[37] D. Falconer, S.L. Ariyavisitakul, A. Benyamin-Seeyar, et al. Frequency domain equalization for single-carrier broadband wireless systems. *IEEE Commun. Mag.*, 40(4):58–66, 2002.

[38] P.A. Haigh, S. Thai Le, S. Zvanovec, et al. Multi-band carrier-less amplitude and phase modulation for band limited visible light communications systems. *IEEE Wireless Commun.*, 22(2):46–53, 2015.

[39] F.M. Wu, C.T. Lin, C.C. Wei, et al. Performance comparison of OFDM signal and CAP signal over high capacity RGB-LED-based WDM visible light communication. *IEEE Photon. J.*, 5(4):7901507, 2013.

[40] Xilinx Inc. *System generator for DSP—User Guide*. http://www.xilinx.com/support/documentation/sw manuals/xilinx11/sysgen user.pdf [accessed October 2015].

[41] Altera Inc. *DSP Builder Introduction*. https://www.altera.com/en us/pdfs/literature/hb/dspb/hbdspbintro.pdf [accessed October 2015].

[42] S. Tam. *Clocking in Modern VLSI Systems*. Springer Science + Business Media, New York, 2009.

[43] *Xilinx User Guide. 7 Series FPGAs Clocking Resources, UG472 (v1.8)*, 2013.

[44] E. Stavinov. *100 Power Tips for FPGA Designers*. 2011.

[45] *Xilinx User Guide. Virtex-6 FPGA DSP48E1 Slice, UG369 (v1.3)*, 2014.

[46] J.E. Volder. The CORDIC trigonometric computing technique. *IRE Trans. Electron. Comput.*, EC-8(3):330–334, 1959.

[47] J.S. Walther. A unified algorithm for elementary functions. In *Proceedings of the Spring Joint Computer Conference*, Pages 379–385, 1971.

[48] K. Chapman. *Multiplexer Design Techniques for Datapath Performance with Minimized Routing Resources*, Xilinx Application Note 522, 2014.

[49] K. Chapman. *Get your Priorities Right Make your Design Up to 50% Smaller*, Xilinx White Paper WP275 (v1.0.1), 2007.

[50] SrikanthErusalagandi. *How do I resetmy FPGA, Xcell Journal, 3rd Quarter 2011*. Available: http://www.eetimes.com/document.asp?docid=1278998 (accessed February 2017).

[51] D. Mills, C.E. Cummings, and S. Golson. Asynchronous & synchronous reset design techniques—Part Deux. *Proceedings of the Synopsys User Group Conference (SNUG)*, 2003.

[52] *Xilinx User Guide. Virtex-6 FPGA Memory Resources, UG363 (v1.8)*, 2014.

[53] Altera White Paper. *Understanding Metastability in FPGAs*, WP-01082-1.2, 2009.

[54] J.J. van de Beek, M. Sandell, and P.O. Borjesson. ML estimation of timing and frequency offset in OFDM systems. *IEEE Transactions on Signal Processing*, 45 (7):1800–1805, 1997.

[55] Xilinx Inc. *LogiCORE IP Fast Fourier Transform v7.1*, Xilinx, 2011.

15

Smart Color-Cluster Indoor VLC Systems

Yeon Ho Chung

CONTENTS

15.1 Introduction

Visible light communication (VLC) systems are applied in various environments such as indoors and outdoors. Among these, an indoor application where lighting is usually provided all the time is one of the most attractive areas for this emerging technology. As conventional fluorescent lamps or incandescent lamps in an office environment are rapidly replaced by light-emitting diodes (LEDs), the VLC technologies based on LEDs can aptly be installed to provide a short-range wireless access via main indoor data networks.

On the other hand, this indoor VLC needs to be considered in the context of existing or potential short-range wireless access schemes. For VLC to be a compelling technology, it needs to be delivered in a smarter way in terms of efficiency, quality of services, and link performance. In particular, it has to offer diverse service applications encompassing indoor users as well as ever-increasing smart devices. This needs to be so, regardless of their locations, for example, in the vicinity or around far corners of an indoor office, and other existing multiple users or networks. To meet this demand, a *color-cluster* based VLC transmission technology is considered and is aptly termed as smart color-cluster indoor VLC schemes.

In this chapter, we consider diverse indoor VLCs that are designed to meet the challenges ahead. First, a color-clustering scheme is considered to provide a relatively high data rate by making use of available colors in LEDs. The scheme also offers a bidirectional transmission method in the color-cluster scheme. Recognizing the fact that these schemes are unable to provide users (or devices) mobility support, an enabling technology for a mobility supporting scheme is discussed.

In a recent trend associated with smart homes, VLC is also a good candidate in the context of smart indoor wireless technology. For the discussion of VLC-based smart home technologies, color-code multiuser schemes are introduced. Considering the optical shadowing effect existent in an indoor environment, a unique design of an optical bidirectional beacon (OBB) is also illustrated. The OBB can ensure seamless coverage over various smart devices present in the smart home environment.

An innovative idea of motion detection on top of VLC-based short-range communications is finally introduced. This technology does not affect existing VLC systems and illumination; instead, it provides detectability of motion by measuring the received signal strength from photodiodes. This technique is anticipated to broaden conventional VLC horizons. In the following section, we provide a detailed description of color-cluster based indoor VLC systems, together with the principle of color-clustering to achieve a multiple access scheme. In addition, mobility support is described in the color-cluster systems.

15.2 Color-Cluster Indoor VLC Systems

In a color-cluster environment, a color cluster is formed by dividing an array of red, green, and blue (RGB) LEDs into three primary colors [1–3]. The motivation behind the idea of a color-cluster scheme is that the entire spectrum of the visible light band is utilized and thus the spectral inefficiency in VLC can be minimized. Also, the modulation capability of the LEDs is exploited to its full extent and thus it achieves a high-speed data transmission in the indoor VLC systems. Figure 15.1a shows the color-clustered indoor VLC environment. Figure 15.1b shows its top view [1,2,4].

15.2.1 Principle of Color Clustering

The main principle of the color-clustering of multiple users is that the users are assigned to the three primary colors: RGB. Therefore, the users are clustered into specific visible spectral bands known as colors. The intensity modulated user data are transmitted via the RGB LEDs. As an example, a red color-clustering is visualized in Figure 15.2. The data are first modulated

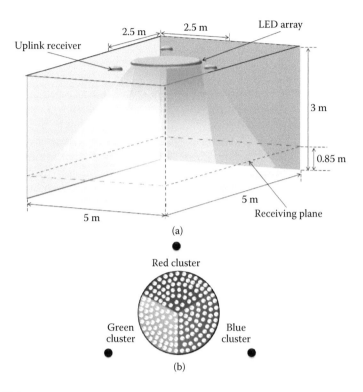

FIGURE 15.1
(a) Color-cluster indoor VLC system and (b) top view. (From Bandara, K., and Chung, Y.-H., *Trans. Emerg. Telecommun. Technol.*, 25, 579–590, 2014. With permission; Sewaiwar, A., et al., *Opt. Commun.*, 339, 153–156, 2015. With permission; Sewaiwar, A., et al., *IEEE Photon. J.*, 7, 7904709(1–9), 2015. With permission.)

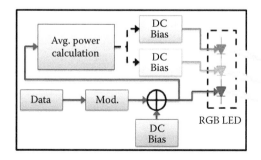

FIGURE 15.2
Red color-cluster data transmission.

and transmitted through the red color and the other two colors are provided with an average DC bias so as to provide white color for illumination.

15.2.2 Color-Cluster Multiuser VLC

In the color-cluster (CC) multiuser VLC system, different users are allocated into three primary colors defined as color clusters; cluster r, cluster g, and cluster b. Using on-off keying (OOK), the data of the users in each cluster are modulated with RGB LEDs individually and simultaneously.

At the receiver, an RGB color sensor is used to detect the intensity of each beam. Clearly, the color sensor provides separate voltages proportional to the detected R, G, and B intensities. Thus, the users of each color cluster can be separated at the receiver. Since there are only three color clusters, only up to three users can transmit data at a time. Therefore, in order to increase the user capacity of the proposed CC multiuser VLC scheme, we allocate more users in one CC by assigning specific intensity to each user within the allocated CC. Since a single LED can produce a single intensity at a time, we use a set of LEDs with the specified intensity for every user in the system. That is, a particular set of LEDs is reserved to transmit the data of a specific user. Therefore, this design can increase the user capacity significantly with an increased set of intensities. Figure 15.3b depicts a summary of the user allocation process. In Figure 15.3b, we assume that the total number of users, N, is three times the number of users within each color cluster, K, so $N = 3K$ [1].

In one transmitter, three sets of LEDs for the three color clusters are used. In each color cluster, a subset of LEDs is allocated for one particular user as depicted in Figure 15.3a. All the LEDs shown in Figure 15.3a are RGB LEDs and the colors indicate the cluster color—the beam used to modulate user data in each color cluster. The number of subsets in one color cluster depends on the number of users connected to the VLC system at a given time. As described previously, the modulation intensity of OOK in a single LED subset is fixed for a given user.

As optical data transmission is often impaired by time dispersion along the transmission path, the transmitted signal would undergo distortion. To compensate for this adverse effect, the receiver needs to employ an equalizer. In the current scheme, a decision feedback equalizer (DFE) could be employed in the receiver. For the DFE equalizer, the recursive least square adaptive algorithm could be used with a training sequence of 500 symbols, so as to achieve the fast convergence of the adaptive filter coefficients. The receiver structure of the MU-VLC is depicted in Figure 15.4.

In the receiver, the first operation is to separate the users. This user separation is performed on the basis of the fact that the users within a single color cluster are assigned with a specific intensity for the *on* state of the OOK, and the intensity of the *off* state is kept the same for all users. It should be noted that the term *user intensity* refers to the intensity of the *on* state of a particular user. The received intensity of each color can be identified using an RGB

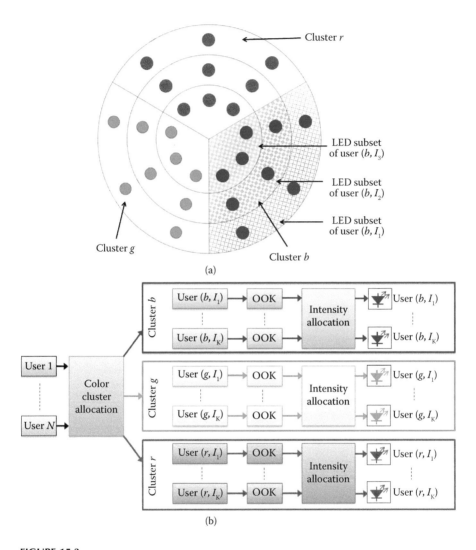

FIGURE 15.3
(a) Transmitter design with multiple LEDs and (b) CC and intensity allocations of N users ($N = 3\,K$). (From Bandara, K., and Chung, Y.-H., *Trans. Emerg. Telecommun. Technol.*, 25, 579–590, 2014. With permission.)

color sensor, which separately converts the RGB intensities into voltages. In this way, the user separation is performed, extracting the composite signal in the particular user color from the received composite color beam.

The second operation to perform is to detect the user data within the color cluster. Figure 15.5a shows the transmitted data of three users with relative intensities of 10, 5, and 2.5 in a cluster b. These intensities are relative to the intensity at the *off* state. An example of the composite signal made by adding

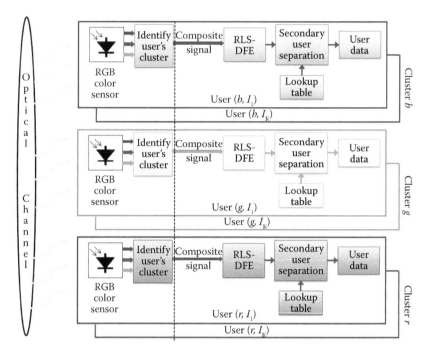

FIGURE 15.4
Structure of the receiver.

individual user intensities of the three users is shown in Figure 15.5b [1]. The composite signal contains eight intensities of 17.5, 15, 12.5, 10, 7.5, 5, 2.5, and 0. In fact, the composite signal resembles an eight-level pulse-amplitude-modulated (8-PAM) signal and each intensity level contains the information of three users. It should be noted that the PAM in this scheme is used as a multiuser signal for each user.

If K number of users is considered within one cluster, the simultaneous transmission of three user signals results in a composite signal that has a maximum of 2^K intensities. As shown in Figure 15.5a and b, if there are three users in the cluster r, there are eight intensities in the composite signal. Since the LEDs are *on* or *off* state continuously, the composite intensity contains either logic 0 or 1 of each user. Therefore, the composite intensity can be obtained, according to the user data. As an example, Table 15.1 shows a look-up table of the composite intensity for the three users. It is apparent that according to Table 15.1, there are specific intensity values for the composite signal where each user's data are in either the *on* or *off* state.

By performing perfect equalization at the receiver, the received signal would be the same as the transmitted composite signal. It is obvious that

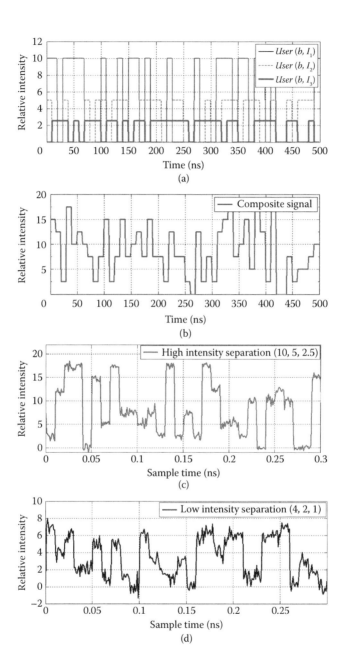

FIGURE 15.5
(a) OOK modulated data of the three users, (b) the composite signal of the three users, (c) received composite signal with high user intensity separation, and (d) low user intensity separation at a signal-to-noise (SNR) value of 5 dB. (From Bandara, K., and Chung, Y.-H., *Trans. Emerg. Telecommun. Technol.*, 25, 579–590, 2014. With permission.)

TABLE 15.1

Lookup Table for User Separation from the Received Signal Intensity (RSI) in One Cluster

User				OOK state				
(b, I_1)	*on*	*on*	*on*	*on*	*off*	*off*	*off*	*off*
(b, I_2)	*on*	*on*	*off*	*off*	*on*	*on*	*off*	*off*
(b, I_3)	*on*	*off*	*on*	*off*	*on*	*off*	*on*	*off*
RSI	17.5	15	12.5	10	7.5	5	2.5	0

after comparing each received composite intensity level with the intensity levels mentioned in the lookup table (see Table 15.1), each user's receiver can estimate the transmitted symbol for that user. The lookup table can readily be generalized to any number of the users in the MU-VLC. Supposing that the number of users connected to each color cluster is known to the user, all the users sharing the same color cluster can generate the lookup table according to the number of users present in order to extract the user data.

In the presence of various noises, such as thermal noise, shot noise, etc., the received composite signal can be distorted so that the intensities cannot be perfectly distinguished at the receiver. In addition to the various noises, the VLC channel will experience multipath-induced intersymbol interference (ISI), which can cause the received signal to be indistinguishable thus leading to performance deterioration. In order to examine the effect of the noise on the composite signal, we assigned two sets of user intensities to the three users: one with high separation between relative user intensities {10, 5, 2.5} and the other with low separation {4, 2, 1}. Figure 15.5c and d show the vulnerability of the composite signal to the noise with the two user intensity sets. Clearly, the composite signal distortion becomes lower if there is a higher intensity separation between user signals, thereby leading to a lower bit error rate (BER).

MU-VLC transmission is simulated under the additive white Gaussian noise (AWGN) channel. Figure 15.6a shows the symbol error rate (SER) performance for three users in the cluster r that have the intensities $I_1 = 10$, $I_2 = 5.0$, $I_3 = 2.5$. Additionally, the average root mean square (rms) delay spread over the entire receiver plane (0.85 m above the floor in a room with dimensions 5 m × 5 m × 3 m) is found to be approximately 1.1 ns [1]. Also, the minimum rms delay spread over the realistic receiver locations in the room is approximately 0.4 ns [1]. These values correspond to the achievable data rates of approximately 90 Mbps and 270 Mbps, respectively, under flat-fading channel transmission environments. The SER performances of the multiuser transmission at these two values of rms delay spreads are also depicted. The simulation results are shown in Figure 15.6b [1].

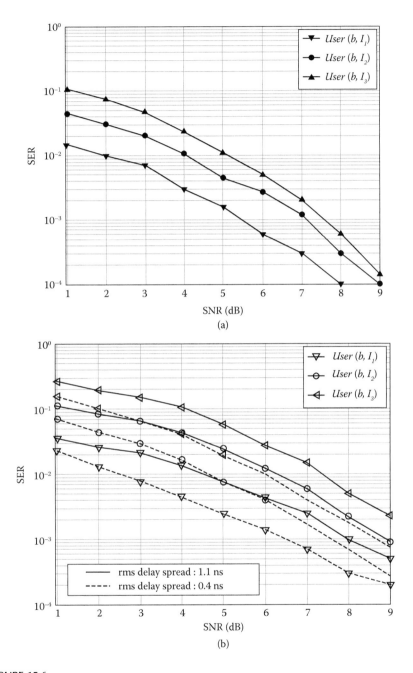

FIGURE 15.6
(a) SER performance of the MU-VLC under AWGN and (b) SER performance with the rms delay spread values of 0.4 ns and 1.1 ns. The relative user intensities (I_1, I_2, I_3) are {10, 5, 2.5}. (From Bandara, K., and Chung, Y.-H., *Trans. Emerg. Telecommun. Technol.*, 25, 579–590, 2014. With permission.)

15.2.3 Mobility-Supported User Allocation in Color-Cluster VLC

As previously noted, the mobility support needs to be considered in the smart color-cluster indoor VLC. This mobility support could have an impact on the user allocation scheme in an indoor full-duplex bidirectional light fidelity (Li-Fi) system. To address this issue more effectively, a user allocation and detection process based on a predefined structure called the "frame" is considered [2]. The key technology of the mobility-supported VLC is CC—three distinctive colors [1]. In a particular CC, the user data are modulated and transmitted using the red or green or blue color beam of RGB LEDs depending on the cluster. At the receiver end, the initial user separation is performed using a color filter that filters respective color. The photodiode (PD), installed behind the color filter, provides the output as individual voltage proportional to the intensity of each color. The uplink data transmission from the receiver is performed by modulating the data using a different color from the one used at the reception.

As shown in Figure 15.1a, the downlink transmitter is an array of LEDs in which each transmitter is composed of three sets of LEDs for the three color clusters. Along with the transmitter, the uplink receiver units with specific color filters for each color cluster are installed. For a particular color cluster, the color used for downlink data transmission is different from the one for uplink data transmission (see Table 15.2), thus resulting in reduced interference between the transmitted data and the received data.

In the mobility-supported CC VLC scheme, three types of frames—*synchronization frame, acknowledgment frame,* and *data frame*—are employed as shown in Figure 15.7 [2]. The functionality of each frame is described. First, the synchronization frame is used by the downlink transmitter (LED array) for broadcast purposes. The synchronization frame is made distinguishable by a 1010 … sequence pattern in the field of data bits and the "available/occupied" bit is always "0", available, and the "uplink/downlink" bit is set to "1", for downlink transmission. This *synchronization* frame is received by the transceiver unit and the *acknowledgment* is sent back to the uplink receiver using the acknowledgment frame. Figure 15.8 shows the simple difference between *synchronization* and *acknowledgment frames* [2].

Data frames represent the actual data bits to transmit (uplink or downlink). The "available/occupied" bit is always "1"—occupied—for data transmission.

TABLE 15.2

Uplink Color Allocation for Each Received Color

Received Color	Uplink Color
Red	Blue
Blue	Green
Green	Red

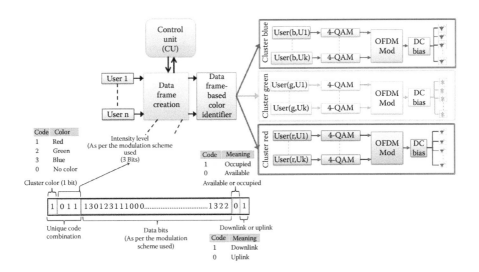

FIGURE 15.7
User allocation process for the CC-based bidirectional VLC network with frame structure. (From Sewaiwar, A., et al., *Opt. Commun.*, 339, 153–156, 2015. With permission.)

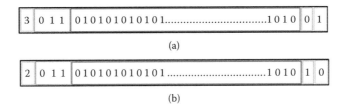

FIGURE 15.8
Structure of (a) received synchronization frame and (b) acknowledgment frame. (From Sewaiwar, A., et al., *Opt. Commun.*, 339, 153–156, 2015. With permission.)

The color bit is set according to the cluster on downlink, while for uplink, it is set according to Table 15.2. The "uplink/downlink" bit will be set to "1" when LED array transmits (downlink), whereas it will be "0" from the user device (uplink). Acknowledgment (ACK) and negative acknowledgment (NACK) for *data frames* can be performed by using a special sequence of bits in the place of data bits in the frame.

As usual in the CC-based VLCs, the scheme operates first by allocating the users with the three primary colors defined as CCs. The data of the users in each cluster are then modulated using 4-QAM. For the multiplexing of the modulated user data, although the OOK-based methods described in 15.2.2 could be applied, we utilize an orthogonal frequency division multiplexing (OFDM)-based scheme, called orthogonal frequency division multiple access (OFDMA), for improved performance and data rate. In the VLC system, since OFDM cannot be applied in its original form, there is a need for providing a DC bias and therefore DCO-OFDMA is applied [5]. Figure 15.7 illustrates the user allocation process. The performance assessment of the CC-based user allocation process is shown in Figure 15.9 [2], in terms of the BER with respect to SNR, and the data rate as a function of the number

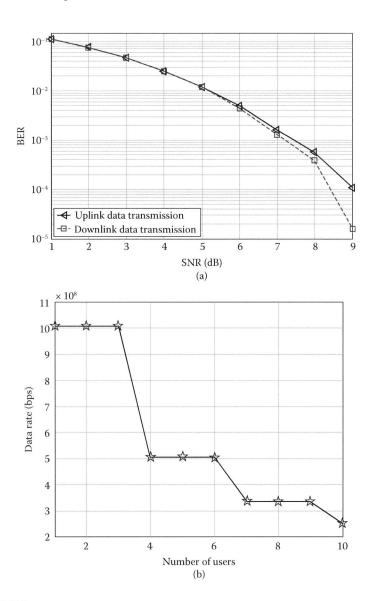

FIGURE 15.9
(a) BER performance and (b) data speed relative to the number of users. (From Sewaiwar, A., et al., *Opt. Commun.*, 339, 153–156, 2015. With permission.)

of users. The downlink BER distribution in a typical indoor environment is also evaluated and shown in Figure 15.10 [2].

In a CC VLC-based network, if the user is mobile, then it may not be able to maintain communication satisfactorily at all times. Figure 15.11a illustrates the user movement from one cluster to another, while Figure 15.11b shows its top view [4]. Apparently, there is a need for mobility support, regardless of whether the user moves in and around the cluster at a certain speed, v.

The user movement can be classified into two categories: *intracluster* and *intercluster* movements. Intracluster movement refers to the user movement in the same cluster in which the link is initially established, shown in Figure 15.11c. Intercluster movement refers to the user movement between clusters. Figure 15.11d depicts the intercluster movement of the user. Although these two categories of movement cause user performance degradation in the multiuser bidirectional VLC system, it is intuitively true that the intercluster movement is more detrimental to the link quality. Therefore, it is justifiable to focus on addressing the intercluster movement. To address this intercluster movement, one can employ a color filter array (CFA). A commonly used CFA is known to be the Bayer filter array [6] and a 3×3 Bayer filter is shown in Figure 15.12a. The CFA typically gives information about the intensity of light in RGB wavelength regions. In order to exploit a more substantial diversity effect from the CFA, a slightly modified form of the CFA shown in Figure 15.12b can be applied [4]. That is, it is modified to form an equal number of color filters.

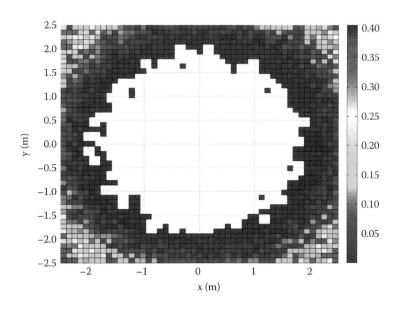

FIGURE 15.10
Downlink BER distribution in an indoor environment. (From Sewaiwar, A., et al., *Opt. Commun.*, 339, 153–156, 2015. With permission.)

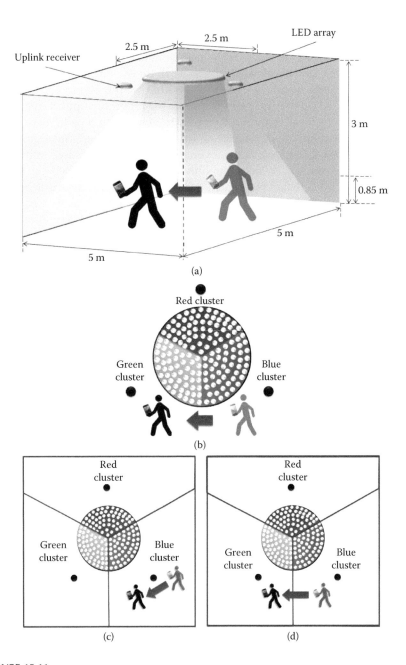

FIGURE 15.11
(a) User movement in a color-cluster environment, (b) its top view, (c) intracluster movement, and (d) intercluster movement. (From Sewaiwar, A., et al., *IEEE Photon. J.*, 7, 7904709(1–9), 2015. With permission.)

Despite this adoption of the CFA, it is apparent that it is not possible to fully filter out the effect of other colors by simply employing RGB LEDs and CFA. This is due to the overlapping relative intensity profiles of CFA and RGB LEDs. In the present scheme, the receiver diversity is implemented via selection combining (SC), in order to compensate for the degradation. The receiver diversity is designed to obtain the most probable signal in the receiver, thereby reducing the error rate. Figure 15.13 shows the block diagram of CFA-based receiver diversity [4].

An algorithm and sequence diagram for mobility support is illustrated in [4].

G	R	G
B	G	B
G	R	G

(a)

G	R	B
B	G	R
R	B	G

(b)

FIGURE 15.12
(a) Bayer's CFA and (b) proposed CFA. (From Sewaiwar, A., et al., *IEEE Photon. J.*, 7, 7904709 (1–9), 2015. With permission.)

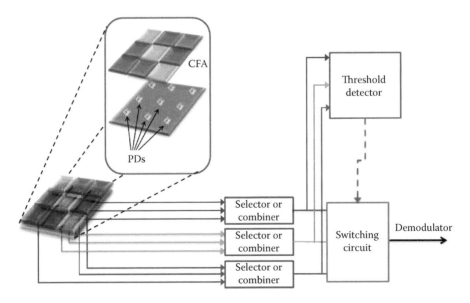

FIGURE 15.13
CFA-based receiver diversity. (From Sewaiwar, A., et al., *IEEE Photon. J.*, 7, 7904709(1–9), 2015. With permission.)

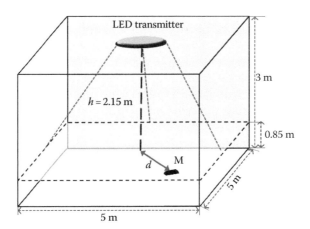

FIGURE 15.14
Indoor VLC environment.

In practice, communication delay would occur, due to the user speed, v. This delay is the time difference caused by the movement of the user over a certain distance. Suppose the communication radius is d with respect to the central point of the receiving plane, as shown in Figure 15.14.

For the intracluster movement, the communication delay, d_{intra}, is computed by [4]

$$d_{intra} = \frac{(\sqrt{d^2 + h^2}) - h}{v} \tag{15.1}$$

where h represents the height of the LED array from the receiving plane.

On the other hand, for the intercluster movement, the communication delay, d_{inter}, can be expressed as the sum of d_{intra} and the duration required for the cluster transfer, T. Therefore, d_{inter} is given by [4]

$$d_{inter} = d_{intra} + T = \frac{(\sqrt{d^2 + h^2}) - h}{v} + T \tag{15.2}$$

The frame loss rate, R_{loss}, can then be obtained as [4]

$$R_{loss} = \frac{N_l}{N_t} = \frac{D_c/D_f}{N_t} \tag{15.3}$$

where N_l, N_t, and D_f denote the number of frames lost, the total number of transmitted frames, and the duration of a single frame, respectively. D_c represents the communication delay, which is either d_{intra} or d_{inter}.

The mobility-supported CC VLC scheme is evaluated in the indoor environment. Figure 15.15a and b show the results in terms of the communication delay and frame loss rate relative to the communication radius, d, over

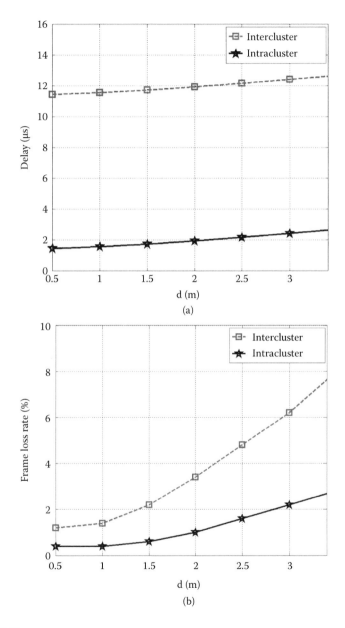

FIGURE 15.15
Effect of communication radius on (a) communication delay and (b) frame loss rate. (From Sewaiwar, A., et al., *IEEE Photon. J.*, 7, 7904709(1–9), 2015. With permission.)

intercluster and intracluster movements with the user speed fixed to 5 km/h [4]. As the communication radius increases, the communication delay and frame loss rate increase linearly for both categories of the movement. This is because the delay and frame loss rate become poorer as the user moves away from the transmitter.

The effect of the user speed is also evaluated with the communication radius fixed to 2 m. As shown in Figure 15.16a and b [4], the delay for the intercluster movement is higher than the one for the intracluster movement. In addition, the frame loss rate increases sharply for the intercluster movement with increasing speed. Therefore, it can be said that the user speed has a significant impact on the frame loss rate in the intercluster movement.

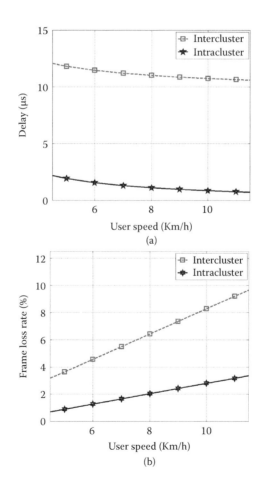

FIGURE 15.16
Effect of user speed on (a) communication delay and (b) frame loss rate. (From Sewaiwar, A., et al., *IEEE Photon. J.*, 7, 7904709(1–9), 2015. With permission.)

In evaluating the BER performance and data rate, we fix the user speed to a constant yet realistic speed of 5 km/h. Figure 15.17a shows that the BER performance of 10^{-3} is attained at a SNR value of approximately merely 5.8 dB for the intercluster downlink transmission. At the identical user speed, Figure 15.17b shows the data rate in terms of the number of users with 4-QAM-DCO-OFDMA with 128 subcarriers employed [2,4]. A minimum data rate of 110 Mbps is found to support when the number of intracluster users in a single cluster is 10. Interestingly, when 10 users are uniformly

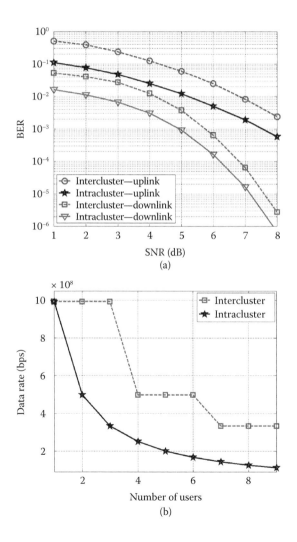

FIGURE 15.17
(a) BER performance relative to SNR values and (b) data rate relative to the number of users. (From Sewaiwar, A., et al., *IEEE Photon. J.*, 7, 7904709(1–9), 2015. With permission.)

distributed over all clusters, an achievable data rate sharply increases up to 250 Mbps at a minimum.

15.3 Smart Home VLC Technologies

Smart home technology has recently attracted much attention as smart devices are ubiquitous at home. The smart home is viewed as a home that has highly advanced automatic systems for lighting, temperature control, security, appliances, and many other functions. That is, household items, such as lamps, thermostats, and locks, are connected wirelessly and are becoming smarter. Up until now, there have been three main smart home technologies—Insteon [7], Z-Wave [8], and ZigBee [9,10]. All these technologies used the radio frequency spectrum that is expeditiously congested with an increasing number of users or devices to support and is, moreover, highly susceptible to hacking.

In the optical domain, infrared (IR)-based device control is common in consumer electronics and equipment. However, the use of IR may limit versatility in smart home technologies compared with VLC, because VLC usually offers illumination plus wireless connections and more diverse control with motion detection [11] for smart home devices. A VLC-based smart home can be conceived in such a way that visible light transmission and reception unit (or transceiver) is assumed to be installed over home appliances such as mobiles, laptops, television, or air conditioners. In addition, the control unit installed at the ceiling provides synchronization command and instructions to the devices on the basis of requests received from the users.

15.3.1 Color-Code Multiuser VLC

As one enabling smart home technology, color-code multiuser VLC is considered. This scheme can be called color-code multiple access (CCMA) [12] for smart home applications in the bidirectional multiuser VLC. The CCMA is designed to transmit the data bidirectionally for multiple devices by utilizing the individual color beams from RGB LEDs, where the red color is used for downlink data transmission, green for synchronization, and blue for uplink transmission. On top of this bidirectional transmission, an orthogonal code is allocated to each device. Therefore, the devices can transmit or receive the data exclusively by the allocated orthogonal codes. It is worth noting that for the smart home applications, the CCMA coupled with RGB LEDs is considered to support bidirectional transmission as the color-cluster multiuser scheme described in 15.2.2 is primarily designed for unidirectional transmission.

The pure visible light-based CCMA model for smart homes is shown in Figure 15.18. Any smart home application can consider the two types of

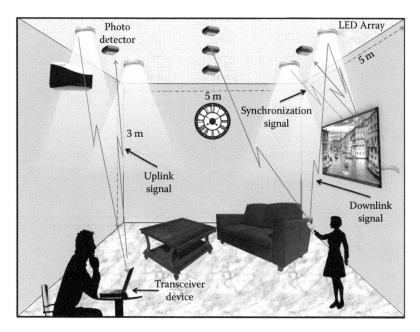

FIGURE 15.18
CCMA VLC-based smart home system.

devices: data user devices and smart home devices. As mentioned previously, the different colors are used for a full-duplex operation and an orthogonal code (Walsh code-based) code division multiple access (OCDMA) [13] is employed to facilitate efficient multiuser transmission. That is, an orthogonal code is assigned to each user and the data are spread for multiple users prior to transmission. In theory, these orthogonal codes will not cause any multiple access interference during data transmission.

The transmission process begins by assigning an orthogonal code sequence to each user where the data bit stream of N individual users are spread, according to the code length. The data of N users in different channels after spreading are multiplexed to form a serial data stream and transmitted simultaneously. The transmitted spread data for kth user is given by:

$$x_k(t) = \sum_{i=1}^{N} a_k(i) \sum_{j=1}^{N} b_k(j) h(t - ijTc) \tag{15.4}$$

where $a_k(i)$ represents the kth channel's information sequence, b_k is the code for kth channel, $h(t - ijT_c)$ is the spreading function. The spread parallel data chips of N channels are multiplexed to form a serial data stream. Figure 15.19 shows the block diagram of the proposed multiuser scheme for downlink transmission. Note that the information related to the length of frame for data

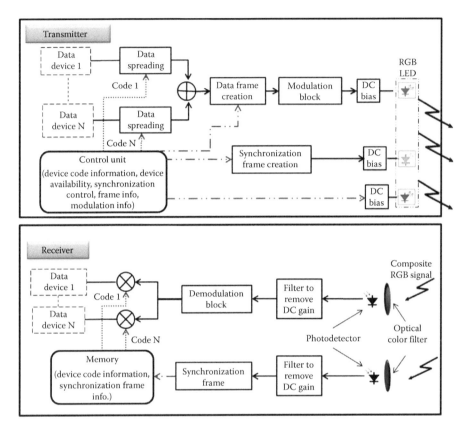

FIGURE 15.19
Block diagram of downlink multiuser transmission.

transmission, the number of devices available in a smart home, and the modulation information are assumed to be stored in the control unit.

As was the case for color (or color-cluster)-based VLC schemes, respective color optical filters are employed at the receiver to filter out for the actual data modulated using specific colors. Then, each user reconstructs the individual data by multiplying the received data with the user specific orthogonal code. The received optical signal can be expressed as:

$$r(t) = \eta x_k(t) + n(t) \tag{15.5}$$

where η is the efficiency of photodetector and $n(t)$ is the AWGN. Each PD measures the light intensity of the signal and provides the corresponding electrical signal. It is assumed that each receiver has knowledge of the orthogonal code used at the transmitter end. The demodulated data chips are demultiplexed by multiplying with their respective codes to recover the original data.

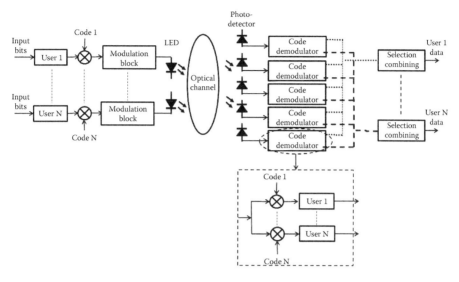

FIGURE 15.20
Block diagram of uplink multiuser transmission.

For the uplink transmission, we consider five receivers, taking into account the design constraints due to limited power (as shown in Figure 15.18) and employ SC to exploit the diversity effect in the indoor environment. Figure 15.20 depicts the operation of the receiver with SC. In principle, the SC is performed at the receiver where replicas of the transmitted signal are often received and then the most probable signal is selected, thereby significantly improving the performance.

The CCMA is evaluated in terms of the BER performance at downlink transmission. The room configuration shown in Figure 15.21a (Case I with single uplink receiver) and b (Case II with receiver diversity) is applied. The evaluation was conducted for all possible locations where devices can be placed within the area. Figure 15.22a shows the average BER distribution for devices at downlink. As anticipated, the locations near the transmitters show better performance as compared to the locations at the corner, due to stronger light intensity and near absence of ISI. Figure 15.22b shows a comparative analysis in terms of the average BER performance of four users for the downlink transmission of the proposed model (Case I and Case II), compared with previous multiuser schemes: color shift keying (4-CSK) [14] and OFDMA [2].

Figure 15.23a shows the average BER distribution measured across the room at uplink, indicating adequate performance at the locations near the center only. Thus, the uplink transmission requires a more rigorous scheme, given the design constraints and limited power availability. When the SC is employed (Case II, Figure 15.21b), the performance improves significantly for nearly all the locations within the room, except the far corners (Figure 15.23b). Note that

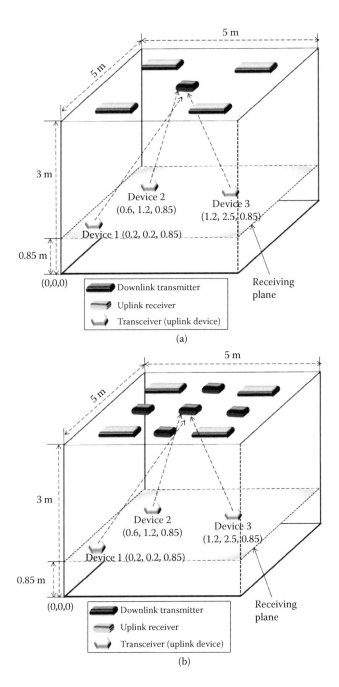

FIGURE 15.21
Indoor environment for the VLC-based smart home with (a) a single uplink receiver (Case I) and (b) receiver diversity (Case II).

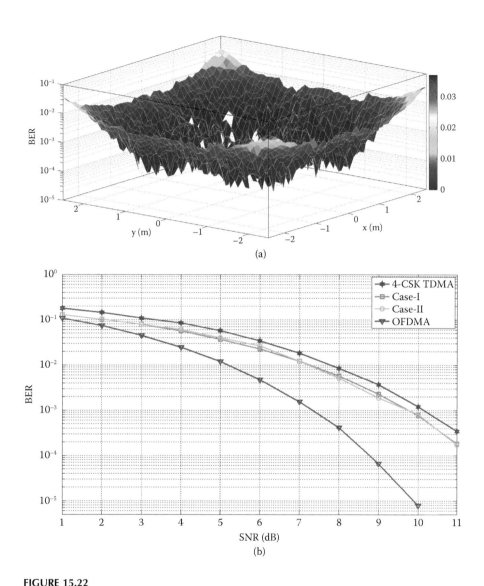

FIGURE 15.22
(a) Distribution of BER of devices for downlink transmission and (b) performance comparison
with conventional transmission schemes at downlink.

the BER distributions of both uplink and downlink transmissions are obtained
on the basis of the fact that the user device is located on the receiver plane
placed 0.85 m above the ground.

When compared with the OFDMA for four users at uplink, a significant
performance improvement is achieved as shown in Figure 15.24.

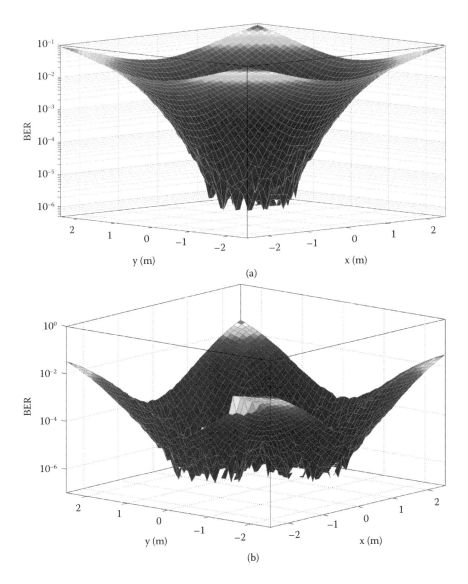

FIGURE 15.23
Distribution of BER of devices for uplink transmission (a) without receiver diversity and (b) with receiver diversity.

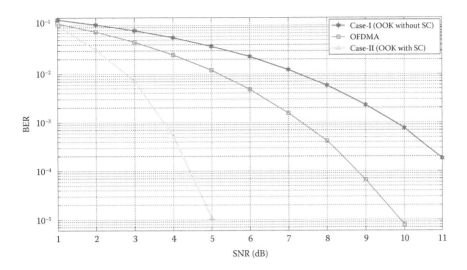

FIGURE 15.24
Performance comparison with conventional transmission schemes at uplink.

15.3.2 Optical Shadowing in Smart Home VLC

In the VLC, the performance depends largely on both the optical power received from the line-of-sight (LOS) transmission path and the distance between the LED transmitter and the receiver. Unfortunately, the LOS does not always exist, due to the obstruction caused by various movements, objects, or man-made structures in an indoor VLC environment. This optical obstruction can be called *optical shadowing*. Intuitively, it is true that the optical shadowing would cause the performance of the VLC system to degrade significantly. At uplink transmission, this degradation becomes even more severe due to design constraints and limited power at uplink devices.

Among various solutions to this optical shadowing, an optical bidirectional beacon (OBB) [15] can be considered. OBB is an independent operating bidirectional transceiver unit consisting of RGB LEDs, PDs, and color filters. Figure 15.25a exhibits a probable scenario for the optical shadowing in an indoor VLC environment [15]. As for the OBB, a conceptual design is shown in Figure 15.26 [15]. That is, the OBB is a transceiver unit composed of N number of RGB LEDs, a PD with a red optical color filter for downlink signal and another PD with a blue optical filter for uplink signal. The number of LEDs deployed in OBB can vary in accordance with practical applications and the dimension of an indoor environment.

Figure 15.27 shows an actual OBB-based VLC transmission environment. Two different colors are used for the bidirectional transmission; that is, the red color of RGB LEDs for downlink and the blue color of a phosphor-based

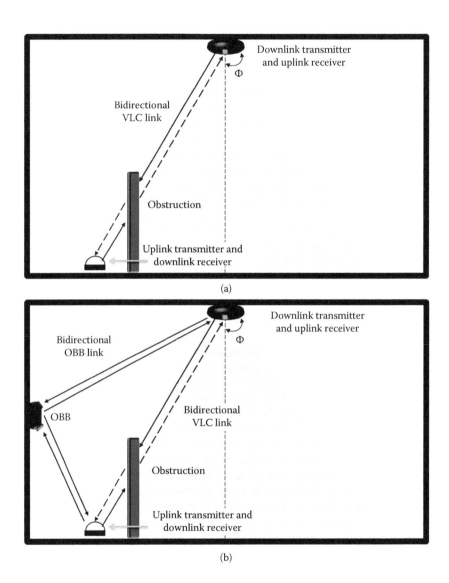

FIGURE 15.25
Optical shadowing scenarios in indoor VLCs (a) without OBB and (b) with OBB. (From Tiwari, S. V., et al., *Opt. Express*, 23, 26551–26564, 2015. With permission.)

white LED for uplink [12]. A single PD installed at the ceiling is used as an uplink receiver for the bidirectional communication. The OBB performance is evaluated by choosing the three representative locations—Device 1, Device 2, and Device 3 in Figure 15.27 [15].

A color filter-based bidirectional transmission scheme described in the previous sections is employed. It is important to note that during the

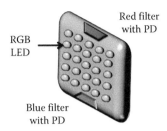

FIGURE 15.26
Illustration of optical bidirectional beacon (OBB) design. (From Tiwari, S. V., et al., *Opt. Express*, 23, 26551–26564, 2015. With permission.)

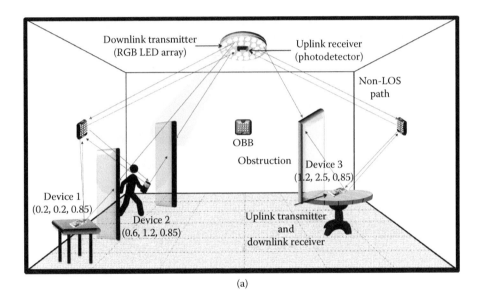

(a)

FIGURE 15.27
Indoor VLC environment with OBB. (From Tiwari, S. V., et al., *Opt. Express*, 23, 26551–26564, 2015. With permission.)

transmission from the OBB, the average power in the other two color components is changed according to the power variation in the blue color, in order to maintain flicker-free white color [1]. The transmitted signal from the OBB is received by the receiver installed at the ceiling that consists of a PD and the blue color filter. The PD installed at the ceiling detects the signal that is considered to be the largest over the bit time span, when the signals from both the device and the OBB are received at the PD. A similar operation is performed for the downlink transmission using the red color of RGB LEDs (transmission) and the red color filter (reception).

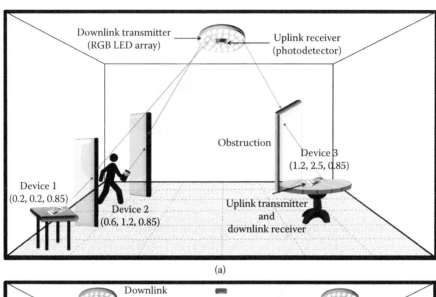

(a)

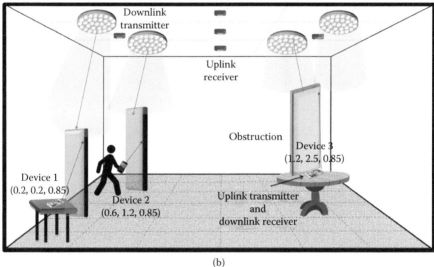

(b)

FIGURE 15.28
Conventional optical shadowing models (a) R1 and (b) R2. (From Tiwari, S. V., et al., *Opt. Express*, 23, 26551–26564, 2015. With permission.)

In order to verify the OBB scheme comprehensively, we introduce two conventional optical shadowing models. Reference 1 (R1) represents a single downlink transmitter and an uplink receiver at the center of the ceiling. It is this model that most VLC studies consider in terms of LED arrangement [2]. Reference 2 (R2) has four transmitters and five PDs. Figure 15.28 shows these two models [15].

Prior to the performance assessment, we consider a measure of the optical shadowing. The optical shadowing indicator (OSI) is introduced. Given an N_{shadow} as an OSI, the received power, P_{shadow}, can be obtained as [15]

$$P_{shadow} = \left[1 - \frac{N_{shadow}}{100}\right] P_{max} \qquad (15.6)$$

where P_{max} is the maximum received power at a particular location in the absence of any shadowing.

Figure 15.29a and b shows the BER variation for R1 and R2 models during downlink transmission, relative to the three locations—L1 (0.2, 0.2, 0.85), L2 (0.6, 1.2, 0.85) and L3 (1.2, 2.5, 0.85) [15]. It is observed that the BER performance degrades severely when the OSI values are higher than 50%, which makes communication nearly impossible, especially for L1. This is due to the fact that the receiver is located in the far corner region and suffers most from the optical shadowing.

Similarly, the impact of the optical shadowing for uplink transmission is analyzed. The results are shown in Figure 15.30a and b [15]. For an uplink transmission, the performance degradation becomes even more severe, due largely to the design constraints and limited power at uplink devices.

When the OBBs are placed on each wall as shown in Figure 15.27, the BER performance exhibits an improvement at the OSI value of 50%, as shown in Figures 15.31 and 15.32 [15].

It is evident from the results that the performance of the bidirectional communication link in a VLC system is significantly enhanced, by virtue of the OBBs in terms of BER performance, while maintaining the illumination level. In addition, this performance improvement is obtained with relatively less complicated circuitry and with no additional LEDs required, thus saving power consumption. In designing the OBBs, however, there would be a concern in terms of the esthetic aspect. The OBBs are supposed to be designed in harmony with their surroundings. A practical and viable design of the OBB compatible with surroundings could thus be required for commercial applications.

15.4 VLC-Based Motion Detection

In the smart indoor VLC, an additional feature can be considered on top of existing functionalities of communication and illumination. It is a novel motion detection based on VLC [11]. In principle, it operates based on an array of PDs, motion is detected by observing the pattern created by intentional obstruction of a VLC link.

The PD array provides not only the motion detection feature but also enhanced VLC performance via receiver diversity obtainable from multiple PDs. Without loss of generality, we assume a total of nine PDs employed

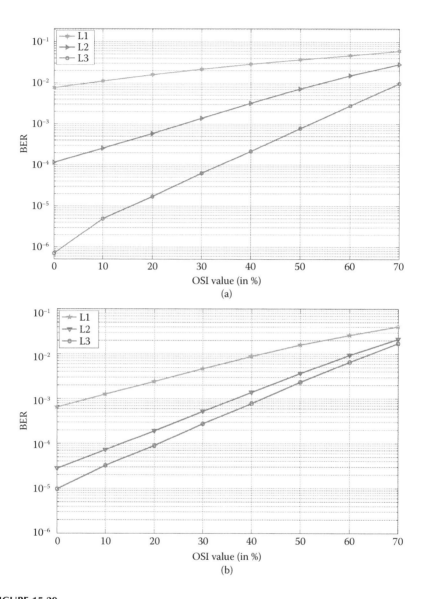

FIGURE 15.29
Downlink BER performance (a) R1 and (b) R2. (From Tiwari, S. V., et al., *Opt. Express*, 23, 26551–26564, 2015. With permission.)

in the present motion detection. Figure 15.33 shows the principle of the proposed motion detection technique [11]. For the data detection, we define two threshold levels, Th_1 and Th_0. That is, the intensity detected above Th_1 is considered "1" and the intensity detected between Th_1 and Th_0 is deemed "0." If the intensity goes below Th_0, it is "no data" (ND) condition.

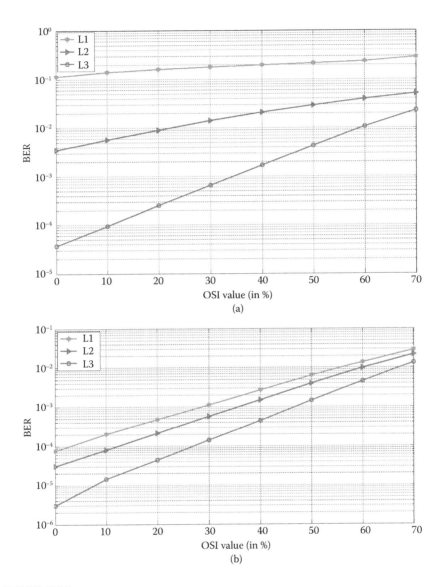

FIGURE 15.30
Uplink BER performance (a) R1 and (b) R2. (From Tiwari, S. V., et al., *Opt. Express*, 23, 26551–26564, 2015. With permission.)

This ND condition occurs as a result of the obstruction created in the VLC link. In other words, the ND detection observed for a period of time in a predefined fashion provides the pattern detection created by the motion. As an example, the pattern for *on* command can be created by intentionally making a straight line over any of the three sets of PDs. Likewise, the pattern for *off*

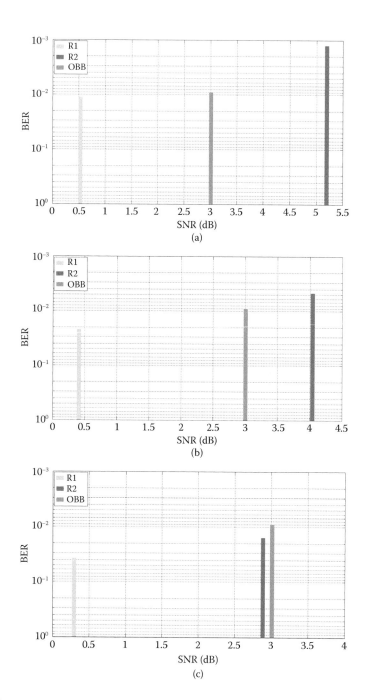

FIGURE 15.31
Comparative downlink BER performance for L1 with OSI values (a) 10%, (b) 30%, and (c) 50%.
(From Tiwari, S. V., et al., *Opt. Express*, 23, 26551–26564, 2015. With permission.)

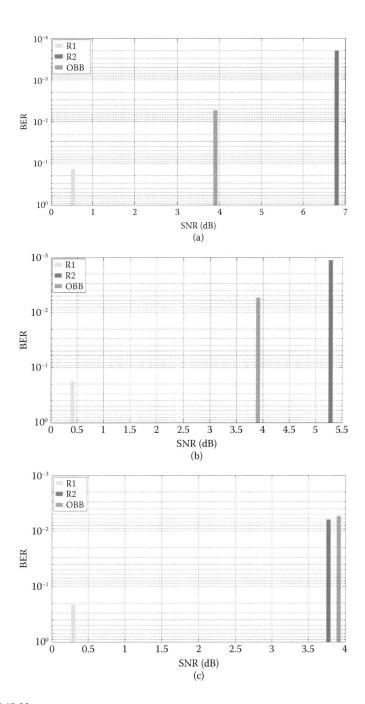

FIGURE 15.32
Comparative uplink BER performance for L1 with OSI values: (a) 10%, (b) 30%, and (c) 50%.
(From Tiwari, S. V., et al., *Opt. Express*, 23, 26551–26564, 2015. With permission.)

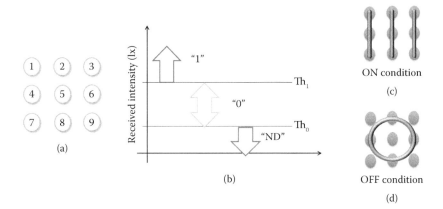

FIGURE 15.33
Principle of motion detection (a) array of PDs, (b) thresholds for data detection, (c) *on* condition, and (d) *off* condition. (From Sewaiwar, A., et al., *Opt. Express*, 23, 18769–18776, 2015. With permission.)

command can be created by making a circle. Figure 15.33c and d shows these patterns. It is important to note that the motion detection technique operates while the VLC link performs data transmission. That is, the motion detection should not hinder the VLC link. In order to ensure an adequate level of illumination and data communication, an array of LEDs is used at the transmitter. The modulation format employed for the data transmission is NRZ-OOK, which is the simplest modulation scheme mentioned in PHY I of IEEE standards [16]. Prior to the NRZ-OOK modulation, Manchester code is applied to remove a long trail of 0 and 1.

Figure 15.34 shows the structure of the VLC-based motion detection technique where there are two paths—the VLC data transmission and the motion detection [11]. For the data transmission, the three groups are formed from all PDs with each group having three PDs. This grouping will facilitate efficient decoding of the transmitted data from the PDs, even when the motion detection technique is in operation. The received signals from the three groups of the PDs are first fed into the threshold detector and demodulator. The threshold detector and demodulator block estimates the received intensity, detects the transmitted symbols, and converts the symbol into a bitstream. Since the OOK modulation is employed, symbols are interchangeable with bits. The binary data from each PD through the threshold detector and demodulator block is then passed to the SC blocks where the most probable bit is detected [3]. Note that when the obstruction occurs, the detected bits from one or two particular SC blocks may not be accurate. For the reliable detection, therefore, the decision circuit is employed.

On the other hand, for the motion detection, the signals from all PDs are fed into the motion detection circuit so as to detect the pattern created by the user as shown in Figure 15.34. As described earlier, the motion detection circuit detects

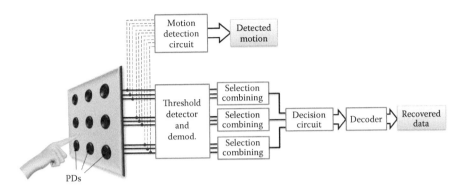

FIGURE 15.34
Block diagram of motion detection technique. (From Sewaiwar, A., et al., *Opt. Express*, 23, 18769–18776, 2015. With permission.)

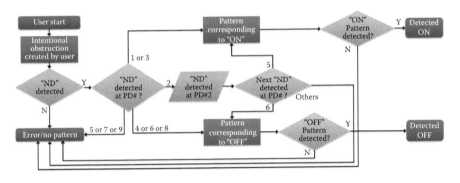

FIGURE 15.35
Motion detection algorithm for *on* and *off* signals. (From Sewaiwar, A., et al., *Opt. Express*, 23, 18769–18776, 2015. With permission.)

the ND condition and subsequently identifies the pattern created by the user. Then, this detected motion will eventually initiate the intended control of the devices. Since the motion detection is based on the obstruction created by the user, the detection of the ND condition for a specific period of time, Δt, at a PD is important. Clearly, the value of interval Δt needs to be determined on the basis of the speed of hand movement of the user. Generally, an empirical value obtained from the experiments for Δt can be used. It is found to be 100 μs with a tolerance of ±10 μs. It should be noted that if the motion is faster or slower than Δt, it is not considered as any signal. Likewise, if the movement is irregular, the pattern is not recognized. Nonetheless, the interval can readily be calibrated for a particular user prior to practical applications.

The algorithm for the motion detection is illustrated in Figure 15.35 [11]. The pattern for the *on* signal can be drawn, starting from either PD 1, PD 2, or PD 3,

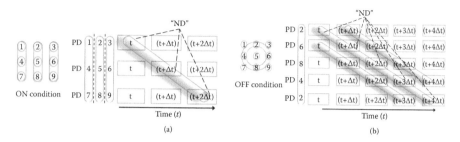

FIGURE 15.36

Principle of pattern detection (a) *on* condition and (b) *off* condition. (From Sewaiwar, A., et al., *Opt. Express*, 23, 18769–18776, 2015. With permission.)

not in the reverse order. Similarly, for the pattern of the *off* signal, the user can start from any of PD 2, 4, 6, and 8.

Figure 15.36a shows the pattern detection starting from any of the three PDs—PD 1, 2, or 3 for the *on* condition [11]. It can be observed that the obstruction occurs at the designated PDs after an interval of Δt. Likewise, Figure 15.36b shows the circular pattern for the *off* condition with the assumption that the pattern begins at PD 2. It also shows flexible pattern detection capability for the *off* signal when the user does not complete a full circle or the user starts from PD 6. However, the *on* signal requires at least three PDs for accuracy and reliability of the motion detection.

For demonstration purposes, experiments were performed with an array of LEDs comprised of 20 RGB LEDs, each having a modulation bandwidth of 120 MHz with an optical output power of 60 mW each. This optical output power is considered adequate in fulfilling the need of illumination. A line encoding scheme mentioned in [16] and the NRZ-OOK modulation scheme as described previously were employed. The data transmission was performed at a data rate of 10 kbps, based on an Arduino ATMEGA 2560. The actual setup for the experiments is visualized in [11]. It shows an array of nine PDs with a field of view of 60°, a physical area of 1.0 cm^2 and responsivity equal to 1.

Experimental results are shown in Figure 15.37. Figure 15.37a shows the detection of the *on* signal based on PD 2, 5, and 8. It can be observed that the sequential ND occurrence is detected for a period of Δt from PD 2, 5, and 8. Therefore, this pattern is identified as the *on* signal according to the proposed algorithm. For the *off* condition, Figure 15.37b shows the received signal [11]. The sequential ND measurement is also observed from PD 2, 6, 8, and 4, which is interpreted as the *off* control signal.

In the experiments, over a distance of up to 60 cm, no bit errors were observed from 21,000 bits transmitted, while maintaining a data rate of 10 kbps. Moreover, a high level of accuracy for the motion detection of the *on* and *off* conditions was obtained from the proposed technique. To identify the threshold values, these are determined by performing the transmission of

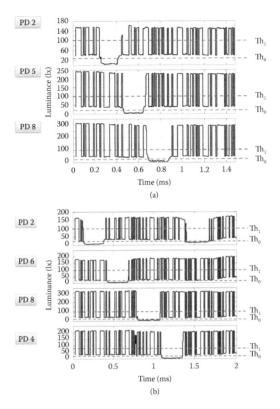

FIGURE 15.37
Received signals (a) *on* condition and (b) *off* condition. (From Sewaiwar, A., et al., *Opt. Express*, 23, 18769–18776, 2015. With permission.)

a known trail of 1s and 0s prior to actual transmission of the data. The values of Th_0 and Th_1 are experimentally found to be 25 lx and 100 lx, respectively.

Performance evaluation of the VLC link was also conducted. Note that the simulation parameters are identical to the parameters of the experiment, except for the data rate, distance, and number of transmitted bits. A data rate of 96 Mbps is utilized and transmitted 10^8 bits. The distance of the data transmission ranges from 40 cm to 90 cm. The simulation results are shown in Figure 15.38 [11]. It was found that the conventional OOK with SC is superior to the conventional OOK without SC [16,17]. This performance gain is due to the receiver diversity in the form of SC, obtained from the PD array. A further analysis was conducted for the effect of the motion detection on the VLC link performance. As shown in Figure 15.38, it is observed that the effect of the proposed motion detection on the existing VLC link is negligible in terms of the BER performance, regardless of

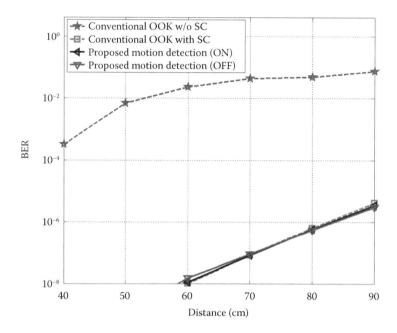

FIGURE 15.38
BER performance comparison. (From Sewaiwar, A., et al., *Opt. Express*, 23, 18769–18776, 2015. With permission.)

whether it is in the *on* or *off* condition. This finding indicates that the communication quality remains unchanged due to SC.

For more advanced motion gestures, the current motion detection technique can be further extended by defining the patterns and subsequently the detection algorithm with a denser PD array having a large number of PDs or the use of imaging receivers.

In this chapter, we have considered the indoor VLC systems that are mainly based on the color-clustering methods. In order to support multiuser or multidevice bidirectionality in indoor environments, the colors of RGB LEDs combined with orthogonal codes are efficiently applied. In addition, from a practical implementation point of view, optical shadowing and mobility support have been discussed. These techniques could pave the way for a smart indoor VLC multiuser (or multidevice) system.

References

[1] K. Bandara and Y.-H. Chung, Novel color-clustered multiuser visible light communication, *Trans. Emerg. Telecommun. Technol.*, vol. 25, no. 6, pp. 579–590, 2014.

[2] A. Sewaiwar, S. V. Tiwari and Y. H. Chung, Novel user allocation scheme for full duplex multiuser bidirectional Li-Fi network, *Opt. Commun.*, vol. 339, pp. 153–156, 2015.

[3] P. P. Han, A. Sewaiwar, S. V. Tiwari and Y. H. Chung, Color clustered multiple-input multiple-output visible light communication, *J. Opt. Soc. Korea*, vol. 19, no. 1, pp. 74–79, 2015.

[4] A. Sewaiwar, S. V. Tiwari and Y.-H. Chung, Mobility support for full-duplex multiuser bidirectional VLC networks, *IEEE Photon. J.*, vol. 7, no. 6, pp. 7904709(1–9), 2015.

[5] J. Fakidis, D. Tsonev and H. Haas, A comparison between DCO-OFDMA and synchronous one-dimensional OCDMA for optical wireless communications, in *IEEE 24th International. Symposium on Personal Indoor and Mobile Radio Communications (PIMRC)*, London, United Kingdom, 2013.

[6] B. Bayer, Color imaging array. United States of America Patent US3971065 A, 3 February 1976.

[7] J. E. Kim, G. Boulos, J. Yackovich, T. Barth, C. Beckel and D. Mosse, Seamless integration of heterogeneous devices and access control in smart homes, in *International Conference on Intelligent Environment*, Guanajuato, 2012.

[8] W. M. T. Vijayananda, K. Samarakoon and J. Ekanayake, Development of a demonstration rig for providing primary frequency response through smart meters, in *International Universities Power Engineering Conference*, Cardiff, Wales, 2010.

[9] K. Gill, S. H. Yang, F. Yao, Y. Liu, Y. L. Liu and H. K. Tsang, A ZigBee-based home automation system, *IEEE Trans. Consum. Electron.*, vol. 55, no. 2, pp. 422–430, 2009.

[10] J. Han, C. S. Choi, W. K. Park, I. Lee and S. H. Kim, Smart home energy management system including renewable energy based on ZigBee and PLC, *IEEE Trans. Consum. Electron.*, vol. 60, no. 2, pp. 198–202, 2014.

[11] A. Sewaiwar, S. V. Tiwari and Y.-H. Chung, Visible light communication based motion detection, *Opt. Express*, vol. 23, no. 14, pp. 18769–18776, 2015.

[12] S. V. Tiwari, A. Sewaiwar and Y. H. Chung, Color coded multiple access scheme for bidirectional multiuser visible light communications in smart home technologies, *Opt. Commun.*, vol. 353, pp. 1–5, 2015.

[13] K. W. Henderson, Some notes on the walsh functions, *IEEE Trans. Electron. Comput.*, vol. 50, pp. EC 13–14, 1964.

[14] J. M. Luna-Rivera, R. Perez-Jimenez, J. Rabadan-Borjes, J. Rufo-Torres, V. Guerra and C. Suarez-Rodriguez, Multiuser CSK scheme for indoor visible light communications, *Opt. Express*, vol. 22, no. 20, pp. 24256–24267, 2014.

[15] S. V. Tiwari, A. Sewaiwar and Y. H. Chung, Optical bidirectional beacon based visible light communication, *Opt. Express*, vol. 23, no. 20, pp. 26551–26564, 2015.

[16] S. Rajagopal, R. D. Roberts and S. K. Lim, IEEE 802.15.7 visible light communication: Modulation schemes and dimming support, *IEEE Commun. Mag.*, vol. 50, no. 3, pp. 72–82, 2012.

[17] T. Komine and M. Nakagawa, Fundamental analysis for visible-light communication system using LED lights, *IEEE Trans. Consum. Electron.*, vol. 50, no. 1, pp. 100–107, 2004.

16

VLC with Organic Photonic Components

Paul Anthony Haigh, Zabih Ghassemlooy, Stanislav Zvánovec,
and Matěj Komanec

CONTENTS

16.1 Introduction

In recent years, organic small molecule and polymer light-emitting diodes (LEDs) and photodetectors (PDs) have been used as optoelectronic components in visible light communications (VLC). The first study appeared in [1] which demonstrated that data transmission rates in the hundreds of kb/s region are possible. This was further improved by using advanced modulation formats such as orthogonal frequency division multiplexing (OFDM) [2]. Ethernet transmission speeds were reported for the first time in [3] and were achieved using the multilayer perceptron artificial neural network equalization technique. The current state-of-the-art transmission speeds available in organic VLC (OVLC) transmission is 55 Mb/s using aggregated wavelength multiplexed data streams [4]. This chapter gives an overview of organic-based VLC focusing on the LED technology trends, organic LED (OLED)-based devices, the organic semiconductors, and visible light PDs. To enhance the OLED-based VLC links blue filtering and a number of equalization schemes including artificial neural network equalizer, decision feedback equalizer, and linear equalizer are discussed and their performance are compared and contrasted. Finally an experimental all-organic VLC system employing both OLED and organic PDs employing artificial neural network base equalizer is introduced and its performance is evaluated. The chapter is completed with concluding remarks.

16.2 OLED Technology and Future Lighting Devices

A substantial problem with using either white phosphor LEDs (WPLEDs) or red, green, and blue (RGB) LEDs (RGBLEDs) as the transmitter in VLC systems is the scalability. LEDs produced with metal alloys such as galium nitride (GaN) by epitaxial thermal evaporation methods result in brittle crystals that cannot easily be fashioned into large area panels, which are desirable for VLCs, solid state lighting or other applications such as screens and displays. One possible solution to this is small molecule and polymer organic optoelectronic devices as a direct replacement for WPLEDs and RGBLEDs. Organic devices offer lower heat dissipation, mechanical flexibility, reduced production cost, and arbitrarily large photoactive areas in complete contrast to the inorganic devices.

The organic electronics sector is now large enough to be considered as a separate industry (the so-called printed electronics industry). According to the market forecasters IDTechEx, the printed electronics industry will be valued at $330 billion as early as 2027 more than the gross value of the Si market today ($225 billion) [5]. Organic electroluminescent polymers were first discovered by Burroughes in 1990 [6] and are now commonly known as polymer LEDs (PLEDs). Alternatively, small molecule–based organic electroluminescent devices known as small molecule organic LEDs (SMOLEDs) were proposed prior to PLEDs in 1987 by Tang and Van Slyke [7]. Aside from the size of molecules used in the semiconductor, PLEDs are more complex and based on long chains of π-conjugated polymers, the main difference between PLEDs and SMOLEDs is the processing method. SMOLEDs are generally thermal-vacuum evaporated while PLEDs can be solution processed, which is the cheaper (and thus more desirable) method.

The idea of organic photonics for communications had been conceived previously and the first postulation of organic photonic devices for optical fiber communications was in 1992 [8]. A summary of the potential of organic communication systems has been outlined in [9]. Perhaps the most striking comment is that organic devices should not be taken as a direct replacement for inorganic devices, as such a transition may never take place due to the strong presence and dominance of inorganic devices in the market as well as the overall cost of replacing the existing infrastructure. However, in [9] it has been identified that organic devices should be seen as a viable alternative technology in applications where inorganic devices are not suitable. A good example of this is an OVLC system where thin films and large area panels are extremely desirable such as deployment in laptop computers, mobile phones, and multifunctional displays.

To date, research and development in organic communications systems has been limited even though reports are starting to emerge that allude to the prospect of organic optical communication systems but they do not explicitly test important performance metrics such as the bit error rate (BER) [9–14]. The root of this could be down to the fact that the device structure has not yet been optimized. For example, the semiconductor interlayer influences the device wavelength and the charge transport characteristics, while the layer structure and organization can affect the efficiency characteristics [15,16].

Organic devices are based on the thin film technology; the general structure for a photonic device is two or more organic semiconductor materials sandwiched by oppositely polarized electrodes, as shown in Figure 16.1, which also includes a charge transport layer made using the polymer poly (3,4-ethylenedioxythiophene)-poly(styrenesulfonate) (PEDOT:PSS) and some possible emissive layer polymers based [2,3,17–20].

The most important manufacturing processes are the solution processing [21], spray coating [22], doctor blading [23], spin coating [24], and inkjet printing [25] all of which are wet processed techniques that can offer potentially ultra-low cost mass production in the future. The total stack thickness for any OLED (either SMOLED or PLED) produced with any manufacturing process is between 100 and 200 nm, which is a very exciting prospect for future displays, considering the common desire to miniaturize electronics as far as possible.

Most of the research focus in VLC systems is on the transmitter and not the receiver. It should be noted that organic photodetectors (OPDs), which are polymer-based, have emerged as an attractive prospect for VLCs not only due to lower materials costs (< £0.20/cm^2, \$0.33/cm^2, EUR 0.25/cm^2) [1] but also due to the fact that OPDs can be spray deposited with higher efficiencies compared to than silicon (Si) [22] in various shapes with a wide range of photoactive area devices [26]. Furthermore, due to the band gap energies of conjugated polymers (in the region of 1–4 eV, encompassing the entire visible range); OPDs can be tailored for the visible light spectrum band while rejecting the entire infrared (IR) region by careful selection of the semiconductor materials.

The empirically measured normalized azimuth emission profile of the SMOLED is shown in Figure 16.2 along with the theoretical normalized Lambertian emission profile [27].

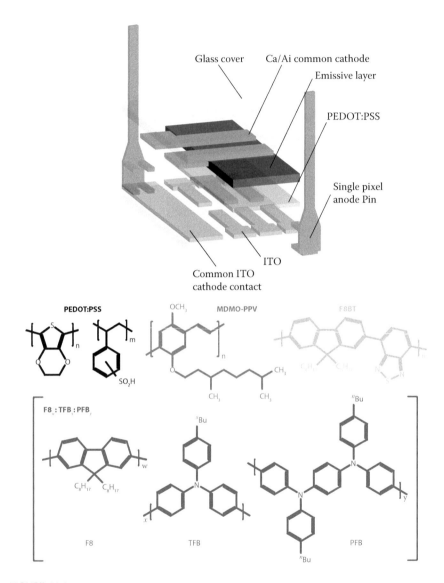

FIGURE 16.1
OLED structure.

The light output versus the current and voltage plots for SMOLED is shown in Figure 16.3a [27]. Note that the OLED goes through a transition phase where the optical power and voltage both drop as a function of time in the first few hours of operation. This could be due to the thermal destruction of unstable molecules and processing defects such as short circuits. This is a common feature of the new SMOLEDs and PLEDs and is colloquially known as burning-in. The device then reaches a steady-state region.

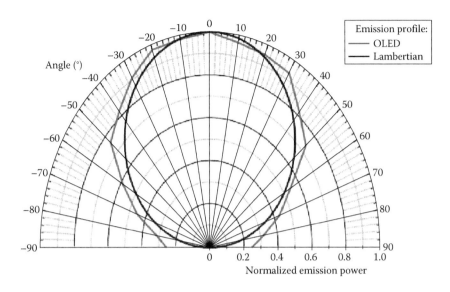

FIGURE 16.2
Polar plot showing the normalized measured emission profile of the SMOLED, which is in close agreement to the normalized Lambertian emission profile. (From Haigh, P.A., Using Equalizers to Increase Data Rates in Organic Photonic Devices for Visible Light Communications Systems, PhD thesis, Northumbria University, 2009. With permission.)

The measured normalized frequency spectrum of SMOLED is depicted in Figure 16.3b for a range of bias voltage. The SMOLED bandwidth is dependent on the injection current and this is a phenomenon that has never been reported for SMOLED devices. At high injection currents (and therefore the bias voltages) the bandwidth extends to 96 kHz while for low injection currents the bandwidth decreases to 26 kHz; a difference of 72 kHz.

16.3 Visible Light Organic Photodetectors

The receiver in VLC systems is generally an individual positive-intrinsic negative (PIN) silicon (Si) PD [28], or less commonly, a Si avalanche PD (APD) [29]. Si-based PDs have responsivity in the range of 200–1100 nm and are very well-established optical wireless communications particularly in free space optics (FSO) communications operating in the near-infrared (NIR) wavelengths, where they offer higher responsivity [30]. On the other hand, the responsivity is very low in the visible range, which is undesirable for VLC links where the information is mostly carried on the blue wavelength. Thus, additional optical power would be required in order to achieve a useful signal voltage level. It is not surprising that a dedicated material has

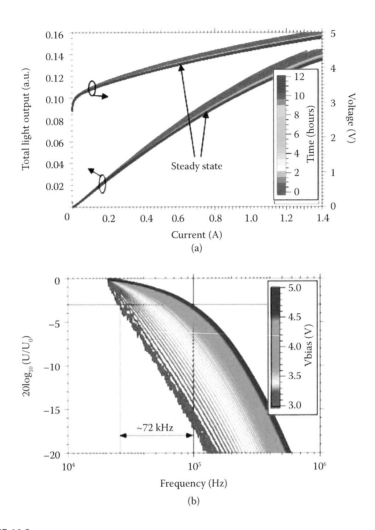

FIGURE 16.3
Measured SMOLED characteristics (a) light output, current, and voltage curve and (b) the normalized frequency response. (Adapted from Haigh, P.A., Using Equalizers to Increase Data Rates in Organic Photonic Devices for Visible Light Communications Systems, PhD thesis, Northumbria University, 2009.)

not emerged for high speed and high responsivity PDs in the visible range. This is because previously no communications technology has utilized this region of the electromagnetic spectrum.

OPDs offer a solution to this problem, with enhanced blue-light absorption using an interpenetrated blend of electron donor [poly(3-hexylthiophene)] and an electron acceptor ([6,6]-phenyl-C61-butyric acid methyl ester [PCBM]), P3HT:PCBM. In order to separate excitons into individual charge carriers, larger energy is required than inorganics-based devices due to the high binding energy as mentioned in the previous section. Electron donors have lower electron affinity (the difference between the band edge and the vacuum energy) than electron acceptors. It should be noted that a high electron affinity is desirable for electron acceptors and vice-versa for electron donors. It is possible to disassociate the exciton at the interface of an electron donor/acceptor configuration due to the unbalanced electron affinities (i.e., the unbalanced energy levels). The bulk heterojunction (BHJ) is an interpenetrated blend of electron donor and electron acceptor that provides such an interface that is distributed across the entire photoactive area of the organic photonic device as illustrated in Figure 16.4 [27]. The reason for interpenetration is due to the fact that the radiative decay of excitons depends on the distance of exciton generation from the electron acceptor–electron donor border (for distances >10 nm the exciton will not offer radiative decay [26]). BHJs were first introduced in [31] and are popular in OPDs due to the fact that they are soluble and offer extremely low-cost processing [22].

As illustrated in Figure 16.4, the materials selected for the electron acceptor and electron donor are [6,6]-phenyl-C61-butyric acid methyl ester (PCBM) and P3HT, respectively. PCBM is a Buckminsterfullerene derivative (the 1996 Nobel Prize in Chemistry was awarded for the discovery of Buckminsterfullerene) that offers the advantage of having a high electron affinity to

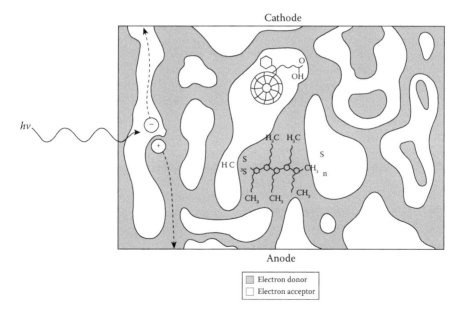

FIGURE 16.4
The bulk heterojunction concept based on PCBM:P3HT. (From Haigh, P.A., *Using Equalizers to Increase Data Rates in Organic Photonic Devices for Visible Light Communications Systems*, PhD thesis, Northumbria University, 2009. With permission.)

produce efficient electron transfer. PCBM is also soluble. P3HT is a conductive polymer (the 2000 Nobel Prize in Chemistry was awarded for its discovery) consisting of π-conjugated orbitals, which are advantageous for photoactive devices.

The OPDs share the aluminum electrode (the cathode) and have an individually structured anode that allows each diode to read out an independent datastream. The anode is made from indium tin oxide (ITO), which is a transparent conductive metal. The fact that the ITO is structured leads to an arbitrary number of photoactive sections on the substrate. This is an important feature of the device as it facilitates many applications that Si-based PDs cannot provide in such a simple manner. The most important application for a communications system is multiple-input multiple-output (MIMO), a highly parallel transmission scheme adopted in optical communications [32].

Another notable application is position sensing. The BHJ is deposited using the spray coating technique proposed in [22] where the materials are dissolved in a solvent and sprayed onto the substrate, thus offering a significant cost reduction at the expense of the surface roughness, which can lead to increased dark currents.

Each BHJ interface can be considered as a miniature P–N junction leading to an expanded Shockley equation that defines the I–V relationship given by [26]:

$$J_{MPN} = J_0(e^{q^{V/n_{ID}k_E T}} - 1) \tag{16.1}$$

where J_0 is the saturation current density, q is the electron charge, V is the voltage, T is the temperature (K), and k_E is the Boltzmann constant. Notice that there is an extra term in the denominator of the exponential term. The additional term is the so-called ideality factor n_{ID} that takes into account bulk morphology. Clearly as $n_{ID} \rightarrow 0$ the diode reaches the saturation current at $V \rightarrow 0$, which is advantageous since a lower bias voltage is required.

Organic semiconductors are typically vertical devices and therefore some insight into the device structure must be given. The substrate can be almost anything in organics including paper [33], plastic [34,35], and glass [22]. The anode is generally made from transparent ITO although there is a growing argument for using graphene due to the emergence of high-efficiency devices with the graphene anodes [36]. The next layers are the organic layers. In state-of-the-art OLED devices, the organic layers are made up of (from bottom to top) a hole injection layer, a hole transport layer, an emissive layer, an electron transport layer, and an electron injection layer followed by the cathode, which is generally aluminum since it is cheap and not necessary to be transparent. There are many devices that offer an increase in performance at the cost of increased complexity such as multiple photon emitters that are not covered here but are referred to in [37]. In BHJ OPDs, the stack structure is significantly less complex, requiring only the two electrodes, the BHJ and an optional interlayer, and selected as P3HT because it offers the highest bandwidth [38]. The interlayer is not covered here; however, it can have a profound effect on the performance of critical parameters of the device such as bandwidth; for a detailed analysis, refer to [38]. As mentioned, the BHJ is an interpenetrated blend of P3HT:PCBM, which are extremely popular materials in BHJ devices due to their relatively high efficiency and solubility.

The band gap energy of P3HT:PCBM is ~ 2 eV that is ideal for VLC applications as the cutoff wavelength is ~ 650 nm, which cuts a portion of the red wavelength that would possibly be useful for wavelength division multiplexing (WDM). By introducing a further, low-band gap material into the BHJ blend such as poly[2,6-(4,4-bis-(2-ethylhexyl)-4H-cyclopenta[2,1-b;3,4-b] dithiphene)-alt-4,7-(2,1,3-benzothiadiazole)] (PCPDTBT) the BHJ band gap can be reduced so the absorption spectrum extends into the NIR region and allows the absorption of such wavelengths. The working principles of P3HT:PCBM and similar BHJs are well covered in the literature and the reader is encouraged to refer to [22,26,31] since no details are given here.

16.4 Organic VLC with Equalization

The BER performance of 2 and 4 pulse-position modulation (PPM) using a SMOLED and Si PD with hard decision decoding using a threshold detector is illustrated in Figure 16.5. It demonstrates data transmission rates of 150 and 50 kb/s along with the 250 kb/s ON-OFF keying (OOK) link for the reference. In comparison to the ~90 kHz system, bandwidth bottleneck

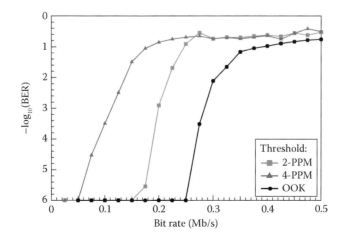

FIGURE 16.5
Unequalized BER performance of PPM and OOK modulation schemes.

introduced by the SMOLED it was expected that OOK would offer the best performance with no equalization as it has half the bandwidth requirement of 2-PPM and four times less than 4-PPM. Furthermore, as expected 2-PPM also outperforms 4-PPM and this is also attributed to the lower bandwidth requirements [39].

Equalization is a well-established subject that has been extensively studied and is widely covered in the literature [40]. Equalizers can undo the effects of intersymbol interference (ISI) mostly caused by multipath and limited channel transmission bandwidth. There are a range of equalizers that can be adopted in organic VLC systems as illustrated in Figure 16.6. Analog equalizers consist of passive components such as resistors, capacitors, and inductors, or active components that add power into the system such as operational amplifiers. Passive analog equalizers are low in complexity but typically offer a limited improvement over the modulation bandwidth due to the associated power penalties.

Digital equalizers can be separated into linear and nonlinear methods. Linear equalizers are less complex than nonlinear equalizers at the cost of reduced BER performance (but more complex than the analog with improved BER performance).

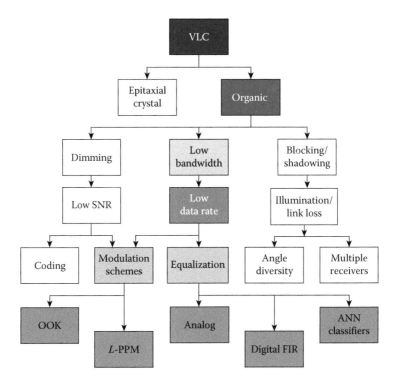

FIGURE 16.6
The most popular equalizers.

There is one additional type of equalizer that has different functionality compared to others, which is not shown in Figure 16.6, and is based on the artificial neural network (ANN). These can be thought of as classifiers as opposed to a traditional equalizer as they classify a signal based on highly nonlinear boundaries that are formed by an adaptive learning sequence. The overall aim of an equalizer in its simplest form is to inverse the undesirable effects of the system response, generally expressed in consideration of the overall system frequency response as follows [1]:

$$H(f) = \frac{1}{Y(f)} \tag{16.2}$$

where $Y(f)$ is the Fourier transform of the system response $y(t)$ at the receiver (i.e., $Y(f) = \Im\{R(t) \otimes g'(t) \otimes h(t) \otimes r(t)\} = R(f)G'(f)H(f)r(f))$, considering the generic VLC schematic block diagram shown in Figure 16.7, where $R(t)$ and $r(t)$ are the transmit and receive filters, respectively.

Equalizers are typically used to equalize the channel response, which can be dispersive or have fading properties in the outdoor environment. The channel response is not being equalized as it is independent of the wavelength and frequency. Therefore $H(f)$ is just a constant. The detailed mathematical analysis can be referred to in [42].

Considering the system response $p(t)$, the factor that is deteriorating the overall system response and introducing ISI is the low-pass transfer functions of the organic devices. Equalizing the low-pass response will allow the data rate to be increased significantly in the presence of a high signal-to-noise ratio (SNR). It should also be noted that the equalizers do not equalize the effects of noise. Detailed mathematical theory of the various equalizers, mentioned in Figure 16.6, is not described here because it is well

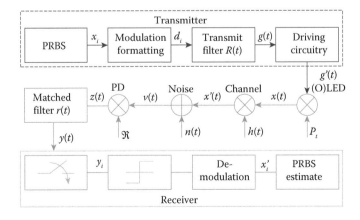

FIGURE 16.7
Example VLC block diagram where $h(t)$ is the inverse Fourier transform of the equalizer.

known and widely available in the literature [27]. The system response is found by the pilot signal and careful filter design is required in order to maximize the effectiveness of the equalizer. There are two main types of filter: analog and digital. In the analog domain, a high-pass filter RC equalizer is the only real choice while in the digital domain the zero forcing (ZF) and the decision feedback (DF) equalizers are the most popular.

Bearing in mind that OOK is the most commonly used modulation scheme in VLCs and is compatible with digital equalizers, there is a noticeable lack of research in this area and the only major reports are based on the analog equalization as previously discussed. An increase in performance can be expected using digital equalizers but there are no reports to provide any further results for a WPLED VLC system aside from [43]. Further, there are no reports that provide any comparison between an adaptive discrete multitone (DMT) link and OOK with equalization, or an RGBLED with digital equalization.

16.5 Type of Equalizers

16.5.1 RC Equalizer

An RC equalizer is the simplest to implement, and consists of a resistor and capacitor arranged into a high-pass filter that is placed between the data source and the optical source (pre-equalizer), or the receiver and the terminal (post-equalizer).

16.5.2 Zero-Forcing Equalizer

The ZF equalizer selects its transfer function as $H(f) = 1/Y(f)$—it tries to force a flat magnitude response by removing the ISI. The ZF is a linear equalizer that has a number of adjustable tap coefficients $\{c_n\}$, as illustrated in Figure 16.8.

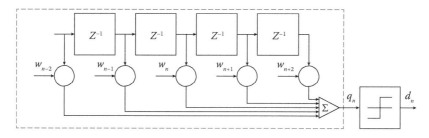

FIGURE 16.8
Zero-forcing equalizer in linear transversal filter format.

The delay given by Z^{-1} is inversely proportional to the filter oversampling period ξ and is either selected equal to the symbol period (symbol spaced) or at a frequency higher than the symbol rate, typically $\xi = T_b/2$ (fractionally spaced). In a fractionally spaced configuration, the output of the filter is also sampled at this rate, as opposed to the symbol rate.

The impulse response of the ZF is given by [44]:

$$q(t) = \sum_{n=-N}^{N} c_n Y(t - n\xi) \tag{16.3}$$

where $Y(t)$ is the incoming training sequence and is built up into an $N \times N$ matrix in order to find the transfer function of the system. The number of taps must be selected in order to span the entire length of the ISI. It must be symmetrical around the current sample to take into account the previous and next samples, so $L \leq 2N + 1$, where L is the number of samples that the ISI spans and N is introduced as a factor in order to make the number of taps symmetrical around the current sample. The condition to force zero ISI is given can be equated to $q(t)$.

Sampling the output at the symbol rate $t = mT_b$ leads to:

$$q(mT_b) = \sum_{n=-N}^{N} c_n Y(mT_b - n\xi) = \begin{cases} 1, & m = 0 \\ 0, & m = \pm 1, \pm 2, \ldots, \pm N \end{cases} \tag{16.4}$$

The filter coefficients are then convolved into the system and periodically updated in case the system response has been modified in some way. Clearly a training sequence is required here in order to build up the impulse response of the system; the longer the training sequence, the better the representation of the system response becomes. It is crucial to notice that the ZF is clearly highly susceptible to the effects of noise as any random receiver noise causing any noise in the channel estimation will be amplified, regardless of the channel estimation accuracy. VLC systems generally exhibit very large SNRs; however, the power penalty for exceeding the system bandwidth is significant, and thus the ZF is not the optimal equalizer for VLC systems.

16.5.3 Adaptive Linear Equalizer

An increase in performance can be obtained if an adaptive algorithm is introduced to find the tap weights as illustrated in Figure 16.9. There are several adaptive algorithms, most notable are the least mean squares (LMS) and recursive least squares (RLS); the others are typically variations of these algorithms. In order to find the tap weights the adaptive algorithm requires training against a header sequence of data symbols that is known at the receiver. The quality of an equalizer is defined by how fast it converges on the target error. As illustrated in Figure 16.9a and b, for a generic OOK VLC link with a five-tap linear equalizer, the RLS algorithm is much faster to converge than

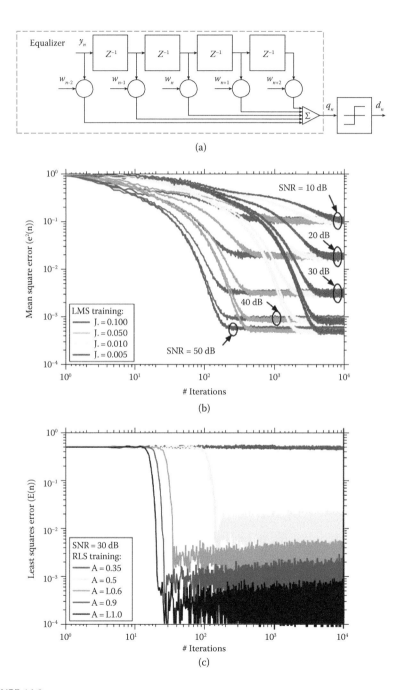

FIGURE 16.9
Adaptive linear transversal equalizer (a) for OOK with five-tap and the mean square error for (b) LMS and (c) RLS.

the LMS, which comes at the cost of increased complexity. As shown in Figure 16.9b, decreasing the step-size parameter results in an improved convergence to the minimum error at the cost of increased convergence time. However, setting the step-size parameter excessively means that the filter becomes unstable and will not convergence on the optimum filter weights. The RLS algorithm with exponentially-weighted forgetting factor demonstrates much faster convergence than LMS, see Figure 16.9c, although not always to a lower error in the case of small forgetting factors (note the difference in range on the y-axis). An increasing forgetting factor offers a faster convergence to the least squares error than the LMS equalizer at a lower error value and there is little difference in performance in each SNR case.

16.5.4 DF Equalization

The performance of an equalizer is directly related to the severity of the ISI experienced in the system. In heavy ISI, linear equalizers will fail due to their inability to produce nonlinear relationships between input and output. Further, if a system transfer function exhibits a deep spectral null, a linear equalizer will struggle to compensate as it will set some of the tap coefficients to be excessively high [44]. Therefore, it is necessary to introduce the nonlinear DF equalizer which works on the principle of estimating the influence of ISI in the current symbol based upon the previously detected symbol. Two filters are required—the feedforward and feedback filters. The feedforward filter is exactly the same as the adaptive linear filters in the previous section and operates in the same way, while the feedback filter is made up of past symbols in order to estimate the contribution of ISI on the current symbol. The output of each filter is subtracted and a decision is made as follows [41]:

$$q_m = \sum_{i=0}^{N_1} c_n y_{m-n} - \sum_{i=0}^{N_2} b_n \hat{d}_{m-n} \qquad (16.5)$$

where c_n is coefficient value of the ith feedforward tap and y_m is the current symbol. The estimate of the previous symbol is given by d_{m-n} and the feedback filter tap coefficients are given by b_n. The block diagram of a DF equalizer is outlined in Figure 16.10.

16.5.5 ANN Equalizer

Although traditional equalizers such as the ones shown in the previous section are very popular, they do not offer the best performance. The major difference between ANNs and transversal equalizers is the structure; the former are arranged into a highly parallel form that allows nonlinear mapping as each input is connected to each neuron. The latter are obviously highly linear (not considering DF) since each input is connected only to its corresponding weight. There are many ANN architectures that can be used as equalizers

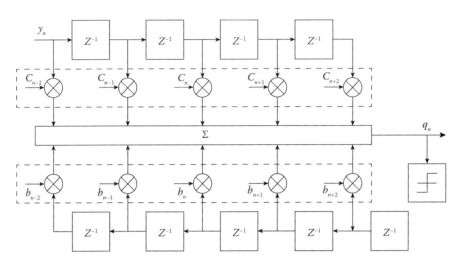

FIGURE 16.10
Decision feedback equalizer with algorithm to update tap coefficients applied at dashed boxes.

in communication systems, including single and multilayer feedforward networks, and feedback networks. For equalization using classification, ANNs require training similar to transversal equalizers. The training sequence simply allows the ANN to adjust the neuron weights according to a gradient descent until the error cost function is satisfied. There are a number of training methods including LMS and RLS and scaled conjugate gradient (SCG) learning but the most popular is the Levenberg–Marquardt back propagation (LMBP) algorithm because it is simple to implement in hardware due to low complexity but requires the most memory. SCG training should converge to a lower error value but requires a longer training period and is more complex for hardware implementation so is not examined here. Having a short training sequence is of paramount importance because it reduces the amount of redundancy in the system, especially if the system is nonstationary and therefore requires frequent retraining to update the input–output map.

There are many variations of ANN that can be used for equalization. The most common are multilayer perceptrons (MLPs), radial basis function (RBF) ANNs, functional link ANNs (FLANNs), and support vector machines (SVMs). Literature has demonstrated that two-layer MLPs offer similar performance to RBFs and SVMs [45] with the advantage of having less complexity and hence are used as the feedforward ANN since any gain obtained using other feedforward ANNs would be marginal with increased hardware complexity. The error convergence is shown in Figure 16.11 for several different MLPs; the one (1H) and two hidden (2H) layer feedforward and one hidden layer DF structures are considered using the same setup as previous equalizers (Figure 16.9a) for an SNR of 30 dB, a number of input taps and neurons of 5 and a training length of 1000, ten times less than in the linear transversal equalizer case.

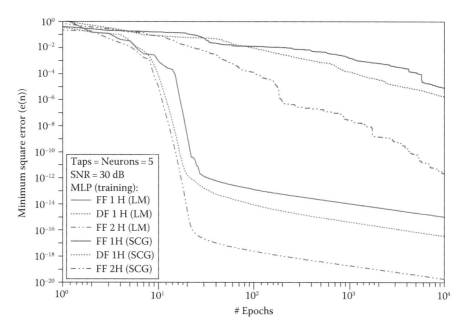

FIGURE 16.11

Comparison of different ANN structures (1H = 1 hidden layer, 2H = 2 hidden layers) and training schemes with SNR = 30 dB; the training length is 1000.

Note that the LMBP training reaches a lower minimum error value than the SCG training for all cases. DF–MLP offers an improvement over the 1H feedforward-MLP using both training methods of a few orders of magnitude. The 2H feedforward-MLP offers a significant improvement in each case. However, this improvement has largely been shown to be theoretical and experimental results have shown that there is little difference between single and two hidden layer structures [46,47] provided an appropriate number of neurons are selected.

16.5.6 ANN Equalizer Performance

The test setup is illustrated by the schematic block diagram in Figure 16.12a. A 2^{10}-1 pseudorandom binary sequence (PRBS-10) is generated using MATLAB® and loaded into the memory of the arbitrary function generator (AFG) using a LabVIEW script. The PRBS-10 data are passed through a unity height rectangular pulse shaping filter $p(t)$ and output to a single pixel via a current mirror driving circuit. Each pixel has an individual driving circuit that is tailored to the respective L-I-V relationships to enable the best possible performance. Individual RGB PLEDs were developed with bandwidths of 350, 110, and 600 kHz, respectively.

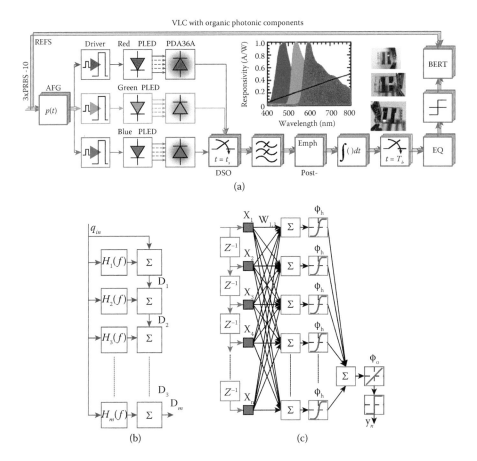

FIGURE 16.12
Schematic block diagrams of (a) the system under test as described in the test insets are top-view photographs of the three RGB PLEDs and the responsivity of the PD in the context of the emission spectra), (b) the postemphasis module, and (c) the equalizer under test.

The transmission distance is set to 0.05 m, consistent with the litera-ture [3,18,20], because singular pixels of ~3.5 mm^2 were used which have relatively low luminance. The optical power is absorbed and the photocur-rent is amplified by a ThorLabs PDA36A packaged silicon PD with inbuilt transimpedance amplifier (10 dB, 5.5 MHz bandwidth). The continuous time photovoltage is sampled by a real-time Agilent DSO9254A oscilloscope at a rate of $t = t_s$. The sampling rate $f_s = 1/t_s$ is set to a maximum of 10 S/sym and at least 10^7 samples were recorded leading to an uncoded BER target of 10^{-6} in line with ITU specifications [48]. The discrete time signal is passed through a fourth-order low-pass filter and the Q-factor is measured (not shown in Figure 16.12a) to give a fast estimate of system performance. The signal is passed through a postemphasis circuit that consists of several high-pass filters of increasing order in parallel, added to the received signal path as shown in Figure 16.12b. The overall frequency response of the post-emphasis circuit can be represented with respect to the normalized cut-on frequency as follows:

$$yf = q(f) + \sum_{\rho=1}^{m} \left(\frac{1}{\left[\sqrt{1 + (f/f_c)^2} \right]} \right)^{\rho} \tag{16.6}$$

where m is the maximum order filter under test and f_c is the filter cut-on frequency and $q(f)$ is the frequency response of the received signal. Extensive studies not shown in this work have demonstrated theoretically and experimentally that a substantial improvement in the Q-factor can be yielded by setting $m = 4$. While for $m > 4$ the improvement obtained is marginal. The filter weights are found by sweeping through a predetermined range of cut-on frequencies and measuring the Q-factor until a maximum value is found. The Q-factor is given by:

$$Q = \frac{v_H - v_L}{\sigma_H + \sigma_L} \tag{16.7}$$

where v_H and v_L are the high and low mean received voltages, respectively, and σ_H and σ_H are the high and low level standard deviations, respectively. Once the maximum value is found for each of the filters in the postemphasis module, the filter coefficients are locked and the transmission occurs.

The postemphasis circuit attempts to restore the attenuated high-frequency components of the signal. The principle of operation of the proposed postemphasis circuit is based on iteratively adjusting the weights of the filter based on the measured Q-factor at the output. Once the optimal filter is found, the signal is downsampled using the integrate/dump method to maximize SNR. The system is then equalized, as will be discussed later and sliced by an average level threshold before comparison with the transmitted bits, symbol-by-symbol in a BER tester (BERT).

Inset in Figure 16.12a represents photographs of the RGB devices, and the responsivity of the photodetector in the context of the normalized emission intensities of the three devices. The responsivities (peak wavelengths) are 0.14 A/W (480 nm), 0.21 A/W (538 nm), and 0.27 A/W (598 nm), respectively. Obviously, there is a significant difference between the responsivities of the blue/red components (around two times) and this will affect the link SNR. This is slightly compensated by the fact that the blue component has a large spectral peak, centered at 624 nm (0.48 A/W). The equalizer used in this work is as outlined in the previous sections and the neurons are selected as $N = \{5; 10; 20; 30; 40\}$.

The results are be presented in terms of individual wavelengths. The first type of devices to be discussed are the F8:TFB:PFB (blue) PLEDs because they offer the best performance, as will be shown. Subsequently, MDMO-PPV (red) devices will be discussed followed by F8BT (green) devices. The unequalized raw BER and the Q-factor performance before and after the postemphasis module for the blue devices are shown in Figure 16.13.

It is clear that the postemphasis module has a clear impact on the BER performance of the link as expected, improving the error-free (at a BER of 10^{-6}) transmission speed from 3 Mb/s up to 7 Mb/s using four filters in parallel. This is also represented by a substantial ~9 dB gain in the Q-factor from 3.8 dB in the raw case to 12.2 dB in the postemphasized case, as is reflected in the eye diagrams inset, which illustrate this improvement with a clear eye opening. The improvement can also be seen at 20 Mb/s, a transmission speed well outside of the modulation bandwidth, reducing the BER from ~0.4 in the raw case to ~4×10^{-2}; or one order of magnitude. On the other hand, this BER is not sufficiently low to successfully apply forward error correction (FEC) codes, which require a maximum BER of 3.6×10^{-3} or 2×10^{-2} considering a 7% or 20% overhead respectively; therefore, error-free communications were not possible [18]. It should be noted that this is a remarkable transmission speed to achieve, representing a net gain of 133% in comparison with the state-of-the-art unequalized error-free data rate reported in the literature (3 Mb/s), which was achieved using both MDMO-PPV and F8:TFB:PFB PLEDs [3,18,20].

The ANN-equalized BER performance before (solid lines) and after (dashed lines) the postemphasis module are both shown in Figure 16.14.

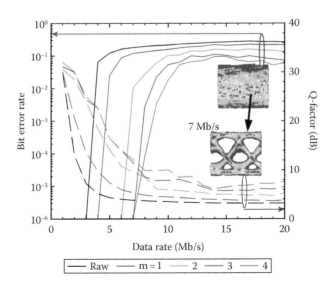

FIGURE 16.13

Unequalized BER and Q-factor performance of the blue PLED; up to 7 Mb/s can be achieved when using the postemphasis module in comparison to 3 Mb/s without it.

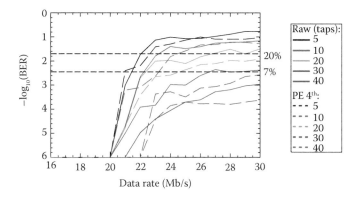

FIGURE 16.14

Equalized BER performance of the blue PLEDs; up to 22 Mb/s can be achieved error free while up to 27.9 Mb/s can be sustained considering a 7% FEC limit.

Recall that the number of taps is set equal to the number of neurons in the equalizer for full efficiency and ranges within the set $N = \{5; 10; 20; 30; 40\}$. The two most important results in Figure 16.14 are as follows; (i) error-free transmission (BER = 10^{-6}) can be supported using $N = \{30; 40\}$ taps if the pre-emphasis module is included. If it is not, transmission speeds are limited to 21 Mb/s for $N = 40$ and 20 Mb/s for $N = \{5; 10; 20; 30\}$; (ii) considering the 7% FEC limit, gross transmission speeds up to 30 Mb/s are readily available in this link either using the postemphasis module and $N = \{30; 40\}$ or without the postemphasis module and $N = 40$, leading to a net transmission speed of 27.9 Mb/s after deduction of the 7% overhead.

For $N = \{5; 10; 20\}$, the system only sustains an error-free transmission speed up to 20 Mb/s as mentioned. For $N = 20$ taps and considering the 7% FEC with and without the postemphasis module, gross (net) transmission speeds of 25 (23.25) and 22 (20.36) Mb/s can be achieved, respectively, which demonstrate a considerable reduction over $N = \{30; 40\}$ and are thus suboptimal. Consider the 20% FEC, gross (net) transmission speeds of 30 (24) and 26 (20.8) Mb/s can be sustained with and without the postemphasis module, respectively. For $N = 10$ and considering the 7% FEC limit, gross (net) transmission speeds of 23 (21.39) Mb/s and 22 (20.46) Mb/s can be supported, offering slight gains of up to ~1.4 Mb/s over the uncoded error rate. The available gross (net) transmission speeds considering the 20% FEC are 23 (18.4) Mb/s and 24 (19.2) Mb/s, actually causing a reduction in available capacity over the uncoded rate. Similarly, for $N = 5$ the available transmission speeds drop below those of the uncoded data rates in all configurations. To provide an illustrative example of the improvement between the worst and best cases, the first 1×10^5 samples at the output of the ANN equalizer are shown for a transmission speed of 30 Mb/s in Figure 16.15, for the system with an ANN equalizer using 5 taps (blue) and 40 taps with the postemphasis module (black).

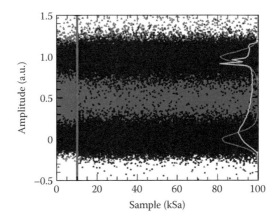

FIGURE 16.15
The equalizer output for the first 10^5 samples (including 10^4 training samples) indicating the substantial classification improvement between $N = 40$ (black) and 5 (blue) taps/neurons.

The vertical red line indicates the length of the training sequence (1×10^4) and hence the pattern slightly degrades after this point due to the high noise level at 30 Mb/s. It is clearly possible to determine two definite signal levels for thresholding in the best case of $N = 40$ taps and the postemphasis, and in the worst case, with $N = 5$ taps, impossible to define any level. The details of the green and blue wavelengths are not shown here but can be referred to in [4].

16.6 All-Organic VLC System

To date a VLC link employing exclusively organic optoelectronic components has not been demonstrated, despite enormous interest in both the organic-based devices [11,14,35] and VLC [3,49,50] in the research community. This can be attributed to two main reasons: first, there is a lack of commercially available organic devices—just a handful of SMOLEDs are available to purchase off the shelf while no OPDs are commercially available and must be custom made. Secondly, OVLC has only recently emerged as a serious topic for research and all of the reports so far have focused on either the transmitter [50] or receiver [49], as opposed to a full system implementation and evaluation. In spite of this, it is necessary to perform such an evaluation because organics have outstanding properties that are ideally suited to the VLC domain. For instance, they can be processed into mechanically flexible, arbitrarily shaped panels with large photoactive areas. Such devices are processed by solution-based processing at room temperature offering a real cost reduction, unlike inorganics which must be processed with epitaxial methods thus resulting in brittle crystals that do not scale well. Further, by careful selection of the semiconducting polymer it is possible to tune the emission or absorption wavelength to visible light as polymers and small molecules with band gap energies of 1–4 eV are abundant.

Even considering a forecasted market value of $330 billion by 2027 [5], it is not anticipated that organic devices will become dominant over inorganics in optical communications and there are several reasons why. Organics-based devices have the potential for applications in areas where inorganics are not perfectly suited or no optimized infrastructure exists. This could be in the screens and chassis of future mobile devices for device-to-device communications where OLEDs have already started appearing. The charge transport characteristics are lower orders of magnitude in organics and the direct result of this is that the bandwidths available for organics are in the kilohertz region (in comparison to MHz for inorganics). This is an open and timely challenge for OVLC links because the bandwidth is the most important factor for increasing capacity. The other is the SNR, which has an upper bound limit caused by the lighting requirements in VLC (max ~400 lux) and also the quantum efficiencies and noise performance of the devices. However, having a low bandwidth is not necessarily a fatal perturbation for OVLC. In [3] a 10 Mb/s link was achieved using an organic polymer LED with 270 kHz bandwidth and an LMS equalizer. Although this report is a significant landmark for OVLC, a silicon PD was used instead of an OPD. Digital equalization techniques are an attractive option to increase transmission speeds as they restore link performance in the presence of ISI caused by the bandwidth limitation. The modulation format selected in this work is OOK due to its simplicity and popularity in the VLC domain [29,49,50].

An experimental test setup is illustrated in Figure 16.16. An AFG (TEK AFG3022B) is loaded with a 2^{10}-1 (PRBS-10) generated in MATLAB and passed through a unit height rectangular pulse shaping filter $p(t)$. The rectangular signal is then mixed with a DC current using a bias tee to ensure operation in the linear region of the transmitter before the DC-biased signal then intensity modulates the OLED. The OPD substrate consists of four independent PDs of 1 cm^2 each as shown inset in Figure 16.16. The incident signal on each detector is sampled by a real-time Tektronix MDO4104-6 oscilloscope, with 10^6 samples acquired with a maximum sampling frequency of 10 samples-per-symbol for further processing offline, meaning a BER target is 10^{-5}. The experiment was conducted in a controlled dark laboratory environment to minimize the ambient noise, and electrical low-pass filters were used in MATLAB to limit the other out-of-band noise sources.

In Figure 16.16b, the ANN-equalized performance of each link is shown. Since there are four OPDs on the substrate, four data streams are recovered, each with an individual data rate. Therefore in Figure 16.16b, the upper and lower BER values are shown for each bandwidth, with the average value

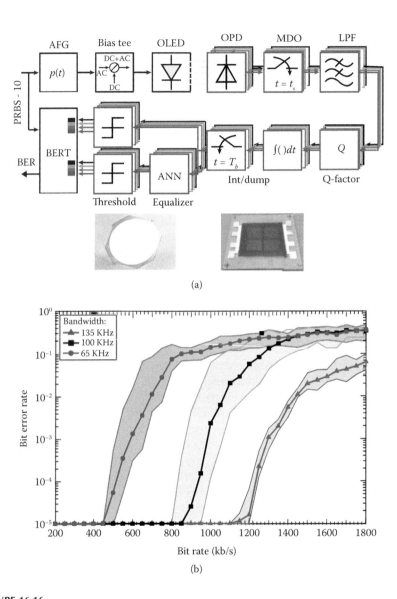

FIGURE 16.16
(a) Block diagram of the experimental setup used with ANN equalizer and (b) equalized BER performance for three bandwidths.

on top. A data rate of 1100 kb/s can be supported at an average BER of 10^{-5} (1150 kb/s at 1.15×10^{-5} BER and 1200 kb/s at 1.6×10^{-5}). This is approximately a threefold improvement over the unequalized case of 350 kb/s. This is due to the ANN's ability to map any input–output sequence given a sufficient SNR and number of neurons. For the 100 kHz case, a reduced equalized transmission speed of 850 kb/s is observed. Similar to the 135 kHz case, the level of performance shows an approximately threefold improvement in transmission speed over the unequalized case (250 kb/s). Finally in the 65 kHz case, an equalized data rate of 450 kb/s can be achieved; once more offering similar performance improvement statistics as the previous two cases.

16.7 Conclusions

VLC is a green technology with multiple functionalities, which is suitable for the future last-meter access networks. Most VLC systems have been using inorganic WPLEDs/RGBLEDs as the transmitters and Si PDs as the receivers due to several advantages: (i) high optical power output (transmitters) and (ii) reasonable bandwidths in the megahertz region. On the other hand, such devices also have drawbacks such as scalability due to brittle crystals produced using epitaxial high vacuum processing methods (transmitters) and low responsivity in the visible range (receivers). Consider these disadvantages with the fact that PLEDs and SMOLEDs are emerging as serious candidates for future lighting systems due to extremely low cost solution-based processing methods and high electrical efficiencies. In order to fully appreciate the proposed organic VLC, this chapter outlined the fundamental principles of organic-based VLC focusing on the LED technology trends, OLED-based devices, the organic semiconductors, and visible light PDs. To enhance the OLED-based VLC links, blue filtering and a number of equalization schemes including the ANN equalizer, DF equalizer, and linear equalizer are discussed and their performances were discussed and compared. It was shown that ANNs with MLP implementation offered a universal classifier between input and output sequences. The MLP was established as the best performing equalizer while being the most complex to implement. Finally, an experimental all-organic VLC system employing both an SMOLED and an OPD employing an ANN-based equalizer offering ~1 Mbps was introduced and its performance was evaluated.

References

[1] P. A. Haigh, Z. Ghassemlooy, H. Le Minh, S. Rajbhandari, F. Arca, S. F. Tedde, O. Hayden, and I. Papakonstantinou. Exploiting equalization techniques for improving data rates in organic optoelectronic devices for visible light communications. *J. Lightwave Technol.*, 30(19):3081–3088, 2012.

[2] S. T. Le, T. Kanesan, F. Bausi, P. A. Haigh, S. Rajbhandari, Z. Ghassemlooy, I. Papakonstantinou, et al. 10 Mb/s visible light transmission system using a

polymer light-emitting diode with orthogonal frequency division multiplexing. *Opt. Lett.*, 39(13):3876–3879, 2014.

[3] P. A. Haigh, F. Bausi, Z. Ghassemlooy, I. Papakonstantinou, H. Le Minh, Ch. Fléchon, and F. Cacialli. Visible light communications: Real time 10 Mb/s link with a low bandwidth polymer light-emitting diode. *Opt. Express*, 22(3):2830–2838, 2014.

[4] P. A. Haigh, F. Bausi, H. Le Minh, I. Papakonstantinou, W. O. Popoola, A. Burton, and F. Cacialli. Wavelength-multiplexed polymer LEDs: Towards 55 mb/s organic visible light communications. *IEEE J. Sel. Areas Commun.*, 33(9):1819–1828, 2015.

[5] R. Das and P. Harrop. *Organic & printed electronicsforecasts, players & opportunities 2007–2027*, IDTechEx, Cambridge, UK, 2010.

[6] J. H. Burroughes, D. D. C. Bradley, A. R. Brown, R. N. Marks, K. Mackay, and R. H. Friend. Light-emitting diodes based on conjugated polymers. *Nature*, 347:539–541, 1990.

[7] C. W. Tang and S. A. Van Slyke. Organic electroluminescent diodes. *Appl. Phys. Lett.*, 51(12):913–915, 1987.

[8] T.-M. Lu. Organic photonics: Materials and devices strategy for computational and communication systems. In *NTC-92 National Telesystems Conference, 1992*, pages 9/7–915, May 1992.

[9] J. Clark and G. Lanzani. Organic photonics for communications. *Nat. Photon.*, vol. 4, 438–446, 2010.

[10] S. Valouch, M. Nintz, S. W. Kettlitz, N. S. Christ, and U. Lemmer. Thickness-dependent transient photocurrent response of organic photodiodes. *IEEE Photon. Technol. Lett.*, 24(7):596–598, 2012.

[11] B. Arredondo, C. de Dios, R. Vergaz, G. del Pozo, and B. Romero. High-bandwidth organic photodetector analyzed by impedance spectroscopy. *IEEE Photon. Technol. Lett.*, 24(20):1868–1871, 2012.

[12] L. Salamandra, G. Susanna, S. Penna, F. Brunetti, and A. Reale. Time-resolved response of polymer bulk-heterojunction photodetectors. *IEEE Photon. Technol. Lett.*, 23(12):780–782, 2011.

[13] E. S. Zaus, S. Tedde, J. Fürst, D. Henseler, and G. H. Döhler. Dynamic and steady state current response to light excitation of multilayered organic photodiodes. *J. Appl. Phys.*, 101:04450, 2007.

[14] I. A. Barlow, T. Kreouzis and D. G. Lidzey. High-speed electroluminescence modulation of a conjugated-polymer light emitting diode. *Appl. Phys. Lett.*, 94(24):243301–243303, 2009.

[15] H. Sasabe, J. Takamatsu, T. Motoyama, S. Watanabe, G. Wagenblast, N. Langer, O. Molt, E. Fuchs, Ch. Lennartz, and J. Kido. High-efficiency blue and white organic light-emitting devices incorporating a blue iridium carbene complex. *Adv. Mater.*, 22(44):5003–5007, 2010.

[16] T. Chiba, Y.-J. Pu, R. Miyazaki, K. Nakayama, H. Sasabe, and J. Kido. Ultra-high efficiency by multiple emission from stacked organic light-emitting devices. *Organ. Electron.*, 12(4):710–715, 2011.

[17] P. A. Haigh, Z. Ghassemlooy, I. Papakonstantinou, F. Arca, S. F. Tedde, O. Hayden, and E. Leitgeb. A 1-Mb/s visible light communications link with low bandwidth organic components. *IEEE Photon. Technol. Lett.*, 26(13):1295–1298, 2014.

[18] P. A. Haigh, F. Bausi, T. Kanesan, S. T. Le, S. Rajbhandari, Z. Ghassem- looy, I. Papakonstantinou, et al. A 10 Mb/s visible light communication system using

a low bandwidth polymer light-emitting diode. In *2014 9th International Symposium on Communication Systems, Networks Digital Signal Processing (CSNDSP)*, pages 999–1004, July 2014.

[19] P. A. Haigh, F. Bausi, T. Kanesan, S. T. Le, S. Rajbhandari, Z. Ghassemlooy, I. Papakonstantinou, et al. A 20-Mb/s VLC link with a polymer LED and a multilayer perceptron equalizer. *IEEE Photon. Technol. Lett.*, 26(19):1975–1978, 2014.

[20] P. A. Haigh, F. Bausi, Z. Ghassemlooy, I. Papakonstantinou, H. Le Minh, C. Flechon, and F. Cacialli. Next generation visible light communications: 10 Mb/s with polymer light-emitting diodes. In *Optical Fiber Communications Conference and Exhibition (OFC), 2014*, pages 1–3, March 2014.

[21] C. D. Muller, A. Falcou, N. Reckefuss, M. Rojahn, V. Wiederhirn, and P. Rudati. Multi-colour organic light-emitting displays by solution processing. *Nature*, 421:829–833, 2003.

[22] S. F. Tedde, J. Kern, T. Sterzl, J. Fürst, P. Lugli, and O. Hayden. Fully spray coated organic photodiodes. *Nano Lett.*, 9(3):980–983, 2009.

[23] J. Shinar. *Organic Light-Emitting Devices: A Survey*. Springer-Verlag, New York, 2003.

[24] Y. Zhao, L. Duan, D. Zhang, L. Hou, J. Qiao, L. Wang, and Y. Qiu. Small molecular phosphorescent organic light-emitting diodes using a spin-coated hole blocking layer. *Appl. Phys. Lett.*, 100(8):083304, 2012.

[25] P. V. Fulvia Villani, G. Nenna, O. Valentino, G. Burrasca, T. Fasolino, C. Minarini, and D. della Sala. Inkjet printed polymer layer on flexible substrate for OLED applications. *J. Phys. Chem. C*, 113(30):13398–13402, 2009.

[26] S. F. Tedde. *Design, Fabrication and Characterization of Organic Photodiodes for Industrial and Medical Applications*. Walter Schottky Institut, Technische Universitt Mnchen, Munchen, Germany, 2009.

[27] P. A. Haigh. *Using Equalizers to Increase Data Rates in Organic Photonic Devices for Visible Light Communications Systems*. PhD thesis, Northumbria University, 2009.

[28] H. Le Minh, D. O'Brien, G. Faulkner, L. Zeng, K. Lee, D. Jung, Y. Oh, and E. T. Won. 100-Mb/s NRZ visible light communications using a postequalized white LED. *IEEE Photon. Technol. Lett.*, 21(15):1063–1065, 2009.

[29] J. Vucic, C. Kottke, S. Nerreter, K. Habel, A. Buttner, K.-D. Langer, and J. W. Walewski. 230 Mbit/s via a wireless visible-light link based on OOK modulation of phosphorescent white LEDs. In *2010 Conference on (OFC/NFOEC) Optical Fiber Communication (OFC), Collocated National Fiber Optic Engineers Conference*, pages 1–3, March 2010.

[30] W. Popoola. *Subcarrier Intensity Modulated Free-Space Optical Communication Systems*. PhD thesis, Northumbria University, 2009.

[31] C. J. Brabec, N. S. Sariciftci, and J. C. Hummelen. Plastic solar cells. *Adv. Funct. Mater.*, 11(1):15–26, 2001.

[32] P. A. Haigh, Z. Ghassemlooy, I. Papakonstantinou, F. Tedde, S. F. Tedde, O. Hayden, and S. Rajbhandari. A MIMO-ANN system for increasing data rates in organic visible light communications systems. In *2013 IEEE International Conference on Communications (ICC)*, pages 5322–5327, June 2013.

[33] F. Eder, H. Klauk, M. Halik, U. Zschieschang, G. Schmid, and Ch. Dehm. Organic electronics on paper. *Appl. Phys. Lett.*, 84(14):2673, 2004.

[34] T. Someya. Flexible electronics: Tiny lamps to illuminate the body. *Nat. Mater.*, 9:879–880, 2010.

[35] Z. B. Wang, M. G. Helander, J. Qiu, D. P. Puzzo, M. T. Greiner, Z. M. Hudson, S. Wang, Z. W. Liu, and Z. H. Lu. Unlocking the full potential of organic light-emitting diodes on flexible plasticing. *Nat. Photon.*, 5:753–757, 2011.

[36] T.-H. Han, Y. Lee, M.-R. Choi, S.-H. Woo, S.-H. Bae, B. H. Hong, J.-H. Ahn, and T.-W. Lee. Extremely efficient flexible organic light-emitting diodes with modified graphene anode. *Nat. Photon.*, 6:105–110, 2012.

[37] H. Sasabe, K. Minamoto, Y.-J. Pu, M. Hirasawa, and J. Kido. Ultra high-efficiency multi-photon emission blue phosphorescent OLEDs with external quantum efficiency exceeding 40. *Org. Electron.*, 13(11):2615–2619, 2012.

[38] F. Arca. Interface trap states in organic photodiodes. *Sci. Rep.*, 3:1324, 2013.

[39] P. A. Haigh, Z. Ghassemlooy, I. Papakonstantinou, and H. Le Minh. 2.7 Mb/s with a 93-kHz white organic light emitting diode and real time ANN equalizer. *IEEE Photon. Technol. Lett.*, 25(17):1687–1690, 2013.

[40] E. Biglieri, J. Proakis, and S. Shamai. Fading channels: Information theoretic and communications aspects. *IEEE Trans. Inform. Theor.*, 44(6):2619–2692, 1998.

[41] S. Haykin. *Adaptive Filter Theory*. Upper Saddle River, NJ: Prentice Hall International, 2001.

[42] S. Haykin. *Communication Systems,* 5th Ed., Wiley Publishing, Chichester, NY, 2009.

[43] P. A. Haigh, Z. Ghassemlooy, S. Rajbhandari, I. Papakonstantinou, and W. Popoola. Visible light communications: 170 Mb/s using an artificial neural network equalizer in a low bandwidth white light configuration. *J. Lightwave Technol.*, 32(9):1807–1813, 2014.

[44] J. G. Proakis and M. Salehi. *Fundamentals of Communication Systems*. Pearson Prentice Hall, New York, 2005.

[45] S. Rajbhandari, J. Faith, Z. Ghassemlooy, and M. Angelova. Comparative study of classifiers to mitigate intersymbol interference in diffuse indoor optical wireless communication links. *Optik—Int. J. Light Electron Opt.*, 124(20):4192–4196, 2013.

[46] S. Trenn. Multilayer perceptrons: Approximation order and necessary number of hidden units. *IEEE Trans. Neural Networks*, 19(5):836–844, 2008.

[47] E. A. Martínez-Rams and V. Garcerán-Hernánde. Assessment of a speaker recognition system based on an auditory model and neural nets. *IWINAC 2009*, 5602:488–498, 2009.

[48] Recommendation ITU-T G. 826: *Error Performance Parameters and Objectives for International Constant Bit Rate Digital Paths at or above the Primary Rate*, International Telecommunication Union, Geneva, Switzerland, 1993.

[49] Z. Ghassemlooy, P. A. Haigh, F. Arca, S. F. Tedde, O. Hayden, I. Papakonstantinou, and S. Rajbhandari. Visible light communications: 3.75 Mbits/s data rate with a 160 kHz bandwidth organic photodetector and artificial neural network equalization [invited]. *Photon. Res.*, 1(2):65–68, 2013.

[50] P. A. Haigh, Z. Ghassemlooy, S. Rajbhandari, and I. Papakonstantinou. Visible light communications using organic light emitting diodes. *IEEE Commun. Mag.*, 51(8):148–154, 2013.

Index